Nonlinear Systems and Complexity

Volume 28

Series editor

Albert C. J. Luo
Southern Illinois University
Edwardsville, IL, USA

Nonlinear Systems and Complexity provides a place to systematically summarize recent developments, applications, and overall advance in all aspects of nonlinearity, chaos, and complexity as part of the established research literature, beyond the novel and recent findings published in primary journals. The aims of the book series are to publish theories and techniques in nonlinear systems and complexity; stimulate more research interest on nonlinearity, synchronization, and complexity in nonlinear science; and fast-scatter the new knowledge to scientists, engineers, and students in the corresponding fields. Books in this series will focus on the recent developments, findings and progress on theories, principles, methodology, computational techniques in nonlinear systems and mathematics with engineering applications. The Series establishes highly relevant monographs on wide ranging topics covering fundamental advances and new applications in the field. Topical areas include, but are not limited to: Nonlinear dynamics; Complexity, nonlinearity, and chaos; Computational methods for nonlinear systems; Stability, bifurcation, chaos and fractals in engineering; Nonlinear chemical and biological phenomena; Fractional dynamics and applications; Discontinuity, synchronization and control.

More information about this series at http://www.springer.com/series/11433

Albert C. J. Luo

Bifurcation and Stability in Nonlinear Dynamical Systems

Springer

Albert C. J. Luo
Southern Illinois University
Edwardsville, IL, USA

ISSN 2195-9994 ISSN 2196-0003 (electronic)
Nonlinear Systems and Complexity
ISBN 978-3-030-22909-2 ISBN 978-3-030-22910-8 (eBook)
https://doi.org/10.1007/978-3-030-22910-8

This Springer imprint is published by the registered company Springer Nature Switzerland AG
The registered company address is: Gewerbestrasse 11, 6330 Cham, Switzerland

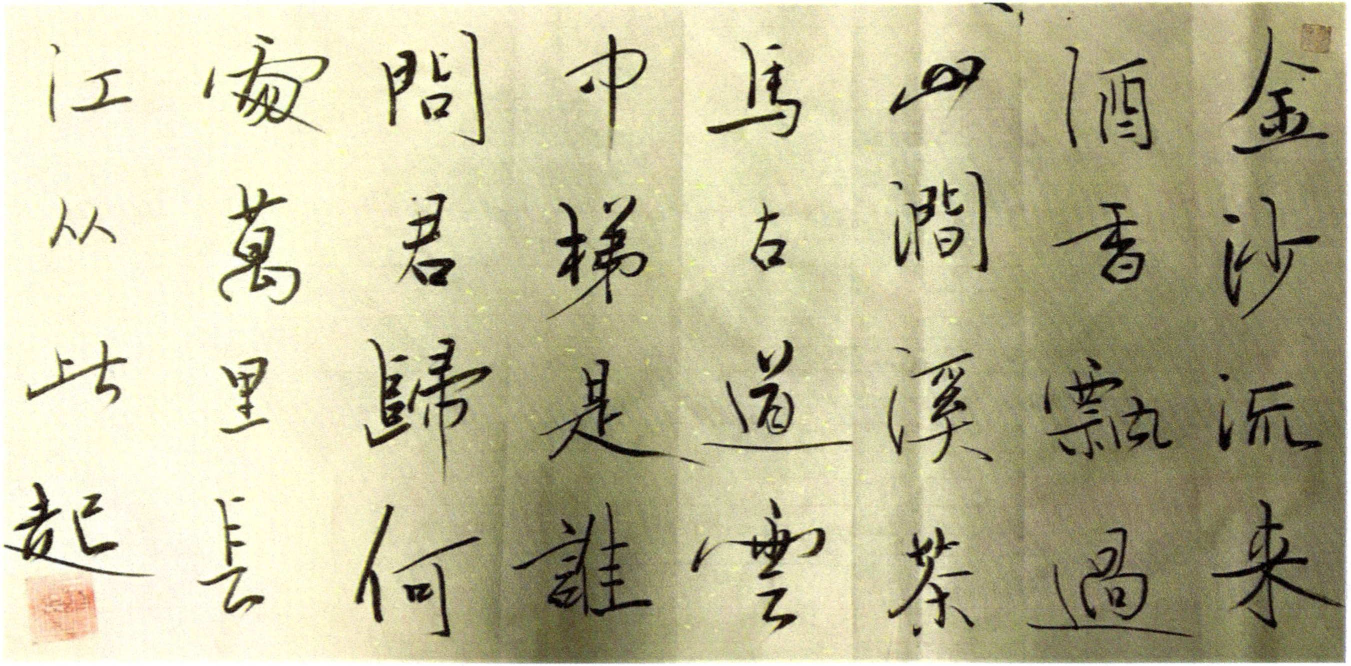
金沙流来
酒香飄過
山澗溪茶
馬古道雲
中梯是誰
問君歸何
處萬里長
江从此起

Preface

This book systematically presents a fundamental theory for the local analysis of bifurcation and stability of equilibriums in nonlinear dynamical systems. Until now, one does not have any efficient way to investigate stability and bifurcation of dynamical systems with higher-order singularity equilibriums. For instance, infinite-equilibrium dynamical systems have higher order singularity, which dramatically changes dynamical behaviors and possesses similar characteristics of discontinuous dynamical systems. The stability and bifurcation of equilibriums on the specific eigenvector are presented, and the spiral stability and Hopf bifurcation of equilibriums in nonlinear systems are presented through the Fourier series transformation. The bifurcation and stability of higher-order singularity equilibriums are presented through the $(2m)^{\text{th}}$ and $(2m+1)^{\text{th}}$-degree polynomial systems. From local analysis, dynamics of infinite-equilibrium systems is discussed. The research on infinite-equilibrium systems will bring us to the new era of dynamical systems and control.

This book consists of eight chapters. The first chapter discusses the local theory of stability of equilibriums in nonlinear dynamical systems. The spiral stability of equilibriums on the eigenvector space is discussed. In addition, based on the Fourier series transformation, the spiral stability of equilibrium is presented. The extended Lyapunov stability theory is also presented. In Chap. 2, the local theory of bifurcations of equilibriums in nonlinear systems is presented on the specific eigenvector, and the Hopf bifurcation of equilibrium is discussed. In Chap. 3, the local analysis of stability and bifurcation for 1-dimensional and 2-dimensional dynamical systems is presented. In Chap. 4, equilibrium higher-order singularity in 1-dimensional systems is discussed globally. In Chap. 5, the global bifurcation theory for low-degree polynomial systems is presented. For a global view of bifurcation and stability of equilibriums in nonlinear dynamical systems, the bifurcation and stability of higher order singularity equilibriums are presented through the $(2m)^{\text{th}}$ and $(2m+1)^{\text{th}}$-degree polynomial systems in Chaps. 6 and 7, respectively. In Chap. 8, infinite-equilibrium systems are discussed through the local analysis of higher-order, singular dynamical systems. The equilibrium computations and normal forms of equilibriums for nonlinear dynamical systems are presented.

Finally, I would like to thank my wife for her support for this research work. The author hopes that the materials presented herein can last long for science and engineering. Such contributions will benefit human beings on their progress and development.

Edwardsville, IL, USA Albert C. J. Luo

Contents

Chapter 1
Stability of Equilibriums

In this chapter, basic concepts of nonlinear dynamical systems are introduced. A local theory of equilibrium stability for nonlinear dynamical systems is discussed. The spiral stability of equilibriums in nonlinear dynamical systems is presented through the Fourier series base. The higher order singularity and stability for nonlinear systems on the specific eigenvectors are developed. The Lyapunov function stability is briefly discussed, and the extended Lyapunov theory for equilibrium stability is also presented.

1.1 Continuous Dynamical Systems

Definition 1.1 For $I \subseteq \mathscr{R}$, $\Omega \subseteq \mathscr{R}^n$, and $\Lambda \subseteq \mathscr{R}^m$, consider a vector function $\mathbf{f} : \Omega \times I \times \Lambda \to \mathscr{R}^n$ which is C^r $(r \geq 1)$-continuous, and there is an ordinary differential equation in a form of

$$\dot{\mathbf{x}} = \mathbf{f}(\mathbf{x}, t, \mathbf{p}) \text{ for } t \in I, \mathbf{x} \in \Omega \text{ and } \mathbf{p} \in \Lambda \tag{1.1}$$

where $\dot{\mathbf{x}} = d\mathbf{x}/dt$ is differentiation with respect to time t, which is simply called the velocity vector of the state variables $\mathbf{x}$. With an initial condition of $\mathbf{x}(t_0) = \mathbf{x}_0$, the solution of Eq. (1.1) is given by

$$\mathbf{x}(t) = \mathbf{\Phi}(\mathbf{x}_0, t - t_0, \mathbf{p}). \tag{1.2}$$

(i) The ordinary differential equation with the initial condition is called a *dynamical system.*
(ii) The vector function $\mathbf{f}(\mathbf{x}, t, \mathbf{p})$ is called a *vector field* on domain Ω.
(iii) The solution $\mathbf{\Phi}(\mathbf{x}_0, t - t_0, \mathbf{p})$ is called the *flow* of dynamical systems.

A. C. J. Luo, *Bifurcation and Stability in Nonlinear Dynamical Systems*, Nonlinear Systems and Complexity 28, https://doi.org/10.1007/978-3-030-22910-8_1

(iv) The projection of the solution $\mathbf{\Phi}(\mathbf{x}_0, t - t_0, \mathbf{p})$ on domain Ω is called the trajectory, phase curve, or orbit of dynamical system, which is defined as

$$\Gamma = \{\mathbf{x}(t) \in \Omega | \mathbf{x}(t) = \mathbf{\Phi}(\mathbf{x}_0, t - t_0, \mathbf{p}) \text{ for } t \in I\} \subset \Omega. \tag{1.3}$$

Definition 1.2 If the vector field of the dynamical system in Eq. (1.1) is independent of time, such a system is called an autonomous dynamical system. Thus, equation (1.1) becomes

$$\dot{\mathbf{x}} = \mathbf{f}(\mathbf{x}, \mathbf{p}) \text{ for } t \in I \subseteq \mathscr{R}, \mathbf{x} \in \Omega \subseteq \mathscr{R}^n \text{ and } \mathbf{p} \in \Lambda \subseteq \mathscr{R}^m \tag{1.4}$$

Otherwise, such a system is called nonautonomous dynamical systems if the vector field of the dynamical system in Eq. (1.1) is dependent on time and state variables.

Definition 1.3 For a vector function $\mathbf{f} \in \mathscr{R}^n$ with $\mathbf{x} \in \mathscr{R}^n$, the operator norm of $\mathbf{f}$ is defined by

$$\|\mathbf{f}\| = \sum\nolimits_{i=1}^{n} \max_{\|\mathbf{x}\| \leq 1, t \in I} | f_i(\mathbf{x}, t) |. \tag{1.5}$$

For an $n \times n$ matrix $\mathbf{f}(\mathbf{x}, \mathbf{p}) = \mathbf{A}\mathbf{x}$ with $\mathbf{A} = (a_{ij})_{n \times n}$, the corresponding norm is defined by

$$\|\mathbf{A}\| = \sum\nolimits_{i,j=1}^{n} | a_{ij} |. \tag{1.6}$$

Definition 1.4 For a vector function $\mathbf{x}(t) = (x_1, x_2, \ldots, x_n)^{\mathrm{T}} \in \mathscr{R}^n$, the derivative and integral of $\mathbf{x}(t)$ are defined by

$$\begin{aligned} \frac{d\mathbf{x}(t)}{dt} &= \left(\frac{dx_1(t)}{dt}, \frac{dx_2(t)}{dt}, \ldots, \frac{dx_n(t)}{dt} \right)^{\mathrm{T}}, \\ \int \mathbf{x}(t)dt &= (\int x_1(t)dt, \int x_2(t)dt, \ldots, \int x_n(t)dt)^{\mathrm{T}}. \end{aligned} \tag{1.7}$$

For an $n \times n$ matrix $\mathbf{A} = (a_{ij})_{n \times n}$, the corresponding derivative and integral are defined by

$$\frac{d\mathbf{A}(t)}{dt} = \left(\frac{da_{ij}(t)}{dt} \right)_{n \times n} \text{ and } \int \mathbf{A}(t)dt = (\int a_{ij}(t)dt)_{n \times n}. \tag{1.8}$$

Definition 1.5 For $I \subseteq \mathscr{R}$, $\Omega \subseteq \mathscr{R}^n$, and $\Lambda \subseteq \mathscr{R}^m$, the vector function $\mathbf{f}(\mathbf{x}, t, \mathbf{p})$ with $\mathbf{f} : \Omega \times I \times \Lambda \rightarrow \mathscr{R}^n$ is differentiable at $\mathbf{x}_0 \in \Omega$ if

$$\frac{\partial \mathbf{f}(\mathbf{x},t,\mathbf{p})}{\partial \mathbf{x}}\Big|_{(\mathbf{x}_0,t,\mathbf{p})} = \lim_{\Delta\mathbf{x}\to\mathbf{0}} \frac{\mathbf{f}(\mathbf{x}_0+\Delta\mathbf{x},t,\mathbf{p})-\mathbf{f}(\mathbf{x}_0,t,\mathbf{p})}{\Delta\mathbf{x}}. \tag{1.9}$$

$\partial\mathbf{f}/\partial\mathbf{x}$ is called the spatial derivative of $\mathbf{f}(\mathbf{x},t,\mathbf{p})$ at $\mathbf{x}_0$, and the derivative is given by the Jacobian matrix:

$$\frac{\partial \mathbf{f}(\mathbf{x},t,\mathbf{p})}{\partial \mathbf{x}} = (\partial f_i/\partial x_j)_{n\times n}. \tag{1.10}$$

Definition 1.6 For $I\subseteq\mathscr{R}$, $\Omega\subseteq\mathscr{R}^n$, and $\Lambda\subseteq\mathscr{R}^m$, consider a vector function $\mathbf{f}(\mathbf{x},t,\mathbf{p})$ with $\mathbf{f}:\Omega\times I\times\Lambda\to\mathscr{R}^n$, $t\in I$, and $\mathbf{x}\in\Omega$ and $\mathbf{p}\in\Lambda$. The vector function $\mathbf{f}(\mathbf{x},t,\mathbf{p})$ satisfies the Lipschitz condition with respect to $\mathbf{x}$ for $I\times\Omega\times\Lambda$,

$$\|\mathbf{f}(\mathbf{x}_2,t,\mathbf{p})-\mathbf{f}(\mathbf{x}_1,t,\mathbf{p})\| \le L\|\mathbf{x}_2-\mathbf{x}_1\| \tag{1.11}$$

with $\mathbf{x}_1$, $\mathbf{x}_2\in\Omega$ and L being a constant. The constant L is called the Lipschitz constant.

Theorem 1.1 *Consider a dynamical system as*

$$\dot{\mathbf{x}} = \mathbf{f}(\mathbf{x},t,\mathbf{p}) \ \text{ with } \mathbf{x}(t_0)=\mathbf{x}_0 \tag{1.12}$$

with $t_0, t\in I=[t_1,t_2]$, $\mathbf{x}\in\Omega=\{\mathbf{x}\,|\,\|\mathbf{x}-\mathbf{x}_0\|\le d\}$, *and* $\mathbf{p}\in\Lambda$. *If the vector function* $\mathbf{f}(\mathbf{x},t,\mathbf{p})$ *is* C^r*-continuous* ($r\ge1$) *in* $G=\Omega\times I\times\Lambda$, *then the dynamical system in Eq. (1.12) has one and only one solution* $\boldsymbol{\Phi}(\mathbf{x}_0,t-t_0,\mathbf{p})$ for

$$|\,t-t_0\,| \le \min\,(t_2-t_1, d/M) \ \text{ with } M=\max_G\|\mathbf{f}\|. \tag{1.13}$$

Proof The proof of this theorem can be referred to the book by Coddington and Levinson (1955). ∎

Theorem 1.2 (Gronwall) *Suppose there is a continuous real valued function g* $(t)\ge0$ *to satisfy*

$$g(t)\le\delta_1\int_{t_0}^{t} g(\tau)d\tau+\delta_2 \tag{1.14}$$

for all $t\in[t_0,t_1]$ and δ_1 and δ_2 *are positive constants. For* $t\in[t_0,t_1]$, *one obtains*

$$g(t)\le\delta_2 e^{\delta_1(t-t_0)}. \tag{1.15}$$

Proof The proof can be referred in Luo (2011). ∎

Theorem 1.3 *Consider a dynamical system as* $\dot{\mathbf{x}} = \mathbf{f}(\mathbf{x}, t, \mathbf{p})$ *with* $\mathbf{x}(t_0) = \mathbf{x}_0$ *in Eq. (1.12) with* $t_0, t \in I = [t_1, t_2]$, $\mathbf{x} \in \Omega = \{\mathbf{x} | \|\mathbf{x} - \mathbf{x}_0\| \leq d\}$, *and* $\mathbf{p} \in \Lambda$. *The vector function* $\mathbf{f}(\mathbf{x}, t, \mathbf{p})$ *is* C^r*-continuous* ($r \geq 1$) *in* $G = \Omega \times I \times \Lambda$. *The solution of* $\dot{\mathbf{x}} = \mathbf{f}(\mathbf{x}, t, \mathbf{p})$ *with* $\mathbf{x}(t_0) = \mathbf{x}_0$ *is* $\mathbf{x}(t)$ *on G and the solution of* $\dot{\mathbf{y}} = \mathbf{f}(\mathbf{y}, t, \mathbf{p})$ *with* $\mathbf{y}(t_0) = \mathbf{y}_0$ *is* $\mathbf{y}(t)$ *on G. For a given* $\varepsilon > 0$, *if* $\|\mathbf{x}_0 - \mathbf{y}_0\| \leq \varepsilon$, *then*

$$\|\mathbf{x}(t) - \mathbf{y}(t)\| \leq \varepsilon e^{L(t-t_0)} \text{ on } I \times \Lambda \tag{1.16}$$

Proof The proof can be referred in Luo (2011). ∎

1.2 Equilibriums and Stability

Definition 1.7 Consider a metric space Ω and $\Omega_\alpha \subseteq \Omega$ ($\alpha = 1, 2, \ldots$).

(i) A map $\mathbf{h}$ is called a homeomorphism of Ω_α onto Ω_β ($\alpha, \beta = 1, 2, \cdots$) if the map $\mathbf{h} : \Omega_\alpha \to \Omega_\beta$ is continuous and one to one, and $\mathbf{h}^{-1} : \Omega_\beta \to \Omega_\alpha$ is continuous.
(ii) Two sets Ω_α and Ω_β are homeomorphic or topologically equivalent if there is a homeomorphism of Ω_α onto Ω_β.

Definition 1.8 A connected, metric space Ω with an open cover $\{\Omega_\alpha\}$ (i.e., $\Omega = \cup_\alpha \Omega_\alpha$) is called an n-dimensional, C^r ($r \geq 1$) differentiable manifold if the following properties exist.

(i) There is an open unit ball $B = \{\mathbf{x} \in \mathscr{R}^n | \|\mathbf{x}\| < 1\}$.
(ii) For all α, there is a homeomorphism $\mathbf{h}_\alpha : \Omega_\alpha \to B$.
(iii) If $\mathbf{h}_\alpha : \Omega_\alpha \to B$ and $\mathbf{h}_\beta : \Omega_\beta \to B$ are homeomorphisms for $\Omega_\alpha \cap \Omega_\beta \neq \varnothing$, then there is a C^r-differentiable map $\mathbf{h} = \mathbf{h}_\alpha \circ \mathbf{h}_\beta^{-1}$ for $h_\alpha(\Omega_\alpha \cap \Omega_\beta) \subset \mathscr{R}^n$ and $\mathbf{h}_\beta(\Omega_\alpha \cap \Omega_\beta) \subset \mathscr{R}^n$ with

$$\mathbf{h} : \mathbf{h}_\beta(\Omega_\alpha \cap \Omega_\beta) \to \mathbf{h}_\alpha(\Omega_\alpha \cap \Omega_\beta), \tag{1.17}$$

and for all $\mathbf{x} \in \mathbf{h}_\beta(\Omega_\alpha \cap \Omega_\beta)$, the Jacobian determinant $\det D\mathbf{h}(\mathbf{x}) \neq 0$.

The manifold Ω is called to be analytic if the maps $\mathbf{h} = \mathbf{h}_\alpha \circ \mathbf{h}_\beta^{-1}$ are analytic.

Definition 1.9 Consider an autonomous, nonlinear dynamical system $\dot{\mathbf{x}} = \mathbf{f}(\mathbf{x}, \mathbf{p})$ in Eq. (1.4). A point $\mathbf{x}^* \in \Omega$ is called an equilibrium point or critical point of a nonlinear system $\dot{\mathbf{x}} = \mathbf{f}(\mathbf{x}, \mathbf{p})$ if

$$\mathbf{f}(\mathbf{x}^*, \mathbf{p}) = \mathbf{0}. \tag{1.18}$$

The linearized system of the nonlinear system $\dot{\mathbf{x}} = \mathbf{f}(\mathbf{x}, \mathbf{p})$ in Eq. (1.4) at the equilibrium point $\mathbf{x}^*$ is given by

$$\dot{\mathbf{y}} = D\mathbf{f}(\mathbf{x}^*, \mathbf{p})\mathbf{y} \ \text{ where } \ \mathbf{y} = \mathbf{x} - \mathbf{x}^*. \tag{1.19}$$

Definition 1.10 Consider an n-dimensional, autonomous, nonlinear dynamical system $\dot{\mathbf{x}} = \mathbf{f}(\mathbf{x}, \mathbf{p})$ in Eq. (1.4) with an equilibrium point $\mathbf{x}^*$. The linearized system of the nonlinear system at the equilibrium point $\mathbf{x}^*$ is $\dot{\mathbf{y}} = D\mathbf{f}(\mathbf{x}^*, \mathbf{p})\mathbf{y}$ ($\mathbf{y} = \mathbf{x} - \mathbf{x}^*$) in Eq. (1.19). The matrix $D\mathbf{f}(\mathbf{x}^*, \mathbf{p})$ possesses n eigenvalues λ_k ($k = 1, 2, \ldots, n$). Set $N = \{1, 2, \ldots, n\}$ and $N_i = \{i_1, i_2, \ldots, i_{n_i}\} \cup \varnothing$ with $i_j \in N$ ($j = 1, 2, \ldots, n_i$; $i = 1, 2, 3$) and $\Sigma_{i=1}^3 n_i = n$. $\cup_{i=1}^3 N_i = N$ and $N_i \cap N_l = \varnothing$ ($l \neq i$). $N_i = \varnothing$ if $n_i = 0$. The corresponding vectors for the negative, positive, and zero eigenvalues of $D\mathbf{f}(\mathbf{x}^*, \mathbf{p})$ are $\{\mathbf{v}_k\}$ ($k \in N_i$, $i = 1, 2, 3$), respectively. The stable, unstable, and invariant subspaces of the linearized nonlinear system in Eq. (1.19) are defined as

$$\begin{aligned}
\mathscr{E}^{\mathrm{s}} &= span\{\mathbf{v}_k | (D\mathbf{f}(\mathbf{x}^*, \mathbf{p}) - \lambda_k \mathbf{I})\mathbf{v}_k = \mathbf{0}, \lambda_k < 0, k \in N_1 \subseteq N \cup \varnothing\}; \\
\mathscr{E}^{\mathrm{u}} &= span\{\mathbf{v}_k | (D\mathbf{f}(\mathbf{x}^*, \mathbf{p}) - \lambda_k \mathbf{I})\mathbf{v}_k = \mathbf{0}, \lambda_k > 0, k \in N_2 \subseteq N \cup \varnothing\}; \\
\mathscr{E}^{\mathrm{i}} &= span\{\mathbf{v}_k | (D\mathbf{f}(\mathbf{x}^*, \mathbf{p}) - \lambda_k \mathbf{I})\mathbf{v}_k = \mathbf{0}, \lambda_k = 0, k \in N_3 \subseteq N \cup \varnothing\}.
\end{aligned} \tag{1.20}$$

Definition 1.11 Consider a $2n$-dimensional, autonomous dynamical system $\dot{\mathbf{x}} = \mathbf{f}(\mathbf{x}, \mathbf{p})$ in Eq. (1.4) with an equilibrium point $\mathbf{x}^*$. The linearized system of the nonlinear system at the equilibrium point $\mathbf{x}^*$ is $\dot{\mathbf{y}} = D\mathbf{f}(\mathbf{x}^*, \mathbf{p})\mathbf{y}$ ($\mathbf{y} = \mathbf{x} - \mathbf{x}^*$) in Eq. (1.19). The matrix $D\mathbf{f}(\mathbf{x}^*, \mathbf{p})$ has complex eigenvalues $\alpha_k \pm \mathbf{i}\beta_k$ with eigenvectors $\mathbf{u}_k \pm \mathbf{i}\mathbf{v}_k$ ($k \in \{1, 2, \ldots, n\}$) and $\mathbf{i} = \sqrt{-1}$, and the base of vector is

$$\mathbf{B} = \{\mathbf{u}_1, \mathbf{v}_1, \ldots, \mathbf{u}_k, \mathbf{v}_k, \ldots, \mathbf{u}_n, \mathbf{v}_n\}. \tag{1.21}$$

The stable, unstable, and center subspaces of Eq. (1.19) are linear subspaces spanned by $\{\mathbf{u}_k, \mathbf{v}_k\}$ ($k \in N_i$, $i = 1, 2, 3$), respectively. $N_i = \{i_1, i_2, \ldots, i_{n_i}\} \cup \varnothing \subseteq N \cup \varnothing$ and $N = \{1, 2, \ldots, n\}$ with $i_j \in N$ ($j = 1, 2, \ldots, n_i$) and $\Sigma_{i=1}^3 n_i = n$. $\cup_{i=1}^3 N_i = N$ and $N_i \cap N_l = \varnothing$ ($l \neq i$). $N_i = \varnothing$ if $n_i = 0$. The stable, unstable, and center subspaces of the linearized nonlinear system in Eq. (1.19) are defined as

$$\mathscr{E}^{\mathrm{s}} = span\left\{ (\mathbf{u}_k, \mathbf{v}_k) \middle| \begin{array}{l} \alpha_k < 0, \beta_k \neq 0, \\ (D\mathbf{f}(\mathbf{x}^*, \mathbf{p}) - (\alpha_k \pm \mathbf{i}\beta_k)\mathbf{I})(\mathbf{u}_k \pm \mathbf{i}\mathbf{v}_k) = \mathbf{0}, \\ k \in N_1 \subseteq \{1, 2, \ldots, n\} \cup \varnothing \end{array} \right\};$$

$$\mathscr{E}^{\mathrm{u}} = span\left\{ (\mathbf{u}_k, \mathbf{v}_k) \middle| \begin{array}{l} \alpha_k > 0, \beta_k \neq 0, \\ (D\mathbf{f}(\mathbf{x}^*, \mathbf{p}) - (\alpha_k \pm \mathbf{i}\beta_k)\mathbf{I})(\mathbf{u}_k \pm \mathbf{i}\mathbf{v}_k) = \mathbf{0}, \\ k \in N_2 \subseteq \{1, 2, \ldots, n\} \cup \varnothing \end{array} \right\};$$

$$\mathscr{E}^c = span\left\{ (\mathbf{u}_k, \mathbf{v}_k) \left| \begin{array}{l} \alpha_k = 0, \beta_k \neq 0, \\ (D\mathbf{f}(\mathbf{x}^*, \mathbf{p}) - (\alpha_k \pm \mathbf{i}\beta_k)\mathbf{I})(\mathbf{u}_k \pm \mathbf{i}\mathbf{v}_k) = \mathbf{0}, \\ k \in N_3 \subseteq \{1, 2, \ldots, n\} \cup \varnothing \end{array} \right. \right\}. \tag{1.22}$$

Theorem 1.4 *Consider an n-dimensional, autonomous, nonlinear dynamical system* $\dot{\mathbf{x}} = \mathbf{f}(\mathbf{x}, \mathbf{p})$ *in Eq. (1.4) with an equilibrium point* $\mathbf{x}^*$. *The linearized system of the nonlinear system at the equilibrium point* $\mathbf{x}^*$ *is* $\dot{\mathbf{y}} = D\mathbf{f}(\mathbf{x}^*, \mathbf{p})\mathbf{y}$ $(\mathbf{y} = \mathbf{x} - \mathbf{x}^*)$ *in Eq. (1.19). The eigenspace of* $D\mathbf{f}(\mathbf{x}^*, \mathbf{p})$ *(i.e.,* $\mathscr{E} \subseteq \mathscr{R}^n$*) in the linearized dynamical system is expressed by direct sum of three subspaces:*

$$\mathscr{E} = \mathscr{E}^s \oplus \mathscr{E}^u \oplus \mathscr{E}^c \tag{1.23}$$

where $\mathscr{E}^s$, $\mathscr{E}^u$, and $\mathscr{E}^c$ *are the stable, unstable, and center spaces, respectively.*

Proof This proof can be referred in Luo (2011). ∎

Definition 1.12 Consider an n-dimensional, autonomous, nonlinear dynamical system $\dot{\mathbf{x}} = \mathbf{f}(\mathbf{x}, \mathbf{p})$ in Eq. (1.4) with an equilibrium point $\mathbf{x}^*$ and $\mathbf{f}(\mathbf{x}, \mathbf{p})$ is C^r-continuous ($r \geq 1$) in a neighborhood of the equilibrium $\mathbf{x}^*$. The corresponding solution is $\mathbf{x}(t) = \mathbf{\Phi}(\mathbf{x}_0, t - t_0, \mathbf{p}) = \mathbf{\Phi}_t(\mathbf{x}_0)$. The linearized system of the nonlinear system at the equilibrium point $\mathbf{x}^*$ is $\dot{\mathbf{y}} = D\mathbf{f}(\mathbf{x}^*, \mathbf{p})\mathbf{y}$ $(\mathbf{y} = \mathbf{x} - \mathbf{x}^*)$ in Eq. (1.19). Suppose there is a neighborhood of the equilibrium $\mathbf{x}^*$ as $U(\mathbf{x}^*) \subset \Omega$, and in the neighborhood

$$\lim_{\|\mathbf{y}\| \to 0} \frac{\|\mathbf{f}(\mathbf{x}^* + \mathbf{y}, \mathbf{p}) - D\mathbf{f}(\mathbf{x}^*, \mathbf{p})\mathbf{y}\|}{\|\mathbf{y}\|} = 0. \tag{1.24}$$

(i) A C^r invariant manifold

$$S_{loc}(\mathbf{x}, \mathbf{x}^*) = \{\mathbf{x} \in U(\mathbf{x}^*) | \lim_{t \to \infty} \mathbf{x}(t) = \mathbf{x}^*, \mathbf{x}(t) \in U(\mathbf{x}^*) \text{ for all } t \geq 0\} \tag{1.25}$$

is called the local stable manifold of $\mathbf{x}^*$, and the corresponding global, stable manifold is defined as

$$\mathscr{S}(\mathbf{x}, \mathbf{x}^*) = \cup_{t \leq 0} \mathbf{\Phi}_t(\mathscr{S}_{loc}(\mathbf{x}, \mathbf{x}^*)). \tag{1.26}$$

(ii) A C^r invariant manifold

$$\mathscr{U}_{loc}(\mathbf{x}, \mathbf{x}^*) = \{\mathbf{x} \in U(\mathbf{x}^*) | \lim_{t \to -\infty} \mathbf{x}(t) = \mathbf{x}^*, \mathbf{x}(t) \in U(\mathbf{x}^*) \text{ for all } t \leq 0\} \tag{1.27}$$

is called the unstable manifold of $\mathbf{x}^*$, and the corresponding global, unstable manifold is defined as

$$\mathscr{U}(\mathbf{x},\mathbf{x}^*) = \cup_{t\geq 0}\mathbf{\Phi}_t(\mathscr{U}_{loc}(\mathbf{x},\mathbf{x}^*)). \tag{1.28}$$

(iii) A C^{r-1} invariant manifold $\mathscr{C}_{loc}(\mathbf{x},\mathbf{x}^*)$ is called the center manifold of $\mathbf{x}^*$ if $\mathscr{C}_{loc}(\mathbf{x},\mathbf{x}^*)$ possesses the same dimension of $\mathscr{E}^c$ for $\mathbf{x}^* \in \mathscr{S}(\mathbf{x},\mathbf{x}^*)$, and the tangential space of $\mathscr{C}_{loc}(\mathbf{x},\mathbf{x}^*)$ is identical to $\mathscr{E}^c$.

The stable and unstable manifolds are unique, but the center manifold is not unique. If the nonlinear vector field $\mathbf{f}$ is C^∞-continuous, then a C^r center manifold can be found for any $r < \infty$.

Theorem 1.5 *Consider an n-dimensional, autonomous, nonlinear dynamical system $\dot{\mathbf{x}} = \mathbf{f}(\mathbf{x},\mathbf{p})$ in Eq. (1.4) with a hyperbolic equilibrium point $\mathbf{x}^*$ and $\mathbf{f}(\mathbf{x},\mathbf{p})$ is C^r-continuous ($r \geq 1$) in a neighborhood of the equilibrium $\mathbf{x}^*$. The corresponding solution is $\mathbf{x}(t) = \mathbf{\Phi}(\mathbf{x}_0, t - t_0, \mathbf{p}) = \mathbf{\Phi}_t(\mathbf{x}_0)$. The linearized system of the nonlinear system at the equilibrium point $\mathbf{x}^*$ is $\dot{\mathbf{y}} = D\mathbf{f}(\mathbf{x}^*,\mathbf{p})\mathbf{y}$ ($\mathbf{y} = \mathbf{x} - \mathbf{x}^*$) in Eq. (1.19). Suppose there is a neighborhood of the hyperbolic equilibrium $\mathbf{x}^*$ as $U(\mathbf{x}^*) \subset \Omega$. If the homeomorphism between the local invariant subspace $E(\mathbf{x},\mathbf{x}^*) \subset U(\mathbf{x}^*)$ under the flow $\mathbf{\Phi}(\mathbf{x}_0, t - t_0, \mathbf{p})$ of $\dot{\mathbf{x}} = \mathbf{f}(\mathbf{x},\mathbf{p})$ in Eq. (1.4) and the eigenspace $\mathscr{E}$ of the linearized system exists with the condition in Eq. (1.24), the local invariant subspace is decomposed by*

$$E(\mathbf{x},\mathbf{x}^*) = \mathscr{S}_{loc}(\mathbf{x},\mathbf{x}^*) \oplus \mathscr{U}_{loc}(\mathbf{x},\mathbf{x}^*). \tag{1.29}$$

(i) *The local stable invariant manifold $\mathscr{S}_{loc}(\mathbf{x},\mathbf{x}^*)$ possesses the following properties:*

  (a) *For $\mathbf{x}^* \in \mathscr{S}_{loc}(\mathbf{x},\mathbf{x}^*)$, $\mathscr{S}_{loc}(\mathbf{x},\mathbf{x}^*)$ possesses the same dimension of $\mathscr{E}^s$ and the tangential space of $\mathscr{S}_{loc}(\mathbf{x},\mathbf{x}^*)$ is identical to $\mathscr{E}^s$.*

  (b) *For $\mathbf{x}_0 \in \mathscr{S}_{loc}(\mathbf{x},\mathbf{x}^*)$, $\mathbf{x}(t) \in \mathscr{S}_{loc}(\mathbf{x},\mathbf{x}^*)$ for all time $t \geq t_0$ and $\lim_{t\to\infty}\mathbf{x}(t) = \mathbf{x}^*$.*

  (c) *For $\mathbf{x}_0 \notin \mathscr{S}_{loc}(\mathbf{x},\mathbf{x}^*)$, $\|\mathbf{x} - \mathbf{x}^*\| \geq \delta$ for $\delta > 0$ with $t \geq t_1 \geq t_0$.*

(ii) *The local unstable invariant manifold $\mathscr{U}_{loc}(\mathbf{x},\mathbf{x}^*)$ possesses the following properties:*

  (a) *For $\mathbf{x}^* \in \mathscr{U}_{loc}(\mathbf{x},\mathbf{x}^*)$, $\mathscr{U}_{loc}(\mathbf{x},\mathbf{x}^*)$ possesses the same dimension of $\mathscr{E}^u$ and the tangential space of $\mathscr{U}_{loc}(\mathbf{x},\mathbf{x}^*)$ is identical to $\mathscr{E}^u$.*

  (b) *For $\mathbf{x}_0 \in \mathscr{U}_{loc}(\mathbf{x},\mathbf{x}^*)$, $\mathbf{x}(t) \in \mathscr{U}_{loc}(\mathbf{x},\mathbf{x}^*)$ for all time $t \leq t_0$ and $\lim_{t\to-\infty}\mathbf{x}(t) = \mathbf{x}^*$.*

  (c) *For $\mathbf{x}_0 \notin \mathscr{U}_{loc}(\mathbf{x},\mathbf{x}^*)$, $\|\mathbf{x} - \mathbf{x}^*\| \geq \delta$ for $\delta > 0$ with $t \leq t_1 \leq t_0$.*

Proof The proof for stable and unstable manifold can be referred in Hartman (1964). The proof for center manifold can be referenced in Marsden and McCracken (1976) or Carr (1981). ∎

Theorem 1.6 *Consider an n-dimensional, autonomous, nonlinear dynamical system $\dot{\mathbf{x}} = \mathbf{f}(\mathbf{x},\mathbf{p})$ in Eq. (1.4) with an equilibrium point $\mathbf{x}^*$. Suppose there is a* *neighborhood of the equilibrium* $\mathbf{x}^*$ *as* $U(\mathbf{x}^*) \subset \Omega$; *then* $\mathbf{f}(\mathbf{x},\mathbf{p})$ *is* C^r*-continuous* $(r \geq 1)$ *in a neighborhood of the equilibrium* $\mathbf{x}^*$. *The corresponding solution is* $\mathbf{x}(t) = \mathbf{\Phi}(\mathbf{x}_0, t - t_0, \mathbf{p})$. *The linearized system of the nonlinear system at the equilibrium point* $\mathbf{x}^*$ *is* $\dot{\mathbf{y}} = D\mathbf{f}(\mathbf{x}^*, \mathbf{p})\mathbf{y}$ $(\mathbf{y} = \mathbf{x} - \mathbf{x}^*)$ *in Eq. (1.19). If the homeomorphism between the local invariant subspace* $E(\mathbf{x},\mathbf{x}^*) \subset U(\mathbf{x}^*)$ *under the flow* $\mathbf{\Phi}(\mathbf{x}_0, t - t_0, \mathbf{p})$ *of* $\dot{\mathbf{x}} = \mathbf{f}(\mathbf{x},\mathbf{p})$ *in Eq. (1.4) and the eigenspace* $\mathscr{E}$ *of the linearized system exists with the condition in Eq. (1.24), in addition to the local stable and unstable invariant manifolds, there is a* C^{r-1} *center manifold* $\mathscr{C}_{loc}(\mathbf{x},\mathbf{x}^*)$. *The center manifold possesses the same dimension of* $\mathscr{E}^c$ *for* $\mathbf{x}^* \in \mathscr{C}_{loc}(\mathbf{x},\mathbf{x}^*)$, *and the tangential space of* $\mathscr{C}_{loc}(\mathbf{x},\mathbf{x}^*)$ *is identical to* $\mathscr{E}^c$. *Thus, the local invariant subspace is decomposed by*

$$E(\mathbf{x},\mathbf{x}^*) = \mathscr{S}_{loc}(\mathbf{x},\mathbf{x}^*) \oplus \mathscr{U}_{loc}(\mathbf{x},\mathbf{x}^*) \oplus \mathscr{C}_{loc}(\mathbf{x},\mathbf{x}^*). \tag{1.30}$$

Proof The proof for stable and unstable manifolds can be referred in Hartman (1964). The proof for center manifolds can be referenced in Marsden and McCracken (1976) or Carr (1981). ∎

Definition 1.13 Consider an n-dimensional, autonomous, nonlinear dynamical system $\dot{\mathbf{x}} = \mathbf{f}(\mathbf{x},\mathbf{p})$ in Eq. (1.4) with an equilibrium point $\mathbf{x}^*$ and $\mathbf{f}(\mathbf{x},\mathbf{p})$ is C^r-continuous $(r \geq 1)$ in a neighborhood of the equilibrium $\mathbf{x}^*$.

(i) The equilibrium $\mathbf{x}^*$ is stable if for all $\varepsilon > 0$ there is a $\delta > 0$ such that for all $\mathbf{x}_0 \in U_\delta(\mathbf{x}^*)$ at $t = t_0$ where $U_\delta(\mathbf{x}^*) = \{\mathbf{x} | \|\mathbf{x} - \mathbf{x}^*\| < \delta\}$ and $t \geq t_0$,

$$\mathbf{x}(t) = \mathbf{\Phi}(\mathbf{x}_0, t - t_0, \mathbf{p}) \in U_\varepsilon(\mathbf{x}^*) = \{\mathbf{x}(t) | \, \|\mathbf{x}(t) - \mathbf{x}^*\|. < \varepsilon, \ \ t \in (t_0, \infty)\}. \tag{1.31}$$

(ii) The equilibrium $\mathbf{x}^*$ is unstable if it is not stable or if for all $\varepsilon > 0$ there is a $\delta > 0$ such that for all $\mathbf{x}_0 \in U_\delta(\mathbf{x}^*)$ at $t = t_0$ where $U_\delta(\mathbf{x}^*) = \{\mathbf{x} | \|\mathbf{x} - \mathbf{x}^*\| < \delta\}$ and $t \geq t_0 > t_1$,

$$\mathbf{x}(t) = \mathbf{\Phi}(\mathbf{x}_0, t - t_0, \mathbf{p}) \notin U_\varepsilon(\mathbf{x}^*) = \{\mathbf{x}(t) | \, \|\mathbf{x}(t) - \mathbf{x}^*\|. < \varepsilon, \ \ t \in (t_0, \infty)\}. \tag{1.32}$$

(iii) The equilibrium $\mathbf{x}^*$ is asymptotically stable if for all $\varepsilon > 0$ there is a $\delta > 0$ such that for all $\mathbf{x}_0 \in U_\delta(\mathbf{x}^*)$ at $t = t_0$ where $U_\delta(\mathbf{x}^*) = \{\mathbf{x} | \|\mathbf{x} - \mathbf{x}^*\| < \delta\}$ and $t \geq t_0$,

$$\lim_{t \to \infty} \mathbf{\Phi}(\mathbf{x}_0, t - t_0, \mathbf{p}) = \mathbf{x}^*. \tag{1.33}$$

(iv) The equilibrium $\mathbf{x}^*$ is asymptotically unstable if for all $\varepsilon > 0$ there is a $\delta > 0$ such that for all $\mathbf{x}_0 \in U_\delta(\mathbf{x}^*)$ at $t = t_0$ where $U_\delta(\mathbf{x}^*) = \{\mathbf{x} | \|\mathbf{x} - \mathbf{x}^*\| < \delta\}$ and $t \leq t_0$,

$$\lim_{t \to -\infty} \mathbf{\Phi}(\mathbf{x}_0, t - t_0, \mathbf{p}) = \mathbf{x}^*. \tag{1.34}$$

Definition 1.14 Consider an n-dimensional, autonomous, nonlinear dynamical system $\dot{\mathbf{x}} = \mathbf{f}(\mathbf{x},\mathbf{p})$ in Eq. (1.4) with an equilibrium point $\mathbf{x}^*$ and $\mathbf{f}(\mathbf{x},\mathbf{p})$ is

C^r-continuous ($r \geq 1$) in a neighborhood of the equilibrium $\mathbf{x}^*$. The corresponding solution is $\mathbf{x}(t) = \boldsymbol{\Phi}(\mathbf{x}_0, t - t_0, \mathbf{p})$. Suppose $U(\mathbf{x}^*) \subset \Omega$ is a neighborhood of equilibrium $\mathbf{x}^*$, and there are n linearly independent vectors $\mathbf{v}_k$ ($k = 1, 2, \ldots, n$). The independent eigenvector $\mathbf{v}_k$ is called the covariant eigenvector. The corresponding contravariant vector of the independent eigenvector $\mathbf{v}_k$ is defined by $\mathbf{v}^k$. For a perturbation of equilibrium $\mathbf{y} = \mathbf{x} - \mathbf{x}^*$, let $\mathbf{y} = c^k(t)\mathbf{v}_k = c_k(t)\mathbf{v}^k$ (tensor notation convention).

$$P = (\mathbf{v}_1, \mathbf{v}_2, \ldots, \mathbf{v}_n) \tag{1.35}$$

is called the eigenvector covariant matrix.

$$Q = (\mathbf{v}^1, \mathbf{v}^2, \ldots, \mathbf{v}^n) \tag{1.36}$$

is called the eigenvector contravariant matrix. The eigenvector contravariant vector is defined as

$$\mathbf{v}^k = (a_{k1}, a_{k2}, \ldots, a_{kn})^{\mathrm{T}} \text{ with } Q^{\mathrm{T}} = P^{-1} = (a_{ij}). \tag{1.37}$$

The component $c^k(t)$ is called the contravariant component. The corresponding vector

$$\overline{\mathbf{c}} = (c^1, c^2, \ldots, c^n)^{\mathrm{T}} \tag{1.38}$$

is called a contravariant component vector. The component c_k is called the covariant component. The corresponding vector is

$$\underline{\mathbf{c}} = (c_1, c_2, \ldots, c_n)^{\mathrm{T}} \tag{1.39}$$

which is called a covariant component vector.

Remark For the Cartesian coordinate, the covariant and contravariant vectors and components are identical.

Since the eigenvectors of the linearized systems of nonlinear systems are not orthogonal, the eigenvector space cannot be used for the Cartesian coordinate. However, the eigenvectors can be used to express the solutions of a dynamical system as $\mathbf{y} = c^k(t)\mathbf{v}_k = c_k(t)\mathbf{v}^k$. The two classes of $\mathbf{v}_k$ and $\mathbf{v}^k$ are the covariant and contravariant eigenvectors.

Definition 1.15 The metric tensors of the covariant and contravariant eigenvectors are defined as for $i, j = 1, 2, \ldots, n$:

$$\begin{aligned} g_{ij} &= (\mathbf{v}_i)^{\mathrm{T}} \cdot \mathbf{v}_j, \;\; g^{ij} = (\mathbf{v}^i)^{\mathrm{T}} \cdot \mathbf{v}^j, \\ g_i^j &= (\mathbf{v}_i)^{\mathrm{T}} \cdot \mathbf{v}^j = \delta_i^j; \end{aligned} \tag{1.40}$$

and the unit vectors of the covariant and contravariant eigenvectors are defined as

$$\mathbf{e}_i = \mathbf{v}_i / \sqrt{g_{ii}}, \;\; \mathbf{e}^i = \mathbf{v}^i / \sqrt{g^{ii}}. \tag{1.41}$$

Two transformations based on the covariant and contravariant eigenvectors are defined as

$$\mathbf{y} = c^i \mathbf{v}_i = c_j \mathbf{v}^j, \tag{1.42}$$

and

$$\mathbf{y} = c^i \sqrt{g_{ii}} \mathbf{e}_i = c_j \sqrt{g^{jj}} \mathbf{e}^j \tag{1.43}$$

with

$$\begin{aligned} (\mathbf{v}^j)^{\mathrm{T}} \cdot \mathbf{y} &= (\mathbf{v}^j)^{\mathrm{T}} \cdot c^i \mathbf{v}_i = c^i (\mathbf{v}^j)^{\mathrm{T}} \cdot \mathbf{v}_i = c^i \delta_i^j = c^j; \\ (\mathbf{v}_j)^{\mathrm{T}} \cdot \mathbf{y} &= (\mathbf{v}_j)^{\mathrm{T}} \cdot c_i \mathbf{v}^i = c_i (\mathbf{v}_j)^{\mathrm{T}} \cdot \mathbf{v}^i = c_i \delta_j^i = c_j. \end{aligned} \tag{1.44}$$

For the orthogonal eigenvectors,

$$\mathbf{v}_i = \mathbf{v}^i, g^{ii} = g_{ii} = 1, g^{ij} = g_{ij} = 0. \tag{1.45}$$

Consider a solution in a 2-dimensional system for $i, j = 1, 2$ as

$$\begin{aligned} \mathbf{y}(t) &= c^i \mathbf{v}_i = c^1 \mathbf{v}_1 + c^2 \mathbf{v}_2, \\ \mathbf{y}(t) &= c_j \mathbf{v}^j = c_1 \mathbf{v}^1 + c_2 \mathbf{v}^2. \end{aligned} \tag{1.46}$$

Based on unit covariant and contravariant eigenvectors,

$$\begin{aligned} \mathbf{y}(t) &= c^1 \sqrt{g_{11}} \mathbf{e}_1 + c^2 \sqrt{g_{22}} \mathbf{e}_2, \\ \mathbf{y}(t) &= c_1 \sqrt{g^{11}} \mathbf{e}^1 + c_2 \sqrt{g^{22}} \mathbf{e}^2. \end{aligned} \tag{1.47}$$

The projected quantities on the covariant and contravariant eigenvectors are

$$\begin{aligned}(\mathbf{e}^i)^{\mathrm{T}} \cdot \mathbf{y} = (\mathbf{e}^i)^{\mathrm{T}} \cdot c^j \mathbf{v}_j = \frac{c^j}{\sqrt{g^{ii}}} (\mathbf{v}^i)^{\mathrm{T}} \cdot \mathbf{v}_j = \frac{c^j}{\sqrt{g^{ii}}} \delta^i_j = \frac{c^i}{\sqrt{g^{ii}}}, \\ (\mathbf{e}_i)^{\mathrm{T}} \cdot \mathbf{y} = (\mathbf{e}_i)^{\mathrm{T}} \cdot c_j \mathbf{v}^j = \frac{c_j}{\sqrt{g_{ii}}} (\mathbf{v}_i)^{\mathrm{T}} \cdot \mathbf{v}^j = \frac{c_j}{\sqrt{g_{ii}}} \delta^j_i = \frac{c_i}{\sqrt{g_{ii}}}.\end{aligned} \tag{1.48}$$

Thus,

$$\begin{aligned}c^i = (\mathbf{v}^i)^{\mathrm{T}} \cdot \mathbf{y} = \sqrt{g^{ii}} (\mathbf{e}^i)^{\mathrm{T}} \cdot \mathbf{y}; \\ c_j = (\mathbf{v}_j)^{\mathrm{T}} \cdot \mathbf{y} = \sqrt{g_{jj}} (\mathbf{e}_j)^{\mathrm{T}} \cdot \mathbf{y}.\end{aligned} \tag{1.49}$$

The covariant and contravariant eigenvectors as the coordinates to express a solution vector of dynamical system are presented in Fig. 1.1. The corresponding geometric interpretation of the relations of components is described. A vector $\mathbf{y}$ is depicted by the covariant and contravariant eigenvectors. The unit vectors and corresponding quantities based on the two eigenvectors are presented.

Definition 1.16 Consider an n-dimensional, autonomous, nonlinear dynamical system $\dot{\mathbf{x}} = \mathbf{f}(\mathbf{x}, \mathbf{p})$ in Eq. (1.4) with an equilibrium point $\mathbf{x}^*$. Suppose there is a neighborhood of the equilibrium $\mathbf{x}^*$ as $U(\mathbf{x}^*) \subset \Omega$; then $\mathbf{f}(\mathbf{x}, \mathbf{p})$ is C^r-continuous ($r \geq 1$) and Eq. (1.24) holds in $U(\mathbf{x}^*) \subset \Omega$. The corresponding solution is $\mathbf{x}(t) = \mathbf{\Phi}(\mathbf{x}_0, t - t_0, \mathbf{p})$. For a linearized dynamical system in Eq. (1.19), consider a real eigenvalue λ_k of matrix $D\mathbf{f}(\mathbf{x}^*, \mathbf{p})$ ($k \in N = \{1, 2, \ldots, n\}$) with an eigenvector $\mathbf{v}_k$. For $\mathbf{y} = c^k \mathbf{v}_k$ (summation of $k = 1, 2, \cdots, n$) and $c^k = (\mathbf{v}^k)^{\mathrm{T}} \cdot \mathbf{y}$, $\dot{c}^k = (\mathbf{v}^k)^{\mathrm{T}} \cdot \dot{\mathbf{y}}$ with $\dot{\mathbf{y}} = D\mathbf{f}(\mathbf{x}^*, \mathbf{p})\mathbf{y} = D\mathbf{f}(\mathbf{x}^*, \mathbf{p})\mathbf{v}_j c^j = \lambda_j c^j \mathbf{v}_j$; thus $\dot{c}^k = \lambda_j c^j \delta^k_j = \lambda_k c^k$.

(i) $\mathbf{x}$ at the equilibrium $\mathbf{x}^*$ on the direction $\mathbf{v}_k$ is stable if

$$\lim_{t\to\infty} c^k = \lim_{t\to\infty} c_0^k e^{\lambda_k t} = 0 \text{ for } \lambda_k < 0. \tag{1.50}$$

(ii) $\mathbf{x}$ at the equilibrium $\mathbf{x}^*$ on the direction $\mathbf{v}_k$ is unstable if

$$\lim_{t\to\infty} | c^k | = \lim_{t\to\infty} | c_0^k e^{\lambda_k t} | = \infty \text{for } \lambda_k > 0. \tag{1.51}$$

(iii) $\mathbf{x}$ at the equilibrium $\mathbf{x}^*$ on the direction $\mathbf{v}_k$ is uncertain (critical) if

$$\lim_{t\to\infty} c^k = \lim_{t\to\infty} e^{\lambda_k t} c_0^k = c_0^k \text{ for } \lambda_k = 0. \tag{1.52}$$

Definition 1.17 Consider a $2n$-dimensional, autonomous, nonlinear dynamical system $\dot{\mathbf{x}} = \mathbf{f}(\mathbf{x}, \mathbf{p})$ in Eq. (1.4) with an equilibrium point $\mathbf{x}^*$. Suppose there is a neighborhood of the equilibrium $\mathbf{x}^*$ as $U(\mathbf{x}^*) \subset \Omega$; then $\mathbf{f}(\mathbf{x}, \mathbf{p})$ is C^r-continuous ($r \geq 1$) and Eq. (1.24) holds in $U(\mathbf{x}^*) \subset \Omega$. The corresponding solution in

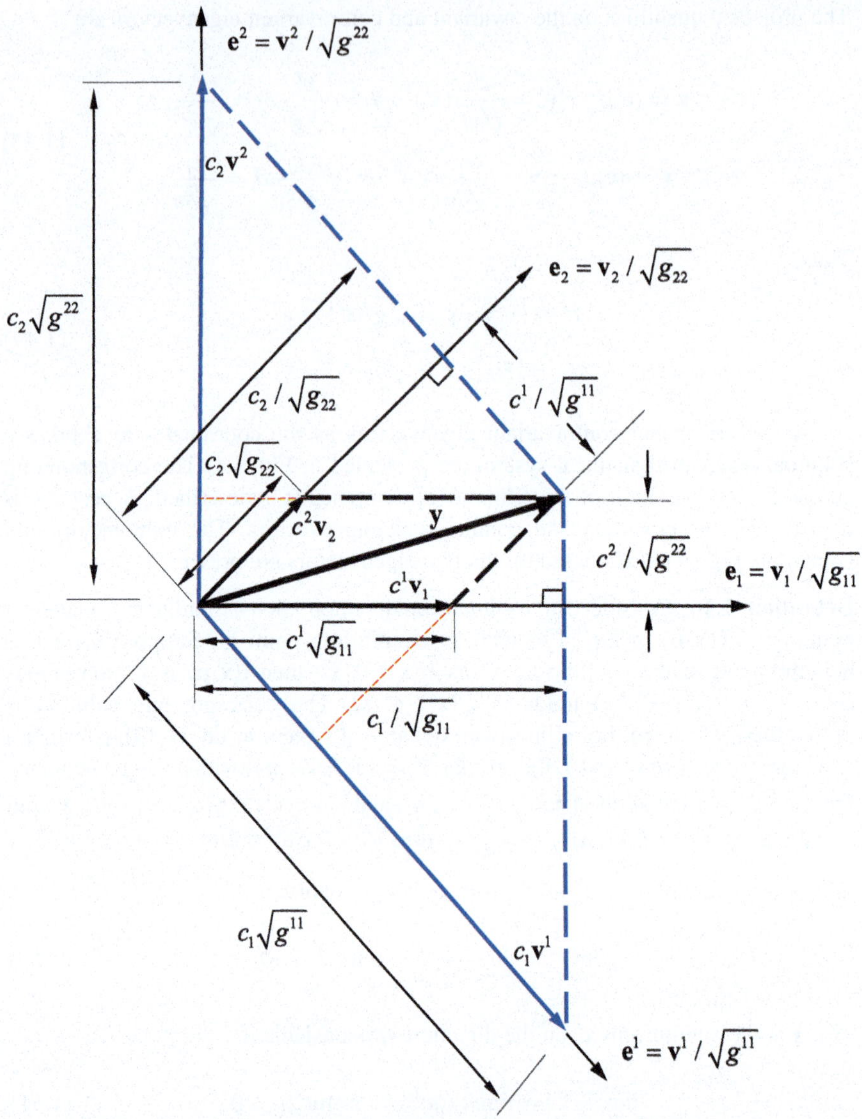

Fig. 1.1 A solution vector based on the covariant and contravariant eigenvectors in a 2-dimensional dynamical system. The thick black and blue colors are used for covariant and contravariant eigenvectors, respectively

$\mathbf{x}(t) = \mathbf{\Phi}(\mathbf{x}_0, t - t_0, \mathbf{p})$. For a linearized dynamical system in Eq. (1.19), consider a pair of complex eigenvalue $\alpha_k \pm \mathbf{i}\beta_k$ ($2k - 1, 2k \in N = \{1, 2, \ldots, n\}$, $\mathbf{i} = \sqrt{-1}$) of matrix $D\mathbf{f}(\mathbf{x}^*, \mathbf{p})$ with a pair of eigenvectors $\mathbf{v}_{2k-1} \pm \mathbf{i}\mathbf{v}_{2k}$. On the invariant plane of $(\mathbf{v}_{2k-1}, \mathbf{v}_{2k})$, for $\dot{\mathbf{y}} = D\mathbf{f}(\mathbf{x}^*, \mathbf{p})\mathbf{y} = D\mathbf{f}(\mathbf{x}^*, \mathbf{p})\mathbf{v}_j c^j$ with j = $2k$–1, $2k$, consider

$$c^{2k-1} = (\mathbf{v}^{2k-1})^{\mathrm{T}} \cdot \mathbf{y}, \;\; c^{2k} = (\mathbf{v}^{2k})^{\mathrm{T}} \cdot \mathbf{y} \tag{1.53}$$

with

$$\begin{aligned} D_{\mathbf{x}}\mathbf{f} \cdot \mathbf{v}_{2k-1} &= \alpha_k \mathbf{v}_{2k-1} - \beta_k \mathbf{v}_{2k}, \\ D_{\mathbf{x}}\mathbf{f} \cdot \mathbf{v}_{2k} &= \beta_k \mathbf{v}_{2k-1} + \alpha_k \mathbf{v}_{2k}. \end{aligned} \tag{1.54}$$

Thus,

$$\begin{aligned} \dot{c}^{2k-1} &= (\mathbf{v}^{2k-1})^{\mathrm{T}} \cdot \dot{\mathbf{y}} = (\mathbf{v}^{2k-1})^{\mathrm{T}} \cdot D\mathbf{f}(\mathbf{x}^*, \mathbf{p}) \mathbf{v}_j c^j = \alpha_k c^{2k-1} + \beta_k c^{2k}, \\ \dot{c}^{2k} &= (\mathbf{v}^{2k})^{\mathrm{T}} \cdot \dot{\mathbf{y}} = (\mathbf{v}^{2k})^{\mathrm{T}} \cdot D\mathbf{f}(\mathbf{x}^*, \mathbf{p}) \mathbf{v}_j c^j = -\beta_k c^{2k-1} + \alpha_k c^{2k}. \end{aligned} \tag{1.55}$$

For $\mathbf{c}^k = (c^{2k-1}, c^{2k})^{\mathrm{T}}$,

$$\dot{\mathbf{c}}^k = \mathbf{E}_k \mathbf{c}^k \Rightarrow \mathbf{c}^k = e^{\alpha_k t} \mathbf{B}_k \mathbf{c}_0^k \tag{1.56}$$

where

$$\mathbf{E}_k = \begin{bmatrix} \alpha_k & \beta_k \\ -\beta_k & \alpha_k \end{bmatrix} \text{ and } \mathbf{B}_k = \begin{bmatrix} \cos \beta_k t & \sin \beta_k t \\ -\sin \beta_k t & \cos \beta_k t \end{bmatrix}. \tag{1.57}$$

(i) $\mathbf{y}$ at the equilibrium $\mathbf{x}^*$ on the plane of $(\mathbf{v}_{2k-1}, \mathbf{v}_{2k})$ is spirally stable if

$$\lim_{t \to \infty} \|\mathbf{c}^k\| = \lim_{t \to \infty} e^{\alpha_k t} \|\mathbf{B}_k\| \times \|\mathbf{c}_0^k\| = 0 \text{ for } \operatorname{Re} \lambda_k = \alpha_k < 0. \tag{1.58}$$

(ii) $\mathbf{y}$ at the equilibrium $\mathbf{x}^*$ on the plane of $(\mathbf{v}_{2k-1}, \mathbf{v}_{2k})$ is spirally unstable if

$$\lim_{t \to \infty} \|\mathbf{c}^k\| = \lim_{t \to \infty} e^{\alpha_k t} \|\mathbf{B}_k\| \times \|\mathbf{c}_0^k\| = \infty \text{ for } \operatorname{Re} \lambda_k = \alpha_k > 0. \tag{1.59}$$

(iii) $\mathbf{y}$ at the equilibrium $\mathbf{x}^*$ on the plane of $(\mathbf{v}_{2k-1}, \mathbf{v}_{2k})$ is on the invariant circle if

$$\lim_{t \to \infty} \|\mathbf{c}^k\| = \lim_{t \to \infty} e^{\alpha_k t} \|\mathbf{B}_k\| \times \|\mathbf{c}_0^k\| = \|\mathbf{c}_0^k\| \text{ for } \operatorname{Re} \lambda_k = \alpha_k = 0. \tag{1.60}$$

(iv) $\mathbf{y}$ at the equilibrium $\mathbf{x}^*$ on the plane of $(\mathbf{v}_{2k-1}, \mathbf{v}_{2k})$ is degenerate in the direction of $\mathbf{v}_{2k-1}$ if $\operatorname{Im}\lambda_k = 0$.

Definition 1.18 Consider an n-dimensional, autonomous, nonlinear dynamical system $\dot{\mathbf{x}} = \mathbf{f}(\mathbf{x}, \mathbf{p})$ in Eq. (1.4) with an equilibrium point $\mathbf{x}^*$. Suppose there is a neighborhood of the equilibrium $\mathbf{x}^*$ as $U(\mathbf{x}^*) \subset \Omega$, and in the neighborhood $\mathbf{f}(\mathbf{x}, \mathbf{p})$ is C^r-continuous ($r \geq 1$) and Eq. (1.24) holds. The corresponding solution is

$\mathbf{x}(t) = \mathbf{\Phi}(\mathbf{x}_0, t - t_0, \mathbf{p})$. The linearized system of the nonlinear system at the equilibrium point $\mathbf{x}^*$ is $\dot{\mathbf{y}} = D\mathbf{f}(\mathbf{x}^*, \mathbf{p})\mathbf{y}$ ($\mathbf{y} = \mathbf{x} - \mathbf{x}^*$) in Eq. (1.19).

(i) The equilibrium $\mathbf{x}^*$ is called *a hyperbolic equilibrium* if none of the eigenvalues of $D\mathbf{f}(\mathbf{x}^*, \mathbf{p})$ is zero real part (i.e., $\mathrm{Re}\lambda_k \neq 0$ ($k = 1, 2, \ldots, n$)).
(ii) The equilibrium $\mathbf{x}^*$ is called *a sink* if all of the eigenvalues of $D\mathbf{f}(\mathbf{x}^*, \mathbf{p})$ have negative real parts (i.e., $\mathrm{Re}\,\lambda_k < 0$ ($k = 1, 2, \ldots, n$)).
(iii) The equilibrium $\mathbf{x}^*$ is called *a source* if all of the eigenvalues of $D\mathbf{f}(\mathbf{x}^*, \mathbf{p})$ have positive real parts (i.e., $\mathrm{Re}\lambda_k > 0$ ($k = 1, 2, \ldots, n$)).
(iv) The equilibrium $\mathbf{x}^*$ is called *a saddle* if it is a hyperbolic equilibrium and $D\mathbf{f}(\mathbf{x}^*, \mathbf{p})$ has at least one eigenvalue with a positive real part (i.e., $\mathrm{Re}\lambda_j > 0$ ($j \in \{1, 2, \ldots, n\}$) and at least one with a negative real part (i.e., $\mathrm{Re}\lambda_k < 0$ ($k \in \{1, 2, \ldots, n\}$).
(v) The equilibrium $\mathbf{x}^*$ is called a center if all of eigenvalues of $D\mathbf{f}(\mathbf{x}^*, \mathbf{p})$ have zero real parts (i.e., $\mathrm{Re}\lambda_j = 0$ ($j = 1, 2, \ldots, n$)) with distinct eigenvalues.
(vi) The equilibrium $\mathbf{x}^*$ is called a stable node if all of eigenvalues of $D\mathbf{f}(\mathbf{x}^*, \mathbf{p})$ are real $\lambda_k < 0$ ($k = 1, 2, \ldots, n$).
(vii) The equilibrium $\mathbf{x}^*$ is called an unstable node if all of eigenvalues of $D\mathbf{f}(\mathbf{x}^*, \mathbf{p})$ are real $\lambda_k > 0$ ($k = 1, 2, \ldots, n$).
(viii) The equilibrium $\mathbf{x}^*$ is called a degenerate case if one of eigenvalues of $D\mathbf{f}(\mathbf{x}^*, \mathbf{p})$ is zero $\lambda_k = 0$ ($k \in \{1, 2, \ldots, n\}$).

The generalized stability and bifurcation of flows in linearized, nonlinear dynamical systems in Eq. (1.4) will be discussed as follows.

Definition 1.19 Consider an n-dimensional, autonomous, nonlinear dynamical system $\dot{\mathbf{x}} = \mathbf{f}(\mathbf{x}, \mathbf{p})$ in Eq. (1.4) with an equilibrium point $\mathbf{x}^*$. Suppose there is a neighborhood of the equilibrium $\mathbf{x}^*$ as $U(\mathbf{x}^*) \subset \Omega$, and in the neighborhood $\mathbf{f}(\mathbf{x}, \mathbf{p})$ is C^r-continuous ($r \geq 1$) and Eq. (1.24) holds. The corresponding solution is $\mathbf{x}(t) = \mathbf{\Phi}(\mathbf{x}_0, t - t_0, \mathbf{p})$. The matrix $D\mathbf{f}(\mathbf{x}^*, \mathbf{p})$ in Eq. (1.19) possesses n eigenvalues λ_k ($k = 1, 2, \ldots, n$). Set $N = \{1, 2, \ldots, m, m{+}1, \ldots, (n{+}m)/2\}$ and $N_i = \{i_1, i_2, \ldots, i_{n_i}\} \cup \varnothing$ with $i_j \in N$ ($j = 1, 2, \ldots, n_i$; $i = 1, 2, \ldots, 6$), $\Sigma_{i=1}^{3} n_i = m$ and $2\Sigma_{i=4}^{6} n_i = n - m$. $\cup_{i=1}^{6} N_i = N$ with $N_i \cap N_l = \varnothing$ ($l \neq i$). $N_i = \varnothing$ if $n_i = 0$. The matrix $D\mathbf{f}(\mathbf{x}^*, \mathbf{p})$ possesses n_1-stable, n_2-unstable, and n_3-invariant real eigenvectors plus n_4-stable, n_5-unstable, and n_6-center pairs of complex eigenvectors. Without repeated complex eigenvalues of $\mathrm{Re}\lambda_k = 0$ ($k \in N_3 \cup N_6$), the flow $\mathbf{\Phi}(t)$ of the nonlinear system $\dot{\mathbf{x}} = \mathbf{f}(\mathbf{x}, \mathbf{p})$ is an $(n_1 : n_2 : [n_3; m_3] | n_4 : n_5 : n_6)$ flow in the neighborhood of $\mathbf{x}^*$. However, with repeated complex eigenvalues of $\mathrm{Re}\lambda_k = 0$ ($k \in N_3 \cup N_6$), the flow $\mathbf{\Phi}(t)$ of the nonlinear system $\dot{\mathbf{x}} = \mathbf{f}(\mathbf{x}, \mathbf{p})$ is an $(n_1 : n_2 : [n_3; \pi_3] | n_4 : n_5 : [n_6, \mathbf{r}; \mathbf{s}])$ flow with $\mathbf{N}^{m_3} = \mathbf{A}^{m_3} = \mathbf{0}$ ($\pi_3 \in \{\varnothing, m_3\}$), and $\mathbf{r} = (r_1, r_2, \ldots, r_l)$, $\mathbf{s} = (s_1, s_2, \ldots, s_l)$ for $l \leq n/2$ and $q_i \in \{\varnothing, m_{6i}\}$ ($i = 1, 2, \ldots, l$). The meanings of notations in the aforementioned structures are defined as follows:

(i) n_1 represents exponential sinks on n_1-directions of $\mathbf{v}_k$ if $\lambda_k < 0$ ($k \in N_1$ and $1 \leq n_1 \leq n$) with distinct or repeated eigenvalues.
(ii) n_2 represents exponential sources on n_2-directions of $\mathbf{v}_k$ if $\lambda_k > 0$ ($k \in N_2$ and $1 \leq n_2 \leq n$) with distinct or repeated eigenvalues.

(iii) $n_3 = 1$ represents an invariant center on 1-direction of $\mathbf{v}_k$ if $\lambda_k = 0$ ($k \in N_3$ and $n_3 = 1$).
(iv) n_4 represents spiral sinks on n_4-pairs of $(\mathbf{u}_k, \mathbf{v}_k)$ if $\mathrm{Re}\lambda_k < 0$ and $\mathrm{Im}\lambda_k \neq 0$ ($k \in N_4$ and $1 \leq n_4 \leq n$) with distinct or repeated eigenvalues.
(v) n_5 represents spiral sources on n_5-pairs of $(\mathbf{u}_k, \mathbf{v}_k)$ if $\mathrm{Re}\lambda_k > 0$ and $\mathrm{Im}\lambda_k \neq 0$ ($k \in N_5$ and $1 \leq n_5 \leq n$) with distinct or repeated eigenvalues.
(vi) n_6 represents invariant centers on n_6-pairs of $(\mathbf{u}_k, \mathbf{v}_k)$ if $\mathrm{Re}\lambda_k = 0$ and $\mathrm{Im}\lambda_k \neq 0$ ($k \in N_6$ and $1 \leq n_6 \leq n$) with distinct eigenvalues.
(vii) $\varnothing$ represents empty or none if $n_i = 0$ ($i \in \{1, 2, \ldots, 6\}$).
(viii) $[n_3; m_3]$ represents invariant centers on $(n_3 - m_3)$-directions of $\mathbf{v}_{k_3}$ ($k_3 \in N_3$) and sources in m_3-directions of $\mathbf{v}_{j_3}$ ($j_3 \in N_3$ and $j_3 \neq k_3$) if $\lambda_k = 0$ ($k \in N_3$ and $n_3 \leq n$) with the $(m_3 + 1)$th-order nilpotent matrix $\mathbf{N}_3^{m_3+1} = \mathbf{0}$ ($0 < m_3 \leq n_2 - 1$).
(ix) $[n_3; \varnothing]$ represents invariant centers on n_3-directions of $\mathbf{v}_k$ if $\lambda_k = 0$ ($k \in N_3$ and $1 < n_3 \leq n$) with a nilpotent matrix $\mathbf{N}_3 = \mathbf{0}$.
(x) $[n_6, \mathbf{r}; \mathbf{s}]$ represents invariant centers on $(r_i - m_{6i})$ pairs of $(\mathbf{u}_{k_{6i}}, \mathbf{v}_{k_{6i}})$ ($k_{6i} \in N_{6i}$), and sources in m_{6i} pairs of $(\mathbf{u}_{j_6}, \mathbf{v}_{j_6})$ ($j_6 \in N_{6i}$ and $j_6 \neq k_{6i}$) if $\mathrm{Re}\lambda_k = 0$ and $\mathrm{Im}\lambda_k \neq 0$ ($k \in N_{6i}$ and $n_6 \leq n/2$) for $(l + 1)$ pairs of repeated eigenvalues with the $(m_{6_i} + 1)$th-order nilpotent matrix $\mathbf{N}_{6i}^{m_{6i}+1} = \mathbf{0}$ ($0 < m_{6i} \leq r_i$) and ($i = 1, 2, \cdots, l$).
(xi) $[n_6, \mathbf{r}; \varnothing]$ represents invariant centers on n_6 pairs of $(\mathbf{u}_k, \mathbf{v}_k)$ if $\mathrm{Re}\lambda_k = 0$ and $\mathrm{Im}\lambda_k \neq 0$ ($k \in N_6$ and $1 \leq n_6 \leq n$) for $(p_i + 1)$ pairs of repeated eigenvalues with a nilpotent matrix $\mathbf{N}_{6i} = \mathbf{0}$ ($i = 1, 2, \cdots, l$).

Definition 1.20 Consider an n-dimensional, autonomous, nonlinear dynamical system $\dot{\mathbf{x}} = \mathbf{f}(\mathbf{x}, \mathbf{p})$ in Eq. (1.4) with an equilibrium point $\mathbf{x}^*$. Suppose there is a neighborhood of the equilibrium $\mathbf{x}^*$ as $U(\mathbf{x}^*) \subset \Omega$, and in the neighborhood $\mathbf{f}(\mathbf{x}, \mathbf{p})$ is C^r-continuous ($r \geq 1$) and Eq. (1.24) holds. The corresponding solution is $\mathbf{x}(t) = \mathbf{\Phi}(\mathbf{x}_0, t - t_0, \mathbf{p})$. The matrix $D\mathbf{f}(\mathbf{x}^*, \mathbf{p})$ in Eq. (1.19) possesses n eigenvalues λ_k ($k = 1, 2, \ldots, n$). Set $N = \{1, 2, \ldots, m, m + 1, \ldots, (n + m)/2\}$ and $N_i = \{i_1, i_2, \ldots, i_{n_i}\} \cup \varnothing$ with $i_j \in N$ ($j = 1, 2, \ldots, n_i$; $i = 1, 2, \ldots, 6$), $\Sigma_{i=1}^{3} n_i = m$ and $2\Sigma_{i=4}^{6} n_i = n - m$. $\cup_{i=1}^{6} N_i = N$ with $N_i \cap N_l = \varnothing$ ($l \neq i$). $N_i = \varnothing$ if $n_i = 0$. The matrix $D\mathbf{f}(\mathbf{x}^*, \mathbf{p})$ possesses n_1-stable, n_2-unstable, and n_3-invariant real eigenvectors plus n_4-stable, n_5-unstable, and n_6-center pairs of complex eigenvectors.

I. Nondegenerate cases

(i) The equilibrium point $\mathbf{x}^*$ is an $(n_1 : n_2 : \varnothing | n_4 : n_5 : \varnothing)$ hyperbolic point (or saddle) for the nonlinear system.
(ii) The equilibrium point $\mathbf{x}^*$ is an $(n_1 : \varnothing : \varnothing | n_4 : \varnothing : \varnothing)$ sink for the nonlinear system.
(iii) The equilibrium point $\mathbf{x}^*$ is an $(\varnothing : n_2 : \varnothing | \varnothing : n_5 : \varnothing)$ source for the nonlinear system.
(iv) The equilibrium point $\mathbf{x}^*$ is an $(\varnothing : \varnothing : \varnothing | \varnothing : \varnothing : n/2)$ center for the nonlinear system.

(v) The equilibrium point $\mathbf{x}^*$ is an $(\varnothing : \varnothing : \varnothing | \varnothing : \varnothing : [n/2, \mathbf{r}; \varnothing])$ center for the nonlinear system.
(vi) The equilibrium point $\mathbf{x}^*$ is an $(\varnothing : \varnothing : \varnothing | \varnothing : \varnothing : [n/2, \mathbf{r}; \mathbf{s}])$ point for the nonlinear system.
(vii) The equilibrium point $\mathbf{x}^*$ is an $(n_1 : \varnothing : \varnothing | n_4 : \varnothing : n_6)$ point for the nonlinear system.
(viii) The equilibrium point $\mathbf{x}^*$ is an $(\varnothing : n_2 : \varnothing | \varnothing : n_5 : n_6)$ point for the nonlinear system.
(ix) The equilibrium point $\mathbf{x}^*$ is an $(n_1 : n_2 : \varnothing | n_4 : n_5 : n_6)$ point for the nonlinear system.

II. Simple degenerate cases

(i) The equilibrium point $\mathbf{x}^*$ is an $(\varnothing : \varnothing : [n; \varnothing] | \varnothing : \varnothing : \varnothing)$-invariant (or static) center for the nonlinear system.
(ii) The equilibrium point $\mathbf{x}^*$ is an $(\varnothing : \varnothing : [n; m_3] | \varnothing : \varnothing : \varnothing)$ point for the nonlinear system.
(iii) The equilibrium point $\mathbf{x}^*$ is an $(\varnothing : \varnothing : [n_3; \varnothing] | \varnothing : \varnothing : n_6)$ point for the nonlinear system.
(iv) The equilibrium point $\mathbf{x}^*$ is an $(\varnothing : \varnothing : [n_3; m_3] | \varnothing : \varnothing : n_6)$ point for the nonlinear system.
(v) The equilibrium point $\mathbf{x}^*$ is an $(\varnothing : \varnothing : [n_3; \varnothing] | \varnothing : \varnothing : [n_6, l; \varnothing])$ point for the nonlinear system.
(vi) The equilibrium point $\mathbf{x}^*$ is an $(\varnothing : \varnothing : [n_3; m_3] | \varnothing : \varnothing : [n_6, l; \varnothing])$ point for the nonlinear system.
(vii) The equilibrium point $\mathbf{x}^*$ is an $(\varnothing : \varnothing : [n_3; \varnothing] | \varnothing : \varnothing : [n_6, l; m_6])$ point for the nonlinear system.
(viii) The equilibrium point $\mathbf{x}^*$ is an $(\varnothing : \varnothing : [n_3; m_3] | \varnothing : \varnothing : [n_6, l; m_6])$ point for the nonlinear system.

III. Complex degenerate cases

(i) The equilibrium point $\mathbf{x}^*$ is an $(n_1 : \varnothing : [n_3; \varnothing] | n_4 : \varnothing : \varnothing)$ point for the nonlinear system.
(ii) The equilibrium point $\mathbf{x}^*$ is an $(n_1 : \varnothing : [n_3; m_3] | n_4 : \varnothing : \varnothing)$ point for the nonlinear system.
(iii) The equilibrium point $\mathbf{x}^*$ is an $(\varnothing : n_2 : [n_3; \varnothing] | \varnothing : n_5 : \varnothing)$ point for the nonlinear system.
(iv) The equilibrium point $\mathbf{x}^*$ is an $(\varnothing : n_2 : [n_3; m_3] | \varnothing : n_5 : \varnothing)$ point for the nonlinear system.
(v) The equilibrium point $\mathbf{x}^*$ is an $(n_1 : \varnothing : [n_3; \varnothing] | n_4 : \varnothing : n_6)$ point for the nonlinear system.
(vi) The equilibrium point $\mathbf{x}^*$ is an $(n_1 : \varnothing : [n_3; m_3] | n_4 : \varnothing : n_6)$ point for the nonlinear system.
(vii) The equilibrium point $\mathbf{x}^*$ is an $(\varnothing : n_2 : [n_3; \varnothing] | \varnothing : n_5 : n_6)$ point for the nonlinear system.
(viii) The equilibrium point $\mathbf{x}^*$ is an $(\varnothing : n_2 : [n_3; m_3] | \varnothing : n_5 : n_6)$ point for the nonlinear system.

Definition 1.21 Consider an n-dimensional, autonomous, nonlinear dynamical system $\dot{\mathbf{x}} = \mathbf{f}(\mathbf{x}, \mathbf{p})$ in Eq. (1.4) with an equilibrium point $\mathbf{x}^*$. Suppose there is a neighborhood of the equilibrium $\mathbf{x}^*$ as $U(\mathbf{x}^*) \subset \Omega$, and in the neighborhood $\mathbf{f}(\mathbf{x}, \mathbf{p})$ is C^r-continuous ($r \geq 1$) and Eq. (1.24) holds. The corresponding solution is $\mathbf{x}(t) = \mathbf{\Phi}(\mathbf{x}_0, t - t_0, \mathbf{p})$. The matrix $D\mathbf{f}(\mathbf{x}^*, \mathbf{p})$ in Eq. (1.19) possesses n eigenvalues λ_k ($k = 1, 2, \ldots, n$). Set $N = \{1, 2, \ldots, n\}$ and $N_i = \{i_1, i_2, \ldots, i_{n_i}\} \cup \varnothing$, with $i_j \in N$ ($j = 1, 2, \ldots, n_i$, $i = 1, 2, 3$) and $\Sigma_{i=1}^{3} n_i = n$. $\cup_{i=1}^{3} N_i = N$ and $N_i \cap N_l = \varnothing$ ($l \neq i$). $N_i = \varnothing$ if $n_i = 0$. The matrix $D\mathbf{f}(\mathbf{x}^*, \mathbf{p})$ possesses n_1-stable, n_2-unstable, and n_3-invariant real eigenvectors. Without repeated eigenvalues of $\lambda_k = 0$ ($k \in N_3$), the flow $\mathbf{\Phi}(t)$ of the nonlinear system $\dot{\mathbf{x}} = \mathbf{f}(\mathbf{x}, \mathbf{p})$ in Eq. (1.4) is a $(n_1 : n_2 : \varnothing \,|$ or $(n_1 : n_2 : 1|$ local flow in the neighborhood of equilibrium point $\mathbf{x}^*$. However, with repeated eigenvalues of $\lambda_k = 0$ ($k \in N_3$), the flow $\mathbf{\Phi}(t)$ of the nonlinear system $\dot{\mathbf{x}} = \mathbf{f}(\mathbf{x}, \mathbf{p})$ in Eq. (1.4) is a $(n_1 : n_2 : [n_3; m_3]|$ local flow in the neighborhood of equilibrium point $\mathbf{x}^*$.

I. Nondegenerate cases

(i) The equilibrium point $\mathbf{x}^*$ is an $(n : \varnothing : \varnothing\,|$-stable node for the nonlinear system.
(ii) The equilibrium point $\mathbf{x}^*$ is an $(\varnothing : n : \varnothing\,|$-unstable node for the nonlinear system.
(iii) The equilibrium point $\mathbf{x}^*$ is an $(n_1 : n_2 : \varnothing\,|$ saddle for the nonlinear system.

II. Degenerate cases

(i) The equilibrium point $\mathbf{x}^*$ is an $(n_1 : n_2 : 1|$-critical state for the nonlinear system.
(ii) The equilibrium point $\mathbf{x}^*$ is an $(n_1 : n_2 : [n_3; \varnothing]|$ point for the nonlinear system.
(iii) The equilibrium point $\mathbf{x}^*$ is an $(n_1 : n_2 : [n_3; m_3]|$ point for the nonlinear system.

Definition 1.22 Consider a $2n$-dimensional, autonomous, nonlinear dynamical system $\dot{\mathbf{x}} = \mathbf{f}(\mathbf{x}, \mathbf{p})$ in Eq. (1.4) with an equilibrium point $\mathbf{x}^*$. Suppose there is a neighborhood of the equilibrium $\mathbf{x}^*$ as $U(\mathbf{x}^*) \subset \Omega$, and in the neighborhood $\mathbf{f}(\mathbf{x}, \mathbf{p})$ is C^r-continuous ($r \geq 1$) and Eq. (1.24) holds. The corresponding solution is $\mathbf{x}(t) = \mathbf{\Phi}(\mathbf{x}_0, t - t_0, \mathbf{p})$. The matrix $D\mathbf{f}(\mathbf{x}^*, \mathbf{p})$ in Eq. (1.19) possesses n pairs of complex eigenvalues ($k = 1, 2, \ldots, n$). Set $N = \{1, 2, \ldots, n\}$ and $N_i = \{i_1, i_2, \ldots, i_{n_i}\} \cup \varnothing$ with $i_j \in N$ ($j = 1, 2, \ldots, n_i$; $i = 4, 5, 6$) and $\Sigma_{i=4}^{6} n_i = n$. $\cup_{i=4}^{6} N_i = N$ and $N_i \cap N_l = \varnothing$ ($l \neq i$). $N_i = \varnothing$ if $n_i = 0$. The matrix $D\mathbf{f}(\mathbf{x}^*, \mathbf{p})$ possesses n_4-stable, n_5-unstable, and n_6-center pairs of complex eigenvectors. Without repeated eigenvalues of $\mathrm{Re}\lambda_k = 0$ ($k \in N_6$), the flow $\mathbf{\Phi}(t)$ of the nonlinear system $\dot{\mathbf{x}} = \mathbf{f}(\mathbf{x}, \mathbf{p})$ in Eq. (1.4) is a $|n_4 : n_5 : n_6)$ local flow in the neighborhood of equilibrium point $\mathbf{x}^*$. However, with repeated eigenvalues of $\mathrm{Re}\lambda_k = 0$ ($k \in N_6$), the flow $\mathbf{\Phi}(t)$ of the nonlinear system $\dot{\mathbf{x}} = \mathbf{f}(\mathbf{x}, \mathbf{p})$ in Eq. (1.4) is a $|n_4 : n_5 : [n_6, \mathbf{r}; \mathbf{s}])$ local flow in the neighborhood of equilibrium point $\mathbf{x}^*$.

I. Nondegenerate cases

(i) The equilibrium point $\mathbf{x}^*$ is an $|n : \varnothing : \varnothing)$-spiral sink for the nonlinear system.
(ii) The equilibrium point $\mathbf{x}^*$ is an $|\varnothing : n : \varnothing)$-spiral source for the nonlinear system.
(iii) The equilibrium point $\mathbf{x}^*$ is an $|\varnothing : \varnothing : n)$ center for the nonlinear system.
(iv) The equilibrium point $\mathbf{x}^*$ is an $|n_4 : n_5 : \varnothing)$-spiral saddle for the nonlinear system.

II. Quasi-degenerate cases

(i) The equilibrium point $\mathbf{x}^*$ is an $|n_4 : \varnothing : n_6)$ point for the nonlinear system.
(ii) The equilibrium point $\mathbf{x}^*$ is an $|\varnothing : n_5 : n_6)$ point for the nonlinear system.
(iii) The equilibrium point $\mathbf{x}^*$ is an $|n_4 : \varnothing : [n_6, \mathbf{r}; \varnothing])$ point for the nonlinear system.
(iv) The equilibrium point $\mathbf{x}^*$ is an $|n_4 : \varnothing : [n_6, \mathbf{r}; \mathbf{s}])$ point for the nonlinear system.
(v) The equilibrium point $\mathbf{x}^*$ is an $|\varnothing : n_5 : [n_6, \mathbf{r}; \varnothing])$ point for the nonlinear system.
(vi) The equilibrium point $\mathbf{x}^*$ is an $|\varnothing : n_5 : [n_6, \mathbf{r}; \mathbf{s}])$ point for the nonlinear system.

1.3 Stability and Singularity

To extend the idea of Definitions 1.16 and 1.17, a new function will be defined to determine the stability and the stability-state switching as in Luo (2019).

1.3.1 *Hyperbolic Stability on Eigenvectors*

Definition 1.24 Consider an n-dimensional, autonomous, nonlinear dynamical system $\dot{\mathbf{x}} = \mathbf{f}(\mathbf{x}, \mathbf{p})$ in Eq. (1.4) with an equilibrium point $\mathbf{x}^*$ and $\mathbf{f}(\mathbf{x}, \mathbf{p})$ is C^r-continuous ($r \geq 1$) in a neighborhood of the equilibrium $\mathbf{x}^*$. The corresponding solution is $\mathbf{x}(t) = \mathbf{\Phi}(\mathbf{x}_0, t - t_0, \mathbf{p})$. Suppose $U(\mathbf{x}^*) \subset \Omega$ is a neighborhood of equilibrium $\mathbf{x}^*$, and there are n linearly independent vectors $\mathbf{v}_k$ ($k = 1, 2, \ldots, n$). For a perturbation of equilibrium ($\mathbf{y} = \mathbf{x} - \mathbf{x}^*$), let $\mathbf{y} = c^k\mathbf{v}_k = c_k\mathbf{v}^k$ and $\dot{\mathbf{y}} = \dot{c}^k\mathbf{v}_k$ (summation with $k = 1, 2, \ldots, n$):

$$c^k = (\mathbf{v}^k)^{\mathrm{T}} \cdot \mathbf{y} = (\mathbf{v}^k)^{\mathrm{T}} \cdot (\mathbf{x} - \mathbf{x}^*). \tag{1.61}$$

Thus

$$\dot{c}^k = G_k(\mathbf{x},\mathbf{p}) = (\mathbf{v}^k)^{\mathrm{T}} \cdot \dot{\mathbf{x}} = (\mathbf{v}^k)^{\mathrm{T}} \cdot \mathbf{f}(\mathbf{x},\mathbf{p})$$
$$= \sum_{r_k=1}^{\infty} \frac{1}{r_k!} G^{(m)}_{k(j_1 j_2 \cdots j_m)}(\mathbf{x}^*,\mathbf{p}) c^{j_1} c^{j_2} \cdots c^{j_{r_k}} \tag{1.62}$$
$$(\text{summation of } j_r = 1,2,\cdots,n \text{ and } r_k = 1,2,\cdots,m\ ),$$

where

$$\begin{aligned} G^{(1)}_{k(j_1)}(\mathbf{x}^*,\mathbf{p}) &= (\mathbf{v}^k)^{\mathrm{T}} \cdot D_{c^{j_1}} \mathbf{f}(\mathbf{x}^*(c^1,\ldots,c^n),\mathbf{p}) \\ &= (\mathbf{v}^k)^{\mathrm{T}} \cdot D_{\mathbf{x}} \mathbf{f}(\mathbf{x}^*(c^1,\ldots,c^n),\mathbf{p}) \partial_{c^{j_1}} \mathbf{x} \\ &= (\mathbf{v}^k)^{\mathrm{T}} \cdot D_{\mathbf{x}} \mathbf{f}(\mathbf{x}^*(c^1,\ldots,c^n),\mathbf{p}) \mathbf{v}_{j_1} \\ (j_1 &= 1,2,\ldots,n), \end{aligned} \tag{1.63}$$

$$\begin{aligned} G^{(m)}_{k(j_1 j_2 \cdots j_m)}(\mathbf{x}^*,\mathbf{p}) &= (\mathbf{v}^k)^{\mathrm{T}} \cdot D^{(m)}_{c^{j_1} c^{j_2} \cdots c^{j_m}} \mathbf{f}(\mathbf{x}^*(c^1,\ldots,c^n),\mathbf{p}) \\ &= (\mathbf{v}^k)^{\mathrm{T}} \cdot D^{(m)}_{\mathbf{x}} \mathbf{f}(\mathbf{x}^*(c^1,\ldots,c^n),\mathbf{p})(\mathbf{v}_{j_1}\mathbf{v}_{j_2}\cdots\mathbf{v}_{j_m}) \\ (j_1, j_2, \ldots, j_m &= 1,2,\ldots,n), \end{aligned} \tag{1.64}$$

with $G_k^{(0)}(\mathbf{x},\mathbf{p}) = G_k(\mathbf{x},\mathbf{p})$ if $m = 0$, and $D_{\mathbf{x}}^{(m)}\mathbf{f}(\mathbf{x}(c_1,\cdots,c_2)$ is the $(m+1)^{th}$-order tensor.

Definition 1.25 Consider an n-dimensional, autonomous, nonlinear dynamical system $\dot{\mathbf{x}} = \mathbf{f}(\mathbf{x},\mathbf{p})$ in Eq. (1.4) with an equilibrium point $\mathbf{x}^*$ and $\mathbf{f}(\mathbf{x},\mathbf{p})$ is C^r-continuous ($r \geq 1$) in a neighborhood of the equilibrium $\mathbf{x}^*$. The corresponding solution is $\mathbf{x}(t) = \mathbf{\Phi}(\mathbf{x}_0, t - t_0, \mathbf{p})$. Suppose $U(\mathbf{x}^*) \subset \Omega$ is a neighborhood of equilibrium $\mathbf{x}^*$, and there are n linearly independent vectors $\mathbf{v}_k$ ($k = 1, 2, \ldots, n$). For a perturbation of equilibrium $\mathbf{y} = \mathbf{x} - \mathbf{x}^*$, let $\mathbf{y} = c^k\mathbf{v}_k$ and $\dot{\mathbf{y}} = \dot{c}^k\mathbf{v}_k$ (summation of k).

(i) $\mathbf{x}$ at the equilibrium $\mathbf{x}^*$ on the direction $\mathbf{v}_k$ is stable if

$$\begin{aligned} &(\mathbf{v}^k)^{\mathrm{T}} \cdot (\mathbf{x}(t+\varepsilon) - \mathbf{x}(t)) < 0 \ \text{ for } (\mathbf{v}^k)^{\mathrm{T}} \cdot (\mathbf{x}(t) - \mathbf{x}^*) > 0, \\ &(\mathbf{v}^k)^{\mathrm{T}} \cdot (\mathbf{x}(t+\varepsilon) - \mathbf{x}(t)) > 0 \ \text{ for } (\mathbf{v}^k)^{\mathrm{T}} \cdot (\mathbf{x}(t) - \mathbf{x}^*) < 0 \end{aligned} \tag{1.65}$$

for all $\mathbf{x} \in U(\mathbf{x}^*) \subset \Omega$ and all $t \in [t_0, \infty)$. The equilibrium $\mathbf{x}^*$ is called a sink (or a stable node) on the direction $\mathbf{v}_k$.

(ii) $\mathbf{x}$ at the equilibrium $\mathbf{x}^*$ on the direction $\mathbf{v}_k$ is unstable if

$$\begin{aligned}(\mathbf{v}^k)^{\mathrm{T}} \cdot (\mathbf{x}(t+\varepsilon) - \mathbf{x}(t)) > 0 \text{ for } (\mathbf{v}^k)^{\mathrm{T}} \cdot (\mathbf{x}(t) - \mathbf{x}^*) > 0,\\ (\mathbf{v}^k)^{\mathrm{T}} \cdot (\mathbf{x}(t+\varepsilon) - \mathbf{x}(t)) < 0 \text{ for } (\mathbf{v}^k)^{\mathrm{T}} \cdot (\mathbf{x}(t) - \mathbf{x}^*) < 0\end{aligned} \tag{1.66}$$

for all $\mathbf{x} \in U(\mathbf{x}^*) \subset \Omega$ and all $t \in [t_0, \infty)$. The equilibrium $\mathbf{x}^*$ is called a source (or an unstable node) on the direction $\mathbf{v}_k$.

(iii) $\mathbf{x}$ at the equilibrium $\mathbf{x}^*$ on the direction $\mathbf{v}_k$ is increasingly unstable if

$$\begin{aligned}(\mathbf{v}^k)^{\mathrm{T}} \cdot (\mathbf{x}(t+\varepsilon) - \mathbf{x}(t)) > 0 \text{ for } (\mathbf{v}^k)^{\mathrm{T}} \cdot (\mathbf{x}(t) - \mathbf{x}^*) > 0,\\ (\mathbf{v}^k)^{\mathrm{T}} \cdot (\mathbf{x}(t+\varepsilon) - \mathbf{x}(t)) > 0 \text{ for } (\mathbf{v}^k)^{\mathrm{T}} \cdot (\mathbf{x}(t) - \mathbf{x}^*) < 0\end{aligned} \tag{1.67}$$

for all $\mathbf{x} \in U(\mathbf{x}^*) \subset \Omega$ and all $t \in [t_0, \infty)$. The equilibrium $\mathbf{x}^*$ is called an increasing saddle (or an *upper saddle*) on the direction $\mathbf{v}_k$.

(iv) $\mathbf{x}$ at the equilibrium $\mathbf{x}^*$ on the direction $\mathbf{v}_k$ is decreasingly unstable if

$$\begin{aligned}(\mathbf{v}^k)^{\mathrm{T}} \cdot (\mathbf{x}(t+\varepsilon) - \mathbf{x}(t)) < 0 \text{ for } (\mathbf{v}^k)^{\mathrm{T}} \cdot (\mathbf{x}(t) - \mathbf{x}^*) > 0,\\ (\mathbf{v}^k)^{\mathrm{T}} \cdot (\mathbf{x}(t+\varepsilon) - \mathbf{x}(t)) < 0 \text{ for } (\mathbf{v}^k)^{\mathrm{T}} \cdot (\mathbf{x}(t) - \mathbf{x}^*) < 0\end{aligned} \tag{1.68}$$

for all $\mathbf{x} \in U(\mathbf{x}^*) \subset \Omega$ and all $t \in [t_0, \infty)$. The equilibrium $\mathbf{x}^*$ is called a decreasing saddle (or a *lower saddle*) on the direction $\mathbf{v}_k$.

(v) $\mathbf{x}$ at the equilibrium $\mathbf{x}^*$ on the direction $\mathbf{v}_k$ is invariant if

$$\begin{aligned}(\mathbf{v}^k)^{\mathrm{T}} \cdot (\mathbf{x}(t+\varepsilon) - \mathbf{x}(t)) = 0\\ \text{for } (\mathbf{v}^k)^{\mathrm{T}} \cdot (\mathbf{x}(t) - \mathbf{x}^*) \neq 0\end{aligned} \tag{1.69}$$

for all $\mathbf{x} \in U(\mathbf{x}^*) \subset \Omega$ and all $t \in [t_0, \infty)$. The equilibrium $\mathbf{x}^*$ is called to be degenerate on the direction $\mathbf{v}_k$.

Theorem 1.7 *Consider an n-dimensional, autonomous, nonlinear dynamical system* $\dot{\mathbf{x}} = \mathbf{f}(\mathbf{x}, \mathbf{p})$ *in Eq. (1.4) with an equilibrium point* $\mathbf{x}^*$ *and* $\mathbf{f}(\mathbf{x}, \mathbf{p})$ *is* C^r $(r \geq 1)$*-continuous in a neighborhood of the equilibrium* $\mathbf{x}^*$ *(i.e.,* $U(\mathbf{x}^*) \subset \Omega$*). The corresponding solution is* $\mathbf{x}(t) = \mathbf{\Phi}(\mathbf{x}_0, t - t_0, \mathbf{p})$*. Suppose Eq. (1.24) holds in* $U(\mathbf{x}^*) \subset \Omega$*. For a linearized dynamical system in Eq. (1.19), consider a real eigenvalue* λ_k *of matrix* $D\mathbf{f}(\mathbf{x}^*, \mathbf{p})$ $(k \in N = \{1, 2, \ldots, n\})$ *with an eigenvector* $\mathbf{v}_k$*. Let* $\mathbf{y} = c^k \mathbf{v}_k$ *and* $\dot{\mathbf{y}} = \dot{c}^k \mathbf{v}_k$ *(summation of* $k = 1, 2, \ldots, n$*). Define*

$$\dot{c}^k = (\mathbf{v}^k)^{\mathrm{T}} \cdot \dot{\mathbf{y}} = (\mathbf{v}^k)^{\mathrm{T}} \cdot \dot{\mathbf{x}} = (\mathbf{v}^k)^{\mathrm{T}} \cdot \mathbf{f}(\mathbf{x}, \mathbf{p}). \tag{1.70}$$

(i) **x** *at the equilibrium* $\mathbf{x}^*$ *on the direction* $\mathbf{v}_k$ *is stable if and only if*

$$
\begin{aligned}
G_k(\mathbf{x}, \mathbf{p}) &= (\mathbf{v}^k)^{\mathrm{T}} \cdot \mathbf{f}(\mathbf{x}, \mathbf{p}) < 0 \;\text{ for } c^k = (\mathbf{v}^k)^{\mathrm{T}} \cdot (\mathbf{x}(t) - \mathbf{x}^*) > 0, \\
G_k(\mathbf{x}, \mathbf{p}) &= (\mathbf{v}^k)^{\mathrm{T}} \cdot \mathbf{f}(\mathbf{x}, \mathbf{p}) > 0 \;\text{ for } c^k = (\mathbf{v}^k)^{\mathrm{T}} \cdot (\mathbf{x}(t) - \mathbf{x}^*) < 0
\end{aligned}
\tag{1.71}
$$

for all $\mathbf{x} \in U(\mathbf{x}^*) \subset \Omega$ *and all* $t \in [t_0, \infty)$.

(ii) **x** *at the equilibrium* $\mathbf{x}^*$ *on the direction* $\mathbf{v}_k$ *is unstable if and only if*

$$
\begin{aligned}
G_k(\mathbf{x}, \mathbf{p}) &= (\mathbf{v}^k)^{\mathrm{T}} \cdot \mathbf{f}(\mathbf{x}, \mathbf{p}) > 0 \;\text{ for } c^k = (\mathbf{v}^k)^{\mathrm{T}} \cdot (\mathbf{x}(t) - \mathbf{x}^*) > 0, \\
G_k(\mathbf{x}, \mathbf{p}) &= (\mathbf{v}^k)^{\mathrm{T}} \cdot \mathbf{f}(\mathbf{x}, \mathbf{p}) < 0 \;\text{ for } c^k = (\mathbf{v}^k)^{\mathrm{T}} \cdot (\mathbf{x}(t) - \mathbf{x}^*) < 0
\end{aligned}
\tag{1.72}
$$

for all $\mathbf{x} \in U(\mathbf{x}^*) \subset \Omega$ *and all* $t \in [t_0, \infty)$.

(iii) **x** *at the equilibrium* $\mathbf{x}^*$ *on the direction* $\mathbf{v}_k$ *is increasingly unstable if and only if*

$$
\begin{aligned}
G_k(\mathbf{x}, \mathbf{p}) &= (\mathbf{v}^k)^{\mathrm{T}} \cdot \mathbf{f}(\mathbf{x}, \mathbf{p}) > 0 \;\text{ for } s_k = (\mathbf{v}^k)^{\mathrm{T}} \cdot (\mathbf{x}(t) - \mathbf{x}^*) > 0, \\
G_k(\mathbf{x}, \mathbf{p}) &= (\mathbf{v}^k)^{\mathrm{T}} \cdot \mathbf{f}(\mathbf{x}, \mathbf{p}) > 0 \;\text{ for } s_k = (\mathbf{v}^k)^{\mathrm{T}} \cdot (\mathbf{x}(t) - \mathbf{x}^*) < 0
\end{aligned}
\tag{1.73}
$$

for all $\mathbf{x} \in U(\mathbf{x}^*) \subset \Omega$ *and all* $t \in [t_0, \infty)$.

(iv) **x** *at the equilibrium* $\mathbf{x}^*$ *on the direction* $\mathbf{v}_k$ *is decreasingly unstable if and only if*

$$
\begin{aligned}
G_k(\mathbf{x}, \mathbf{p}) &= (\mathbf{v}^k)^{\mathrm{T}} \cdot \mathbf{f}(\mathbf{x}, \mathbf{p}) < 0 \;\text{ for } c^k = (\mathbf{v}^k)^{\mathrm{T}} \cdot (\mathbf{x}(t) - \mathbf{x}^*) > 0, \\
G_k(\mathbf{x}, \mathbf{p}) &= (\mathbf{v}^k)^{\mathrm{T}} \cdot \mathbf{f}(\mathbf{x}, \mathbf{p}) < 0 \;\text{ for } c^k = (\mathbf{v}^k)^{\mathrm{T}} \cdot (\mathbf{x}(t) - \mathbf{x}^*) < 0
\end{aligned}
\tag{1.74}
$$

for all $\mathbf{x} \in U(\mathbf{x}^*) \subset \Omega$ *and all* $t \in [t_0, \infty)$.

(v) **x** *at the equilibrium* $\mathbf{x}^*$ *on the direction* $\mathbf{v}_k$ *is invariant if*

$$
G_k(\mathbf{x}, \mathbf{p}) = (\mathbf{v}^k)^{\mathrm{T}} \cdot \mathbf{f}(\mathbf{x}, \mathbf{p}) = 0 \tag{1.75}
$$

for all $\mathbf{x} \in U(\mathbf{x}^*) \subset \Omega$ *and all* $t \in [t_0, \infty)$.

Proof Because

$$
\begin{aligned}
(\mathbf{v}^k)^{\mathrm{T}} \cdot (\mathbf{x}(t+\varepsilon) - \mathbf{x}(t)) &= (\mathbf{v}^k)^{\mathrm{T}} \cdot (\mathbf{x}(t) + \dot{\mathbf{x}}(t)\varepsilon + o(\varepsilon) - \mathbf{x}(t)) \\
&= (\mathbf{v}^k)^{\mathrm{T}} \cdot \dot{\mathbf{x}}(t)\varepsilon + o(\varepsilon),
\end{aligned}
$$

and $\dot{\mathbf{x}} = \mathbf{f}(\mathbf{x}, \mathbf{p})$, we have

$$
\begin{aligned}
(\mathbf{v}^k)^{\mathrm{T}} \cdot (\mathbf{x}(t+\varepsilon) - \mathbf{x}(t)) &= (\mathbf{v}^k)^{\mathrm{T}} \cdot \mathbf{f}(\mathbf{x}, \mathbf{p})\varepsilon + o(\varepsilon) \\
&= G_k(\mathbf{x}, \mathbf{p})\varepsilon + o(\varepsilon).
\end{aligned}
$$

(i) Due to any selection of $\varepsilon > 0$, for $c^k = (\mathbf{v}^k)^{\mathrm{T}} \cdot (\mathbf{x}(t) - \mathbf{x}^*) < 0$,

$$(\mathbf{v}^k)^{\mathrm{T}} \cdot (\mathbf{x}(t+\varepsilon) - \mathbf{x}(t)) < 0 \text{ if } G_k(\mathbf{x}, \mathbf{p}) > 0,$$

and vice versa; and for $c^k = (\mathbf{v}^k)^{\mathrm{T}} \cdot (\mathbf{x}(t) - \mathbf{x}^*) > 0$,

$$(\mathbf{v}^k)^{\mathrm{T}} \cdot (\mathbf{x}(t+\varepsilon) - \mathbf{x}(t)) > 0 \text{ if } G_k(\mathbf{x}, \mathbf{p}) < 0,$$

and vice versa.

(ii) For $c^k = (\mathbf{v}^k)^{\mathrm{T}} \cdot (\mathbf{x}(t) - \mathbf{x}^*) > 0$,

$$(\mathbf{v}^k)^{\mathrm{T}} \cdot (\mathbf{x}(t+\varepsilon) - \mathbf{x}(t)) > 0 \text{ if } G_k(\mathbf{x}, \mathbf{p}) > 0,$$

and vice versa; and for $c^k = (\mathbf{v}^k)^{\mathrm{T}} \cdot (\mathbf{x}(t) - \mathbf{x}^*) < 0$,

$$(\mathbf{v}^k)^{\mathrm{T}} \cdot (\mathbf{x}(t+\varepsilon) - \mathbf{x}(t)) < 0 \text{ if } G_k(\mathbf{x}, \mathbf{p}) < 0,$$

and vice versa.

(iii) For $c^k = (\mathbf{v}^k)^{\mathrm{T}} \cdot (\mathbf{x}(t) - \mathbf{x}^*) > 0$,

$$(\mathbf{v}^k)^{\mathrm{T}} \cdot (\mathbf{x}(t+\varepsilon) - \mathbf{x}(t)) > 0 \text{ if } G_k(\mathbf{x}, \mathbf{p}) > 0,$$

and vice versa; and for $s_k = (\mathbf{v}^k)^{\mathrm{T}} \cdot (\mathbf{x}(t) - \mathbf{x}^*) < 0$,

$$(\mathbf{v}^k)^{\mathrm{T}} \cdot (\mathbf{x}(t+\varepsilon) - \mathbf{x}(t)) < 0 \text{ if } G_k(\mathbf{x}, \mathbf{p}) > 0,$$

and vice versa.

(iv) For $c^k = (\mathbf{v}^k)^{\mathrm{T}} \cdot (\mathbf{x}(t) - \mathbf{x}^*) > 0$,

$$(\mathbf{v}^k)^{\mathrm{T}} \cdot (\mathbf{x}(t+\varepsilon) - \mathbf{x}(t)) > 0 \text{ if } G_k(\mathbf{x}, \mathbf{p}) < 0,$$

and vice versa; and for $c^k = (\mathbf{v}^k)^{\mathrm{T}} \cdot (\mathbf{x}(t) - \mathbf{x}^*) < 0$,

$$(\mathbf{v}^k)^{\mathrm{T}} \cdot (\mathbf{x}(t+\varepsilon) - \mathbf{x}(t)) < 0 \text{ if } G_k(\mathbf{x}, \mathbf{p}) < 0,$$

and vice versa.

(v) For $c^k = (\mathbf{v}^k)^{\mathrm{T}} \cdot (\mathbf{x}(t) - \mathbf{x}^*) > 0$,

$$(\mathbf{v}^k)^{\mathrm{T}} \cdot (\mathbf{x}(t+\varepsilon) - \mathbf{x}(t)) = 0 \text{ if } G_k(\mathbf{x}, \mathbf{p}) = 0,$$

and vice versa. Similarly, for $c^k = (\mathbf{v}^k)^{\mathrm{T}} \cdot (\mathbf{x}(t) - \mathbf{x}^*) > 0$,

$$(\mathbf{v}^k)^{\mathrm{T}} \cdot (\mathbf{x}(t+\varepsilon) - \mathbf{x}(t)) = 0 \text{ if } G_k(\mathbf{x}, \mathbf{p}) = 0,$$

and vice versa. The theorem is proved. ∎

Theorem 1.8 *Consider an n-dimensional, autonomous, nonlinear dynamical system* $\dot{\mathbf{x}} = \mathbf{f}(\mathbf{x}, \mathbf{p})$ *in Eq. (1.4) with an equilibrium point* $\mathbf{x}^*$ *and* $\mathbf{f}(\mathbf{x}, \mathbf{p})$ *is* C^r*-continuous* $(r \geq 1)$ *in a neighborhood of the equilibrium* $\mathbf{x}^*$ *(i.e.,* $U(\mathbf{x}^*) \subset \Omega$*). The corresponding solution is* $\mathbf{x}(t) = \mathbf{\Phi}(\mathbf{x}_0, t - t_0, \mathbf{p})$*. Suppose Eq. (1.24) holds in* $U(\mathbf{x}^*) \subset \Omega$*. For a linearized dynamical system in Eq. (1.19), consider a real eigenvalue* λ_k *of matrix* $D\mathbf{f}(\mathbf{x}^*, \mathbf{p})$ $(k \in N = \{1, 2, \ldots, n\})$ *with an eigenvector* $\mathbf{v}_k$. Let $\mathbf{y} = c^k \mathbf{v}_k$ and $\dot{\mathbf{y}} = \dot{c}^k \mathbf{v}_k$ (summation with $k = 1, 2, \ldots, n$). $c^k = (\mathbf{v}^k)^{\mathrm{T}} \cdot \mathbf{y}$ *and* $\dot{c}^k = (\mathbf{v}^k)^{\mathrm{T}} \cdot \dot{\mathbf{y}} = (\mathbf{v}^k)^{\mathrm{T}} \cdot \mathbf{f}(\mathbf{x}, \mathbf{p})$. *Suppose* $\|G_k^{(2)}(\mathbf{x}^*, \mathbf{p})\| < \infty$ *and* $D\mathbf{f}(\mathbf{x}^*, \mathbf{p})\mathbf{v}_j = \lambda_j \mathbf{v}_j$ $(j = 1, 2, \ldots, n)$.

(i) $\mathbf{x}$ *at the equilibrium* $\mathbf{x}^*$ *on the direction* $\mathbf{v}_k$ *is stable if and only if*

$$\lambda_k < 0 \tag{1.76}$$

for all $\mathbf{x} \in U(\mathbf{x}^*) \subset \Omega$ *and all* $t \in [t_0, \infty)$.

(ii) $\mathbf{x}$ *at the equilibrium* $\mathbf{x}^*$ *on the direction* $\mathbf{v}_k$ *is unstable if and only if*

$$\lambda_k > 0 \tag{1.77}$$

for all $\mathbf{x} \in U(\mathbf{x}^*) \subset \Omega$ *and all* $t \in [t_0, \infty)$.

(iii) $\mathbf{x}$ *at the equilibrium* $\mathbf{x}^*$ *on the direction* $\mathbf{v}_k$ *is increasingly unstable if and only if*

$$\lambda_k = 0, \text{and } G_{k(j_1 j_2)}^{(2)}(\mathbf{x}^*, \mathbf{p}) c^{j_1} c^{j_2} > 0 \tag{1.78}$$

for all $\mathbf{x} \in U(\mathbf{x}^*) \subset \Omega$ *and all* $t \in [t_0, \infty)$.

(iv) $\mathbf{x}$ *at the equilibrium* $\mathbf{x}^*$ *on the direction* $\mathbf{v}_k$ *is decreasingly unstable if and only if*

$$\lambda_k = 0, \text{and } G_{k(j_1 j_2)}^{(2)}(\mathbf{x}^*, \mathbf{p}) c^{j_1} c^{j_2} < 0 \tag{1.79}$$

for all $\mathbf{x} \in U(\mathbf{x}^*) \subset \Omega$ *and all* $t \in [t_0, \infty)$.

(v) $\mathbf{x}$ *at the equilibrium* $\mathbf{x}^*$ *on the direction* $\mathbf{v}_k$ *is invariant if and only if*

$$G_{k(j_1 j_2 \cdots j_m)}^{(m)}(\mathbf{x}^*, \mathbf{p}) = 0 \quad (m = 0, 1, 2, \ldots) \tag{1.80}$$

for all $\mathbf{x} \in U(\mathbf{x}^*) \subset \Omega$ *and all* $t \in [t_0, \infty)$.

Proof For $\mathbf{x} = \mathbf{x}^*$, $c^k = 0$. With $D\mathbf{f}(\mathbf{x}^*, \mathbf{p})\mathbf{v}_j = \lambda_j \mathbf{v}_j$, the Taylor series expansion gives

$$\begin{aligned}
\dot{c}^k &= (\mathbf{v}^k)^{\mathrm{T}} \cdot \mathbf{f}(\mathbf{x}, \mathbf{p}) \\
&= (\mathbf{v}^k)^{\mathrm{T}} \cdot (\mathbf{f}(\mathbf{x}^*, \mathbf{p}) + D_{\mathbf{x}}\mathbf{f}(\mathbf{x}^*, \mathbf{p})\mathbf{v}_i c^i) + o(\|\bar{\mathbf{c}}\|) \\
&= (\mathbf{v}^k)^{\mathrm{T}} \cdot (D_{\mathbf{x}}\mathbf{f}(\mathbf{x}^*, \mathbf{p})\mathbf{v}_i c^i) + o(\|\bar{\mathbf{c}}\|) \\
&= G^{(1)}_{k(i)}(\mathbf{x}^*, \mathbf{p}) c^i + o(\|\bar{\mathbf{c}}\|) \\
&= (\mathbf{v}^k)^{\mathrm{T}} \cdot \lambda_i \mathbf{v}_i c^i + o(\|\bar{\mathbf{c}}\|) \\
&= (\lambda_i \delta_i^k) c^i + o(\|\bar{\mathbf{c}}\|) \\
&= \lambda_k c^k + o(\|\bar{\mathbf{c}}\|) \\
&= G^{(1)}_{k(k)}(\mathbf{x}^*, \mathbf{p}) c^k + o(\|\bar{\mathbf{c}}\|),
\end{aligned}$$

and

$$G^{(1)}_{k(k)}(\mathbf{x}^*, \mathbf{p}) = (\mathbf{v}^k)^{\mathrm{T}} \cdot (D_{\mathbf{x}}\mathbf{f}(\mathbf{x}^*, \mathbf{p})\mathbf{v}_k) = \lambda_k.$$

(i) For $c^k > 0$,

$$\dot{c}^k = \lambda_k c^k < 0,$$

and for $c^k < 0$,

$$\dot{c}^k = \lambda_k c^k > 0.$$

Thus, $G^{(1)}_{k(k)}(\mathbf{x}^*, \mathbf{p}) = \lambda_k < 0$.

(ii) For $c^k > 0$,

$$G_k(\mathbf{x}, \mathbf{p}) = \dot{c}^k = \lambda_k c^k > 0,$$

and for $c^k < 0$,

$$G_k(\mathbf{x}, \mathbf{p}) = \dot{c}^k = \lambda_k c^k < 0.$$

Thus, $G^{(1)}_{k(k)}(\mathbf{x}^*, \mathbf{p}) = \lambda_k > 0$.

(iii) For $c^k > 0$,

$$G_k(\mathbf{x}, \mathbf{p}) = \dot{c}^k = \lambda_k c^k > 0,$$

and for $c^k < 0$,

$$G_k(\mathbf{x}, \mathbf{p}) = \dot{c}^k = \lambda_k c^k > 0.$$

Thus, $G^{(1)}_{k(k)}(\mathbf{x}^*, \mathbf{p}) = \lambda_k = 0$ and the higher order should be considered. The higher order Taylor series expansion gives

$$\begin{aligned}
\dot{c}^k &= (\mathbf{v}^k)^{\mathrm{T}} \cdot \mathbf{f}(\mathbf{x}, \mathbf{p}) \\
&= (\mathbf{v}^k)^{\mathrm{T}} \cdot ([\mathbf{f}(\mathbf{x}^*, \mathbf{p}) + D_{c^{j_1}} \mathbf{f}(\mathbf{x}^*, \mathbf{p}) c^{j_1}. \\
&\quad + \frac{1}{2!} D^{(2)}_{c^{j_1} c^{j_2}} (\mathbf{x}^*, \mathbf{p}) c^{j_1} c^{j_2}] + o(\|\bar{\mathbf{c}}\|^2) \\
&= \frac{1}{2!} [(\mathbf{v}^k)^{\mathrm{T}} \cdot D^{(2)}_{\mathbf{x}} (\mathbf{x}^*, \mathbf{p})(\mathbf{v}_{j_1} \mathbf{v}_{j_2}) c^{j_1} c^{j_2} + o(\|\bar{\mathbf{c}}\|^2) \\
&= \frac{1}{2!} G^{(2)}_{k(j_1 j_2)} (\mathbf{x}^*, \mathbf{p}) c^{j_1} c^{j_2} + o(\|\bar{\mathbf{c}}\|^2).
\end{aligned}$$

For $c^k > 0$ with summation of $j_1, j_2 = 1, 2, \ldots, n$,

$$G_k(\mathbf{x}, \mathbf{p}) = \dot{c}^k = \frac{1}{2!} G^{(2)}_{k(j_1 j_2)} (\mathbf{x}^*, \mathbf{p}) c^{j_1} c^{j_2} > 0,$$

and for $c^k < 0$,

$$G_k(\mathbf{x}, \mathbf{p}) = \dot{c}^k = \frac{1}{2!} G^{(2)}_{k(j_1 j_2)} (\mathbf{x}^*, \mathbf{p}) c^{j_1} c^{j_2} > 0.$$

So we have

$$G^{(2)}_{k(j_1 j_2)} (\mathbf{x}^*, \mathbf{p}) c^{j_1} c^{j_2} > 0.$$

(iv) Similar to (iii), we have $G^{(1)}_k(\mathbf{x}^*, \mathbf{p}) = \lambda_k = 0$. In addition, for $c^k > 0$,

$$G_k(\mathbf{x}, \mathbf{p}) = \dot{c}^k = \frac{1}{2!} G^{(2)}_{k(j_1 j_2)} (\mathbf{x}^*, \mathbf{p}) c^{j_1} c^{j_2} < 0,$$

and for $s_k < 0$,

$$G_k(\mathbf{x},\mathbf{p}) = \dot{c}^k = \frac{1}{2!}G^{(2)}_{k(j_1j_2)}(\mathbf{x}^*,\mathbf{p})c^{j_1}c^{j_2} < 0.$$

So

$$G^{(2)}_{k(j_1j_2)}(\mathbf{x}^*,\mathbf{p})c^{j_1}c^{j_2} < 0.$$

(v) Using the Taylor series expansion yields

$$\begin{aligned}\dot{c}^k &= (\mathbf{v}^k)^{\mathrm{T}} \cdot \mathbf{f}(\mathbf{x},\mathbf{p}) \\ &= \sum\nolimits_{m=1}^{N} \frac{1}{m!}G^{(m)}_{k(j_1j_2\cdots j_m)}(\mathbf{x}^*,\mathbf{p})c^{j_1}c^{j_2}\ldots c^{j_m} + o(\|\bar{\mathbf{c}}\|^N) \\ &= 0 \\ &(N = 1,2,\ldots)\end{aligned}$$

Thus only if

$$G^{(m)}_{k(j_1j_2\cdots j_m)}(\mathbf{x}^*,\mathbf{p}) = 0 \ (m = 1,2,\ldots),$$

the above equation of $\dot{c}^k = 0$ holds, and vice versa. The theorem is proved. ∎

Definition 1.26 Consider an n-dimensional, autonomous, nonlinear dynamical system $\dot{\mathbf{x}} = \mathbf{f}(\mathbf{x},\mathbf{p})$ in Eq. (1.4) with an equilibrium point $\mathbf{x}^*$ and $\mathbf{f}(\mathbf{x},\mathbf{p})$ is C^r $(r \geq 1)$-continuous in a neighborhood of the equilibrium $\mathbf{x}^*$ (i.e., $U(\mathbf{x}^*) \subset \Omega$). The corresponding solution is $\mathbf{x}(t) = \mathbf{\Phi}(\mathbf{x}_0, t - t_0, \mathbf{p})$. Suppose Eq. (1.24) holds in $U(\mathbf{x}^*) \subset \Omega$. For a linearized dynamical system in Eq. (1.19), consider a real eigenvalue λ_k of matrix $D\mathbf{f}(\mathbf{x}^*,\mathbf{p})$ $(k \in N = \{1,2,\ldots,n\})$ with an eigenvector $\mathbf{v}_k$ and let $\mathbf{y} = c^k\mathbf{v}_k$. $c^k = (\mathbf{v}^k)^{\mathrm{T}} \cdot \mathbf{y}$ and $\dot{c}^k = (\mathbf{v}^k)^{\mathrm{T}} \cdot \dot{\mathbf{y}} = (\mathbf{v}^k)^{\mathrm{T}} \cdot \mathbf{f}(\mathbf{x},\mathbf{p})$.

(i) $\mathbf{x}$ at the equilibrium $\mathbf{x}^*$ on the direction $\mathbf{v}_k$ is stable of the m_kth order if

$$\begin{aligned}&G^{(r_k)}_{k(j_1j_2\cdots j_{r_k})}(\mathbf{x}^*,\mathbf{p}) = 0, r_k = 0,1,2,\ldots,m_k - 1; \\ &(\mathbf{v}^k)^{\mathrm{T}} \cdot (\mathbf{x}(t+\varepsilon) - \mathbf{x}(t)) < 0 \ \text{ for } (\mathbf{v}^k)^{\mathrm{T}} \cdot (\mathbf{x}(t) - \mathbf{x}^*) > 0, \\ &(\mathbf{v}^k)^{\mathrm{T}} \cdot (\mathbf{x}(t+\varepsilon) - \mathbf{x}(t)) > 0 \ \text{ for } (\mathbf{v}^k)^{\mathrm{T}} \cdot (\mathbf{x}(t) - \mathbf{x}^*) < 0\end{aligned}$$

for all $\mathbf{x} \in U(\mathbf{x}^*) \subset \Omega$ and all $t \in [t_0, \infty)$. The equilibrium $\mathbf{x}^*$ is called a sink (or a stable node) of the m_kth order on the direction $\mathbf{v}_k$.

(ii) $\mathbf{x}$ at the equilibrium $\mathbf{x}^*$on the direction $\mathbf{v}_k$ is unstable of the m_kth order if

$$\begin{aligned} &G^{(r_k)}_{k(j_1j_2\cdots j_{r_k})}(\mathbf{x}^*,\mathbf{p})=0, r_k=0,1,2,\ldots,m_k-1;\\ &(\mathbf{v}^k)^{\mathrm{T}}\cdot(\mathbf{x}(t+\varepsilon)-\mathbf{x}(t))>0 \ \text{ for } (\mathbf{v}^k)^{\mathrm{T}}\cdot(\mathbf{x}(t)-\mathbf{x}^*)>0,\\ &(\mathbf{v}^k)^{\mathrm{T}}\cdot(\mathbf{x}(t+\varepsilon)-\mathbf{x}(t))<0 \ \text{ for } (\mathbf{v}^k)^{\mathrm{T}}\cdot(\mathbf{x}(t)-\mathbf{x}^*)<0 \end{aligned} \tag{1.82}$$

for all $\mathbf{x}\in U(\mathbf{x}^*)\subset\Omega$ and all $t\in[t_0,\infty)$. The equilibrium $\mathbf{x}^*$ is called a source (or an unstable node) of the m_kth order on the direction $\mathbf{v}_k$.

(iii) $\mathbf{x}$ at the equilibrium $\mathbf{x}^*$ on the direction $\mathbf{v}_k$ is increasingly unstable of the m_kth order if

$$\begin{aligned} &G^{(r_k)}_{k(j_1j_2\cdots j_{r_k})}(\mathbf{x}^*,\mathbf{p})=0, r_k=0,1,2,\ldots,m_k-1;\\ &(\mathbf{v}^k)^{\mathrm{T}}\cdot(\mathbf{x}(t+\varepsilon)-\mathbf{x}(t))>0 \ \text{ for } (\mathbf{v}^k)^{\mathrm{T}}\cdot(\mathbf{x}(t)-\mathbf{x}^*)>0,\\ &(\mathbf{v}^k)^{\mathrm{T}}\cdot(\mathbf{x}(t+\varepsilon)-\mathbf{x}(t))>0 \ \text{ for } (\mathbf{v}^k)^{\mathrm{T}}\cdot(\mathbf{x}(t)-\mathbf{x}^*)<0 \end{aligned} \tag{1.83}$$

for all $\mathbf{x}\in U(\mathbf{x}^*)\subset\Omega$ and all $t\in[t_0,\infty)$. The equilibrium $\mathbf{x}^*$ is called an increasing saddle (or an *upper saddle*) of the m_kth order on the direction $\mathbf{v}_k$.

(iv) $\mathbf{x}$ at the equilibrium $\mathbf{x}^*$ on the direction $\mathbf{v}_k$ is decreasingly unstable of the m_kth order if

$$\begin{aligned} &G^{(r_k)}_{k(j_1j_2\cdots j_{r_k})}(\mathbf{x}^*,\mathbf{p})=0, r_k=0,1,2,\ldots,m_k-1;\\ &(\mathbf{v}^k)^{\mathrm{T}}\cdot(\mathbf{x}(t+\varepsilon)-\mathbf{x}(t))<0 \ \text{ for } (\mathbf{v}^k)^{\mathrm{T}}\cdot(\mathbf{x}(t)-\mathbf{x}^*)>0,\\ &(\mathbf{v}^k)^{\mathrm{T}}(\mathbf{x}(t+\varepsilon)-\mathbf{x}(t))<0 \ \text{ for } (\mathbf{v}^k)^{\mathrm{T}}\cdot(\mathbf{x}(t)-\mathbf{x}^*)<0 \end{aligned} \tag{1.84}$$

for all $\mathbf{x}\in U(\mathbf{x}^*)\subset\Omega$ and all $t\in[t_0,\infty)$. The equilibrium $\mathbf{x}^*$ is called a decreasing saddle (or a *lower saddle*) of the m_kth order on the direction $\mathbf{v}_k$.

Theorem 1.9 *Consider an n-dimensional, autonomous, nonlinear dynamical system* $\dot{\mathbf{x}}=\mathbf{f}(\mathbf{x},\mathbf{p})$ *in Eq. (1.4) with an equilibrium point* $\mathbf{x}^*$ *and* $\mathbf{f}(\mathbf{x},\mathbf{p})$ *is* C^r $(r\geq 1)$*-continuous in a neighborhood of the equilibrium* $\mathbf{x}^*$ *(i.e.,* $U(\mathbf{x}^*)\subset\Omega$*). The corresponding solution is* $\mathbf{x}(t)=\mathbf{\Phi}(\mathbf{x}_0,t-t_0,\mathbf{p})$*. Suppose Eq. (1.24) holds in* $U(\mathbf{x}^*)\subset\Omega$*. For a linearized dynamical system in Eq. (1.19), consider a real eigenvalue* λ_k *of matrix* $D\mathbf{f}(\mathbf{x}^*,\mathbf{p})$ $(k\in N=\{1,2,\ldots,n\})$ *with an eigenvector* $\mathbf{v}_k$ *and let* $\mathbf{y}=c^k\mathbf{v}_k$*.* $c^k=(\mathbf{v}^k)^{\mathrm{T}}\cdot\mathbf{y}$ *and* $\dot{c}^k=(\mathbf{v}^k)^{\mathrm{T}}\cdot\dot{\mathbf{y}}=(\mathbf{v}^k)^{\mathrm{T}}\cdot\mathbf{f}(\mathbf{x},\mathbf{p})$.

(i) **x** *at the equilibrium* $\mathbf{x}^*$ *on the direction* $\mathbf{v}_k$ *is stable of the* m_kth *order if and only if*

$$\begin{aligned} &G^{(r_k)}_{k(j_1 j_2 \cdots j_{r_k})}(\mathbf{x}^*, \mathbf{p}) = 0, r_k = 0, 1, 2, \ldots, m_k - 1; \\ &c^k \times G^{(m_k+1)}_{k(j_1 j_2 \cdots j_{2m_k+1})}(\mathbf{x}^*, \mathbf{p})(c^{j_1} c^{j_2} \cdots c^{j_{m_k}}) < 0 \end{aligned} \tag{1.85}$$

for all $\mathbf{x} \in U(\mathbf{x}^*) \subset \Omega$ *and all* $t \in [t_0, \infty)$.

(ii) **x** *at the equilibrium* $\mathbf{x}^*$ *on the direction* $\mathbf{v}_k$ *is unstable of the* m_kth *order if and only if*

$$\begin{aligned} &G^{(r_k)}_{k(j_1 j_2 \cdots j_{r_k})}(\mathbf{x}^*, \mathbf{p}) = 0, r_k = 0, 1, 2, \ldots, m_k - 1; \\ &c^k \times G^{(m_k)}_{k(j_1 j_2 \cdots j_{m_k})}(\mathbf{x}^*, \mathbf{p})(c^{j_1} c^{j_2} \cdots c^{j_{m_k}}) > 0 \end{aligned} \tag{1.86}$$

for all $\mathbf{x} \in U(\mathbf{x}^*) \subset \Omega$ *and all* $t \in [t_0, \infty)$.

(iii) **x** *at the equilibrium* $\mathbf{x}^*$ *on the direction* $\mathbf{v}_k$ *is increasingly unstable of the* m_kth *order if and only if*

$$\begin{aligned} &G^{(r_k)}_{k(j_1 j_2 \cdots j_{r_k})}(\mathbf{x}^*, \mathbf{p}) = 0, r_k = 0, 1, 2, \ldots, m_k - 1; \\ &G^{(m_k)}_{k(j_1 j_2 \cdots j_{m_k})}(\mathbf{x}^*, \mathbf{p})(c^{j_1} c^{j_2} \cdots c^{j_{m_k}}) > 0 \end{aligned} \tag{1.87}$$

for all $\mathbf{x} \in U(\mathbf{x}^*) \subset \Omega$ *and all* $t \in [t_0, \infty)$.

(iv) **x** *at the equilibrium* $\mathbf{x}^*$ *on the direction* $\mathbf{v}_k$ *is decreasingly unstable of the* m_kth *order if and only if*

$$\begin{aligned} &G^{(r_k)}_{k(j_1 j_2 \cdots j_{r_k})}(\mathbf{x}^*, \mathbf{p}) = 0, r_k = 0, 1, 2, \ldots, m_k - 1; \\ &G^{(m_k)}_{k(j_1 j_2 \cdots j_{m_k})}(\mathbf{x}^*, \mathbf{p})(c^{j_1} c^{j_2} \cdots c^{j_{m_k}}) < 0 \end{aligned} \tag{1.88}$$

for all $\mathbf{x} \in U(\mathbf{x}^*) \subset \Omega$ *and all* $t \in [t_0, \infty)$.

Proof For $\mathbf{x} = \mathbf{x}^*$, $c^k = 0$. Using the Taylor series expansion gives

$$\begin{aligned} \dot{c}^k &= (\mathbf{v}^k)^{\mathrm{T}} \cdot \mathbf{f}(\mathbf{x}, \mathbf{p}) \\ &= \sum\nolimits_{r_k=1}^{m_k-1} \frac{1}{r_k!} G^{(r_k)}_{k(j_1 j_2 \cdots j_{r_k})}(\mathbf{x}^*, \mathbf{p})(c^{j_1} c^{j_2} \cdots c^{j_{r_k}}) \\ &\quad + \frac{1}{m_k!} G^{(m_k)}_{k(j_1 j_2 \cdots j_{m_k})}(\mathbf{x}^*, \mathbf{p})(c^{j_1} c^{j_2} \cdots c^{j_{m_k}}) \\ &\quad + o(\|\bar{\mathbf{c}}\|^{m_k}), \end{aligned}$$

and

$$G^{(r_k)}_{k(j_1 j_2 \cdots j_{r_k})}(\mathbf{x}^*, \mathbf{p}) = 0 \text{ for } r_k = 0, 1, 2, \cdots, m_k - 1,$$

$$\dot{c}^k = \frac{1}{m_k!} G^{(m_k)}_{k(j_1 j_2 \cdots j_{m_k})}(\mathbf{x}^*, \mathbf{p})(c^{j_1} c^{j_2} \cdots c^{j_{m_k}}).$$

(i) For $c^k > 0$,

$$G_k(\mathbf{x}, \mathbf{p}) = \dot{c}^k = \frac{1}{m_k!} G^{(m_k)}_{k(j_1 j_2 \cdots j_{m_k})}(\mathbf{x}^*, \mathbf{p})(c^{j_1} c^{j_2} \cdots c^{j_{m_k}}) < 0,$$

and for $c^k < 0$,

$$G_k(\mathbf{x}, \mathbf{p}) = \dot{c}^k = \frac{1}{m_k!} G^{(m_k)}_{k(j_1 j_2 \cdots j_{m_k})}(\mathbf{x}^*, \mathbf{p})(c^{j_1} c^{j_2} \cdots c^{j_{m_k}}) > 0.$$

Thus

$$c^k \times G^{(m_k)}_{k(j_1 j_2 \cdots j_{m_k})}(\mathbf{x}^*, \mathbf{p})(c^{j_1} c^{j_2} \cdots c^{j_{m_k}}) < 0.$$

(ii) For $c^k > 0$,

$$G_k(\mathbf{x}, \mathbf{p}) = \dot{c}^k = \frac{1}{m_k!} G^{(m_k)}_{k(j_1 j_2 \cdots j_{m_k})}(\mathbf{x}^*, \mathbf{p})(c^{j_1} c^{j_2} \cdots c^{j_{m_k}}) > 0,$$

and for $c^k < 0$,

$$G_k(\mathbf{x}, \mathbf{p}) = \dot{c}^k = \frac{1}{m_k!} G^{(m_k)}_{k(j_1 j_2 \cdots j_{m_k})}(\mathbf{x}^*, \mathbf{p})(c^{j_1} c^{j_2} \cdots c^{j_{m_k}}) < 0.$$

Thus

$$c^k \times G^{(m_k)}_{k(j_1 j_2 \cdots j_{m_k})}(\mathbf{x}^*, \mathbf{p})(c^{j_1} c^{j_2} \cdots c^{j_{m_k}}) > 0.$$

(iii) For $c^k > 0$,

$$G_k(\mathbf{x}, \mathbf{p}) = \dot{c}^k = \frac{1}{m_k!} G^{(m_k)}_{k(j_1 j_2 \cdots j_{m_k})}(\mathbf{x}^*, \mathbf{p})(c^{j_1} c^{j_2} \cdots c^{j_{m_k}}) > 0,$$

and for $c^k < 0$,

$$G_k(\mathbf{x},\mathbf{p}) = \dot{c}^k = \frac{1}{m_k!} G^{(m_k)}_{k(j_1 j_2 \cdots j_{m_k})}(\mathbf{x}^*,\mathbf{p})(c^{j_1} c^{j_2} \cdots c^{j_{m_k}}) > 0.$$

So we have

$$G^{(m_k)}_{k(j_1 j_2 \cdots j_{m_k})}(\mathbf{x}^*,\mathbf{p})(c^{j_1} c^{j_2} \cdots c^{j_{m_k}}) > 0.$$

(iv) Similar to (iii), for $c^k > 0$,

$$G_k(\mathbf{x},\mathbf{p}) = \dot{c}^k = \frac{1}{m_k!} G^{(m_k)}_{k(j_1 j_2 \cdots j_{m_k})}(\mathbf{x}^*,\mathbf{p})(c^{j_1} c^{j_2} \cdots c^{j_{m_k}}) < 0,$$

and for $c^k < 0$,

$$G_k(\mathbf{x},\mathbf{p}) = \dot{c}^k = \frac{1}{m_k!} G^{(m_k)}_{k(j_1 j_2 \cdots j_{m_k})}(\mathbf{x}^*,\mathbf{p})(c^{j_1} c^{j_2} \cdots c^{j_{m_k}}) < 0.$$

So

$$G^{(m_k)}_{k(j_1 j_2 \cdots j_{m_k})}(\mathbf{x}^*,\mathbf{p})(c^{j_1} c^{j_2} \cdots c^{j_{m_k}}) < 0.$$

The theorem is proved. ∎

1.3.2 *Spiral Stability on an Invariant Eigenplane*

Definition 1.27 Consider an n-dimensional, autonomous, nonlinear dynamical system $\dot{\mathbf{x}} = \mathbf{f}(\mathbf{x},\mathbf{p})$ in Eq. (1.4) with an equilibrium point $\mathbf{x}^*$ and $\mathbf{f}(\mathbf{x},\mathbf{p})$ is C^r $(r \geq 1)$-continuous in a neighborhood of the equilibrium $\mathbf{x}^*$ (i.e., $U(\mathbf{x}^*) \subset \Omega$). The corresponding solution is $\mathbf{x}(t) = \mathbf{\Phi}(\mathbf{x}_0, t - t_0, \mathbf{p})$. Suppose Eq. (1.24) holds in $U(\mathbf{x}^*) \subset \Omega$. For a linearized dynamical system in Eq. (1.19), consider a pair of complex eigenvalue $\alpha_k \pm \mathbf{i}\beta_k$ $(2k-1, 2k \in N = \{1,2,\ldots,n\}, \mathbf{i} = \sqrt{-1})$ of matrix $D\mathbf{f}(\mathbf{x}^*,\mathbf{p})$ with a pair of eigenvectors $\mathbf{v}_{2k-1} \pm \mathbf{i}\mathbf{v}_{2k}$. On the invariant plane of $(\mathbf{v}_{2k-1}, \mathbf{v}_{2k})$, the contravariants are defined as

$$c^{2k-1} = (\mathbf{v}^{2k-1})^{\mathrm{T}} \cdot \mathbf{y},\ \ c^{2k} = (\mathbf{v}^{2k})^{\mathrm{T}} \cdot \mathbf{y} \tag{1.89}$$

with

$$
\begin{aligned}
D_{\mathbf{x}}\mathbf{f} \cdot \mathbf{v}_{2k-1} &= \alpha_k \mathbf{v}_{2k-1} - \beta_k \mathbf{v}_{2k}, \\
D_{\mathbf{x}}\mathbf{f} \cdot \mathbf{v}_{2k} &= \beta_k \mathbf{v}_{2k-1} + \alpha_k \mathbf{v}_{2k}.
\end{aligned} \tag{1.90}
$$

The time change rates of the contravariants are

$$
\begin{aligned}
\dot{c}^{2k-1} &= G_{2k-1}(\mathbf{x}, \mathbf{p}) = (\mathbf{v}^{2k-1})^{\mathrm{T}} \cdot \mathbf{f}(\mathbf{x}, \mathbf{p}) \\
&= \sum_{r_k=1}^{\infty} \frac{1}{r_k!} G^{(r_k)}_{(2k-1)(j_1 j_2 \cdots j_{r_k})}(\mathbf{x}^*, \mathbf{p}) c^{j_1} c^{j_2} \cdots c^{j_{r_k}}, \\
\dot{c}^{2k} &= G_{2k}(\mathbf{x}, \mathbf{p}) = (\mathbf{v}^{2k})^{\mathrm{T}} \cdot \mathbf{f}(\mathbf{x}, \mathbf{p}) \\
&= \sum_{r_k=1}^{\infty} \frac{1}{r_k!} G^{(r_k)}_{(2k)(j_1 j_2 \cdots j_{r_k})}(\mathbf{x}^*, \mathbf{p}) c^{j_1} c^{j_2} \cdots c^{j_{r_k}},
\end{aligned} \tag{1.91}
$$

where for $j_{r_k} \in \{1, 2, \ldots, n\}, r_k = 1, 2, \ldots$

$$
\begin{aligned}
G^{(r_k)}_{(2k-1)(j_1 j_2 \cdots j_{r_k})}(\mathbf{x}^*, \mathbf{p}) &= (\mathbf{v}^{2k-1})^{\mathrm{T}} \cdot D^{(r_k)}_{c^{j_1} c^{j_2} \cdots c^{j_{r_k}}} \mathbf{f}(\mathbf{x}^*, \mathbf{p}) \\
&= (\mathbf{v}^{2k-1})^{\mathrm{T}} \cdot D^{(r_k)}_{\mathbf{x}} \mathbf{f}(\mathbf{x}^*, \mathbf{p}) \mathbf{v}_{j_1} \mathbf{v}_{j_2} \cdots \mathbf{v}_{j_{r_k}}, \\
G^{(r_k)}_{(2k)(j_1 j_2 \cdots j_{r_k})}(\mathbf{x}^*, \mathbf{p}) &= (\mathbf{v}^{2k})^{\mathrm{T}} \cdot D^{(r_k)}_{c^{j_1} c^{j_2} \cdots c^{j_{r_k}}} \mathbf{f}(\mathbf{x}^*, \mathbf{p}) \\
&= (\mathbf{v}^{2k})^{\mathrm{T}} \cdot D^{(r_k)}_{\mathbf{x}} \mathbf{f}(\mathbf{x}^*, \mathbf{p}) \mathbf{v}_{j_1} \mathbf{v}_{j_2} \cdots \mathbf{v}_{j_{r_k}}.
\end{aligned} \tag{1.92}
$$

Thus, the time change rates of the contravariants in Eq. (1.91) become

$$
\begin{aligned}
\begin{Bmatrix} \dot{c}^{2k-1} \\ \dot{c}^{2k} \end{Bmatrix} &= \begin{bmatrix} \alpha_k & \beta_k \\ -\beta_k & \alpha_k \end{bmatrix} \begin{Bmatrix} c^{2k-1} \\ c^{2k} \end{Bmatrix} \\
&\quad + \sum_{r_k=2}^{\infty} \frac{1}{r_k!} \begin{Bmatrix} G^{(r_k)}_{(2k-1)(j_1 j_2 \cdots j_{r_k})}(\mathbf{x}^*, \mathbf{p}) \\ G^{(r_k)}_{(2k)(j_1 j_2 \cdots j_{r_k})}(\mathbf{x}^*, \mathbf{p}) \end{Bmatrix} c^{j_1} c^{j_2} \cdots c^{j_{r_k}}.
\end{aligned} \tag{1.93}
$$

A radial variable is defined as

$$
\rho_k = \sqrt{(c^{2k-1})^2 + (c^{2k})^2}, \tag{1.94}
$$

with

$$c^{2k-1} = \rho_k \cos\theta_k, c^{2k} = \rho_k \sin\theta_k, \tag{1.95}$$

where we have

$$\begin{aligned} \dot{c}^{2k-1} &= \dot{\rho}_k \cos\theta_k - \dot{\theta}_k \rho_k \sin\theta_k, \\ \dot{c}^{2k} &= \dot{\theta}_k \rho_k \cos\theta_k + \dot{\rho}_k \sin\theta_k. \end{aligned} \tag{1.96}$$

Thus

$$\begin{aligned} \dot{\rho}_k &= \dot{c}^{2k-1} \cos\theta_k + \dot{c}^{2k} \sin\theta_k, \\ \dot{\theta}_k \rho_k &= \dot{c}^{2k-1} \sin\theta_k - \dot{c}^{2k} \cos\theta_k \end{aligned} \tag{1.97}$$

and

$$\dot{\rho}_k = \sum_{r_k=1}^{\infty} \frac{1}{r_k!} G_{\rho_k}^{(r_k)} \text{ and } \dot{\theta}_k = \sum_{r_k=1}^{\infty} \frac{1}{r_k!} G_{\theta_k}^{(r_k)} \tag{1.98}$$

where

$$\begin{aligned} G_{\rho_k}^{(r_k)} &= [G_{(2k-1)(j_1 j_2 \cdots j_{r_k})}^{(r_k)}(\mathbf{x}^*, \mathbf{p}) c^{j_1} c^{j_2} \cdots c^{j_{r_k}} \cos\theta_k \\ &\quad + G_{(2k)(j_1 j_2 \cdots j_{r_k})}^{(r_k)}(\mathbf{x}^*, \mathbf{p}) c^{j_1} c^{j_2} \cdots c^{j_{r_k}} \sin\theta_k], \\ G_{\theta_k}^{(r_k)} &= \frac{1}{\rho_k} [G_{(2k-1)(j_1 j_2 \cdots j_{r_k})}^{(r_k)}(\mathbf{x}^*, \mathbf{p}) c^{j_1} c^{j_2} \cdots c^{j_{r_k}} \sin\theta_k \\ &\quad - G_{(2k)(j_1 j_2 \cdots j_{r_k})}^{(r_k)}(\mathbf{x}^*, \mathbf{p}) c^{j_1} c^{j_2} \cdots c^{j_{r_k}} \cos\theta_k] \\ &\quad (\text{summation of } j_1, j_2, \cdots, j_{r_k}). \end{aligned} \tag{1.99}$$

For $\rho_k \to 0$, the linearized equation is

$$\begin{Bmatrix} \dot{c}^{2k-1} \\ \dot{c}^{2k} \end{Bmatrix} = \begin{bmatrix} \alpha_k & \beta_k \\ -\beta_k & \alpha_k \end{bmatrix} \begin{Bmatrix} c^{2k-1} \\ c^{2k} \end{Bmatrix} + \begin{Bmatrix} o(\rho_k) \\ o(\rho_k) \end{Bmatrix} \tag{1.100}$$

Equation (1.98) becomes

$$\dot{\rho}_k = \alpha_k \rho_k + o(\rho_k), \dot{\theta}_k \rho_k = -\beta_k \rho_k + o(\rho_k). \tag{1.101}$$

For $\rho_k \to 0$, we have

$$\dot{\rho}_k = \alpha_k \rho_k, \dot{\theta}_k = -\beta_k. \tag{1.102}$$

Thus

$$\rho_k = \rho_{k0} e^{\alpha_k}, \theta_k = -\beta_k t + \theta_0. \tag{1.103}$$

So

$$\begin{aligned} c^{2k-1} &= \rho_{k0} e^{\alpha_k t} \cos(\beta_k t - \theta_0) \\ &= e^{\alpha_k t}(c_0^{2k-1} \cos \beta_k t + c_0^{2k} \sin \beta_k t), \\ c^{2k} &= -\rho_{k0} e^{\alpha_k t} \sin(\beta_k t - \theta_0) \\ &= e^{\alpha_k t}(-c_0^{2k-1} \sin \beta_k t + c_0^{2k} \cos \beta_k t) \end{aligned} \tag{1.104}$$

and

$$\begin{Bmatrix} c^{2k-1} \\ c^{2k} \end{Bmatrix} = e^{\alpha_k t} \begin{bmatrix} \cos \beta_k t & \sin \beta_k t \\ -\sin \beta_k t & \cos \beta_k t \end{bmatrix} \begin{Bmatrix} c_0^{2k-1} \\ c_0^{2k} \end{Bmatrix}. \tag{1.105}$$

Letting $\mathbf{c}^k = (c^{2k-1}, c^{2k})^{\mathrm{T}}$, we have

$$\dot{\mathbf{c}}^k = \mathbf{E}_k \mathbf{c}^k \Rightarrow \mathbf{c}^k = e^{\alpha_k t} \mathbf{B}_k \mathbf{c}_0^k \tag{1.106}$$

where

$$\mathbf{E}_k = \begin{bmatrix} \alpha_k & \beta_k \\ -\beta_k & \alpha_k \end{bmatrix} \text{ and } \mathbf{B}_k = \begin{bmatrix} \cos \beta_k t & \sin \beta_k t \\ -\sin \beta_k t & \cos \beta_k t \end{bmatrix}. \tag{1.107}$$

Definition 1.28 Consider an n-dimensional, autonomous, nonlinear dynamical system $\dot{\mathbf{x}} = \mathbf{f}(\mathbf{x}, \mathbf{p})$ in Eq. (1.4) with an equilibrium point $\mathbf{x}^*$ and $\mathbf{f}(\mathbf{x}, \mathbf{p})$ is C^r-continuous ($r \geq 1$) in a neighborhood of the equilibrium $\mathbf{x}^*$. The corresponding solution is $\mathbf{x}(t) = \mathbf{\Phi}(\mathbf{x}_0, t - t_0, \mathbf{p})$. Suppose $U(\mathbf{x}^*) \subset \Omega$ is a neighborhood of equilibrium $\mathbf{x}^*$. For a linearized dynamical system in Eq. (1.19), consider a pair of complex eigenvalue $\alpha_k \pm \mathbf{i}\beta_k$ ($2k - 1, 2k \in N = \{1, 2, \ldots, n\}, \mathbf{i} = \sqrt{-1}$) of matrix $D\mathbf{f}(\mathbf{x}^*, \mathbf{p})$ with a pair of eigenvectors $\mathbf{v}_{2k-1} \pm \mathbf{i}\mathbf{v}_{2k}$. On the invariant plane of $(\mathbf{v}_{2k-1}, \mathbf{v}_{2k})$,

$$c^{2k-1} = (\mathbf{v}^{2k-1})^{\mathrm{T}} \cdot \mathbf{y}, \ c^{2k} = (\mathbf{v}^{2k})^{\mathrm{T}} \cdot \mathbf{y} \tag{1.108}$$

with

$$
\begin{aligned}
D_{\mathbf{x}}\mathbf{f} \cdot \mathbf{v}_{2k-1} &= \alpha_k \mathbf{v}_{2k-1} - \beta_k \mathbf{v}_{2k}, \\
D_{\mathbf{x}}\mathbf{f} \cdot \mathbf{v}_{2k} &= \beta_k \mathbf{v}_{2k-1} + \alpha_k \mathbf{v}_{2k}
\end{aligned} \tag{1.109}
$$

and

$$
\begin{aligned}
\rho_k &= \sqrt{(c^{2k-1})^2 + (c^{2k})^2}, \\
c^{2k-1} &= \rho_k \cos\theta_k, c^{2k} = \rho_k \sin\theta_k.
\end{aligned} \tag{1.110}
$$

For any arbitrarily small $\varepsilon > 0$, the stability of the equilibrium $\mathbf{x}^*$ on the invariant plane of $(\mathbf{v}_{2k-1}, \mathbf{v}_{2k})$ can be determined.

(i) $\mathbf{x}$ at the equilibrium $\mathbf{x}^*$ on the plane of $(\mathbf{v}_{2k-1}, \mathbf{v}_{2k})$ is spirally stable if

$$
\rho_k(t+\varepsilon) - \rho_k(t) < 0. \tag{1.111}
$$

(ii) $\mathbf{x}$ at the equilibrium $\mathbf{x}^*$ on the plane of $(\mathbf{v}_{2k-1}, \mathbf{v}_{2k})$ is spirally unstable if

$$
\rho_k(t+\varepsilon) - \rho_k(t) > 0. \tag{1.112}
$$

(iii) $\mathbf{x}$ at the equilibrium $\mathbf{x}^*$ on the plane of $(\mathbf{v}_{2k-1}, \mathbf{v}_{2k})$ is stable with the m_kth-order singularity if for $\theta_k \in [0, 2\pi]$

$$
\begin{aligned}
&G_{\rho_k}^{(r_k)}(\mathbf{x}^*, \mathbf{p}) = 0 \ \text{ for } r_k = 1, 2, \ldots, m_k - 1 \\
&\rho_k(t+\varepsilon) - \rho_k(t) < 0.
\end{aligned} \tag{1.113}
$$

(iv) $\mathbf{x}$ at the equilibrium $\mathbf{x}^*$ on the plane of $(\mathbf{v}_{2k-1}, \mathbf{v}_{2k})$ is spirally unstable with the m_kth-order singularity if for $\theta_k \in [0, 2\pi]$

$$
\begin{aligned}
&G_{r_k}^{(s_k)}(\theta_k) = 0 \ \text{ for } s_k = 0, 1, 2, \ldots, m_k - 1 \\
&\rho_k(t+\varepsilon) - \rho_k(t) > 0.
\end{aligned} \tag{1.114}
$$

(v) $\mathbf{x}$ at the equilibrium $\mathbf{x}^*$ on the plane of $(\mathbf{v}_{2k-1}, \mathbf{v}_{2k})$ is circular if for $\theta_k \in [0, 2\pi]$

$$
r_k(t+\varepsilon) - r_k(t) = 0. \tag{1.115}
$$

(vi) $\mathbf{x}$ at the equilibrium $\mathbf{x}^*$ on the plane of $(\mathbf{u}_k, \mathbf{v}_k)$ is degenerate in the direction of $\mathbf{u}_k$ if

$$
\beta_k = 0 \text{ and } \theta_k(t+\varepsilon) - \theta_k(t) = 0. \tag{1.116}
$$

Theorem 1.10 *Consider an n-dimensional, autonomous, nonlinear dynamical system* $\dot{\mathbf{x}} = \mathbf{f}(\mathbf{x}, \mathbf{p})$ *in Eq. (1.4) with an equilibrium point* $\mathbf{x}^*$ *and* $\mathbf{f}(\mathbf{x}, \mathbf{p})$ *is* C^r*-continuous* $(r \geq 1)$ *in a neighborhood of the equilibrium* $\mathbf{x}^*$*. The corresponding solution is*

$\mathbf{x}(t) = \mathbf{\Phi}(\mathbf{x}_0, t - t_0, \mathbf{p})$. *Suppose* $U(\mathbf{x}^*) \subset \Omega$ *is a neighborhood of equilibrium* $\mathbf{x}^*$. *For a linearized dynamical system in Eq. (1.19), consider a pair of complex eigenvalue* $\alpha_k \pm \mathbf{i}\beta_k$ *(*$2k-1, 2k \in N = \{1, 2, \ldots, n\}$*,* $\mathbf{i} = \sqrt{-1}$*) of matrix* $D\mathbf{f}(\mathbf{x}^*, \mathbf{p})$ *with a pair of eigenvectors* $\mathbf{v}_{2k-1} \pm \mathbf{i}\mathbf{v}_{2k}$. *On the invariant plane of* $(\mathbf{v}_{2k-1}, \mathbf{v}_{2k})$,

$$c^{2k-1} = (\mathbf{v}^{2k-1})^{\mathrm{T}} \cdot \mathbf{y}, \;\; c^{2k} = (\mathbf{v}^{2k})^{\mathrm{T}} \cdot \mathbf{y} \tag{1.117}$$

with

$$\begin{aligned} D_{\mathbf{x}}\mathbf{f} \cdot \mathbf{v}_{2k-1} &= \alpha_k \mathbf{v}_{2k-1} - \beta_k \mathbf{v}_{2k}, \\ D_{\mathbf{x}}\mathbf{f} \cdot \mathbf{v}_{2k} &= \beta_k \mathbf{v}_{2k-1} + \alpha_k \mathbf{v}_{2k} \end{aligned} \tag{1.118}$$

and

$$\begin{aligned} \rho_k &= \sqrt{(c^{2k-1})^2 + (c^{2k})^2}, \\ c^{2k-1} &= \rho_k \cos\theta_k, c^{2k} = \rho_k \sin\theta_k. \end{aligned} \tag{1.119}$$

For any arbitrarily small $\varepsilon > 0$*, the stability of the equilibrium* $\mathbf{x}^*$ *on the invariant plane of* $(\mathbf{v}_{2k-1}, \mathbf{v}_{2k})$ *can be determined.*

(i) $\mathbf{x}$ *at the equilibrium* $\mathbf{x}^*$ *on the plane of* $(\mathbf{v}_{2k-1}, \mathbf{v}_{2k})$ *is spirally stable if and only if*

$$G_{\rho_k}^{(1)}(\mathbf{x}^*, \mathbf{p}) = \alpha_k < 0. \tag{1.120}$$

(ii) $\mathbf{x}$ *at the equilibrium* $\mathbf{x}^*$ *on the plane of* $(\mathbf{v}_{2k-1}, \mathbf{v}_{2k})$ *is spirally unstable if and only if*

$$G_{\rho_k}^{(1)}(\mathbf{x}^*, \mathbf{p}) = \alpha_k > 0. \tag{1.121}$$

(iii) $\mathbf{x}$ *at the equilibrium* $\mathbf{x}^*$ *on the plane of* $(\mathbf{v}_{2k-1}, \mathbf{v}_{2k})$ *is spirally stable with the* m_k*th-order singularity if and only if for* $\theta_k \in [0, 2\pi]$

$$\begin{aligned} &G_{\rho_k}^{(r_k)}(\mathbf{x}^*, \mathbf{p}) = 0 \;\text{ for } r_k = 1, 2, \ldots, m_k - 1 \\ &\text{and } G_{\rho_k}^{(m_k)}(\mathbf{x}^*, \mathbf{p}) < 0. \end{aligned} \tag{1.122}$$

(iv) $\mathbf{x}$ *at the equilibrium* $\mathbf{x}^*$ *on the plane of* $(\mathbf{v}_{2k-1}, \mathbf{v}_{2k})$ *is spirally unstable with the* m_k*th-order singularity if and only if for* $\theta_k \in [0, 2\pi]$

$$\begin{aligned} &G_{\rho_k}^{(r_k)}(\mathbf{x}^*, \mathbf{p}) = 0 \text{ for } r_k = 1, 2, \ldots, m_k - 1 \\ &\text{and } G_{\rho_k}^{(m_k)}(\mathbf{x}^*, \mathbf{p}) > 0. \end{aligned} \tag{1.123}$$

(v) **x** *at the equilibrium* $\mathbf{x}^*$ *on the plane of* $(\mathbf{v}_{2k-1}, \mathbf{v}_{2k})$ *is circular if and only if for* $\theta_k \in [0, 2\pi]$

$$G_{\rho_k}^{(r_k)}(\mathbf{x}^*, \mathbf{p}) = 0 \text{ for } r_k = 1, 2, \ldots. \tag{1.124}$$

(vi) **x** *at the equilibrium* $\mathbf{x}^*$ *on the plane of* $(\mathbf{v}_{2k-1}, \mathbf{v}_{2k})$ *is degenerate in the direction of* $\mathbf{v}_{2k-1}$ *if and only if*

$$Im\lambda_k = \beta_k = 0 \text{ and } G_{\theta_k}^{(s_k)}(\theta_k) = 0 \text{ for } s_k = 2, 3, \ldots. \tag{1.125}$$

Proof For $\mathbf{x} = \mathbf{x}^*$, $c^{2k-1} \to 0$, and $c^{2k} \to 0$, consider

$$\rho_k = \sqrt{(c^{2k-1})^2 + (c^{2k})^2}.$$

Assuming

$$c^{2k-1} = \rho_k \cos\theta_k, c^{2k} = \rho_k \sin\theta_k,$$

we have

$$\begin{aligned} \dot{c}^{2k-1} &= \dot{\rho}_k \cos\theta_k - \dot{\theta}_k \rho_k \sin\theta_k, \\ \dot{c}^{2k} &= \dot{\theta}_k \rho_k \cos\theta_k + \dot{\rho}_k \sin\theta_k. \end{aligned}$$

Thus

$$\begin{aligned} \dot{\rho}_k &= \dot{c}^{2k-1} \cos\theta_k + \dot{c}^{2k} \sin\theta_k, \\ \dot{\theta}_k \rho_k &= \dot{c}^{2k-1} \sin\theta_k - \dot{c}^{2k} \cos\theta_k, \end{aligned}$$

where

$$\begin{aligned} \dot{c}^{2k-1} &= G_{2k-1}(\mathbf{x}, \mathbf{p}) = (\mathbf{v}^{2k-1})^{\mathrm{T}} \cdot \mathbf{f}(\mathbf{x}, \mathbf{p}) \\ &= \sum_{r_k=1}^{\infty} \frac{1}{r_k!} G_{(2k-1)(j_1 j_2 \cdots j_{r_k})}^{(r_k)}(\mathbf{x}^*, \mathbf{p}) c^{j_1} c^{j_2} \cdots c^{j_{r_k}}, \\ \dot{c}^{2k} &= G_{2k}(\mathbf{x}, \mathbf{p}) = (\mathbf{v}^{2k})^{\mathrm{T}} \cdot \mathbf{f}(\mathbf{x}, \mathbf{p}) \\ &= \sum_{r_k=1}^{\infty} \frac{1}{r_k!} G_{(2k)(j_1 j_2 \cdots j_{r_k})}^{(r_k)}(\mathbf{x}^*, \mathbf{p}) c^{j_1} c^{j_2} \cdots c^{j_{r_k}}, \\ &(j_{r_k} \in \{1, 2, \cdots, n\}, r_k = 1, 2, \cdots). \end{aligned}$$

Therefore,

$$\dot{\rho}_k = \sum_{r_k=1}^{\infty} \frac{1}{r_k!} G_{\rho_k}^{(r_k)} \text{ and } \dot{\theta}_k = \sum_{r_k=1}^{\infty} \frac{1}{r_k!} G_{\theta_k}^{(r_k)},$$

where

$$\begin{aligned} G_{\rho_k}^{(r_k)}(\mathbf{x}^*, \mathbf{p}) &= [G_{(2k-1)(j_1 j_2 \cdots j_{r_k})}^{(r_k)}(\mathbf{x}^*, \mathbf{p}) c^{j_1} c^{j_2} \cdots c^{j_{r_k}} \cos\theta_k \\ &\quad + G_{(2k)(j_1 j_2 \cdots j_{r_k})}^{(r_k)}(\mathbf{x}^*, \mathbf{p}) c^{j_1} c^{j_2} \cdots c^{j_{r_k}} \sin\theta_k], \\ G_{\theta_k}^{(r_k)}(\mathbf{x}^*, \mathbf{p}) &= \frac{1}{\rho_k} [G_{(2k-1)(j_1 j_2 \cdots j_{r_k})}^{(r_k)}(\mathbf{x}^*, \mathbf{p}) c^{j_1} c^{j_2} \cdots c^{j_{r_k}} \sin\theta_k \\ &\quad - G_{(2k)(j_1 j_2 \cdots j_{r_k})}^{(r_k)}(\mathbf{x}^*, \mathbf{p}) c^{j_1} c^{j_2} \cdots c^{j_{r_k}} \cos\theta_k], \\ &\quad (\text{summation of } j_1, j_2, \ldots, j_{r_k}). \end{aligned}$$

To the first-order approximation of $\dot{c}^{2k-1}$ and $\dot{c}^{2k}$,

$$\begin{aligned} \dot{c}^{2k-1} &= \alpha_k c^{2k-1} + \beta_k c^{2k} + o(\rho_k), \\ \dot{c}^{2k} &= -\beta_k c^{2k-1} + \alpha_k c^{2k} + o(\rho_k) \end{aligned}$$

or

$$\begin{bmatrix} \dot{c}^{2k-1} \\ \dot{c}^{2k} \end{bmatrix} = \begin{bmatrix} \alpha_k & \beta_k \\ -\beta_k & \alpha_k \end{bmatrix} \begin{bmatrix} c^{2k-1} \\ c^{2k} \end{bmatrix} + \begin{bmatrix} o(\rho_k) \\ o(\rho_k) \end{bmatrix}.$$

Using

$$\begin{aligned} \dot{\rho}_k &= \dot{c}^{2k-1} \cos\theta_k + \dot{c}^{2k} \sin\theta_k, \\ \dot{\theta}_k &= \rho_k^{-1} (\dot{c}^{2k} \cos\theta_k - \dot{c}^{2k-1} \sin\theta_k), \end{aligned}$$

we obtain

$$\begin{aligned} \dot{\rho}_k &= (\alpha_k c^{2k-1} + \beta_k c^{2k-1}) \cos\theta_k + (-\beta_k c^{2k-1} + \alpha_k c^{2k}) \sin\theta_k \\ &= \alpha_k \rho_k + o(\rho_k), \\ \dot{\theta}_k \rho_k &= (-\beta_k c^{2k-1} + \alpha_k c^{2k}) \cos\theta_k + (\alpha_k c^{2k-1} + \beta_k c^{2k}) \sin\theta_k \\ &= -\beta_k \rho_k + o(\rho_k). \end{aligned}$$

Further

$$\dot{\rho}_k = \alpha_k \rho_k + o(\rho_k),$$
$$\dot{\theta}_k \rho_k = -\beta_k \rho_k + o(\rho_k).$$

So

$$\dot{\rho}_k = \alpha_k \rho_k$$
$$\dot{\theta}_k \rho_k = -\beta_k \rho_k \Rightarrow \dot{\theta}_k = -\beta_k.$$

Therefore,

$$G^{(1)}_{\rho_k}(\mathbf{x}^*, \mathbf{p}) = \alpha_k \rho_k \text{ and } G^{(1)}_{\theta_k}(\mathbf{x}^*, \mathbf{p}) = -\beta_k \theta_k.$$

The corresponding higher order expression is given by

$$\dot{\rho}_k = \sum\nolimits_{r_k=1}^{m_k-1} \frac{1}{r_k!} G^{(s_k)}_{\rho_k}(\mathbf{x}^*, \mathbf{p}) + \frac{1}{m_k!} G^{(m_k)}_{\rho_k}(\mathbf{x}^*, \mathbf{p}) + o(\rho_k^{m_k}).$$

Because for $\varepsilon > 0$ and $\varepsilon \to 0$,

$$\begin{aligned}
\rho_k(t+\varepsilon) - \rho_k(t) &= \dot{\rho}_k \varepsilon \\
&= \varepsilon \sum\nolimits_{r_k=1}^{m_k-1} \frac{1}{r_k!} G^{(s_k)}_{\rho_k}(\mathbf{x}^*, \mathbf{p}) + \varepsilon \frac{1}{m_k!} G^{(m_k)}_{\rho_k}(\mathbf{x}^*, \mathbf{p}) + o(\varepsilon \rho_k^{m_k}).
\end{aligned}$$

(i) For equilibrium stability, $\rho_k > 0$ and $\rho_k \to 0$. If $\alpha_k \neq 0$, we have

$$\dot{\rho}_k = G^{(1)}_{\rho_k}(\mathbf{x}^*, \mathbf{p}) = \alpha_k \rho_k + o(\rho_k).$$

Due to $\rho_k > 0$, if $\alpha_k < 0$, then $\dot{\rho}_k < 0$. Therefore,

$$\rho_k(t+\varepsilon) - \rho_k(t) = \dot{\rho}_k \varepsilon < 0$$

which implies that $\mathbf{x}$ at the equilibrium $\mathbf{x}^*$ on the plane of $(\mathbf{v}_{2k-1}, \mathbf{v}_{2k})$ is spirally stable, and vice versa.

(ii) Due to $\rho_k > 0$, if $\alpha_k > 0$, then $\dot{\rho}_k > 0$. Thus,

$$\rho_k(t+\varepsilon) - \rho_k(t) = \dot{\rho}_k \varepsilon > 0,$$

which implies that $\mathbf{x}$ at the equilibrium $\mathbf{x}^*$ on the plane of $(\mathbf{v}_{2k-1}, \mathbf{v}_{2k})$ is spirally unstable, and vice versa.

(iii) If for $\theta_k \in [0, 2\pi]$ the following conditions exist

$$G_{\rho_k}^{(r_k)}(\mathbf{x}^*, \mathbf{p}) = 0 \ \text{ for r}_k = 1, 2, \ldots, m_k - 1;$$
$$G_{\rho_k}^{(m_k)}(\mathbf{x}^*, \mathbf{p}) \neq 0, \text{and } \mid G_{\rho_k}^{(r_k)}(\mathbf{x}^*, \mathbf{p}) \mid < \infty \ \text{ for } r_k = m_k + 1, m_k + 2, \ldots,$$

then the higher order terms can be ignored, i.e.,

$$\dot{\rho}_k = \frac{1}{m_k!} G_{\rho_k}^{(m_k)}(\mathbf{x}^*, \mathbf{p}) + o(\rho_k^{m_k}).$$

Due to $\rho_k > 0$, if $G_{\rho_k}^{(m_k)}(\mathbf{x}^*, \mathbf{p}) < 0$, then $\dot{\rho}_k < 0$. Therefore,

$$\rho_k(t + \varepsilon) - \rho_k(t) = \dot{\rho}_k \varepsilon < 0.$$

In other words, $\mathbf{x}$ at the equilibrium $\mathbf{x}^*$ on the plane of $(\mathbf{v}_{2k-1}, \mathbf{v}_{2k})$ is spirally stable with the m_kth-order singularity, and vice versa.

(iv) Due to $\rho_k > 0$, if $G_{\rho_k}^{(m_k)}(\mathbf{x}^*, \mathbf{p}) > 0$, then $\dot{\rho}_k > 0$. Therefore,

$$\rho_k(t + \varepsilon) - \rho_k(t) = \dot{\rho}_k \varepsilon > 0.$$

In other words, $\mathbf{x}$ at the equilibrium $\mathbf{x}^*$ on the plane of $(\mathbf{v}_{2k-1}, \mathbf{v}_{2k})$ is spirally unstable with the (m_k-1)th-order singularity, and vice versa.

(v) If for $\theta_k \in [0, 2\pi]$ the following conditions exist

$$G_{\rho_k}^{(r_k)}(\mathbf{x}^*, \mathbf{p}) = 0 \text{ for } r_k = 1, 2, \ldots,$$

then

$$\rho_k(t + \varepsilon) - \rho_k(t) = \dot{\rho}_k \varepsilon = 0$$

and vice versa. Therefore $\rho_k(t)$ is constant. $\mathbf{x}$ at the equilibrium $\mathbf{x}^*$ on the plane of $(\mathbf{v}_{2k-1}, \mathbf{v}_{2k})$ is circular.

(vi) Consider

$$\begin{aligned} &\theta_k(t + \varepsilon) - \theta_k(t) = \dot{\theta}_k \varepsilon \\ &= \varepsilon\{-\beta_k + \frac{1}{\rho_k}[\sum\nolimits_{r_k=2}^{m_k-1} \frac{1}{r_k!} G_{\theta_k}^{(r_k)}(\mathbf{x}^*, \mathbf{p}) + \frac{1}{m_k!} G_{\theta_k}^{(m_k)}(\mathbf{x}^*, \mathbf{p}) + o(\rho_k^{m_k})]\}. \end{aligned}$$

If for $\theta_k \in [0, 2\pi]$ the following conditions exist

$$\beta_k = 0 \text{ and } G_{\theta_k}^{(r_k)}(\mathbf{x}^*, \mathbf{p}) = 0 \text{ for } r_k = 2, 3, \ldots$$

then

$$\theta_k(t+\varepsilon) - \theta_k(t) = \dot{\theta}_k \varepsilon = 0.$$

Therefore, $\mathbf{x}$ at the equilibrium $\mathbf{x}^*$ on the plane of $(\mathbf{v}_{2k-1}, \mathbf{v}_{2k})$ is degenerate in the direction of $\mathbf{v}_{2k-1}$. This theorem is proved. ∎

1.3.3 Spiral Stability Based on the Fourier Series Base

From the previous method, the spiral stability of the equilibrium $\mathbf{x}^*$ is difficult, and such spiral stability is based on the circular assumption as in Eq. (1.119), which is difficult to separate $\rho_k(t)$ and $\theta_k(t)$. For the linear case, it is very easy to do so. However, for the nonlinear case, only special cases can be discussed for the spiral stability. Thus, an alternative approach is presented to determine the spiral stability of the equilibrium $\mathbf{x}^*$ as in Luo (2019).

Definition 1.29 Consider an n-dimensional, autonomous, nonlinear dynamical system $\dot{\mathbf{x}} = \mathbf{f}(\mathbf{x}, \mathbf{p})$ in Eq. (1.4) with an equilibrium point $\mathbf{x}^*$ and $\mathbf{f}(\mathbf{x}, \mathbf{p})$ is C^r-continuous ($r \geq 1$) in a neighborhood of the equilibrium $\mathbf{x}^*$ (i.e., $U(\mathbf{x}^*) \subset \Omega$). The corresponding solution is $\mathbf{x}(t) = \mathbf{\Phi}(\mathbf{x}_0, t-t_0, \mathbf{p})$. Suppose Eq. (1.24) holds in $U(\mathbf{x}^*) \subset \Omega$. For a linearized dynamical system in Eq. (1.19), consider a pair of complex eigenvalue $\alpha_k \pm \mathbf{i}\beta_k$ ($2k-1$, $2k \in N = \{1, 2, \cdots, n\}$,$\mathbf{i} = \sqrt{-1}$) of matrix $D\mathbf{f}(\mathbf{x}^*, \mathbf{p})$ with a pair of eigenvectors $\mathbf{v}_{2k-1} \pm \mathbf{i}\mathbf{v}_{2k}$. In the vicinity of $\mathbf{x}^*$, for $\mathbf{x} = \mathbf{x}^* + \mathbf{y}$, there is a solution in the form of

$$\mathbf{x} = \mathbf{x}^* + \mathbf{y} = \mathbf{x}^* + \mathbf{a}_0(t) + \sum_{j=1}^{\infty} \mathbf{b}_j(t)\cos(j\beta_k t) + \mathbf{c}_j(t)\sin(j\beta_k t) \tag{1.126}$$

with the condition of

$$\lim_{j\to\infty} \|\mathbf{A}_j\| = 0 \tag{1.127}$$

where

$$\begin{aligned}
&\mathbf{a}_0 = (a_{01}, a_{02}, \ldots, a_{0n})^{\mathrm{T}}, \mathbf{b}_j = (b_{j1}, b_{j2}, \ldots, b_{jn})^{\mathrm{T}}, \\
&\mathbf{c}_j = (c_{j1}, c_{j2}, \ldots, c_{jn})^{\mathrm{T}}, \mathbf{A}_j = (A_{j1}, A_{j2}, \ldots, A_{jn})^{\mathrm{T}}, \\
&A_{jl} = \sqrt{b_{jl}^2 + c_{jl}^2}\ (j = 1, 2, \ldots; l = 1, 2, \ldots, n).
\end{aligned} \tag{1.128}$$

Thus, the approximate solution of $\mathbf{x}$ in the vicinity of the equilibrium $\mathbf{x}^*$ is for a large number $N \to \infty$ as

$$\mathbf{x} = \mathbf{x}^* + \mathbf{y} \approx \mathbf{x}^* + \mathbf{a}_0(t) + \sum_{j=1}^{N} \mathbf{b}_j(t)\cos(j\beta_k t) + \mathbf{c}_j(t)\sin(j\beta_k t). \tag{1.129}$$

With the transformation in Eq. (1.129), Eq. (1.4) becomes

$$\begin{aligned} \dot{\mathbf{a}}_0 &= \mathbf{F}_0(\mathbf{a}_0, \mathbf{b}, \mathbf{c}), \\ \dot{\mathbf{b}} &= -\beta_k \mathbf{k}\mathbf{c} + \mathbf{F}^c(\mathbf{a}_0, \mathbf{b}, \mathbf{c}), \\ \dot{\mathbf{c}} &= \beta_k \mathbf{k}\mathbf{b} + \mathbf{F}^s(\mathbf{a}_0, \mathbf{b}, \mathbf{c}) \end{aligned} \tag{1.130}$$

where

$$\begin{aligned} \mathbf{b} &= (\mathbf{b}_1, \mathbf{b}_2, \ldots, \mathbf{b}_N)^{\mathrm{T}}, \mathbf{c} = (\mathbf{c}_1, \mathbf{c}_2, \ldots, \mathbf{c}_N)^{\mathrm{T}}, \\ \mathbf{F}^c &= (\mathbf{F}_1^c, \mathbf{F}_2^c, \ldots, \mathbf{F}_N^c)^{\mathrm{T}}, \mathbf{F}^s = (\mathbf{F}_1^s, \mathbf{F}_2^s, \ldots, \mathbf{F}_N^s)^{\mathrm{T}}, \\ \mathbf{k} &= \mathit{diag}\,(1, 2, \ldots, N), \mathbf{A} = (\mathbf{A}_1, \mathbf{A}_2, \ldots, \mathbf{A}_N)^{\mathrm{T}}; \end{aligned} \tag{1.131}$$

$$\begin{aligned} \mathbf{F}_0(\mathbf{a}_0, \mathbf{b}, \mathbf{c}) &= \frac{1}{T}\int_0^T \mathbf{f}(\mathbf{x}, \mathbf{p})dt, \\ \mathbf{F}_j^c(\mathbf{a}_0, \mathbf{b}, \mathbf{c}) &= \frac{2}{T}\int_0^T \mathbf{f}(\mathbf{x}, \mathbf{p})\cos(j\beta_k t)dt, \\ \mathbf{F}_j^s(\mathbf{a}_0, \mathbf{b}, \mathbf{c}) &= \frac{2}{T}\int_0^T \mathbf{f}(\mathbf{x}, \mathbf{p})\sin(j\beta_k t)dt, \\ T &= \frac{2\pi}{\beta_k}, \quad (j = 1, 2, \ldots, N). \end{aligned} \tag{1.132}$$

The equilibrium of Eq. (1.130) is given by

$$\begin{aligned} \mathbf{0} &= \mathbf{F}_0(\mathbf{a}_0^*, \mathbf{b}^*, \mathbf{c}^*), \\ \mathbf{0} &= -\beta_k \mathbf{k}\mathbf{c}^* + \mathbf{F}^c(\mathbf{a}_0^*, \mathbf{b}^*, \mathbf{c}^*), \\ \mathbf{0} &= \beta_k \mathbf{k}\mathbf{b}^* + \mathbf{F}^s(\mathbf{a}_0^*, \mathbf{b}^*, \mathbf{c}^*). \end{aligned} \tag{1.133}$$

If Eq. (1.133) has a zero solution, i.e.,

$$\mathbf{a}_0^* = \mathbf{0}, \mathbf{b}^* = \mathbf{0}, \mathbf{c}^* = \mathbf{0}, \tag{1.134}$$

and then the equilibrium $\mathbf{x}^*$ has the following stability.

(i) The equilibrium $\mathbf{x}^*$ is spirally stable if

$$\begin{aligned}&\lim_{t\to\infty}\|\mathbf{a}_0(t)\|=0,\ \lim_{t\to\infty}\|\mathbf{b}(t)\|=0,\ \lim_{t\to\infty}\|\mathbf{c}(t)\|=0\\&(\text{or}\ \lim_{t\to\infty}\|\mathbf{a}_0(t)\|=0\lim_{t\to\infty}\|\mathbf{A}(t)\|=0).\end{aligned}\tag{1.135}$$

(ii) The equilibrium $\mathbf{x}^*$ is spirally unstable if

$$\begin{aligned}&\lim_{t\to\infty}\|\mathbf{a}_0(t)\|=\infty,\text{or}\ \lim_{t\to\infty}\|\mathbf{b}(t)\|=\infty,\text{or}\ \lim_{t\to\infty}\|\mathbf{c}(t)\|=\infty\\&(\ \text{or}\lim_{t\to\infty}\|\mathbf{a}_0(t)\|=\infty,\text{or}\ \lim_{t\to-\infty}\|\mathbf{A}(t)\|=\infty).\end{aligned}\tag{1.136}$$

(iii) The equilibrium $\mathbf{x}^*$ is stable if

$$\begin{aligned}&\lim_{t\to\infty}\|\mathbf{a}_0(t)\|=0,\mathbf{b}(t)=\mathbf{0},\mathbf{c}(t)=\mathbf{0}\\&(\text{or}\lim_{t\to\infty}\|\mathbf{a}_0(t)\|=0,\mathbf{A}(t)=\mathbf{0}).\end{aligned}\tag{1.137}$$

(iv) The equilibrium $\mathbf{x}^*$ is unstable if

$$\begin{aligned}&\lim_{t\to\infty}\|\mathbf{a}_0(t)\|=\infty,\mathbf{b}(t)=\mathbf{0},\mathbf{c}(t)=\mathbf{0}\\&(\text{or}\lim_{t\to\infty}\|\mathbf{a}_0(t)\|=\infty,\mathbf{A}(t)=\mathbf{0}).\end{aligned}\tag{1.138}$$

Definition 1.30 Consider an n-dimensional, autonomous, nonlinear dynamical system $\dot{\mathbf{x}}=\mathbf{f}(\mathbf{x},\mathbf{p})$ in Eq. (1.4) with an equilibrium point $\mathbf{x}^*$ and $\mathbf{f}(\mathbf{x},\mathbf{p})$ is C^r-continuous ($r\geq 1$) in a neighborhood of the equilibrium $\mathbf{x}^*$ (i.e.,$U(\mathbf{x}^*)\subset\Omega$). The corresponding solution is $\mathbf{x}(t)=\boldsymbol{\Phi}(\mathbf{x}_0,t-t_0,\mathbf{p})$. Suppose Eq. (1.24) holds in $U(\mathbf{x}^*)\subset\Omega$. For a linearized dynamical system in Eq. (1.19), consider a pair of complex eigenvalue $\alpha_k\pm\mathbf{i}\beta_k$ ($2k-1$, $2k\in N=\{1,2,\ldots,n\}$,$\mathbf{i}=\sqrt{-1}$) of matrix $D\mathbf{f}(\mathbf{x}^*,\mathbf{p})$ with a pair of eigenvectors $\mathbf{v}_{2k-1}\pm\mathbf{i}\mathbf{v}_{2k}$. In the vicinity of $\mathbf{x}^*$, there exists a transformation of

$$\mathbf{x}=\mathbf{x}^*+\mathbf{y}=\mathbf{x}^*+\mathbf{a}_0(t)+\sum_{j=1}^{N}\mathbf{b}_j(t)\cos(j\beta_k t)+\mathbf{c}_j(t)\sin(j\beta_k t)\tag{1.139}$$

under $\lim_{j\to\infty}\|\mathbf{A}_j\|=0$ with Eq. (1.128). The corresponding dynamical system of coefficients is

$$\begin{aligned}\dot{\mathbf{a}}_0 &= \mathbf{F}_0(\mathbf{a}_0, \mathbf{b}, \mathbf{c}),\\ \dot{\mathbf{b}} &= -\beta_k \mathbf{k}\mathbf{c} + \mathbf{F}^c(\mathbf{a}_0, \mathbf{b}, \mathbf{c}),\\ \dot{\mathbf{c}} &= \beta_k \mathbf{k}\mathbf{b} + \mathbf{F}^s(\mathbf{a}_0, \mathbf{b}, \mathbf{c})\end{aligned} \tag{1.140}$$

with Eqs. (1.131) and (1.132). The equivalent equation of Eq. (1.140) is

$$\dot{\mathbf{z}} = \mathbf{g}(\mathbf{z}) \tag{1.141}$$

where

$$\mathbf{z} = (\mathbf{a}_0, \mathbf{b}, \mathbf{c})^{\mathrm{T}}, \mathbf{g} = (\mathbf{F}_0, \mathbf{F}^c, \mathbf{F}^s)^{\mathrm{T}}. \tag{1.142}$$

If Eq. (1.141) has an equilibrium of $\mathbf{z}^* = \mathbf{0}$ (i.e., $\mathbf{a}_0^* = \mathbf{0}, \mathbf{b}^* = \mathbf{0}, \mathbf{c}^* = \mathbf{0}$), and in a neighborhood of $\mathbf{z}^* = \mathbf{0}$, the linearized equation with $\Delta\mathbf{z} = \mathbf{z} - \mathbf{z}^*$ is

$$\Delta\dot{\mathbf{z}} = D\mathbf{g}(\mathbf{z}^*)\Delta\mathbf{z}, \tag{1.143}$$

and then the eigenvalues of $D\mathbf{g}(\mathbf{z}^*)$determined by

$$| D\mathbf{g}(\mathbf{z}^*) - \lambda \mathbf{I}_{n(2N+1)\times n(2N+1)} | = 0 \tag{1.144}$$

yield the stability of the equilibrium $\mathbf{x}^*$ with a kind of

$$(n_1 : n_2 : n_3 | n_4 : n_5 : n_6) \tag{1.145}$$

with $\Sigma_{r=1}^{6} n_r = n(2N+1)$.

(i) If $\mathrm{Re}\lambda_l < 0$ $(l = 1, 2, \ldots, n(2N+1))$ with at least $\mathbf{b} \neq \mathbf{0}$ or $\mathbf{c} \neq \mathbf{0}$, the equilibrium $\mathbf{x}^*$ is spirally stable.
(ii) If $\mathrm{Re}\lambda_l > 0$ $(l \in \{1, 2, \ldots, n(2N+1)\})$ with at least $\mathbf{b} \neq \mathbf{0}$ or $\mathbf{c} \neq \mathbf{0}$, the equilibrium $\mathbf{x}^*$ is spirally unstable.
(iii) The boundary between the stable and unstable solutions of equilibrium $\mathbf{x}^*$ is determined by the higher singularity.

Remark 1.1 (i) For $\mathbf{A}(t) = \mathbf{0}$, $\mathbf{a}_0(t) \neq \mathbf{0}$, Eq. (1.139) is recalled to Eq. (1.19) with n-eigenvalues, and the stability of the equilibrium $\mathbf{x}^*$ is determined identically by Eq. (1.19). (ii) For $\mathbf{A}(t) \neq \mathbf{0}$, $\mathbf{a}_0(t) = \mathbf{0}$, Eq. (1.144) gives the spiral stability of the equilibrium $\mathbf{x}^*$ without the shifted center. (iii) From eigenvectors of Eq. (1.144), the stability on each eigenvector direction for $\mathbf{z}$ in the vicinity of $\mathbf{z}^* = \mathbf{0}$ can be determined in the similar fashion.

1.4 Spiral Stability in Second-Order Nonlinear Systems

The stability of equilibriums in the second-order dynamical systems can be discussed through the first-order dynamical system. Since second-order dynamical systems are used extensively in physics and engineering, the spiral stability of equilibrium is specially discussed as in Luo (2019).

Definition 1.31 Consider a second-order, autonomous, nonlinear dynamical system with $2n$-dimensions as

$$\ddot{\mathbf{x}} = \mathbf{f}(\mathbf{x}, \dot{\mathbf{x}}, \mathbf{p}) \in R^n. \tag{1.146}$$

There is an equilibrium point $(\mathbf{x}^*, \mathbf{0})$ determined by

$$\mathbf{f}(\mathbf{x}^*, \mathbf{0}, \mathbf{p}) = \mathbf{0}, \tag{1.147}$$

and $\mathbf{f}(\mathbf{x}, \dot{\mathbf{x}}, \mathbf{p})$ is C^r-continuous ($r \geq 1$) in a neighborhood of the equilibrium $(\mathbf{x}^*, \mathbf{0})$ (i.e., $U(\mathbf{x}^*, \mathbf{0}) \subset \Omega$). The corresponding solution is $\mathbf{x}(t) = \mathbf{\Phi}(\mathbf{x}_0, t-t_0, \mathbf{p})$. Suppose

$$\lim_{\|\Delta\mathbf{x}\|+\|\Delta\dot{\mathbf{x}}\|\to 0} \frac{\|\mathbf{f}(\mathbf{x}, \dot{\mathbf{x}}, \mathbf{p}) - D_{\mathbf{x}}\mathbf{f}(\mathbf{x}^*, \mathbf{0}, \mathbf{p})\Delta\mathbf{x} - D_{\dot{\mathbf{x}}}\mathbf{f}(\mathbf{x}^*, \mathbf{0}, \mathbf{p})\Delta\dot{\mathbf{x}}\|}{\|\Delta\mathbf{x}\| + \|\Delta\dot{\mathbf{x}}\|} = 0 \tag{1.148}$$

in $U(\mathbf{x}^*, \mathbf{0}) \subset \Omega$. The linearization of Eq. (1.146) is

$$\Delta\ddot{\mathbf{x}} = D_{\mathbf{x}}\mathbf{f}(\mathbf{x}^*, \mathbf{0}, \mathbf{p})\Delta\mathbf{x} + D_{\dot{\mathbf{x}}}\mathbf{f}(\mathbf{x}^*, \mathbf{0}, \mathbf{p})\Delta\dot{\mathbf{x}}. \tag{1.149}$$

For the linearized system, consider a pair of complex eigenvalues $\alpha_k \pm \mathbf{i}\beta_k$ ($2k-1$, $2k \in N = \{1, 2, \ldots, 2n\}, \mathbf{i} = \sqrt{-1}$) with a pair of eigenvectors $\mathbf{v}_{2k-1} \pm \mathbf{i}\mathbf{v}_{2k}$. In the vicinity of $(\mathbf{x}^*, \mathbf{0})$, for $\mathbf{x} = \mathbf{x}^* + \mathbf{y}$, there is a solution in the form of

$$\mathbf{x} = \mathbf{x}^* + \mathbf{y} = \mathbf{x}^* + \mathbf{a}_0(t) + \sum_{j=1}^{\infty} \mathbf{b}_j(t)\cos(j\beta_k t) + \mathbf{c}_j(t)\sin(j\beta_k t) \tag{1.150}$$

with the condition of

$$\lim_{j\to\infty} \|\mathbf{A}_j\| = 0 \tag{1.151}$$

where

$$\begin{aligned} &\mathbf{a}_0 = (a_{01}, a_{02}, \ldots, a_{0n})^{\mathrm{T}}, \mathbf{b}_j = (b_{j1}, b_{j2}, \ldots, b_{jn})^{\mathrm{T}}, \\ &\mathbf{c}_j = (c_{j1}, c_{j2}, \ldots, c_{jn})^{\mathrm{T}}, \mathbf{A}_j = (A_{j1}, A_{j2}, \ldots, A_{jn})^{\mathrm{T}}, \\ &A_{jl} = \sqrt{b_{jl}^2 + c_{jl}^2}\ (j = 1, 2, \ldots; l = 1, 2, \ldots, n). \end{aligned} \tag{1.152}$$

Thus, the approximate solution of $\mathbf{x}$ in the vicinity of the equilibrium $(\mathbf{x}^*, \mathbf{0})$ is for a large number $N \to \infty$ as

$$\mathbf{x} = \mathbf{x}^* + \mathbf{y} \approx \mathbf{x}^* + \mathbf{a}_0(t) + \sum_{j=1}^{N} \mathbf{b}_j(t)\cos(j\beta_k t) + \mathbf{c}_j(t)\sin(j\beta_k t). \tag{1.153}$$

With the transformation in Eq. (1.153), Eq. (1.146) becomes

$$\begin{aligned}
\ddot{\mathbf{a}}_0 &= \mathbf{F}_0(\mathbf{a}_0, \mathbf{b}, \mathbf{c}, \dot{\mathbf{a}}_0, \dot{\mathbf{b}}, \dot{\mathbf{c}}),\\
\ddot{\mathbf{b}} &= -2\beta_k \mathbf{k}_1 \dot{\mathbf{c}} + \beta_k^2 \mathbf{k}_2 \mathbf{b} + \mathbf{F}^c(\mathbf{a}_0, \mathbf{b}, \mathbf{c}, \dot{\mathbf{a}}_0, \dot{\mathbf{b}}, \dot{\mathbf{c}}),\\
\ddot{\mathbf{c}} &= 2\beta_k \mathbf{k}_1 \dot{\mathbf{b}} + \beta_k^2 \mathbf{k}_2 \mathbf{c} + \mathbf{F}^s(\mathbf{a}_0, \mathbf{b}, \mathbf{c}, \dot{\mathbf{a}}_0, \dot{\mathbf{b}}, \dot{\mathbf{c}})
\end{aligned} \tag{1.154}$$

where

$$\begin{aligned}
&\mathbf{b} = (\mathbf{b}_1, \mathbf{b}_2, \ldots, \mathbf{b}_N)^{\mathrm{T}}, \mathbf{c} = (\mathbf{c}_1, \mathbf{c}_2, \ldots, \mathbf{c}_N)^{\mathrm{T}},\\
&\mathbf{F}^c = (\mathbf{F}_1^c, \mathbf{F}_2^c, \ldots, \mathbf{F}_N^c)^{\mathrm{T}}, \mathbf{F}^s = (\mathbf{F}_1^s, \mathbf{F}_2^s, \ldots, \mathbf{F}_N^s)^{\mathrm{T}},\\
&\mathbf{k}_1 = \mathit{diag}\,(1, 2, \ldots, N), \mathbf{k}_2 = \mathit{diag}\,(1^2, 2^2, \ldots, N^2),\\
&\mathbf{A} = (\mathbf{A}_1, \mathbf{A}_2, \ldots, \mathbf{A}_N)^{\mathrm{T}};
\end{aligned} \tag{1.155}$$

$$\begin{aligned}
\mathbf{F}_0(\mathbf{a}_0, \mathbf{b}, \mathbf{c}, \dot{\mathbf{a}}_0, \dot{\mathbf{b}}, \dot{\mathbf{c}}) &= \frac{1}{T}\int_0^T \mathbf{f}(\mathbf{x}, \dot{\mathbf{x}}, \mathbf{p})dt,\\
\mathbf{F}_j^c(\mathbf{a}_0, \mathbf{b}, \mathbf{c}, \dot{\mathbf{a}}_0, \dot{\mathbf{b}}, \dot{\mathbf{c}}) &= \frac{2}{T}\int_0^T \mathbf{f}(\mathbf{x}, \dot{\mathbf{x}}, \mathbf{p})\cos(j\beta_k t)dt,\\
\mathbf{F}_j^s(\mathbf{a}_0, \mathbf{b}, \mathbf{c}, \dot{\mathbf{a}}_0, \dot{\mathbf{b}}, \dot{\mathbf{c}}) &= \frac{2}{T}\int_0^T \mathbf{f}(\mathbf{x}, \dot{\mathbf{x}}, \mathbf{p})\sin(j\beta_k t)dt,\\
T &= \frac{2\pi}{\beta_k},\ (j = 1, 2, \ldots, N).
\end{aligned} \tag{1.156}$$

The equilibrium of Eq. (1.154) is given by

$$\begin{aligned}
\mathbf{0} &= \mathbf{F}_0(\mathbf{a}_0^*, \mathbf{b}^*, \mathbf{c}^*, \mathbf{0}, \mathbf{0}, \mathbf{0}),\\
\mathbf{0} &= \beta_k^2 \mathbf{k}_2 \mathbf{b}^* + \mathbf{F}^c(\mathbf{a}_0^*, \mathbf{b}^*, \mathbf{c}^*, \mathbf{0}, \mathbf{0}, \mathbf{0}),\\
\mathbf{0} &= \beta_k^2 \mathbf{k}_2 \mathbf{c}^* + \mathbf{F}^s(\mathbf{a}_0^*, \mathbf{b}^*, \mathbf{c}^*, \mathbf{0}, \mathbf{0}, \mathbf{0}).
\end{aligned} \tag{1.157}$$

If Eq. (1.157) has a zero solution, i.e.,

$$\mathbf{a}_0^* = \mathbf{0}, \mathbf{b}^* = \mathbf{0}, \mathbf{c}^* = \mathbf{0}, \tag{1.158}$$

then the equilibrium $\mathbf{x}^*$ has the following stability:

(i) The equilibrium $\mathbf{x}^*$ is spirally stable if

$$\begin{aligned} &\lim_{t\to\infty}\|\mathbf{a}_0(t)\| = 0,\ \lim_{t\to\infty}\|\mathbf{b}(t)\| = 0,\ \lim_{t\to\infty}\|\mathbf{c}(t)\| = 0\\ &(\text{or } \lim_{t\to\infty}\|\mathbf{a}_0(t)\| = 0,\ \lim_{t\to\infty}\|\mathbf{A}(t)\| = 0). \end{aligned} \tag{1.159}$$

(ii) The equilibrium $\mathbf{x}^*$ is spirally unstable if

$$\begin{aligned} &\lim_{t\to\infty}\|\mathbf{a}_0(t)\| = \infty, \text{or } \lim_{t\to\infty}\|\mathbf{b}(t)\| = \infty, \text{or } \lim_{t\to\infty}\|\mathbf{c}(t)\| = \infty\\ &(\text{or } \lim_{t\to\infty}\|\mathbf{a}_0(t)\| = \infty, \text{or } \lim_{t\to\infty}\|\mathbf{A}(t)\| = \infty). \end{aligned} \tag{1.160}$$

(iii) The equilibrium $\mathbf{x}^*$ is asymptotically stable if

$$\begin{aligned} &\lim_{t\to\infty}\|\mathbf{a}_0(t)\| = 0, \mathbf{b}(t) = \mathbf{0}, \mathbf{c}(t) = \mathbf{0}\\ &(\text{or} \lim_{t\to\infty}\|\mathbf{a}_0(t)\| = 0, \mathbf{A}(t) = \mathbf{0}). \end{aligned} \tag{1.161}$$

(iv) The equilibrium $\mathbf{x}^*$ is asymptotically unstable if

$$\begin{aligned} &\lim_{t\to\infty}\|\mathbf{a}_0(t)\| = 0, \mathbf{b}(t) = \mathbf{0}, \mathbf{c}(t) = \mathbf{0}\\ &(\text{or } \lim_{t\to-\infty}\|\mathbf{a}_0(t)\| = 0, \mathbf{A}(t) = \mathbf{0}). \end{aligned} \tag{1.162}$$

Definition 1.32 Consider a $2n$-dimensional, autonomous, nonlinear dynamical system $\ddot{\mathbf{x}} = \mathbf{f}(\mathbf{x}, \dot{\mathbf{x}}, \mathbf{p})$ in Eq. (1.146) with an equilibrium point $(\mathbf{x}^*, \mathbf{0})$ and $\mathbf{f}(\mathbf{x}, \dot{\mathbf{x}}, \mathbf{p})$ is C^r-continuous ($r \geq 1$) in a neighborhood of the equilibrium $(\mathbf{x}^*, \mathbf{0})$ (i.e., $U(\mathbf{x}^*, \mathbf{0}) \subset \Omega$). The corresponding solution is $\mathbf{x}(t) = \boldsymbol{\Phi}(\mathbf{x}_0, t-t_0, \mathbf{p})$. Suppose Eq. (1.148) holds in $U(\mathbf{x}^*, \mathbf{0}) \subset \Omega$. For a linearized dynamical system in Eq. (1.149), consider a pair of complex eigenvalues $\alpha_k \pm \mathbf{i}\beta_k$ ($2k-1$, $2k \in N = \{1, 2, \ldots, 2n\}$, $\mathbf{i} = \sqrt{-1}$) with a pair of eigenvectors $\mathbf{v}_{2k-1} \pm \mathbf{i}\mathbf{v}_{2k}$. In the vicinity of $(\mathbf{x}^*, \mathbf{0})$, there exists a transformation of

$$\mathbf{x} = \mathbf{x}^* + \mathbf{y} = \mathbf{x}^* + \mathbf{a}_0(t) + \sum_{j=1}^{N} \mathbf{b}_j(t)\cos(j\beta_k t) + \mathbf{c}_j(t)\sin(j\beta_k t) \tag{1.163}$$

under $\lim_{j\to\infty}\|\mathbf{A}_j\| = 0$ with Eq. (1.152). The corresponding dynamical system of coefficients is

$$\begin{aligned}
\ddot{\mathbf{a}}_0 &= \mathbf{F}_0(\mathbf{a}_0, \mathbf{b}, \mathbf{c}, \dot{\mathbf{a}}_0, \dot{\mathbf{b}}, \dot{\mathbf{c}}),\\
\ddot{\mathbf{b}} &= -2\beta_k \mathbf{k}_1 \dot{\mathbf{c}} + \beta_k^2 \mathbf{k}_2 \mathbf{b} + \mathbf{F}^c(\mathbf{a}_0, \mathbf{b}, \mathbf{c}, \dot{\mathbf{a}}_0, \dot{\mathbf{b}}, \dot{\mathbf{c}}),\\
\ddot{\mathbf{c}} &= 2\beta_k \mathbf{k}_1 \dot{\mathbf{b}} + \beta_k^2 \mathbf{k}_2 \mathbf{c} + \mathbf{F}^s(\mathbf{a}_0, \mathbf{b}, \mathbf{c}, \dot{\mathbf{a}}_0, \dot{\mathbf{b}}, \dot{\mathbf{c}})
\end{aligned} \tag{1.164}$$

with Eqs. (1.155) and (1.156). The equivalent equation of Eq. (1.164) is

$$\ddot{\mathbf{z}} = \mathbf{g}(\mathbf{z}, \dot{\mathbf{z}}) \tag{1.165}$$

where

$$\mathbf{z} = (\mathbf{a}_0, \mathbf{b}, \mathbf{c})^{\mathrm{T}}, \dot{\mathbf{z}} = (\dot{\mathbf{a}}_0, \dot{\mathbf{b}}, \dot{\mathbf{c}})^{\mathrm{T}}, \mathbf{g} = (\mathbf{F}_0, \mathbf{F}^c, \mathbf{F}^s)^{\mathrm{T}}. \tag{1.166}$$

The equivalent first-order equation of Eq. (1.165) is

$$\dot{\mathbf{u}} = \bar{\mathbf{f}}(\mathbf{u}) \tag{1.167}$$

where

$$\mathbf{u} = (\mathbf{z}, \dot{\mathbf{z}})^{\mathrm{T}}; \bar{\mathbf{f}} = (\dot{\mathbf{z}}, \mathbf{g})^{\mathrm{T}} \tag{1.168}$$

If Eq. (1.165) has an equilibrium of $(\mathbf{z}^*, \mathbf{0})$ (i.e., $\mathbf{u}^* = \mathbf{0}$), and in a neighborhood of $\mathbf{u}^* = \mathbf{0}$ the linearized equation with $\Delta\mathbf{u} = \mathbf{u} - \mathbf{u}^*$ is

$$\Delta\dot{\mathbf{u}} = D\mathbf{f}(\mathbf{u}^*)\Delta\mathbf{u} \tag{1.169}$$

with

$$D\bar{\mathbf{f}}(\mathbf{u}^*) = \begin{bmatrix} \mathbf{0}_{n(2N+1)\times n(2N+1)} & \mathbf{I}_{n(2N+1)\times n(2N+1)} \\ \mathbf{G}_{n(2N+1)\times n(2N+1)} & \mathbf{H}_{n(2N+1)\times n(2N+1)} \end{bmatrix} \tag{1.170}$$

and

$$\mathbf{G} = \left.\frac{\partial \mathbf{g}}{\partial \mathbf{z}}\right|_{\mathbf{u}^* = \mathbf{0}}, \mathbf{H} = \left.\frac{\partial \mathbf{g}}{\partial \dot{\mathbf{z}}}\right|_{\mathbf{u}^* = \mathbf{0}} \tag{1.171}$$

then the eigenvalues of $D\bar{\mathbf{f}}(\mathbf{u}^*)$ determined by

$$| D\bar{\mathbf{f}}(\mathbf{u}^*) - \lambda \mathbf{I}_{2n(2N+1)\times 2n(2N+1)} | = 0 \tag{1.172}$$

yield the stability of the equilibrium $\mathbf{x}^*$ with a kind of

$$(n_1 : n_2 : n_3 | n_4 : n_5 : n_6) \tag{1.173}$$

with $\Sigma_{r=1}^{6} n_r = 2n(2N + 1)$.

(i) If $\mathrm{Re}\lambda_l < 0$ (l = 1, 2,..., $2n(2N + 1)$) with at least $\mathbf{b} \neq \mathbf{0}$ and $\mathbf{c} \neq \mathbf{0}$, the equilibrium $\mathbf{x}^*$ is spirally stable.
(ii) If $\mathrm{Re}\lambda_l > 0$ ($l \in \{1, 2, \ldots, 2n(2N + 1)\}$) with at least $\mathbf{b} \neq \mathbf{0}$ and $\mathbf{c} \neq \mathbf{0}$, the equilibrium $\mathbf{x}^*$ is spirally unstable.
(iii) The boundary between the stable and unstable solutions of the equilibrium $\mathbf{x}^*$ is determined by the higher singularity.

1.5 Lyapunov Functions and Stability

Consider an n-dimensional, autonomous, nonlinear dynamical system $\dot{\mathbf{x}} = \mathbf{f}(\mathbf{x}, \mathbf{p})$ in Eq. (1.4). Let $V : U \to \mathscr{R}$ be a differentiable function defined in a neighborhood of equilibrium $\mathbf{x}^*$ on $U/\{\mathbf{x}^*\}$. A function $\dot{V} : U \to \mathscr{R}$ is defined by

$$\dot{V}(\mathbf{x}) = DV(\mathbf{x}) = \mathbf{n}^{\mathrm{T}} \cdot \mathbf{f}(\mathbf{x}, t) \tag{1.175}$$

where $\mathbf{n}^{\mathrm{T}} = (\partial_{x_1} V, \ldots, \partial_{x_n} V)$ and $\mathbf{f}(\mathbf{x}, t) = (f_1, \ldots, f_n)^{\mathrm{T}}$.

Definition 1.33 Consider an n-dimensional, autonomous, nonlinear dynamical system $\dot{\mathbf{x}} = \mathbf{f}(\mathbf{x}, \mathbf{p})$ in Eq. (1.4) with an equilibrium point $\mathbf{x}^*$ and $\mathbf{f}(\mathbf{x}, \mathbf{p})$ is C^r-continuous ($r \geq 1$) in a neighborhood of the equilibrium $\mathbf{x}^*$. Suppose $U(\mathbf{x}^*) \subset \Omega$ is a neighborhood of equilibrium $\mathbf{x}^*$. There is a continuous function $V : U \to \mathscr{R}$ which is differentiable on $U/\{\mathbf{x}^*\}$, such that $V(\mathbf{x}^*) = 0$ and $V(\mathbf{x}) > 0$ if $\mathbf{x} \neq \mathbf{x}^*$.

(i) If $\dot{V} \leq 0$ in $U/\{\mathbf{x}^*\}$, the function V is called *a Lyapunov function* for equilibrium $\mathbf{x}^*$.
(ii) If $\dot{V} < 0$ in $U/\{\mathbf{x}^*\}$, the function V is called a *strict Lyapunov function* for equilibrium $\mathbf{x}^*$.

Theorem 1.11 *Consider an n-dimensional, autonomous, nonlinear dynamical system* $\dot{\mathbf{x}} = \mathbf{f}(\mathbf{x}, \mathbf{p})$ *in Eq. (1.4) with an equilibrium point* $\mathbf{x}^*$ *and* $\mathbf{f}(\mathbf{x}, \mathbf{p})$ *is* C^r*-continuous* ($r \geq 1$) *in a neighborhood of the equilibrium* $\mathbf{x}^*$. *Suppose* $U(\mathbf{x}^*) \subset \Omega$ *is a neighborhood of equilibrium* $\mathbf{x}^*$. *There is a continuous function* $V : U \to \mathscr{R}$ *which is differentiable on* $U/\{\mathbf{x}^*\}$, *such that* $V(\mathbf{x}^*) = 0$ *and* $V(\mathbf{x}) > 0$ *if* $\mathbf{x} \neq \mathbf{x}^*$.

(i) *If* $\dot{V} \leq 0$ *in* $U/\{\mathbf{x}^*\}$, *the equilibrium* $\mathbf{x}^*$ *is stable.*
(ii) *If* $\dot{V} < 0$ *in* $U/\{\mathbf{x}^*\}$, *the equilibrium* $\mathbf{x}^*$ *is asymptotically stable.*

Proof The proof can be referred in Hirsch et al. (2004). ∎

As in Luo (2012), to investigate the stability of the continuous systems in the vicinity of equilibrium $\mathbf{x}^*$, a measuring function should be introduced through the relative position vector to equilibrium $\mathbf{x}^*$. The relative position vector is given by

$$\mathbf{r} = \mathbf{x} - \mathbf{x}^*. \tag{1.176}$$

Definition 1.34 For an equilibrium point $\mathbf{x}^*$, consider a flow $\mathbf{x} \in U(\mathbf{x}^*)$ in dynamical system of Eq. (1.4). A relative distance function for the flow $\mathbf{x}$ to the equilibrium $\mathbf{x}^*$is defined by

$$d(\mathbf{x}, \mathbf{x}^*) = \|\mathbf{x} - \mathbf{x}^*\|. \tag{1.177}$$

If $d(\mathbf{x}, \mathbf{x}_p) = C = const$, there is a surface given by

$$\|\mathbf{x} - \mathbf{x}^*\| = C \tag{1.178}$$

which is called *the equi-distance surface* of equilibrium $\mathbf{x}^*$. Further, if there is a monotonically increasing or decreasing function of the relative distance $d(\mathbf{x}, \mathbf{x}^*)$,

$$E = V(\mathbf{x}, \mathbf{x}^*) \equiv f(d(\mathbf{x}, \mathbf{x}^*)) \tag{1.179}$$

with the following property:

$$V(\mathbf{x}, \mathbf{x}^*) = f(d(\mathbf{x}, \mathbf{x}^*)) = \min \text{ (or } \max\text{) if } d(\mathbf{x}, \mathbf{x}^*) = 0. \tag{1.180}$$

Such a monotonic function $V(\mathbf{x}, \mathbf{x}^*)$ is called a generalized *measuring function* of dynamical system in a neighborhood of the equilibrium $\mathbf{x}^*$. If $E = C =$ const, there is a surface given by

$$V(\mathbf{x}, \mathbf{x}^*) = C \tag{1.181}$$

which is called the *equi-measuring function* surface.

To explain the concept of the measuring function, consider a domain $U(\mathbf{x}^*)$ in phase space in the vicinity of the equilibrium $\mathbf{x}^*$ with big circular symbol, as shown in Fig. 1.2. Suppose a point $\mathbf{x}$ with a small circular symbol is the solution of the subsystem. The relative location vector to the equilibrium $\mathbf{x}^*$ is expressed by a vector $\mathbf{r} = \mathbf{x} - \mathbf{x}^*$ and the corresponding relative distance is expressed by $d(\mathbf{x}, \mathbf{x}^*)$. Such a point $\mathbf{x}$ is on the equi-measuring function surface of $E = V(d(\mathbf{x}, \mathbf{x}^*))$. For different values of $E = E_i$ $(i = 1, 2, \cdots)$, a set of the equi-measuring function surfaces will fill the entire domain of $U(\mathbf{x})$, which are depicted by the thick curves in Fig. 1.2. Based on the equi-measuring function surface, there is a dynamical system.

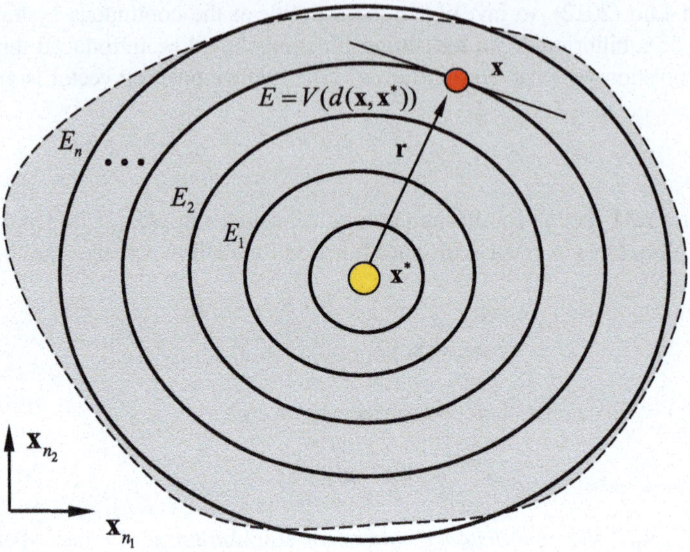

Fig. 1.2 A relative position vector from $\mathbf{x}_p$ to $\mathbf{x}$ and a measuring function with $n_1 + n_2 = n$

Definition 1.35 For any *equi-measuring function* surface in Eq. (1.181), there is a dynamical system as

$$\dot{\mathbf{x}}^m = \mathbf{f}^m(\mathbf{x}^m) \tag{1.182}$$

with the initial condition $(\mathbf{x}_0^m, t_0)$ and the equi-measuring function surface can be expressed by

$$V(\mathbf{x}^m, \mathbf{x}^*) = V(\mathbf{x}_0^m, \mathbf{x}^*) = E. \tag{1.183}$$

The dynamical system given in Eq. (1.182) is invariant in the sense of the measuring function in Eq. (1.181). The subscript or superscript "m" represents the flow on the "equi-measuring function surface." To measure dynamical behaviors of dynamical system in Eq. (1.4) to the equi-measuring function surface, from Luo (2008a,b), the following functions are introduced.

Definition 1.36 Consider a flow $\mathbf{x}(t)$ of the dynamical system with a vector field $\mathbf{f}(\mathbf{x}, t, \mathbf{p})$ in Eq. (1.4). At time t, if the flow $\mathbf{x}(t)$ arrives to the *equi-measuring function* surface in Eq. (1.181), the kth-order, G-functions at the constant measuring function level are defined as

$$
\begin{aligned}
G_m^{(k)}(\mathbf{x}(t),\mathbf{x}^*) &= (k+1)!\lim_{\varepsilon\to 0}\frac{1}{\varepsilon^{k+1}}\{[\mathbf{n}(\mathbf{x}^m(t+\varepsilon),\mathbf{x}^*)]^{\mathrm{T}}\cdot\mathbf{x}(t+\varepsilon) \\
&\quad - [\mathbf{n}^m(\mathbf{x}(t),\mathbf{x}^*)]^{\mathrm{T}}\cdot\mathbf{x}(t) - \sum_{q=1}^{k}\frac{1}{q!}G_m^{(k)}(\mathbf{x}(t),\mathbf{x}^*)\varepsilon^q\} \\
&= \sum\nolimits_{r=1}^{k+1} C_{k+1}^r D^{(k+1-r)}[\mathbf{n}(\mathbf{x}^m(t),\mathbf{x}^*)]^{\mathrm{T}}\cdot[D^{(r-1)}\mathbf{f}(\mathbf{x}(t),\mathbf{p}) \\
&\quad -D^{(r-1)}\mathbf{f}^m(\mathbf{x}^m,\mathbf{x}^*)]|_{\mathbf{x}^m=\mathbf{x}}
\end{aligned}
\tag{1.184}
$$

for $k = 0,1,2,\ldots$. The normal vector of the equi-measuring function surface is

$$
\mathbf{n}(\mathbf{x}^m,\mathbf{x}^*) = \frac{\partial V(\mathbf{x}^m,\mathbf{x}^*)}{\partial \mathbf{x}^m} = \left(\frac{\partial V}{\partial x_1^m},\frac{\partial V}{\partial x_2^m},\ldots,\frac{\partial V}{\partial x_n^m}\right)^{\mathrm{T}}
\tag{1.185}
$$

where the total differential operator is given by

$$
\begin{aligned}
&D = \frac{\partial(\cdot)}{\partial \mathbf{x}}\dot{\mathbf{x}} + \frac{\partial(\cdot)}{\partial t}, \\
&D^{(r)}(\cdot) = D\circ D^{(r-1)}(\cdot) = D(D^{(r-1)}(\cdot)), \\
&\text{and } D^{(0)}(\cdot) = \mathbf{I}
\end{aligned}
\tag{1.186}
$$

with

$$
C_{k+1}^r = \frac{(k+1)!}{r!(k+1-r)!} \text{ and } r! = 1\times 2\times 3\cdots\times r.
\tag{1.187}
$$

From Eq. (1.181), the following relation holds:

$$
0 = \dot{\mathbf{x}}^{\mathrm{T}}\cdot\frac{\partial V}{\partial \mathbf{x}} = [\mathbf{n}(\mathbf{x}^m,\mathbf{x}^*)]^{\mathrm{T}}\cdot\mathbf{f}^m(\mathbf{x}^m,\mathbf{x}^*).
\tag{1.188}
$$

For a zero-order G-function ($k = 0$), one obtains

$$
G_m^{(0)}(\mathbf{x},\mathbf{x}^*) = [\mathbf{n}(\mathbf{x}^m,\mathbf{x}^*)]^{\mathrm{T}}\cdot[\mathbf{f}(\mathbf{x},\mathbf{p}) - \mathbf{f}^m(\mathbf{x}^m,\mathbf{x}^*)]|_{\mathbf{x}^m=\mathbf{x}} = [\mathbf{n}(\mathbf{x},\mathbf{x}^*)]^{\mathrm{T}}\cdot\mathbf{f}(\mathbf{x},\mathbf{p}).
\tag{1.189}
$$

The zero-order G-function is the dot product of the vector field $\mathbf{f}(\mathbf{x},\mathbf{p})$ for the dynamical system and the normal vector $\mathbf{n}(\mathbf{x}^m,\mathbf{x}^*)$. Consider an instantaneous value of the equi-measuring function at time t. In other words, letting $\mathbf{x}^m = \mathbf{x}$, Eq. (1.179) gives

$$E(t) = V(\mathbf{x}, \mathbf{x}^*). \tag{1.190}$$

The corresponding time change ratio of the measuring function is

$$\begin{aligned}
\frac{dE(t)}{dt} &= \frac{\partial V(\mathbf{x}, \mathbf{x}^*)}{\partial \mathbf{x}} \cdot \dot{\mathbf{x}} \\
&= [\mathbf{n}(\mathbf{x}, \mathbf{x}^*)]^{\mathrm{T}} \cdot \mathbf{f}(\mathbf{x}, \mathbf{p}) \\
&= G_m^{(0)}(\mathbf{x}, \mathbf{x}^*).
\end{aligned} \tag{1.191}$$

From the foregoing equation, the change of the equi-measuring function for the dynamical system for time $t \in [t_k, t_{k+1}]$ can be defined as in Luo (2008a,b).

Definition 1.37 For a flow $\mathbf{x}(t)$ of the dynamical system with a vector field $\mathbf{f}(\mathbf{x}, \mathbf{p})$ in Eq. (1.4), consider the equi-measuring function $V(\mathbf{x}, \mathbf{x}^*)$ in Eq. (1.179) to monotonically increase to a metric function $d(\mathbf{x}, \mathbf{x}_p)$ in Eq. (1.177). The total change of the equi-measuring function for the time interval $[t_k, t]$ is defined as

$$\begin{aligned}
L(\mathbf{x}^*, t_k, t) &= \int_{t_k}^{t} \frac{dE(t)}{dt} = \int_{t_k}^{t} G_m^{(0)}(\mathbf{x}, \mathbf{x}^*)dt \\
&= \int_{t_k}^{t} [\mathbf{n}(\mathbf{x}, \mathbf{x}^*)]^{\mathrm{T}} \cdot \mathbf{f}(\mathbf{x}, \mathbf{p})dt \\
&= V(\mathbf{x}(t), \mathbf{x}^*) - V(\mathbf{x}_k, \mathbf{x}^*)
\end{aligned} \tag{1.192}$$

where $\mathbf{x}_k = \mathbf{x}(t_k)$. For a given $t = t_{k+1} > t_k$, the increment of the equi-measuring function to dynamical system in Eq. (1.4) for $t \in [t_k, t_{k+1}]$ is

$$\begin{aligned}
L(\mathbf{x}^*, t_k, t_{k+1}) &= \int_{t_k}^{t_{k+1}} \frac{dE(t)}{dt} = \int_{t_k}^{t_{k+1}} G_m^{(0)}(\mathbf{x}, \mathbf{x}^*)dt \\
&= \int_{t_k}^{t_{k+1}} [\mathbf{n}(\mathbf{x}, \mathbf{x}^*)]^{\mathrm{T}} \cdot \mathbf{f}(\mathbf{x}, \mathbf{p})dt \\
&= V(\mathbf{x}_{k+1}, \mathbf{x}^*) - V(\mathbf{x}_k, \mathbf{x}^*).
\end{aligned} \tag{1.193}$$

From Eq. (1.193), the equi-measuring function quantity is used to measure changes of the dynamical system. Thus such a function can be used to investigate the stability of dynamical systems.

Definition 1.38 For a dynamical system in Eq. (1.4), consider the equi-measuring function $V(\mathbf{x}, \mathbf{x}^*)$ in Eq. (1.179) to monotonically increase to a metric function $d(\mathbf{x}, \mathbf{x}_p)$ in Eq. (1.177) in $U(\mathbf{x}^*) \subset \mathscr{R}^n$. A flow $\mathbf{x}(t)$ at $\mathbf{x}_k$ for $t = t_k$ is:

(i) locally decreasing to the equi-measuring function surface in $U(\mathbf{x}^*) \subset \mathscr{R}^n$ if

$$\begin{aligned} V(\mathbf{x}_k, \mathbf{x}^*) - V(\mathbf{x}_{k-\varepsilon}, \mathbf{x}^*) &< 0, \\ V(\mathbf{x}_{k+\varepsilon}, \mathbf{x}^*) - V(\mathbf{x}_k, \mathbf{x}^*) &< 0; \end{aligned} \tag{1.194}$$

(ii) locally increasing to the equi-measuring function surface in $U(\mathbf{x}^*) \subset \mathscr{R}^n$ if

$$\begin{aligned} V(\mathbf{x}_k, \mathbf{x}^*) - V(\mathbf{x}_{k-\varepsilon}, \mathbf{x}^*) &> 0, \\ V(\mathbf{x}_{k+\varepsilon}, \mathbf{x}^*) - V(\mathbf{x}_k, \mathbf{x}^*) &> 0; \end{aligned} \tag{1.995}$$

(iii) locally tangential to the equi-measuring function surface in $U(\mathbf{x}^*) \subset \mathscr{R}^n$ if

$$\begin{aligned} &\text{either} \quad \left.\begin{aligned} V(\mathbf{x}_k, \mathbf{x}^*) - V(\mathbf{x}_{k-\varepsilon}, \mathbf{x}^*) &< 0, \\ V(\mathbf{x}_{k+\varepsilon}, \mathbf{x}^*) - V(\mathbf{x}_k, \mathbf{x}^*) &> 0 \end{aligned}\right\} \text{for the external surface;} \\ &\text{or} \quad \begin{aligned} V(\mathbf{x}_k, \mathbf{x}^*) - V(\mathbf{x}_{k-\varepsilon}, \mathbf{x}^*) &> 0, \\ V(\mathbf{x}_{k+\varepsilon}, \mathbf{x}^*) - V(\mathbf{x}_k, \mathbf{x}^*) &< 0 \end{aligned} \}\text{for the interior surface.} \end{aligned} \tag{1.196}$$

From the previous definitions, the locally increasing and decreasing of a flow $\mathbf{x}(t)$ at $\mathbf{x}_k$ to the measuring function surface are depicted in Fig. 1.3 in the vicinity of the point $\mathbf{x}^*$. A flow $\mathbf{x}(t)$ at $\mathbf{x}_k$, locally tangential to the equi-measuring function surface, can be similarly sketched.

Theorem 1.12 *For a dynamical system in Eq. (1.4), consider the equi-measuring function $V(\mathbf{x}, \mathbf{x}^*)$in Eq. (1.179) to be monotonically increased to a metric function d $(\mathbf{x}, \mathbf{x}^*)$ in Eq. (1.177). A flow $\mathbf{x}(t)$ at $\mathbf{x}_k$ for $t{=}t_k$ in the domain $U(\mathbf{x}^*) \subset \mathscr{R}^n$ is:*

(i) *locally decreasing to the equi-measuring function surface in $U(\mathbf{x}^*) \subset \mathscr{R}^n$ if and only if*

$$G_m^{(0)}(\mathbf{x}_k, \mathbf{x}^*) = [\mathbf{n}(\mathbf{x}_k, \mathbf{x}^*)]^{\mathrm{T}} \cdot \mathbf{f}(\mathbf{x}_k, \mathbf{p}) < 0; \tag{1.197}$$

(ii) *locally increasing to the equi-measuring function surface in $U(\mathbf{x}^*) \subset \mathscr{R}^n$ if and only if*

$$G_m^{(0)}(\mathbf{x}_k, \mathbf{x}^*) = [\mathbf{n}(\mathbf{x}_k, \mathbf{x}^*)]^{\mathrm{T}} \cdot \mathbf{f}(\mathbf{x}_k, \mathbf{p}) < 0; \tag{1.198}$$

(iii) *locally tangential to the equi-measuring function surface in $U(\mathbf{x}^*) \subset \mathscr{R}^n$ if and only if*

$$\begin{aligned} &G_m^{(0)}(\mathbf{x}_k, \mathbf{x}^*, t_k) = [\mathbf{n}(\mathbf{x}_k, \mathbf{x}^*)]^{\mathrm{T}} \cdot \mathbf{f}(\mathbf{x}_k, \mathbf{p}) = 0 \\ &G_m^{(1)}(\mathbf{x}_k, \mathbf{x}^*, t_k) > 0 \text{ for the external surface;} \\ &G_m^{(1)}(\mathbf{x}_k, \mathbf{x}^*, t_k) < 0 \text{ for the interior surface.} \end{aligned} \tag{1.199}$$

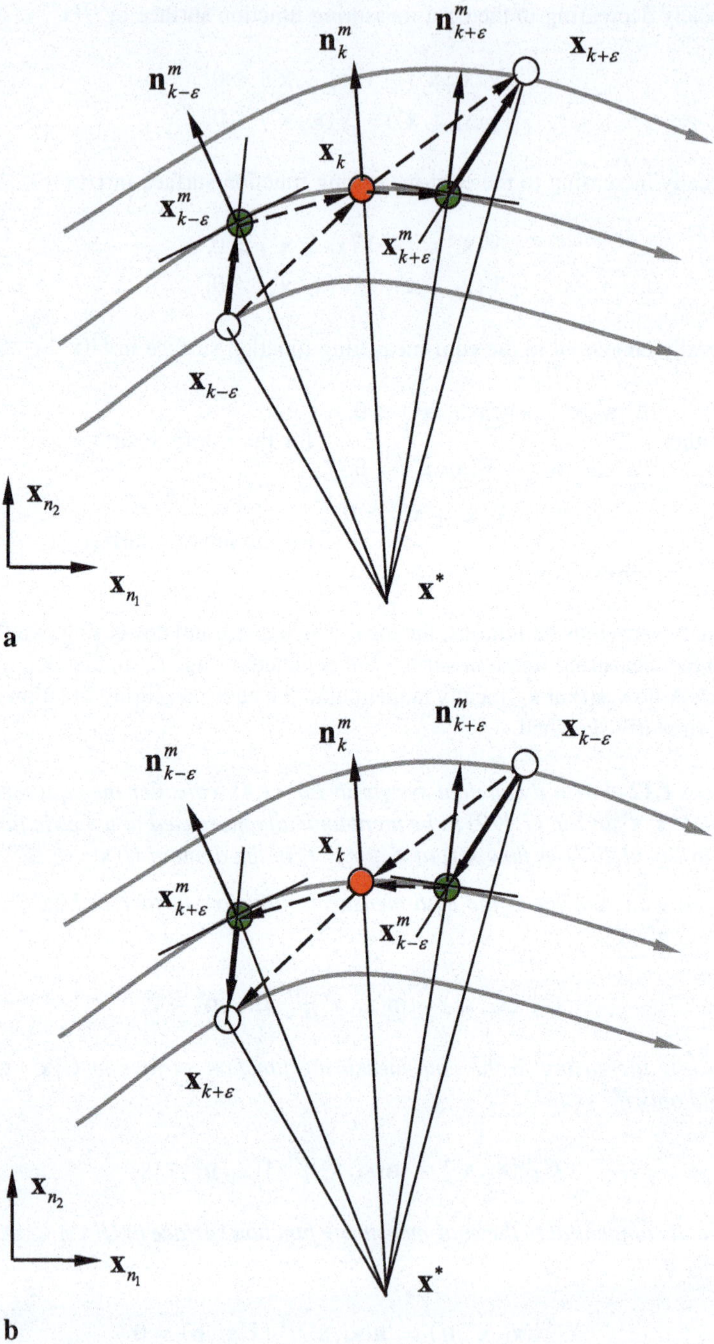

Fig. 1.3 (**a**) A locally increasing flow and (**b**) a locally decreasing flow to a measuring function with $n_1 + n_2 = n$

Proof Using G-function and the Taylor series expansion, the theorem can be proved directly. ∎

Definition 1.39 For a dynamical system in Eq. (1.4), consider the equi-measuring function $V(\mathbf{x},\mathbf{x}^*)$ in Eq. (1.179) to monotonically increase to a metric function $d(\mathbf{x},\ \mathbf{x}_p)$ in Eq. (1.177) in $U(\mathbf{x}^*) \subset \mathscr{R}^n$. A flow $\mathbf{x}(t)$ for $\mathbf{x}_k$ to $\mathbf{x}_{k+1}$ in $U(\mathbf{x}^*)$ for $t = t_s \in (t_k, t_{k+1})$ is:

(i) uniformly decreasing to the equi-measuring function surface in $U(\mathbf{x}^*) \subset \mathscr{R}^n$ if

$$\begin{aligned} V(\mathbf{x}_s,\mathbf{x}^*) - V(\mathbf{x}_{s-\varepsilon},\mathbf{x}^*) < 0, \\ V(\mathbf{x}_{s+\varepsilon},\mathbf{x}^*) - V(\mathbf{x}_s,\mathbf{x}^*) < 0; \end{aligned} \tag{1.200}$$

(ii) uniformly increasing to the equi-measuring function surface in $U(\mathbf{x}^*) \subset \mathscr{R}^n$ if

$$\begin{aligned} V(\mathbf{x}_s,\mathbf{x}^*) - V(\mathbf{x}_{s-\varepsilon},\mathbf{x}^*) < 0, \\ V(\mathbf{x}_{s+\varepsilon},\mathbf{x}^*) - V(\mathbf{x}_s,\mathbf{x}^*) < 0; \end{aligned} \tag{1.201}$$

(iii) uniformly invariant to the equi-measuring function surface in $U(\mathbf{x}^*) \subset \mathscr{R}^n$ if

$$V(\mathbf{x}_{s-\varepsilon},\mathbf{x}^*) = V(\mathbf{x}_s,\mathbf{x}^*) = V(\mathbf{x}_{s+\varepsilon},\mathbf{x}^*). \tag{1.202}$$

Theorem 1.13 *For a dynamical system in Eq. (1.4), consider the equi-measuring function $V(\mathbf{x},\mathbf{x}^*)$ in Eq. (1.179) to monotonically increase to a metric function $d(\mathbf{x},\ \mathbf{x}_p)$ in Eq. (1.177) in $U(\mathbf{x}^*) \subset \mathscr{R}^n$. A flow $\mathbf{x}(t)$ for $\mathbf{x}_k$ to $\mathbf{x}_{k+1}$ for $t \in (t_k, t_{k+1})$ in $U(\mathbf{x}^*) \subset \mathscr{R}^n$ is:*

(i) *uniformly decreasing to the equi-measuring function surface in $U(\mathbf{x}^*) \subset \mathscr{R}^n$ if and only if all points $\mathbf{x}(t)$ for $t \in [t_k,\ t_{k+1}]$ on the flow γ satisfy the following condition:*

$$G_m^{(0)}(\mathbf{x},\mathbf{x}^*) = [\mathbf{n}(\mathbf{x},\mathbf{x}^*)]^{\mathrm{T}} \cdot \mathbf{f}(\mathbf{x},\mathbf{p}) < 0; \tag{1.203}$$

(ii) *uniformly increasing to the equi-measuring function surface in $U(\mathbf{x}^*) \subset \mathscr{R}^n$ if and only if all points $\mathbf{x}(t)$ for $t \in [t_k,\ t_{k+1}]$ on the flow γ satisfy the following condition:*

$$G_m^{(0)}(\mathbf{x},\mathbf{x}^*) = [\mathbf{n}(\mathbf{x},\mathbf{x}^*)]^{\mathrm{T}} \cdot \mathbf{f}(\mathbf{x},\mathbf{p}) > 0; \tag{1.204}$$

(iii) *uniformly invariant to the equi-measuring function surface in $U(\mathbf{x}^*) \subset \mathscr{R}^n$ if and only if all points $\mathrm{x}^{(i)}(t)$ for $t \in [t_k,\ t_{k+1}]$ on the flow γ satisfy the following condition:*

$$G_m^{(k)}(\mathbf{x},\mathbf{x}^*) = 0 \ \ k = 0, 1, 2, \ldots. \tag{1.205}$$

Proof Using G-function and the Taylor series expansion, the theorem can be proved directly. ∎

Definition 1.40 For a dynamical system in Eq. (1.4), consider the equi-measuring function $V(\mathbf{x}, \mathbf{x}^*)$ in Eq. (1.179) to monotonically increase to a metric function $d(\mathbf{x}, \mathbf{x}_p)$ in Eq. (1.177) in $U(\mathbf{x}^*) \subset \mathscr{R}^n$. A flow $\mathbf{x}(t)$ at $\mathbf{x}_k$ for $t=t_k$ is:

(i) locally decreasing with the $(2s)^{\text{th}}$ order to the equi-measuring function surface in $U(\mathbf{x}^*) \subset \mathscr{R}^n$ if

$$\begin{aligned} &G_m^{(r)}(\mathbf{x}_k, \mathbf{x}^*) = 0, \text{for } r = 0,1,2,\ldots,2s-1; \\ &\left.\begin{aligned} &V(\mathbf{x}_k, \mathbf{x}^*) - V(\mathbf{x}_{k-\varepsilon}, \mathbf{x}^*) < 0, \\ &V(\mathbf{x}_{k+\varepsilon}, \mathbf{x}^*) - V(\mathbf{x}_k, \mathbf{x}^*) < 0; \end{aligned}\right\} \end{aligned} \tag{1.206}$$

(ii) locally increasing with the $(2s)^{\text{th}}$ order to the equi-measuring function surface in $U(\mathbf{x}^*) \subset \mathscr{R}^n$ if

$$\begin{aligned} &G_m^{(r)}(\mathbf{x}_k, \mathbf{x}^*) = 0, \text{for } r = 0,1,2,\ldots,2s-1 \\ &\left.\begin{aligned} &V(\mathbf{x}_k, \mathbf{x}^*) - V(\mathbf{x}_{k-\varepsilon}, \mathbf{x}^*) > 0, \\ &V(\mathbf{x}_{k+\varepsilon}, \mathbf{x}^*) - V(\mathbf{x}_k, \mathbf{x}^*) > 0; \end{aligned}\right\} \end{aligned} \tag{1.207}$$

(iii) locally tangential with the $(2s+1)^{\text{th}}$ order to the equi-measuring function surface in $U(\mathbf{x}^*) \subset \mathscr{R}^n$ if

$$\begin{aligned} &G_m^{(r)}(\mathbf{x}_k, \mathbf{x}^*) = 0, \text{for } r = 0,1,2,\ldots,2s; \\ &\left.\begin{aligned} &V(\mathbf{x}_k, \mathbf{x}^*) - V(\mathbf{x}_{k-\varepsilon}, \mathbf{x}^*) < 0, \\ &V(\mathbf{x}_{k+\varepsilon}, \mathbf{x}^*) - V(\mathbf{x}_k, \mathbf{x}^*) > 0 \end{aligned}\right\} \text{for the external surface,} \\ &\left.\begin{aligned} &V(\mathbf{x}_k, \mathbf{x}^*) - V(\mathbf{x}_{k-\varepsilon}, \mathbf{x}^*) > 0, \\ &V(\mathbf{x}_{k+\varepsilon}, \mathbf{x}^*) - V(\mathbf{x}_k, \mathbf{x}^*) < 0 \end{aligned}\right\} \text{for the interior surface.} \end{aligned} \tag{1.208}$$

Theorem 1.14 *For a dynamical system in Eq. (1.4), consider the equi-measuring function* $V(\mathbf{x}, \mathbf{x}^*)$ *in Eq. (1.179) to monotonically increase to a metric function* $d(\mathbf{x}, \mathbf{x}_p)$ *in Eq. (1.177) in* $U(\mathbf{x}^*) \subset \mathscr{R}^n$. *A flow* $\mathbf{x}(t)$ *at* $\mathbf{x}_k$ *for* $t=t_k$ *is:*

(i) *locally decreasing with the* $(2s)^{\text{th}}$ *order to the equi-measuring function surface in* $U(\mathbf{x}^*) \subset \mathscr{R}^n$ *if and only if*

$$\begin{aligned} &G_m^{(r)}(\mathbf{x}_k, \mathbf{x}^*) = 0, \text{ for } r = 0, 1, 2, \ldots, 2s-1; \\ &G_m^{(2s)}(\mathbf{x}_k, \mathbf{x}^*) < 0; \end{aligned} \tag{1.209}$$

(ii) *locally increasing with the* $(2s)^{\text{th}}$ *order to the equi-measuring function surface in* $U(\mathbf{x}^*) \subset \mathscr{R}^n$ *if and only if*

$$\begin{aligned} &G_m^{(r)}(\mathbf{x}_k, \mathbf{x}^*) = 0, \text{for } r = 0, 1, 2, \ldots, 2s-1; \\ &G_m^{(2s)}(\mathbf{x}_k, \mathbf{x}^*) > 0; \end{aligned} \tag{1.210}$$

(iii) *locally tangential with the* $(2s+1)^{\text{th}}$ *order to the equi-measuring function surface in* $U(\mathbf{x}^*) \subset \mathscr{R}^n$ *if and only if*

$$\begin{aligned} &G_m^{(r)}(\mathbf{x}_k, \mathbf{x}^*) = 0, \text{for } r = 0, 1, 2, \ldots, 2s; \\ &G_m^{(2s+1)}(\mathbf{x}_k, \mathbf{x}^*) > 0 \ \text{ for the external surface,} \\ &G_m^{(2s+1)}(\mathbf{x}_k, \mathbf{x}^*) < 0 \ \text{ for the interior surface.} \end{aligned} \tag{1.211}$$

Proof Using G-function and the Taylor series expansion, the theorem can be proved directly. ∎

References

Carr, J.,1981, *Applications of Center Manifold Theory*, Applied Mathematical Science **35**, Springer-Verlag, New York.

Coddington, E.A. and Levinson, N., 1955, *Theory of Ordinary Differential Equations*, New York: McGraw-Hill.

Hartman, P., *Ordinary Differential Equations*, Wiley, New York. (2nd ed. Birkhauser, Boston Basel Stuttgart, 1964).

Hirsch, M.W., Smale, S. and Devaney, R.L., 2004, *Differential Equations, Dynamical Systems, and An Introduction to Chaos*, Amsterdam: Elsevier.

Luo, A.C.J., 2008a, A theory for flow switchability in discontinuous dynamical systems, *Nonlinear Analysis: Hybrid Systems*, **2**(4), pp. 1030–1061.

Luo, A.C.J., 2008b, *Global Transversality, Resonance and Chaotic Dynamics*, Singapore: World Scientific.

Luo, A.C.J., 2011, *Regularity and Complexity in Dynamical Systems*, New York: Springer.

Luo, A.C.J., 2012, *Continuous Dynamical Systems*, HEP/L&H Scientific, Beijing/Glen Carbon.

Luo, A.C.J., 2019, On stability and bifurcation of equilibriums in nonlinear systems, *Journal of Vibration Testing and System Dynamics*, **3**(2), pp. 147–232.

Marsden, J.E. and McCracken, M.F., 1976, *The Hopf Bifurcation and Its Applications*, Applied Mathematical Science 19, Springer-Verlag, New York.

Chapter 2
Bifurcations of Equilibrium

In this chapter, the hyperbolic bifurcations of equilibriums on the eigenvectors in nonlinear dynamical systems are discussed, and the Hopf bifurcation of an equilibrium on a specific eigenvector plane is presented. Based on the Fourier series base, the transformation for the spiral stability is introduced for the Hopf bifurcation of equilibriums. The Hopf bifurcation of equilibriums in the second-order nonlinear dynamical systems is discussed from the Fourier series transformation.

2.1 Bifurcations

The dynamical characteristics of equilibriums in nonlinear dynamical systems in Eq. (1.4) are based on the given parameters. With varying parameters in dynamical systems, the corresponding dynamical behaviors will change qualitatively. The qualitative switching of dynamical behaviors in dynamical systems is called *bifurcation* and the corresponding parameter values are called *bifurcation values*. To understand the qualitative changes of dynamical behaviors of nonlinear systems with parameters in the neighborhood of equilibriums, the bifurcation theory for equilibrium of nonlinear dynamical system in Eq. (1.4) will be presented. $D_{\mathbf{x}}(\cdot) = \partial(\cdot)/\partial\mathbf{x}$ and $D_{\mathbf{p}}(\cdot) = \partial(\cdot)/\partial\mathbf{p}$ will be adopted from now on. For no specific notice, $D \equiv D_{\mathbf{x}}$.

Definition 2.1 Consider an n-dimensional, autonomous, nonlinear dynamical system $\dot{\mathbf{x}} = \mathbf{f}(\mathbf{x}, \mathbf{p})$ in Eq. (1.4) with an equilibrium point $(\mathbf{x}^*, \mathbf{p})$. Suppose there is a neighborhood of the equilibrium $\mathbf{x}^*$ as $U(\mathbf{x}^*) \subset \Omega$, and in the neighborhood Eq. (1.24) holds. The linearized system of the nonlinear system at the equilibrium point $(\mathbf{x}^*, \mathbf{p})$ is $\dot{\mathbf{y}} = D_{\mathbf{x}}\mathbf{f}(\mathbf{x}^*, \mathbf{p})\mathbf{y}$ $(\mathbf{y} = \mathbf{x} - \mathbf{x}^*)$ in Eq. (1.19).

(i) The equilibrium point $(\mathbf{x}_0^*, \mathbf{p}_0)$ is called *the switching point* of equilibrium solutions if $D_{\mathbf{x}}\mathbf{f}(\mathbf{x}^*, \mathbf{p})$ at $(\mathbf{x}_0^*, \mathbf{p}_0)$ possesses at least one real eigenvalue (or one pair of complex eigenvalues) with zero real part.

A. C. J. Luo, *Bifurcation and Stability in Nonlinear Dynamical Systems*, Nonlinear Systems and Complexity 28, https://doi.org/10.1007/978-3-030-22910-8_2

(ii) The value $\mathbf{p}_0$ in Eq. (1.4) is called *a switching value* of $\mathbf{p}$ if the dynamical characteristics at point $(\mathbf{x}_0^*, \mathbf{p}_0)$ change from one state into another state.
(iii) The equilibrium point $(\mathbf{x}_0^*, \mathbf{p}_0)$ is called the *bifurcation point* of equilibrium solutions if $D_{\mathbf{x}}\mathbf{f}(\mathbf{x}^*, \mathbf{p})$ at $(\mathbf{x}_0^*, \mathbf{p}_0)$ possesses at least one real eigenvalue (or one pair of complex eigenvalues) with zero real part, and more than one branches of equilibrium solutions appear or disappear.
(iv) The value $\mathbf{p}_0$ in Eq. (1.4) is called a *bifurcation value* of $\mathbf{p}$ if the dynamical characteristics at point $(\mathbf{x}_0^*, \mathbf{p}_0)$ change from one stable state into another unstable state.

2.2 Hyperbolic Bifurcations on Eigenvectors

Definition 2.2 Consider an n-dimensional, autonomous, nonlinear dynamical system $\dot{\mathbf{x}} = \mathbf{f}(\mathbf{x}, \mathbf{p})$ in Eq. (1.4) with an equilibrium point $\mathbf{x}^*$ and $\mathbf{f}(\mathbf{x}, \mathbf{p})$ is C^r-continuous $(r \geq 1)$ in a neighborhood of the equilibrium $\mathbf{x}^*$ (i.e., $U(\mathbf{x}^*) \subset \Omega$). The corresponding solution is $\mathbf{x}(t) = \mathbf{\Phi}(\mathbf{x}_0, t - t_0, \mathbf{p})$. Suppose Eq. (1.24) holds in $U(\mathbf{x}^*) \subset \Omega$. For a linearized dynamical system in Eq. (1.19), consider a real eigenvalue λ_k of matrix $D\mathbf{f}(\mathbf{x}^*, \mathbf{p}^*)$ $(k \in N = \{1, 2, \ldots, n\})$ with an eigenvector $\mathbf{v}_k$. Suppose one of n independent solutions $\mathbf{y} = c^k \mathbf{v}_k$ and $\dot{\mathbf{y}} = \dot{c}^k \mathbf{v}_k$,

$$c^k = (\mathbf{v}^k)^{\mathrm{T}} \cdot \mathbf{y} = (\mathbf{v}^k)^{\mathrm{T}} \cdot (\mathbf{x} - \mathbf{x}^*) \tag{2.1}$$

and

$$\dot{c}^k = (\mathbf{v}^k)^{\mathrm{T}} \cdot \dot{\mathbf{y}} = (\mathbf{v}^k)^{\mathrm{T}} \cdot \dot{\mathbf{x}} = (\mathbf{v}^k)^{\mathrm{T}} \cdot \mathbf{f}(\mathbf{x}, \mathbf{p}). \tag{2.2}$$

In the vicinity of point $(\mathbf{x}_0^*, \mathbf{p}_0)$, $(\mathbf{v}^k)^{\mathrm{T}} \cdot \mathbf{f}(\mathbf{x}, \mathbf{p})$ can be expanded for $(0 < \theta_i < 1)$ and $(0 < \gamma_j < 1)$ with $\boldsymbol{\gamma} = (\gamma_1, \gamma_2, \ldots, \gamma_m)^{\mathrm{T}}$ as

$$\begin{aligned}
(\mathbf{v}^k)^{\mathrm{T}} \cdot \mathbf{f}(\mathbf{x}, \mathbf{p}) &= (\mathbf{v}^k)^{\mathrm{T}} \cdot \mathbf{f}(\mathbf{x}_0^* + \mathbf{y}^*, \mathbf{p}_0 + \Delta\mathbf{p}) \\
&= a_{k(k)} z^k + \mathbf{b}_k^{\mathrm{T}} \cdot (\mathbf{p} - \mathbf{p}_0) \\
&\quad + \sum_{q=2}^{m_k} \sum_{r=0}^{q} \frac{1}{q!} C_q^r \mathbf{a}_{k(j_1 j_2 \cdots j_{q-r})}^{(q-r,\, r)} z^{j_1 *} z^{j_2 *} \cdots z^{j_{q-r} *} (\mathbf{p} - \mathbf{p}_0)^r \\
&\quad + \sum_{r=0}^{m_k+1} \frac{1}{(m_k+1)!} C_{m_k+1}^r \mathbf{a}_{k(j_1 j_2 \cdots j_{m_k+1-r})}^{(m_k+1-r,\, r)} (\theta_1 z^{j_1 *})(\theta_2 z^{j_2 *}) \\
&\quad \cdots (\theta_{m_k+1-r} z^{j_{m_k+1-r} *})(\boldsymbol{\gamma}^{\mathrm{T}} \cdot (\mathbf{p} - \mathbf{p}_0))^r,
\end{aligned} \tag{2.3}$$

where

$$
\begin{aligned}
z^{s*} &= (\mathbf{v}^s)^{\mathrm{T}} \cdot \mathbf{y}^* = c^{s*} - c_0^{s*} \quad (s = 1, 2, \ldots, n),\\
\Delta\mathbf{p} &= \mathbf{p} - \mathbf{p}_0;\\
\mathbf{b}_k^{\mathrm{T}} &= (\mathbf{v}^k)^{\mathrm{T}} \cdot \partial_{\mathbf{p}} \mathbf{f}(\mathbf{x}, \mathbf{p})|_{(\mathbf{x}_0^*, \mathbf{p}_0)},\\
a_{k(j_1)} &= (\mathbf{v}^k)^{\mathrm{T}} \cdot \partial_{c^{j_1}} \mathbf{f}(\mathbf{x}, \mathbf{p})|_{(\mathbf{x}_0^*, \mathbf{p}_0)}\\
&= (\mathbf{v}^k)^{\mathrm{T}} \cdot \partial_{\mathbf{x}} \mathbf{f}(\mathbf{x}, \mathbf{p})|_{(\mathbf{x}_0^*, \mathbf{p}_0)} \mathbf{v}_{j_1}\\
&= G_{k(j_1)}^{(1)}(\mathbf{x}_0^*, \mathbf{p}_0),\\
\mathbf{a}_{k(j_1 j_2 \cdots j_r)}^{(r,s)} &= (\mathbf{v}^k)^{\mathrm{T}} \cdot \partial_{c^{j_1} c^{j_2} \cdots c^{j_r}}^{(r)} \partial_{\mathbf{p}}^{(s)} \mathbf{f}(\mathbf{x}, \mathbf{p})|_{(\mathbf{x}_0^*, \mathbf{p}_0)}\\
&= (\mathbf{v}^k)^{\mathrm{T}} \cdot \partial_{\mathbf{x}}^{(r)} \partial_{\mathbf{p}}^{(s)} \mathbf{f}(\mathbf{x}, \mathbf{p})|_{(\mathbf{x}_0^*, \mathbf{p}_0)} \mathbf{v}_{j_1} \mathbf{v}_{j_2} \cdots \mathbf{v}_{j_r};\\
\mathbf{a}_{k(j_1 j_2 \cdots j_r)}^{(r,0)} &= (\mathbf{v}^k)^{\mathrm{T}} \cdot \partial_{c^{j_1} c^{j_2} \cdots c^{j_r}}^{(r)} \mathbf{f}(\mathbf{x}, \mathbf{p})|_{(\mathbf{x}_0^*, \mathbf{p}_0)}\\
&= (\mathbf{v}^k)^{\mathrm{T}} \cdot \partial_{\mathbf{x}}^{(r)} \mathbf{f}(\mathbf{x}, \mathbf{p})|_{(\mathbf{x}_0^*, \mathbf{p}_0)} \mathbf{v}_{j_1} \mathbf{v}_{j_2} \cdots \mathbf{v}_{j_r}\\
&= G_{k(j_1 j_2 \cdots j_r)}^{(r)}(\mathbf{x}_0^*, \mathbf{p}_0).
\end{aligned}
\tag{2.4}
$$

Definition 2.3 Consider an n-dimensional, autonomous, nonlinear dynamical system $\dot{\mathbf{x}} = \mathbf{f}(\mathbf{x}, \mathbf{p})$ in Eq. (1.4) with an equilibrium point $\mathbf{x}^*$ and $\mathbf{f}(\mathbf{x}, \mathbf{p})$ is C^r-continuous $(r \geq 1)$ in a neighborhood of the equilibrium $\mathbf{x}^*$ (i.e., $U(\mathbf{x}^*) \subset \Omega$). The corresponding solution is $\mathbf{x}(t) = \mathbf{\Phi}(\mathbf{x}_0, t - t_0, \mathbf{p})$. Suppose Eq. (1.24) holds in $U(\mathbf{x}^*) \subset \Omega$. For a linearized dynamical system in Eq. (1.19), consider a real eigenvalue λ_k of matrix $D\mathbf{f}$ $(\mathbf{x}^*, \mathbf{p}^*)$ $(k \in N = \{1, 2, \ldots, n\})$ with an eigenvector $\mathbf{v}_k$. Suppose one of n independent solutions $\mathbf{y} = c^k \mathbf{v}_k$ and $\dot{\mathbf{y}} = \dot{c}^k \mathbf{v}_k$.

For a specific $j \neq k$, equilibrium equation in the vicinity of $(\mathbf{x}_0, \mathbf{p}_0)$ is

$$
\begin{aligned}
&(\mathbf{v}^j)^{\mathrm{T}} \cdot \mathbf{f}(\mathbf{x}_0^* + \mathbf{y}^*, \mathbf{p}_0 + \Delta\mathbf{p}) = 0\\
&(\text{or } \lambda_{0j} z^{j*} + \mathbf{b}_j^{\mathrm{T}} \cdot (\mathbf{p} - \mathbf{p}_0) \approx 0),\\
&(j = 1, 2, \ldots, n \text{ but } j \neq k), \lambda_{0j} \neq 0.
\end{aligned}
\tag{2.5}
$$

For a specific k, if $a_{k(k)} = \lambda_{0k} = 0$ at $(\mathbf{x}_0, \mathbf{p}_0)$, equilibrium equation is

$$\begin{aligned}
&(\mathbf{v}^k)^{\mathrm{T}} \cdot \mathbf{f}(\mathbf{x}_0^* + \mathbf{y}^*, \mathbf{p}_0 + \Delta\mathbf{p}) = 0 \,\text{with}\, \lambda_{0k} = 0; \\
&(\text{or}\, G^{(1)}_{k(j_1)}(\mathbf{x}_0^*, \mathbf{p}_0) z^{j_1 *} + \mathbf{b}_k^{\mathrm{T}} \cdot (\mathbf{p} - \mathbf{p}_0) \\
&+ \sum_{q=2}^{s_k} \sum_{r=0}^{q} \frac{1}{q!} C_q^r \mathbf{a}^{(q-r,r)}_{k(j_1 j_2 \cdots j_{q-r})} z^{j_1 *} z^{j_2 *} \ldots z^{j_{q-r} *} (\mathbf{p} - \mathbf{p}_0)^r \approx 0) \\
&\text{equivalent to} \\
&\dot{z}^k = A_{k0}(z^k)^{s_k} + A_{k1}(z^k)^{s_k - 1} + \cdots + A_{ks_k} = 0 \\
&A_{ki} = A_{ki}(z^{1*}, z^{2*}, \cdots, z^{k-1*}, z^{k+1*}, \ldots, z^{n*}, \mathbf{p}), \\
&(i = 0, 1, 2, \ldots, s_k).
\end{aligned} \tag{2.6}$$

Equations (2.5) and (2.6) give l_k-equilibrium z^{k*} $(0 < l_k \le s_k)$ with $\sum_{i=1}^{l_k} \alpha_i = m_k$ on the eigenvector $\mathbf{v}_k$ direction.

(i) If

$$\left.\begin{aligned}
&\lambda_{0k} = 0, G^{(r_k)}_{k(j_1 j_2 \cdots j_{r_k})}(\mathbf{x}_0^*, \mathbf{p}_0) = 0 \; (r_k = 1, 2, 3, \ldots, m_k - 1), \\
&G^{(m_k)}_{k(j_1 j_2 \cdots j_{m_k})}(\mathbf{x}_0^*, \mathbf{p}_0) z^{j_1} z^{j_2} \ldots z^{j_{m_k}} > 0,
\end{aligned}\right\} \tag{2.7}$$

the bifurcation of equilibrium $\mathbf{x}^*$ at point $(\mathbf{x}_0^*, \mathbf{p}_0)$ is called an increasing saddle-node (or an *upper-saddle-node*) bifurcation of the $(m_k)^{\text{th}}$ order on the eigenvector $\mathbf{v}_k$. The bifurcation point $(\mathbf{x}_0^*, \mathbf{p}_0)$ on the eigenvector $\mathbf{v}_k$ is an increasing saddle (or an upper saddle) of the $(m_k)^{\text{th}}$ order.

(ii) If

$$\left.\begin{aligned}
&\lambda_{0k} = 0, G^{(r_k)}_{k(j_1 j_2 \cdots j_{r_k})}(\mathbf{x}_0^*, \mathbf{p}_0) = 0 \; (r_k = 1, 2, 3, \ldots, m_k - 1), \\
&G^{(m_k)}_{k(j_1 j_2 \cdots j_{m_k})}(\mathbf{x}_0^*, \mathbf{p}_0) z^{j_1} z^{j_2} \ldots z^{j_{m_k}} < 0,
\end{aligned}\right\} \tag{2.8}$$

the bifurcation of equilibrium $\mathbf{x}^*$ at point $(\mathbf{x}_0^*, \mathbf{p}_0)$ is called a decreasing saddle-node (or a *lower-saddle-node*) bifurcation of the m_kth order on the eigenvector $\mathbf{v}_k$. The bifurcation point $(\mathbf{x}_0^*, \mathbf{p}_0)$ on the eigenvector $\mathbf{v}_k$ is a decreasing saddle (or a lower saddle) of the m_kth order.

(iii) If

$$\left.\begin{aligned}
&\lambda_{0k} = 0, G^{(r_k)}_{k(j_1 j_2 \cdots j_{r_k})}(\mathbf{x}_0^*, \mathbf{p}_0) = 0 \; (r_k = 1, 2, 3, \ldots, m_k - 1), \\
&z^k \times G^{(m_k)}_{k(j_1 j_2 \cdots j_{m_k})}(\mathbf{x}_0^*, \mathbf{p}_0) z^{j_1} z^{j_2} \ldots z^{j_{m_k}} < 0,
\end{aligned}\right\} \tag{2.9}$$

the bifurcation of equilibrium $\mathbf{x}^*$ at point $(\mathbf{x}_0^*, \mathbf{p}_0)$ is called an $(m_k)^{\text{th}}$-order sink bifurcation on the eigenvector $\mathbf{v}_k$. The bifurcation point $(\mathbf{x}_0^*, \mathbf{p}_0)$ on the eigenvector $\mathbf{v}_k$ is an $(m_k)^{\text{th}}$-order sink.

(iv) If

$$\left.\begin{array}{l}\lambda_{0k}=0, G^{(r_k)}_{k(j_1j_2\cdots j_{r_k})}(\mathbf{x}_0^*,\mathbf{p}_0)=0\ (r_k=1,2,3,\ldots,m_k-1),\\ z^k\times G^{(m_k)}_{k(j_1j_2\cdots j_{m_k})}(\mathbf{x}_0^*,\mathbf{p}_0)z^{j_1}z^{j_2}\ldots z^{j_{m_k}}>0,\end{array}\right\}\tag{2.10}$$

the bifurcation of equilibrium $\mathbf{x}^*$ at point $(\mathbf{x}_0^*, \mathbf{p}_0)$ is called an $(m_k)^{\text{th}}$-order source bifurcation of the eigenvector $\mathbf{v}_k$. The bifurcation point $(\mathbf{x}_0^*, \mathbf{p}_0)$ on the eigenvector $\mathbf{v}_k$ is an $(m_k)^{\text{th}}$-order source.

Definition 2.5 Consider an n-dimensional, autonomous, nonlinear dynamical system $\dot{\mathbf{x}} = \mathbf{f}(\mathbf{x}, \mathbf{p})$ in Eq. (1.4) with an equilibrium point $\mathbf{x}^*$ and $\mathbf{f}(\mathbf{x}, \mathbf{p})$ is C^r $(r \geq 1)$-continuous in a neighborhood of the equilibrium $\mathbf{x}^*$ (i.e., $U(\mathbf{x}^*) \subset \Omega$). The corresponding solution is $\mathbf{x}(t) = \mathbf{\Phi}(\mathbf{x}_0, t - t_0, \mathbf{p})$. Suppose Eq. (1.24) holds in U-$(\mathbf{x}^*) \subset \Omega$. For a linearized dynamical system in Eq. (1.19), consider a real eigenvalue λ_k of matrix $D\mathbf{f}(\mathbf{x}^*, \mathbf{p}^*)$ $(k \in N = \{1, 2, \ldots, n\})$ with an eigenvector $\mathbf{v}_k$. Suppose one of n independent solutions $\mathbf{y} = c^k\mathbf{v}_k$ and $\dot{\mathbf{y}} = \dot{c}^k\mathbf{v}_k$. Consider $a_{l_\alpha(l_\alpha)} = \lambda_{l_\alpha} \neq 0$ $(l \in S = \{l_1, l_2, \ldots, l_p\}$ and $S \subset N = \{1, 2, \ldots, n\})$; the higher order term is ignorable in Eq. (2.3). The p equations for the corresponding directions are

$$\begin{array}{l}(\mathbf{v}^{l_\alpha})^{\mathrm{T}}\cdot\mathbf{f}(\mathbf{x}_0^*+\mathbf{y}^*,\mathbf{p}_0+\Delta\mathbf{p})=0,\\ (\text{or}\ \lambda_{0l_\alpha}z^{l_\alpha *}+\mathbf{b}_{l_\alpha}^{\mathrm{T}}\cdot(\mathbf{p}-\mathbf{p}_0)\approx 0)\\ (l_\alpha=l_1,l_2,\ldots,l_p),\lambda_{0l_\alpha}\neq 0\end{array}\tag{2.10}$$

and at $\mathbf{p} = \mathbf{p}_0$, for a specific k_α, if $a_{k_\alpha(k_\alpha)} = \lambda_{k_\alpha} = 0$ $(k_\alpha \in N\backslash S = \{l_{p+1}, l_{p+2}, \ldots, l_n\} \subseteq N)$, the s equations $(s = n - p)$ on the corresponding directions are given by

$$\begin{array}{l}(\mathbf{v}^{k_\alpha})^{\mathrm{T}}\cdot\mathbf{f}(\mathbf{x}_0^*+\mathbf{y}^*,\mathbf{p}_0+\Delta\mathbf{p})=0\ \text{with}\ \lambda_{0k_\alpha}=0\\ (\text{or}\ G^{(1)}_{k_\alpha(j_1)}z^{j_1 *}+\mathbf{b}_k^{\mathrm{T}}\cdot(\mathbf{p}-\mathbf{p}_0)\\ +\sum_{q=2}^{s_k}\sum_{r=0}^{q}\frac{1}{q!}C_q^r\mathbf{a}^{(q-r,r)}_{k_\alpha(j_1j_2\cdots j_{q-r})}z^{j_1 *}z^{j_2 *}\ldots z^{j_{q-r}*}(\mathbf{p}-\mathbf{p}_0)^r\approx 0);\end{array}\tag{2.11}$$

equivalent to

$$\begin{array}{l}\dot{z}^{k_\alpha}=A_{k_\alpha 0}(z^{k_\alpha})^{s_{k_\alpha}}+A_{k_\alpha 1}(z^{k_\alpha})^{s_k-1}+\cdots+A_{k_\alpha s_{k_\alpha}}=0,\\ A_{k_e i}=A_{k_\alpha i}(z^{1*},z^{2*},\ldots,z^{k-1*},z^{k+1*},\ldots,z^{n*},\mathbf{p}),\\ (i=0,1,2,\ldots,s_{k_\alpha};k_\alpha=l_{p+1},l_{p+2},\ldots,l_n).\end{array}$$

Equations (2.10) and (2.11) give l_{k_α}-equilibrium $z^{k_\alpha *}$ with $\sum_{i=1}^{l_{k_\alpha}} \alpha_i^{k_\alpha} = m_{k_\alpha}$ $(0 < m_{k_\alpha} \leq s_{k_\alpha})$ on the eigenvector $\mathbf{v}_{k_\alpha}$ direction. Such a bifurcation at equilibrium point $(\mathbf{x}_0^*, \mathbf{p}_0)$ is called the *hyperbolic bifurcation* of $(\beta_{l_{p+1}} : \beta_{l_{p+2}} : \cdots : \beta_{l_n})$ order on the eigenvectors of $\mathbf{v}_{k_\alpha}$ $(k_\alpha = l_{p+1}, l_{p+2}, \ldots, l_n)$ with $\beta_\alpha \in \{2m_\alpha, 2m_\alpha + 1\}$.

(i) If

$$\left.\begin{array}{l} \lambda_{0k_\alpha} = 0, G^{(r_{k_\alpha})}_{k_\alpha(j_1 j_2 \cdots j_{r_{k_\alpha}})}(\mathbf{x}_0^*, \mathbf{p}_0) = 0 \; (r_{k_\alpha} = 1, 2, 3, \ldots, m_{k_\alpha} - 1), \\ G^{(m_{k_\alpha})}_{k_\alpha(j_1 j_2 \cdots j_{2m_{k_\alpha}})}(\mathbf{x}_0^*, \mathbf{p}_0) z^{j_1} z^{j_2} \cdots z^{j_{m_{k_\alpha}}} > 0 \\ (k_\alpha \in \{l_{p+1}, l_{p+2}, \ldots, l_n\}), \end{array}\right\} \quad (2.12)$$

the bifurcation of equilibrium $\mathbf{x}^*$ at point $(\mathbf{x}_0^*, \mathbf{p}_0)$ is called an increasing saddle-node (or an upper-saddle-node) bifurcation of the $(m_{k_\alpha})^{\text{th}}$ order on the eigenvector $\mathbf{v}_{k_\alpha}$. The bifurcation point $(\mathbf{x}_0^*, \mathbf{p}_0)$ on the eigenvector $\mathbf{v}_{k_\alpha}$ is an increasing saddle (or an *upper saddle*) of the $(m_k)^{\text{th}}$ order.

(ii) If

$$\left.\begin{array}{l} \lambda_{0k_\alpha} = 0, G^{(r_{k_\alpha})}_{k_\alpha(j_1 j_2 \cdots j_{r_{k_\alpha}})}(\mathbf{x}_0^*, \mathbf{p}_0) = 0 \; (r_{k_\alpha} = 1, 2, 3, \ldots, m_{k_\alpha} - 1), \\ G^{(m_{k_\alpha})}_{k_\alpha(j_1 j_2 \cdots j_{m_{k_\alpha}})}(\mathbf{x}_0^*, \mathbf{p}_0) z^{j_1} z^{j_2} \cdots z^{j_{m_{k_\alpha}}} < 0 \\ (k_\alpha \in \{l_{p+1}, l_{p+2}, \ldots, l_n\}), \end{array}\right\} \quad (2.13)$$

the bifurcation of equilibrium $\mathbf{x}^*$ at point $(\mathbf{x}_0^*, \mathbf{p}_0)$ is called a decreasing saddle-node (or a *lower-saddle-node*) bifurcation of the $(m_k)^{\text{th}}$ order on the eigenvector $\mathbf{v}_{k_\alpha}$. The bifurcation point $(\mathbf{x}_0^*, \mathbf{p}_0)$ on the eigenvector $\mathbf{v}_{k_\alpha}$ is a decreasing saddle (or a *lower saddle*) of the $(m_k)^{\text{th}}$ order

(iii) If

$$\left.\begin{array}{l} \lambda_{0k_\alpha} = 0, G^{(r_{k_\alpha})}_{k_\alpha(j_1 j_2 \cdots j_{r_{k_\alpha}})}(\mathbf{x}_0^*, \mathbf{p}_0) = 0 \; (r_{k_\alpha} = 1, 2, 3, \ldots, m_{k_\alpha} - 1), \\ z^{k_\alpha} \times G^{(m_{k_\alpha})}_{k_\alpha(j_1 j_2 \cdots j_{m_{k_\alpha}})}(\mathbf{x}_0^*, \mathbf{p}_0) z^{j_1} z^{j_2} \ldots z^{j_{m_{k_\alpha}}} < 0 \\ (k_\alpha \in \{l_{p+1}, l_{p+2}, \ldots, l_n\}), \end{array}\right\} \quad (2.14)$$

the bifurcation of equilibrium $\mathbf{x}^*$ at point $(\mathbf{x}_0^*, \mathbf{p}_0)$ is called an (m_{k_α})th-*order sink* bifurcation on the eigenvector $\mathbf{v}_{k_\alpha}$. The bifurcation point $(\mathbf{x}_0^*, \mathbf{p}_0)$ on the eigenvector $\mathbf{v}_{k_\alpha}$ is an $(m_{k_\alpha})^{\text{th}}$-order sink.

(iv) If

$$\left.\begin{array}{l}\lambda_{0k_a}=0, G^{(r_{k_a})}_{k_a(j_1j_2\cdots j_{r_{k_a}})}(\mathbf{x}_0^*,\mathbf{p}_0)=0\ (r_{k_a}=2,3,\ldots,m_{k_a}-1),\\ z^{k_a}\times G^{(m_{k_a})}_{k_a(j_1j_2\cdots j_{m_{k_a}})}(\mathbf{x}_0^*,\mathbf{p}_0)z^{j_1}z^{j_2}\ldots z^{j_{m_{k_a}}}<0\\ (k_a\in\{l_{p+1},l_{p+2},\ldots,l_n\}),\end{array}\right\}\tag{2.15}$$

the bifurcation of equilibrium $\mathbf{x}^*$ at point $(\mathbf{x}_0^*,\mathbf{p}_0)$ is called an $(m_k)^{\text{th}}$-order source bifurcation on the eigenvector $\mathbf{v}_{k_a}$. The bifurcation point $(\mathbf{x}_0^*,\mathbf{p}_0)$ on the eigenvector $\mathbf{v}_{k_a}$ is an $(m_k)^{\text{th}}$-order source.

Definition 2.6 Consider an n-dimensional, autonomous, nonlinear dynamical system $\dot{\mathbf{x}}=\mathbf{f}(\mathbf{x},\mathbf{p})$ in Eq. (1.4) with an equilibrium point $\mathbf{x}^*$ and $\mathbf{f}(\mathbf{x},\mathbf{p})$ is C^r-continuous $(r\geq 1)$ in a neighborhood of the equilibrium $\mathbf{x}^*$ (i.e., $U(\mathbf{x}^*)\subset\Omega$). The corresponding solution is $\mathbf{x}(t)=\mathbf{\Phi}(\mathbf{x}_0,t-t_0,\mathbf{p})$. Suppose Eq. (1.24) holds in $U(\mathbf{x}^*)\subset\Omega$. For a linearized dynamical system in Eq. (1.19), consider a real eigenvalue λ_k of matrix $D\mathbf{f}(\mathbf{x}^*,\mathbf{p}^*)$ $(k\in N=\{1,2,\ldots,n\})$ with an eigenvector $\mathbf{v}_k$. Suppose one of n independent solutions $\mathbf{y}=c^k\mathbf{v}_k$ and $\dot{\mathbf{y}}=\dot{c}^k\mathbf{v}_k$. Three special cases of bifurcations are defined as follows.

(i) Consider

$$\lambda_s\neq 0\ (s=1,2,\ldots,n\text{ but }s\neq k)\tag{2.16}$$

with

$$\left.\begin{array}{l}\lambda_s z^{s*}+\mathbf{b}_s^{\mathrm{T}}\cdot(\mathbf{p}-\mathbf{p}_0)=0,\\ s=1,2,\ldots,n\text{ but }s\neq k;\end{array}\right\}$$

$$\left.\begin{array}{l}\mathbf{b}_k^{\mathrm{T}}\cdot(\mathbf{p}-\mathbf{p}_0)+\frac{1}{2!}\mathbf{a}_k^{(0,2)}(\mathbf{p}-\mathbf{p}_0)^2+\mathbf{a}_{k(j_1)}^{(1,1)}(\mathbf{p}-\mathbf{p}_0)z^{j_1*}\\ +\frac{1}{2!}a^{(2,0)}_{k(j_1j_2)}z^{j_1*}z^{j_2*}=0,\\ \text{equivalent to}\\ \dot{z}^k=A_{k0}(z^k)^2+A_{k1}z^k+A_{k2}=0;\\ A_{ki}=A_{ki}(z^{1*},z^{2*},\ldots,z^{k-1*},z^{k+1*},\ldots,z^{n*},\mathbf{p})\\ (i=0,1,2);\\ \Delta_k=B_{k1}^2-4C_{k1}\geq 0\text{ with }B_{k1}=\frac{A_{k1}}{A_{k0}},C_{k1}=\frac{A_{k2}}{A_{k0}},\\ A_{k1}=0,A_{k0}=a^{(2,0)}_{k(kk)}\neq 0,C_{1k}\leq 0,\end{array}\right\}\tag{2.17}$$

where

$$
\begin{aligned}
a_{k(j_1j_2)}^{(2,0)} &= (\mathbf{v}^k)^{\mathrm{T}} \cdot \partial_{c^{j_1}c^{j_2}}^{(2)} \partial_{\mathbf{p}}^{(0)} \mathbf{f}(\mathbf{x},\mathbf{p})|_{(\mathbf{x}_0^*,\mathbf{p}_0)} = (\mathbf{v}^k)^{\mathrm{T}} \cdot \partial_{c^{j_1}c^{j_2}}^{(2)} \mathbf{f}(\mathbf{x},\mathbf{p})|_{(\mathbf{x}_0^*,\mathbf{p}_0)} \\
&= (\mathbf{v}^k)^{\mathrm{T}} \cdot \partial_{\mathbf{x}}^{(2)} \mathbf{f}(\mathbf{x},\mathbf{p})(\mathbf{v}_{j_1}\mathbf{v}_{j_1})|_{(\mathbf{x}_0^*,\mathbf{p}_0)} = G_{k(j_1j_2)}^{(2)}(\mathbf{x}_0^*,\mathbf{p}_0), \\
\mathbf{a}_{k(j_1)}^{(1,1)} &= (\mathbf{v}^k)^{\mathrm{T}} \cdot \partial_{c^{j_1}}^{(1)} \partial_{\mathbf{p}}^{(1)} \mathbf{f}(\mathbf{x},\mathbf{p})|_{(\mathbf{x}_0^*,\mathbf{p}_0)} = (\mathbf{v}^k)^{\mathrm{T}} \cdot \partial_{\mathbf{x}} \partial_{\mathbf{p}} \mathbf{f}(\mathbf{x},\mathbf{p}) \mathbf{v}_{j_1}|_{(\mathbf{x}_0^*,\mathbf{p}_0)}, \\
\mathbf{b}_k^{\mathrm{T}} &= (\mathbf{v}^k)^{\mathrm{T}} \cdot \partial_{\mathbf{p}} \mathbf{f}(\mathbf{x},\mathbf{p})|_{(\mathbf{x}_0^*,\mathbf{p}_0)}, \mathbf{a}_k^{(0,2)} = (\mathbf{v}^k)^{\mathrm{T}} \cdot \partial_{\mathbf{p}}^{(2)} \mathbf{f}(\mathbf{x},\mathbf{p}), \\
a_{k(kk)}^{(2,0)} &= G_{k(kk)}^{(2)}(\mathbf{x}_0^*,\mathbf{p}_0) \neq 0.
\end{aligned}
\tag{2.18}
$$

If Eq. (2.17) has two equilibrium solutions, such a bifurcation at point $(\mathbf{x}_0^*, \mathbf{p}_0)$ is called the *saddle-node* bifurcation on the eigenvector $\mathbf{v}_k$.

(i$_1$) If

$$
\left.
\begin{array}{l}
\lambda_k = 0, a_{k(j_1j_2)}^{(2,0)} z^{j_1} z^{j_2} = G_{k(j_1j_2)}^{(2)}(\mathbf{x}_0^*,\mathbf{p}_0) z^{j_1} z^{j_2} > 0 \\
\text{for } j_1, j_2 \in \{1, 2, \ldots, n\}
\end{array}
\right\}
\tag{2.19}
$$

such a bifurcation at point $(\mathbf{x}_0^*, \mathbf{p}_0)$ is called an increasing *saddle-node* (or an upper-saddle-node) switching/appearing bifurcation on the eigenvector $\mathbf{v}_k$. The bifurcation point at $(\mathbf{x}_0^*, \mathbf{p}_0)$ is an increasing saddle (or an upper saddle) on the eigenvector $\mathbf{v}_k$.

(i$_2$) If

$$
\left.
\begin{array}{l}
\lambda_k = 0, a_{k(j_1j_2)}^{(2,0)} z^{j_1} z^{j_2} = G_{k(j_1j_2)}^{(2)}(\mathbf{x}_0^*,\mathbf{p}_0) z^{j_1} z^{j_2} < 0 \\
\text{for } j_1, j_2 \in \{1, 2, \ldots, n\};
\end{array}
\right\}
\tag{2.20}
$$

such a bifurcation at point $(\mathbf{x}_0^*, \mathbf{p}_0)$ is called a decreasing *saddle-node (or a lower-saddle-node) switching/appearing* bifurcation of the second order on the eigenvector $\mathbf{v}_k$. The bifurcation point at $(\mathbf{x}_0^*, \mathbf{p}_0)$ is a decreasing saddle (or a lower saddle) of the second order on the eigenvector $\mathbf{v}_k$.

(ii) Consider

$$
\lambda_s \neq 0 \; (s = 1, 2, \ldots, n \text{ but } s \neq k)
\tag{2.21}
$$

with

$$
\left.
\begin{array}{l}
\lambda_s z^{s*} + \mathbf{b}_s^{\mathrm{T}} \cdot (\mathbf{p} - \mathbf{p}_0) = 0, \\
s = 1, 2, \ldots, n \text{ but } s \neq k;
\end{array}
\right\}
\tag{2.22a}
$$

$$\left.\begin{array}{l}\mathbf{b}_k^{\mathrm{T}} \cdot (\mathbf{p}-\mathbf{p}_0) + \frac{1}{2!}\mathbf{a}_k^{(0,2)}(\mathbf{p}-\mathbf{p}_0)^2 + \mathbf{a}_{k(j_1)}^{(1,1)}(\mathbf{p}-\mathbf{p}_0)z^{j_1*} \\ + \frac{1}{2!}a_{k(j_1j_2)}^{(2,0)} z^{j_1*}z^{j_2*} = 0, \\ \text{equivalent to} \\ \dot{z}^k = A_{k0}(z^k)^2 + A_{k1}z^k + A_{k2} = 0; \\ A_{ki} = A_{ki}(z^{1*}, z^{2*}, \ldots, z^{k-1*}, z^{k+1*}, \ldots, z^{n*}, \mathbf{p}) \\ (i = 0, 1, 2); \end{array}\right\} \quad (2.22\text{b})$$

$$\Delta_k = B_{k1}^2 - 4C_{k1} \geq 0 \text{ with } B_{k1} = \frac{A_{k1}}{A_{k0}}, C_{k1} = \frac{A_{k2}}{A_{k0}},$$
$$A_{k2} = 0, A_{k0} = a_{k(kk)}^{(2,0)} \neq 0, A_{k1} \in (-\infty, \infty),$$

where

$$\begin{aligned} a_{k(j_1j_2)}^{(2,0)} &= (\mathbf{v}^k)^{\mathrm{T}} \cdot \partial_{c^{j_1}c^{j_2}}^{(2)} \partial_{\mathbf{p}}^{(0)} \mathbf{f}(\mathbf{x},\mathbf{p})\big|_{(\mathbf{x}_0^*,\mathbf{p}_0)} = (\mathbf{v}^k)^{\mathrm{T}} \cdot \partial_{c^{j_1}c^{j_2}}^{(2)} \mathbf{f}(\mathbf{x},\mathbf{p})\big|_{(\mathbf{x}_0^*,\mathbf{p}_0)} \\ &= (\mathbf{v}^k)^{\mathrm{T}} \cdot \partial_{\mathbf{x}}^{(2)} \mathbf{f}(\mathbf{x},\mathbf{p})(\mathbf{v}_{j_1}\mathbf{v}_{j_2})\big|_{(\mathbf{x}_0^*,\mathbf{p}_0)} = G_{k(j_1j_2)}^{(2)}(\mathbf{x}_0^*,\mathbf{p}_0), \\ \mathbf{a}_{k(j_1)}^{(1,1)} &= (\mathbf{v}^k)^{\mathrm{T}} \cdot \partial_{c^{j_1}}^{(1)} \partial_{\mathbf{p}}^{(1)} \mathbf{f}(\mathbf{x},\mathbf{p})\big|_{(\mathbf{x}_0^*,\mathbf{p}_0)} = (\mathbf{v}^k)^{\mathrm{T}} \cdot \partial_{c^{j_1}} \partial_{\mathbf{p}} \mathbf{f}(\mathbf{x},\mathbf{p})\big|_{(\mathbf{x}_0^*,\mathbf{p}_0)} \\ &= (\mathbf{v}^k)^{\mathrm{T}} \cdot \partial_{\mathbf{x}} \partial_{\mathbf{p}} \mathbf{f}(\mathbf{x},\mathbf{p}) \mathbf{v}_{j_1}\big|_{(\mathbf{x}_0^*,\mathbf{p}_0)} \neq \mathbf{0}, \\ a_{k(kk)}^{(2,0)} &= G_{k(kk)}^{(2)}(\mathbf{x}_0^*,\mathbf{p}_0) \neq 0. \end{aligned} \quad (2.23)$$

If Eq. (2.20) has two solutions of equilibriums, such a bifurcation at point $(\mathbf{x}_0^*, \mathbf{p}_0)$ is called a saddle-node *transcritical (or switching)* bifurcation of the second order on the eigenvector of $\mathbf{v}_k$.

(ii$_1$) If

$$\left.\begin{array}{l}\lambda_k = 0, a_{k(j_1j_2)}^{(2,0)} z^{j_1}z^{j_2} = G_{k(j_1j_2)}^{(2)}(\mathbf{x}_0^*,\mathbf{p}_0) z^{j_1}z^{j_2} < 0 \\ \text{for } j_1, j_2 \in \{1, 2, \cdots, n\}, \end{array}\right\} \quad (2.24)$$

such a bifurcation at point $(\mathbf{x}_0^*, \mathbf{p}_0)$ is called an increasing *transcritical (or an upper-saddle-node) switching* bifurcation of the second order on the eigenvector $\mathbf{v}_k$. The bifurcation point at $(\mathbf{x}_0^*, \mathbf{p}_0)$ is an increasing saddle (or an upper saddle) of the second order on the eigenvector $\mathbf{v}_k$.

(ii$_2$) If

$$\left.\begin{array}{l}\lambda_k = 0, a^{(2,0)}_{k(j_1 j_2)} z^{j_1} z^{j_2} = G^{(2)}_{k(j_1 j_2)}(\mathbf{x}_0^*, \mathbf{p}_0) z^{j_1} z^{j_2} > 0 \\ \text{for } j_1, j_2 \in \{1, 2, \ldots, n\};\end{array}\right\} \tag{2.25}$$

such a bifurcation at point $(\mathbf{x}_0^*, \mathbf{p}_0)$ is called a decreasing *transcritical* (or a lower-saddle-node) switching bifurcation of the second order on the eigenvector $\mathbf{v}_k$. The bifurcation point at $(\mathbf{x}_0^*, \mathbf{p}_0)$ is a decreasing saddle (or a lower saddle) of the second order on the eigenvector $\mathbf{v}_k$.

(iii) Consider

$$\lambda_s \neq 0 \; (s = 1, 2, \ldots, n \text{ but } s \neq k) \tag{2.26}$$

with

$$\left.\begin{array}{l}\lambda_s z^{s*} + \mathbf{b}_s^{\mathrm{T}} \cdot (\mathbf{p} - \mathbf{p}_0) = 0, \\ s = 1, 2, \ldots, n \text{ but } s \neq k;\end{array}\right\}$$

$$\left.\begin{array}{l}(\mathbf{v}^k)^{\mathrm{T}} \cdot \mathbf{f}(\mathbf{x}_0^* + \mathbf{y}^*, \mathbf{p}_0 + \Delta\mathbf{p}) = 0 \text{ with } \lambda_{0k} = 0; \\ \sum_{q=2}^{3} \sum_{r=0}^{q} \frac{1}{q!} C_q^r \mathbf{a}^{(q-r,r)}_{k(j_1 j_2 \cdots j_{q-r})} z^{j_1 *} z^{j_2 *} \ldots z^{j_{q-r} *} (\mathbf{p} - \mathbf{p}_0)^r + \mathbf{b}_k^{\mathrm{T}} \cdot (\mathbf{p} - \mathbf{p}_0) \approx 0 \\ \text{equivalent to} \\ \dot{z}^k = A_{k0} (z^k)^3 + A_{k1} (z^k)^2 + A_{k2} z^k + A_{k3} = 0, \\ A_{ki} = A_{ki}(z^{1*}, z^{2*}, \ldots, z^{k-1*}, z^{k+1*}, \ldots, z^{n*}, \mathbf{p}) \; (i = 0, 1, 2, 3);\end{array}\right\} \tag{2.27}$$

$$A_{k3} = A_{k1} = 0, A_{k0} = a^{(3,0)}_{k(kkk)} \neq 0, A_{k2} \times A_{k0} < 0,$$

where

$$\begin{aligned} a^{(3,0)}_{k(j_1 j_2 j_3)} &= (\mathbf{v}^k)^{\mathrm{T}} \cdot \partial^{(3)}_{c^{j_1} c^{j_2} c^{j_3}} \partial^{(0)}_{\mathbf{p}} \mathbf{f}(\mathbf{x}, \mathbf{p})|_{(\mathbf{x}_0^*, \mathbf{p}_0)} = (\mathbf{v}^k)^{\mathrm{T}} \cdot \partial^{(3)}_{c^{j_1} c^{j_2} c^{j_3}} \mathbf{f}(\mathbf{x}, \mathbf{p})|_{(\mathbf{x}_0^*, \mathbf{p}_0)} \\ &= (\mathbf{v}^k)^{\mathrm{T}} \cdot \partial^{(3)}_{\mathbf{x}} \mathbf{f}(\mathbf{x}, \mathbf{p})(\mathbf{v}_{j_1} \mathbf{v}_{j_2} \mathbf{v}_{j_3})|_{(\mathbf{x}_0^*, \mathbf{p}_0)} = G^{(3)}_{k(j_1 j_2 j_3)}(\mathbf{x}_0^*, \mathbf{p}_0), \\ \mathbf{a}^{(1,1)}_{k(j_1)} &= (\mathbf{v}^k)^{\mathrm{T}} \cdot \partial^{(1)}_{c^{j_1}} \partial^{(1)}_{\mathbf{p}} \mathbf{f}(\mathbf{x}, \mathbf{p})|_{(\mathbf{x}_0^*, \mathbf{p}_0)} = (\mathbf{v}^k)^{\mathrm{T}} \cdot \partial_{c^{j_1}} \partial_{\mathbf{p}} \mathbf{f}(\mathbf{x}, \mathbf{p})|_{(\mathbf{x}_0^*, \mathbf{p}_0)} \\ &= (\mathbf{v}^k)^{\mathrm{T}} \cdot \partial_{\mathbf{x}} \partial_{\mathbf{p}} \mathbf{f}(\mathbf{x}, \mathbf{p}) \mathbf{v}_{j_1}|_{(\mathbf{x}_0^*, \mathbf{p}_0)}, \\ a^{(3,0)}_{k(kkk)} &= G^{(3)}_{k(kkk)}(\mathbf{x}_0^*, \mathbf{p}_0) \neq 0. \end{aligned} \tag{2.28}$$

If Eq. (2.27) has three solutions, such a bifurcation at point $(\mathbf{x}_0^*, \mathbf{p}_0)$ is called the *pitchfork*-appearing bifurcation on the eigenvector of $\mathbf{v}_k$.

(iii$_1$) If

$$\left.\begin{array}{l}\lambda_k = 0, z^k \times a^{(3,0)}_{k(j_1 j_2 j_3)} z^{j_1} z^{j_2} z^{j_3} = z^k \times G^{(3)}_{k(j_1 j_2 j_3)}(\mathbf{x}_0^*, \mathbf{p}_0) z^{j_1} z^{j_2} z^{j_3} < 0 \\ \text{for } j_1, j_2, j_3 \in \{1, 2, \ldots, n\},\end{array}\right\} \tag{2.29}$$

such a bifurcation at point $(\mathbf{x}_0^*, \mathbf{p}_0)$ is called a stable *pitchfork* bifurcation of the third order (*or a third-order sink*) on the eigenvector $\mathbf{v}_k$. The bifurcation point at $(\mathbf{x}_0^*, \mathbf{p}_0)$ is a *third-order sink* on the eigenvector $\mathbf{v}_k$.

(iii$_2$) If

$$\left.\begin{array}{l}\lambda_k = 0, z^k \times a^{(3,0)}_{k(j_1 j_2 j_3)} z^{j_1} z^{j_2} z^{j_3} = z^k \times G^{(3)}_{k(j_1 j_2 j_3)}(\mathbf{x}_0^*, \mathbf{p}_0) z^{j_1} z^{j_2} z^{j_3} > 0 \\ \text{for } j_1, j_2, j_3 \in \{1, 2, \ldots, n\},\end{array}\right\} \tag{2.30}$$

such a bifurcation at point $(\mathbf{x}_0^*, \mathbf{p}_0)$ is called an unstable *pitchfork* bifurcation of the third order (or a *third-order source*) on the eigenvector $\mathbf{v}_k$. The bifurcation point at $(\mathbf{x}_0^*, \mathbf{p}_0)$ is a *third-order source* on the eigenvector $\mathbf{v}_k$.

2.3 Hopf Bifurcation on an Eigenvector Plane

Definition 2.7 Consider an n-dimensional, autonomous, nonlinear dynamical system $\dot{\mathbf{x}} = \mathbf{f}(\mathbf{x}, \mathbf{p})$ in Eq. (1.4) with an equilibrium point $\mathbf{x}^*$ and $\mathbf{f}(\mathbf{x}, \mathbf{p})$ is C^r $(r \geq 1)$-continuous in a neighborhood of the equilibrium $\mathbf{x}^*$. The corresponding solution is $\mathbf{x}(t) = \boldsymbol{\Phi}(\mathbf{x}_0, t - t_0, \mathbf{p})$. Suppose $U(\mathbf{x}^*) \subset \Omega$ is a neighborhood of equilibrium $\mathbf{x}^*$. For a linearized dynamical system in Eq. (1.19), consider a pair of complex eigenvalue $\alpha_k \pm \mathbf{i}\beta_k$ $(2k-1, 2k \in N = \{1, 2, \ldots, n\}, \mathbf{i} = \sqrt{-1})$ of matrix $D\mathbf{f}(\mathbf{x}^*, \mathbf{p})$ with a pair of eigenvectors $\mathbf{v}_{2k-1} \pm \mathbf{i}\mathbf{v}_{2k}$. On the invariant plane of $(\mathbf{v}_{2k-1}, \mathbf{v}_{2k})$,

$$c^{2k-1} = (\mathbf{v}^{2k-1})^{\mathrm{T}} \cdot \mathbf{y}, \;\; c^{2k} = (\mathbf{v}^{2k})^{\mathrm{T}} \cdot \mathbf{y}, \tag{2.31}$$

with

$$\begin{array}{l} D_{\mathbf{x}}\mathbf{f} \cdot \mathbf{v}_{2k-1} = \alpha_k \mathbf{v}_{2k-1} - \beta_k \mathbf{v}_{2k}, \\ D_{\mathbf{x}}\mathbf{f} \cdot \mathbf{v}_{2k} = \beta_k \mathbf{v}_{2k-1} + \alpha_k \mathbf{v}_{2k}, \end{array} \tag{2.32}$$

and

$$\begin{aligned} \rho_k &= \sqrt{(c^{2k-1})^2 + (c^{2k})^2}, \\ c^{2k-1} &= \rho_k \cos\theta_k, c^{2k} = \rho_k \sin\theta_k, \end{aligned} \tag{2.33}$$

and

$$\begin{aligned} \dot{c}^{2k-1} &= G_{(2k-1)}(\mathbf{x},\mathbf{p}) = (\mathbf{v}^{2k-1})^{\mathrm{T}} \cdot \mathbf{f}(\mathbf{x},\mathbf{p}), \\ \dot{c}^{2k} &= G_{(2k)}(\mathbf{x},\mathbf{p}) = (\mathbf{v}^{2k})^{\mathrm{T}} \cdot \mathbf{f}(\mathbf{x},\mathbf{p}), \end{aligned} \tag{2.34}$$

where

$$\begin{aligned} & G_{(2k-1)}(\mathbf{x},\mathbf{p}) \\ & = (\mathbf{v}^{2k-1})^{\mathrm{T}} \cdot \mathbf{f}(\mathbf{x},\mathbf{p}) \\ & = \mathbf{a}^{\mathrm{T}}_{(2k-1)} \cdot (\mathbf{p}-\mathbf{p}_0) + a_{(2k-1)(2k-1)} z^{2k-1} + a_{(2k-1)(2k)} z^{2k} \\ & + \sum_{q=2}^{m} \sum_{r=0}^{q} \frac{1}{q!} C_q^r \mathbf{a}^{(q-r,r)}_{(2k-1)(j_1 j_2 \cdots j_{q-r})} (\mathbf{x}^*,\mathbf{p}_0) z^{j_1} z^{j_2} \ldots z^{j_{q-r}} (\mathbf{p}-\mathbf{p}_0)^r \\ & + \frac{1}{(m+1)!} \sum_{r=0}^{m+1} C_{m+1}^r \mathbf{a}^{(m+1-r,r)}_{(2k-1)(j_1 j_2 \cdots j_{m+1-r})} (\mathbf{x}^*,\mathbf{p}_0) z^{j_1} z^{j_2} \ldots z^{j_{m+1-r}} (\mathbf{p}-\mathbf{p}_0)^r, \\ & G_{(2k)}(\mathbf{x},\mathbf{p}) \\ & = (\mathbf{v}^{2k})^{\mathrm{T}} \cdot \mathbf{f}(\mathbf{x},\mathbf{p}) \\ & = \mathbf{a}^{\mathrm{T}}_{(2k)} \cdot (\mathbf{p}-\mathbf{p}_0) + a_{(2k)(2k-1)} z^{2k-1} + a_{(2k)(2k)} z^{2k} \\ & + \sum_{q=2}^{m} \sum_{r=0}^{q} \frac{1}{q!} C_q^r \mathbf{a}^{(q-r,r)}_{(2k)(j_1 j_2 \cdots j_{q-r})} (\mathbf{x}^*,\mathbf{p}_0) z^{j_1} z^{j_2} \ldots z^{j_{q-r}} (\mathbf{p}-\mathbf{p}_0)^r \\ & + \frac{1}{(m+1)!} \sum_{r=0}^{m+1} C_{m+1}^r \mathbf{a}^{(m+1-r,r)}_{(2k)(j_1 j_2 \cdots j_{m+1-r})} (\mathbf{x}^*,\mathbf{p}_0) z^{j_1} z^{j_2} \ldots z^{j_{m+1-r}} (\mathbf{p}-\mathbf{p}_0)^r, \end{aligned} \tag{2.35}$$

and

$$\begin{aligned} \mathbf{a}^{(r_k,s_k)}_{(2k-1)(j_1 j_2 \cdots j_{r_k})}(\mathbf{x}^*,\mathbf{p}) &= (\mathbf{v}^{2k-1})^{\mathrm{T}} \cdot \partial^{(r_k)}_{c^{j_1} c^{j_2} \ldots c^{j_{r_k}}} \partial^{(s_k)}_{\mathbf{p}} \mathbf{f}(\mathbf{x}^*,\mathbf{p}) \\ &= (\mathbf{v}^{2k-1})^{\mathrm{T}} \cdot \partial^{(r_k)}_{\mathbf{x}} \partial^{(s_k)}_{\mathbf{p}} \mathbf{f}(\mathbf{x}^*,\mathbf{p}) \mathbf{v}_{j_1} \mathbf{v}_{j_2} \ldots \mathbf{v}_{j_{r_k}}, \\ \mathbf{a}^{(r_k,s_k)}_{(2k)(j_1 j_2 \cdots j_{r_k})}(\mathbf{x}^*,\mathbf{p}) &= (\mathbf{v}^{2k})^{\mathrm{T}} \cdot \partial^{(r_k)}_{c^{j_1} c^{j_2} \ldots c^{j_{r_k}}} \partial^{(s_k)}_{\mathbf{p}} \mathbf{f}(\mathbf{x}^*,\mathbf{p}) \\ &= (\mathbf{v}^{2k})^{\mathrm{T}} \cdot \partial^{(r_k)}_{\mathbf{x}} \partial^{(s_k)}_{\mathbf{p}} \mathbf{f}(\mathbf{x}^*,\mathbf{p}) \mathbf{v}_{j_1} \mathbf{v}_{j_2} \ldots \mathbf{v}_{j_{r_k}}; \\ \mathbf{a}^{\mathrm{T}}_{(2k-1)} &= (\mathbf{v}^{2k-1})^{\mathrm{T}} \cdot \partial_{\mathbf{p}} \mathbf{f}(\mathbf{x},\mathbf{p}), \\ \mathbf{a}^{\mathrm{T}}_{(2k)} &= (\mathbf{v}^{2k})^{\mathrm{T}} \cdot \partial_{\mathbf{p}} \mathbf{f}(\mathbf{x},\mathbf{p}); \end{aligned} \tag{2.36}$$

$$
\begin{aligned}
a_{(2k-1)(2k-1)} &= (\mathbf{v}^{2k-1})^{\mathrm{T}} \cdot \partial_{\mathbf{x}}\mathbf{f}(\mathbf{x},\mathbf{p})\mathbf{v}_{2k-1},\\
a_{(2k-1)(2k)} &= (\mathbf{v}^{2k-1})^{\mathrm{T}} \cdot \partial_{\mathbf{x}}\mathbf{f}(\mathbf{x},\mathbf{p})\mathbf{v}_{2k};\\
a_{(2k)(2k-1)} &= (\mathbf{v}^{2k})^{\mathrm{T}} \cdot \partial_{\mathbf{x}}\mathbf{f}(\mathbf{x},\mathbf{p})\mathbf{v}_{2k-1},\\
a_{(2k)(2k)} &= (\mathbf{v}^{2k})^{\mathrm{T}} \cdot \partial_{\mathbf{x}}\mathbf{f}(\mathbf{x},\mathbf{p})\mathbf{v}_{2k}.
\end{aligned}
\tag{2.37}
$$

Thus

$$
\begin{aligned}
\dot{\rho}_k &= \dot{c}^{2k-1}\cos\theta_k + \dot{c}^{2k}\sin\theta_k = \sum_{q_k=1}^{\infty}\sum_{r_k=0}^{q_k}\frac{1}{q_k!}G_{\rho_k}^{(q_k-r_k,\,r_k)}\\
&= (\alpha_k + \mathbf{a}_{\rho_k(k)}^{(1,1)}(\theta_k,\mathbf{p}_0)\cdot(\mathbf{p}-\mathbf{p}_0))\rho_k\\
&\quad + \sum_{q=2}^{\infty}\sum_{r=0}^{q}\frac{1}{q!}\mathbf{a}_{\rho_k(j_1\cdots j_{q-r})}^{(q-r,\,r)}(\theta_k,\mathbf{p}_0)z^{j_1}\ldots z^{j_{q-r}}(\mathbf{p}-\mathbf{p}_0)^r,\\
\dot{\theta}_k &= \rho_k^{-1}(\dot{c}^{2k}\cos\theta_k - \dot{c}^{2k-1}\sin\theta_k) = \sum_{q_k=1}^{\infty}\sum_{r_k=0}^{q}\frac{1}{q_k!}G_{\theta_k}^{(q_k-r_k,\,r_k)}\\
&= \beta_k + \mathbf{a}_{\theta_k(k)}^{(1,1)}(\theta_k,\mathbf{p}_0)\cdot(\mathbf{p}-\mathbf{p}_0)\\
&\quad + \frac{1}{\rho_k}\sum_{q=2}^{3}\sum_{r=0}^{q}\frac{1}{q!}\mathbf{a}_{\theta_k(j_1\cdots j_{q-r})}^{(q-r,\,r)}(\theta_k,\mathbf{p}_0)z^{j_1}\ldots z^{j_{q-r}}(\mathbf{p}-\mathbf{p}_0)^r,
\end{aligned}
\tag{2.38}
$$

where

$$
\begin{aligned}
G_{\rho_k}^{(s,r)} &= \mathbf{a}_{\rho_k(j_1 j_2\cdots j_s)}^{(s,r)}z^{j_1}z^{j_2}\ldots z^{j_s}(\mathbf{p}-\mathbf{p}_0)^r,\\
G_{\theta_k}^{(s,r)} &= \frac{1}{\rho_k}\mathbf{a}_{\theta_k(j_1 j_2\cdots j_s)}^{(s,r)}z^{j_1}z^{j_2}\ldots z^{j_s}(\mathbf{p}-\mathbf{p}_0)^r\\
&\qquad (\text{summation of } j_1, j_2, \ldots, j_s),
\end{aligned}
\tag{2.39}
$$

and

$$
\begin{aligned}
\mathbf{a}_{\rho_k(j_1 j_2\cdots j_s)}^{(s,r)} &= \cos\theta_k\mathbf{a}_{(2k-1)(j_1 j_2\cdots j_s)}^{(s,r)}(\mathbf{x}^*,\mathbf{p}) + \sin\theta_k(\mathbf{a}_{(2k)(j_1 j_2\cdots j_s)}^{(s,r)}(\mathbf{x}^*,\mathbf{p}),\\
\mathbf{a}_{\theta_k(j_1 j_2\cdots j_s)}^{(s,r)} &= -\sin\theta_k\mathbf{a}_{(2k-1)(j_1 j_2\cdots j_s)}^{(s,r)}(\mathbf{x}^*,\mathbf{p}) + \cos\theta_k(\mathbf{a}_{(2k)(j_1 j_2\cdots j_s)}^{(s,r)}(\mathbf{x}^*,\mathbf{p}).
\end{aligned}
\tag{2.40}
$$

Definition 2.8 Consider an n-dimensional, autonomous, nonlinear dynamical system $\dot{\mathbf{x}} = \mathbf{f}(\mathbf{x},\mathbf{p})$ in Eq. (1.4) with an equilibrium point $\mathbf{x}^*$ and $\mathbf{f}(\mathbf{x},\mathbf{p})$ is C^r-continuous $(r \geq 1)$ in a neighborhood of the equilibrium $\mathbf{x}^*$ (i.e., $U(\mathbf{x}^*) \subset \Omega$). The corresponding solution is $\mathbf{x}(t) = \mathbf{\Phi}(\mathbf{x}_0, t-t_0, \mathbf{p})$. Suppose Eq. (1.24) holds in $U(\mathbf{x}^*) \subset \Omega$. For a linearized dynamical system in Eq. (1.19), consider a real eigenvalue λ_k of matrix

$D\mathbf{f}(\mathbf{x}^*,\mathbf{p}^*)$ ($k \in N = \{1,2,\ldots,n\}$) with an eigenvector $\mathbf{v}_k$. Suppose one of n independent solutions $\mathbf{y} = c^k \mathbf{v}_k$ and $\dot{\mathbf{y}} = \dot{c}^k \mathbf{v}_k$. Consider $a_{s(s)} = \lambda_s \neq 0$ ($s \in S = \{1,2,\ldots,p\}$ and $S \subset N = \{1,2,\ldots,n\}$); the higher order term is ignorable in Eq. (2.3). The p equations for the corresponding directions are

$$\begin{aligned} &\lambda_{s_\alpha} z^{s_\alpha} + \mathbf{b}_{s_\alpha}^{\mathrm{T}} \cdot (\mathbf{p} - \mathbf{p}_0) = 0 \\ &\lambda_{s_\alpha} \neq 0 \; (s_\alpha = 1, 2, \ldots, p). \end{aligned} \tag{2.41}$$

For a pair of complex eigenvalues $\alpha_{k_\alpha} \pm \mathbf{i}\beta_{k_\alpha}$ ($2k_\alpha - 1, 2k_\alpha \in N = \{1,2,\ldots,n\}$, $\mathbf{i} = \sqrt{-1}$) of matrix $D\mathbf{f}(\mathbf{x}^*,\mathbf{p})$ with a pair of eigenvectors $\mathbf{v}_{2k_\alpha-1} \pm \mathbf{i}\mathbf{v}_{2k_\alpha}$. On the invariant plane of $(\mathbf{v}_{2k_\alpha-1}, \mathbf{v}_{2k_\alpha})$ and at $\mathbf{p} = \mathbf{p}_0$, for a specific k_α, if $a_{k_\alpha(k_\alpha)} = \alpha_{k_\alpha} = 0$ ($k_\alpha \in N/S = \{\frac{1}{2}p+1, \frac{1}{2}p+2, \ldots, \frac{1}{2}n\} \subseteq N$), the s-paired equations ($s = (n-p)/2$) on the corresponding directions are given by

$$\left.\begin{aligned} &0 = (\alpha_{k_\alpha} + \mathbf{a}^{(1,1)}_{\rho_{k\alpha}(k_\alpha)}(\theta_{k_\alpha}, \mathbf{p}_0) \cdot (\mathbf{p} - \mathbf{p}_0))\rho_{k_\alpha} \\ &\quad + \sum_{q=2}^{m_{k\alpha}} \sum_{r=0}^{q} \frac{1}{q!} \mathbf{a}^{(q-r,r)}_{\rho_{k\alpha}(j_1 \cdots j_{q-r})}(\theta_{k_\alpha}, \mathbf{p}_0) z^{j_1} \ldots z^{j_{q-r}} (\mathbf{p} - \mathbf{p}_0)^r, \\ &\quad \text{equivalent to} \\ &\dot{\rho}_{k_\alpha} = A_{k_\alpha 0}(\rho_{k_\alpha})^{m_{k\alpha}} + A_{k_\alpha 1}(\rho_{k_\alpha})^{m_{k\alpha}-1} + \cdots + A_{k_\alpha m_\alpha} = 0, \\ &A_{k_\alpha i} = A_{k_\alpha i}(z^1, z^2, \ldots, z^{2k_\alpha - 2}, z^{2k_\alpha+1}, \ldots, z^n, \theta_{k_\alpha}, \mathbf{p}_0) \\ &\quad (i = 0, 1, 2, \ldots, m_{k\alpha}), \end{aligned}\right\}$$

$$\left.\begin{aligned} &\dot{\theta}_{k_\alpha} = \beta_{k_\alpha} + \mathbf{a}^{(1,1)}_{\theta_k(k_\alpha)}(\theta_{k_\alpha}, \mathbf{p}_0) \cdot (\mathbf{p} - \mathbf{p}_0) \\ &\quad + \frac{1}{\rho_{k_\alpha}} \sum_{q=2}^{m_{k\alpha}} \sum_{r=0}^{q} \frac{1}{q!} \mathbf{a}^{(q-r,r)}_{\theta_{k\alpha}(j_1 \cdots j_{q-r})}(\theta_{k_\alpha}, \mathbf{p}_0) z^{j_1} \ldots z^{j_{q-r}} (\mathbf{p} - \mathbf{p}_0)^r, \\ &\quad \text{equivalent to} \\ &\dot{\theta}_{k_\alpha} = \frac{1}{\rho_{k_\alpha}} [B_{k_\alpha 0}(\rho_{k_\alpha})^{m_{k\alpha}} + B_{k_\alpha 1}(\rho_{k_\alpha})^{m_{k\alpha}} + \cdots + B_{k_\alpha m_\alpha}], \\ &B_{k_\alpha i} = B_{k_\alpha i}(z^1, z^2, \ldots, z^{2k_\alpha - 2}, z^{2k_\alpha+1}, \ldots, z^n, \theta_{k_\alpha}, \mathbf{p}_0) \\ &\quad (i = 0, 1, 2, \ldots, m_{k\alpha}), \end{aligned}\right\} \tag{2.42}$$

$$(k_\alpha = \frac{1}{2}p + 1, \frac{1}{2}p + 2, \ldots, \frac{1}{2}n)$$

with

$$G_{\rho_{k_\alpha}}^{(m_{k_\alpha})} = \mathbf{a}_{\rho_{k_\alpha}(j_1 \cdots j_{m_{k_\alpha}})}^{(m_{k_\alpha},0)}(\theta_{k_\alpha}, \mathbf{p}_0) z^{j_1} \ldots z^{j_{m_{k_\alpha}}} \neq 0. \tag{2.43}$$

Equations (2.41)–(2.43) possess l_{k_α}-branch spiral equilibriums $\mathbf{x}^*$ with $\sum_{i=1}^{l_{k_\alpha}} \alpha_i^{k_\alpha} = m_{k_\alpha}$ $(0 < l_{k_\alpha} \leq m_{k_\alpha})$. Such l_{k_α}-branch solutions are called the bifurcation solutions of equilibrium $\mathbf{x}^*$ on the eigenvector of $\mathbf{v}_k$ in the neighborhood of $(\mathbf{x}_0^*, \mathbf{p}_0)$. Such a bifurcation at point $(\mathbf{x}_0^*, \mathbf{p}_0)$ is called the *Hopf bifurcation* of $(m_{\frac{1}{2}p+1} : m_{\frac{1}{2}p+2} : \cdots : m_{\frac{1}{2}n})$ order on the eigenvector of $(\mathbf{v}_{2k_\alpha-1}, \mathbf{v}_{2k_\alpha})$.

(i) If

$$\left.\begin{array}{l} \alpha_{k_\alpha} = 0, G_{\rho_{k_\alpha}}^{(r_{k_\alpha})} = 0 \ (r_{k_\alpha} = 2, 3, \ldots, m_{k_\alpha} - 1), \\ G_{\rho_{k_\alpha}}^{(m_{k_\alpha})} = \mathbf{a}_{\rho_{k_\alpha}(j_1 \cdots j_{m_\alpha})}^{(m_{k_\alpha},0)}(\theta_{k_\alpha}, \mathbf{p}_0) z^{j_1} \cdots z^{j_{m_{k_\alpha}}} > 0 \\ (k_\alpha \in \{\frac{1}{2}p + 1, \frac{1}{2}p + 2, \ldots, \frac{1}{2}n\}), \end{array}\right\} \tag{2.44}$$

(i$_1$) The bifurcation of equilibrium $\mathbf{x}^*$ at point $(\mathbf{x}_0^*, \mathbf{p}_0)$ is called the source Hopf bifurcation of the $(2l_{k_\alpha} + 1)$th order on the eigenvector plane of $(\mathbf{v}_{2k_\alpha-1}, \mathbf{v}_{2k_\alpha})$ if $m_{k_\alpha} = 2l_{k_\alpha} + 1$. The bifurcation point $(\mathbf{x}_0^*, \mathbf{p}_0)$ on the eigenvector plane of $(\mathbf{v}_{2k_\alpha-1}, \mathbf{v}_{2k_\alpha})$ is the $(2l_{k_\alpha} + 1)$th-order source.

(i$_2$) The bifurcation of equilibrium $\mathbf{x}^*$ at point $(\mathbf{x}_0^*, \mathbf{p}_0)$ is called the upper-saddle Hopf bifurcation of the $(2l_{k_\alpha})$th order on the eigenvector plane of $(\mathbf{v}_{2k_\alpha-1}, \mathbf{v}_{2k_\alpha})$ if $m_{k_\alpha} = 2l_{k_\alpha}$. The bifurcation point $(\mathbf{x}_0^*, \mathbf{p}_0)$ on the eigenvector plane of $(\mathbf{v}_{2k_\alpha-1}, \mathbf{v}_{2k_\alpha})$ is the $(2l_{k_\alpha})$th-order upper saddle.

(ii) If

$$\left.\begin{array}{l} \alpha_{k_\alpha} = 0, G_{\rho_{k_\alpha}}^{(r_{k_\alpha})} = 0 \ (r_{k_\alpha} = 2, 3, \ldots, m_{k_\alpha} - 1) \\ G_{\rho_{k_\alpha}}^{(m_{k_\alpha})} = \mathbf{a}_{\rho_{k_\alpha}(j_1 \cdots j_{m_\alpha})}^{(m_{k_\alpha},0)}(\theta_{k_\alpha}, \mathbf{p}_0) z^{j_1} \cdots z^{j_{m_{k_\alpha}}} < 0 \\ (k_\alpha \in \{\frac{1}{2}p + 1, \frac{1}{2}p + 2, \ldots, \frac{1}{2}n\}), \end{array}\right\} \tag{2.45}$$

(ii$_1$) The bifurcation of equilibrium $\mathbf{x}^*$ at point $(\mathbf{x}_0^*, \mathbf{p}_0)$ is called the sink Hopf bifurcation of the $(2l_{k_\alpha} + 1)$th order on the eigenvector plane of $(\mathbf{v}_{2k_\alpha-1}, \mathbf{v}_{2k_\alpha})$ if $m_{k_\alpha} = 2l_{k_\alpha} + 1$. The bifurcation point $(\mathbf{x}_0^*, \mathbf{p}_0)$ on the eigenvector plane of $(\mathbf{v}_{2k_\alpha-1}, \mathbf{v}_{2k_\alpha})$ is the $(2l_{k_\alpha} + 1)$th-order sink.

(ii$_2$) The bifurcation of equilibrium $\mathbf{x}^*$ at point $(\mathbf{x}_0^*, \mathbf{p}_0)$ is called the lower saddle Hopf bifurcation of the $(2l_{k_\alpha})$th order on the eigenvector plane of

$(\mathbf{v}_{2k_\alpha-1}, \mathbf{v}_{2k_\alpha})$ if $m_{k_\alpha} = 2l_{k_\alpha}$. The bifurcation point $(\mathbf{x}_0^*, \mathbf{p}_0)$ on the eigenvector plane of $(\mathbf{v}_{2k_\alpha-1}, \mathbf{v}_{2k_\alpha})$ is the $(2l_{k_\alpha})$th-order lower saddle.

Definition 2.9 Consider an n-dimensional, autonomous, nonlinear dynamical system $\dot{\mathbf{x}} = \mathbf{f}(\mathbf{x}, \mathbf{p})$ in Eq. (1.4) with an equilibrium point $\mathbf{x}^*$ and $\mathbf{f}(\mathbf{x}, \mathbf{p})$ is C^r-continuous $(r \geq 1)$ in a neighborhood of the equilibrium $\mathbf{x}^*$ (i.e., $U(\mathbf{x}^*) \subset \Omega$). The corresponding solution is $\mathbf{x}(t) = \mathbf{\Phi}(\mathbf{x}_0, t - t_0, \mathbf{p})$. Suppose Eq. (1.24) holds in $U(\mathbf{x}^*) \subset \Omega$. For a linearized dynamical system in Eq. (1.19), consider a real eigenvalue λ_s of matrix $D\mathbf{f}(\mathbf{x}^*, \mathbf{p}^*)$ $(s = 1, 2, \ldots, n$ but $s \neq 2k - 1, 2k)$ with an eigenvector $\mathbf{v}_s$. Suppose one of n independent solutions $\mathbf{y} = c^j \mathbf{v}_j$ and $\dot{\mathbf{y}} = \dot{c}^j \mathbf{v}_j$. Suppose

$$\left.\begin{array}{l} \lambda_s z^s + \mathbf{b}_s^{\mathrm{T}} \cdot (\mathbf{p} - \mathbf{p}_0) = 0 \\ \lambda_s \neq 0 \ (s = 1, 2, \ldots, n \text{ but } s \neq 2k - 1, 2k). \end{array}\right. \tag{2.46}$$

For a pair of complex eigenvalue $\alpha_k \pm \mathbf{i}\beta_k$ $(2k - 1, 2k \in N = \{1, 2, \ldots, n\}, \mathbf{i} = \sqrt{-1})$ of matrix $D\mathbf{f}(\mathbf{x}^*, \mathbf{p})$ with a pair of eigenvectors $\mathbf{v}_{2k-1} \pm \mathbf{i}\mathbf{v}_{2k}$. On the invariant plane of $(\mathbf{v}_{2k-1}, \mathbf{v}_{2k})$ at $\mathbf{p} = \mathbf{p}_0$,

$$\left.\begin{array}{l} \dot{\rho}_k = (\alpha_k + \mathbf{a}_{\rho_k(k)}^{(1,1)}(\theta_k, \mathbf{p}_0) \cdot (\mathbf{p} - \mathbf{p}_0))\rho_k \\ \qquad + \displaystyle\sum_{q=2}^{3} \sum_{r=0}^{q} \frac{1}{q!} \mathbf{a}_{\rho_k(j_1 \cdots j_{q-r})}^{(q-r,r)}(\theta_k, \mathbf{p}_0) z^{j_1} \cdots z^{j_{q-r}} (\mathbf{p} - \mathbf{p}_0)^r + o(\rho_k^3), \\ \quad \text{equivalent to} \\ \quad \dot{\rho}_k = A_{k0}(\rho_k)^3 + A_{k1}(\rho_k)^2 + A_{k2}\rho_k + A_{k3} = 0, \\ \quad A_{ki} = A_{ki}(z^1, z^2, \ldots, z^{2k-2}, z^{2k+1}, \ldots, z^n, \theta_k, \mathbf{p}_0) \\ \quad (i = 0, 1, 2, 3), \\ \dot{\theta}_k = \beta_k + \mathbf{a}_{\theta_k(k)}^{(1,1)}(\theta_k, \mathbf{p}_0) \cdot (\mathbf{p} - \mathbf{p}_0) \\ \qquad + \dfrac{1}{\rho_k} \displaystyle\sum_{q=2}^{3} \sum_{r=0}^{q} \frac{1}{q!} \mathbf{a}_{\theta_k(j_1 \cdots j_{q-r})}^{(q-r,r)}(\theta_k, \mathbf{p}_0) z^{j_1} \ldots z^{j_{q-r}} (\mathbf{p} - \mathbf{p}_0)^r + o(\rho_k^3) \\ \quad \text{equivalent to} \\ \rho_k \dot{\theta}_k = B_{k0}(\rho_k)^3 + B_{k1}(\rho_k)^2 + B_{k1}\rho_k + B_{k3}, \\ B_{ki} = B_{ki}(z^1, z^2, \ldots, z^{2k-2}, z^{2k+1}, \ldots, z^n, \theta_k, \mathbf{p}_0) \\ (i = 0, 1, 2, 3). \end{array}\right\} \tag{2.47}$$

Consider a specific case:

$$\begin{aligned} & B_{k3} = B_{k1} = A_{k1} = A_{k3} = 0; \\ & A_{k0}, A_{k2}, B_{k0}, B_{k2} \text{ are independent of } \theta_k, \\ & A_{k0} = \mathrm{a}^{(3,0)}_{\rho_k(kkk)}(\theta_k, \mathbf{p}_0) \neq 0. \end{aligned} \tag{2.48}$$

If $\alpha_k = 0$ and $\mathbf{p} = \mathbf{p}_0$, the stability of current equilibrium $\mathbf{x}^*$ on an eigenvector plane of $(\mathbf{v}_{2k-1}, \mathbf{v}_{2k})$ changes from stable to unstable state (or from unstable to stable state). The bifurcation manifold on the plane of $(\mathbf{v}_{2k-1}, \mathbf{v}_{2k})$ is determined by

$$\begin{aligned} \dot{\rho}_k &= A_{k2}\rho_k + A_{k0}\rho_k^3 + o(\rho_k^3), \\ \dot{\theta}_k &= B_{k2} + B_{k0}\rho_k^2 + o(\rho_k^2), \end{aligned} \tag{2.49}$$

where

$$A_{k0} \times A_{k2} < 0. \tag{2.50}$$

Such a bifurcation at point $(\mathbf{x}_0^*, \mathbf{p}_0)$ is called the Hopf bifurcation on the eigenvector plane of $(\mathbf{v}_{2k-1}, \mathbf{v}_{2k})$.

(i) If

$$G^{(3)}_{\rho_k} = a^{(3,0)}_{\rho_k(j_1 j_2 j_3)}(\theta_k, \mathbf{p}_0) z^{j_1} z^{j_2} z^{j_3} > 0, \tag{2.51}$$

the bifurcation of equilibrium $\mathbf{x}^*$ at point $(\mathbf{x}_0^*, \mathbf{p}_0)$ is called an unstable Hopf bifurcation of the third order on the eigenvector plane of $(\mathbf{v}_{2k-1}, \mathbf{v}_{2k})$. The bifurcation point $(\mathbf{x}_0^*, \mathbf{p}_0)$ on the eigenvector plane of $(\mathbf{v}_{2k-1}, \mathbf{v}_{2k})$ is a third-order source.

(ii) If

$$G^{(3)}_{\rho_k} = a^{(3,0)}_{\rho_k(j_1 j_2 j_3)}(\theta_k, \mathbf{p}_0) z^{j_1} z^{j_2} z^{j_3} < 0, \tag{2.52}$$

the bifurcation of equilibrium $\mathbf{x}^*$ at point $(\mathbf{x}_0^*, \mathbf{p}_0)$ is called a stable Hopf bifurcation of the third order on the eigenvector plane of $(\mathbf{v}_{2k-1}, \mathbf{v}_{2k})$. The bifurcation point $(\mathbf{x}_0^*, \mathbf{p}_0)$ on the eigenvector plane of $(\mathbf{v}_{2k-1}, \mathbf{v}_{2k})$ is a third-order sink.

2.4 Hopf Bifurcation Based on the Fourier Series Base

In the traditional Hopf bifurcation analysis, it is very difficult to determine the limit cycle and stability. As in Luo (2019), an alternative method was presented herein.

Definition 2.10 Consider an n-dimensional, autonomous, nonlinear dynamical system $\dot{\mathbf{x}} = \mathbf{f}(\mathbf{x}, \mathbf{p})$ in Eq. (1.4) with an equilibrium point $\mathbf{x}^*$ and $\mathbf{f}(\mathbf{x}, \mathbf{p})$ is C^r-continuous ($r \geq 1$) in a neighborhood of the equilibrium $\mathbf{x}^*$ (i.e., $U(\mathbf{x}^*) \subset \Omega$). The corresponding solution is $\mathbf{x}(t) = \mathbf{\Phi}(\mathbf{x}_0, t - t_0, \mathbf{p})$. Suppose Eq. (1.24) holds in $U(\mathbf{x}^*) \subset \Omega$. For a linearized dynamical system in Eq. (1.19), consider a pair of complex eigenvalues $\alpha_k \pm \mathbf{i}\beta_k$ ($2k - 1$, $2k \in N = \{1, 2, \ldots, n\}$,$\mathbf{i} = \sqrt{-1}$) of matrix $D\mathbf{f}(\mathbf{x}^*, \mathbf{p})$ with a pair of eigenvectors $\mathbf{v}_{2k-1} \pm \mathbf{i}\mathbf{v}_{2k}$. In the vicinity of $\mathbf{x}^*$, there exists a transformation of

$$\mathbf{x} = \mathbf{x}^* + \mathbf{y} = \mathbf{x}^* + \mathbf{a}_0(t) + \sum_{j=1}^{N} \mathbf{b}_j(t) \cos(j\beta_k t) + \mathbf{c}_j(t) \sin(j\beta_k t) \tag{2.53}$$

under $\lim_{j \to \infty} \|\mathbf{A}_j\| = 0$ with Eq. (1.128). The corresponding dynamical system of coefficients is

$$\begin{aligned} \dot{\mathbf{a}}_0 &= \mathbf{F}_0(\mathbf{a}_0, \mathbf{b}, \mathbf{c}), \\ \dot{\mathbf{b}} &= -\beta_k \mathbf{k}\mathbf{c} + \mathbf{F}^c(\mathbf{a}_0, \mathbf{b}, \mathbf{c}), \\ \dot{\mathbf{c}} &= \beta_k \mathbf{k}\mathbf{b} + \mathbf{F}^s(\mathbf{a}_0, \mathbf{b}, \mathbf{c}) \end{aligned} \tag{2.54}$$

with Eqs. (1.131) and (1.132). If Eq. (2.54) has a nonzero solution, i.e.,

$$(\mathbf{a}_0^*, \mathbf{b}^*, \mathbf{c}^*) \neq \mathbf{0}, \tag{2.55}$$

then in the vicinity of equilibrium $\mathbf{x}^*$, there is a bifurcation solution of equilibrium$\mathbf{x}^*$.

(i) The point $(\mathbf{x}_0^*, \mathbf{p}_0)$ is called a hyperbolic bifurcation of the equilibrium $(\mathbf{x}^*, \mathbf{p})$ if

$$\begin{aligned} &\mathbf{a}_0^*(\mathbf{x}^*, \mathbf{p}) \neq \mathbf{0} \text{ and } \lim_{(\mathbf{x}^*, \mathbf{p}) \to (\mathbf{x}_0^*, \mathbf{p}_0),} \mathbf{a}_0^*(\mathbf{x}^*, \mathbf{p}) = \mathbf{0}; \\ &\mathbf{A}^*(\mathbf{x}^*, \mathbf{p}) = \mathbf{0} \text{ (or } \mathbf{b}^*(\mathbf{x}^*, \mathbf{p}) = \mathbf{0}, \mathbf{c}^*(\mathbf{x}^*, \mathbf{p}) = \mathbf{0}) \\ &\text{or } \beta_k = 0. \end{aligned} \tag{2.56}$$

The new equilibrium

$$\bar{\mathbf{x}}^* = \mathbf{x}^* + \mathbf{a}_0^*(\mathbf{x}^*, \mathbf{p}) \tag{2.57}$$

is called the bifurcation branch of the equilibrium $(\mathbf{x}^*, \mathbf{p})$.

If there are l new branches of equilibrium points $\bar{\mathbf{x}}^*$, such a bifurcation is called the l-branch bifurcation of the equilibrium $\mathbf{x}^*$.

(i_1) For $l = 1$, the bifurcation at the point $(\mathbf{x}_0^*, \mathbf{p}_0)$ is called a transcritical bifurcation of the equilibrium $\mathbf{x}^*$ if $\bar{\mathbf{x}}^*$ is linear with parameter $\mathbf{p}$.

(i_2) For $l = 1$, the bifurcation at the point $(\mathbf{x}_0^*, \mathbf{p}_0)$ is called a saddle-node bifurcation of the equilibrium $\mathbf{x}^*$ if $\overline{\mathbf{x}}^*$ and $\mathbf{x}^*$ are parabolic with parameter $\mathbf{p}$.

(i_3) For $l = 2$, the bifurcation at the point $(\mathbf{x}_0^*, \mathbf{p}_0)$ is called a pitchfork bifurcation of the equilibrium $\mathbf{x}^*$ if two new equilibriums $\overline{\mathbf{x}}^*$ and $\mathbf{x}^*$ form a pitchfork.

(i_4) For $l = 2m - 1$, the bifurcation at the point $(\mathbf{x}_0^*, \mathbf{p}_0)$ is called a $(2m)^{\text{th}}$-order transcritical bifurcation of the equilibrium $\mathbf{x}^*$ if $\overline{\mathbf{x}}^*$ is a $(2m)^{\text{th}}$-order inflexion point at $(\mathbf{x}_0^*, \mathbf{p}_0)$ with parameter $\mathbf{p}$.

(i_5) For $l = 2m - 1$, the bifurcation at the point $(\mathbf{x}_0^*, \mathbf{p}_0)$ is called a $(2m)^{\text{th}}$-order saddle-node bifurcation of the equilibrium $\mathbf{x}^*$ if the $(2m - 1)$ branches of $\overline{\mathbf{x}}^*$ and $\mathbf{x}^*$ are parabolic-alike of the $(2m)^{\text{th}}$ order with parameter $\mathbf{p}$.

(i_6) For $l = 2m$, the bifurcation at the point $(\mathbf{x}_0^*, \mathbf{p}_0)$ is called a $(2m + 1)^{\text{th}}$-order pitchfork bifurcation of the equilibrium $\mathbf{x}^*$ if the $(2m)$ branch of $\overline{\mathbf{x}}^*$ and $\mathbf{x}^*$ forms a $(2m + 1)^{\text{th}}$-order pitchfork with $(2m + 1)$ branches.

(ii) The point $(\mathbf{x}_0^*, \mathbf{p}_0)$ is called a Hopf bifurcation of the equilibrium $(\mathbf{x}^*, \mathbf{p})$ if

$$\begin{aligned} &\mathbf{a}_0^*(\mathbf{x}^*, \mathbf{p}) \neq \mathbf{0} \text{ and } \lim_{(\mathbf{x}^*, \mathbf{p}) \to (\mathbf{x}_0^*, \mathbf{p}_0),} \mathbf{a}_0^*(\mathbf{x}^*, \mathbf{p}) = \mathbf{0}; \\ &\mathbf{A}^*(\mathbf{x}^*, \mathbf{p}) \neq \mathbf{0} \text{ and } \lim_{(\mathbf{x}^*, \mathbf{p}) \to (\mathbf{x}_0^*, \mathbf{p}_0),} \mathbf{A}^*(\mathbf{x}^*, \mathbf{p}) = \mathbf{0}. \end{aligned} \tag{2.58}$$

The periodic motion near the equilibrium $(\mathbf{x}^*, \mathbf{p})$,

$$\overline{\mathbf{x}}(t) = \mathbf{x}^* + \mathbf{a}_0^*(\mathbf{x}^*, \mathbf{p}) + \sum_{j=1}^{N} \mathbf{b}_j^*(\mathbf{x}^*, \mathbf{p}) \cos(j\beta_k t) + \mathbf{c}_j^*(\mathbf{x}^*, \mathbf{p}) \sin(j\beta_k t) \tag{2.59}$$

is called the limit cycle (or periodic motion) near the equilibrium $(\mathbf{x}^*, \mathbf{p})$.

Definition 2.11 Consider an n-dimensional, autonomous, nonlinear dynamical system $\dot{\mathbf{x}} = \mathbf{f}(\mathbf{x}, \mathbf{p})$ in Eq. (1.4) with an equilibrium point $\mathbf{x}^*$ and $\mathbf{f}(\mathbf{x}, \mathbf{p})$ is C^r-continuous ($r \geq 1$) in a neighborhood of the equilibrium $\mathbf{x}^*$ (i.e., $U(\mathbf{x}^*) \subset \Omega$). The corresponding solution is $\mathbf{x}(t) = \mathbf{\Phi}(\mathbf{x}_0, t - t_0, \mathbf{p})$. Suppose Eq. (1.24) holds in $U(\mathbf{x}^*) \subset \Omega$. For a linearized dynamical system in Eq. (1.19), consider a pair of complex eigenvalues $\alpha_k \pm \mathbf{i}\beta_k$ ($2k - 1, 2k \in N = \{1, 2, \ldots, n\}, \mathbf{i} = \sqrt{-1}$) of matrix $D\mathbf{f}(\mathbf{x}^*, \mathbf{p})$ with a pair of eigenvectors $\mathbf{v}_{2k-1} \pm \mathbf{i}\mathbf{v}_{2k}$. In the vicinity of $\mathbf{x}^*$, there exists a transformation of

$$\mathbf{x} = \mathbf{x}^* + \mathbf{y} = \mathbf{x}^* + \mathbf{a}_0(t) + \sum_{j=1}^{N} \mathbf{b}_j(t) \cos(j\beta_k t) + \mathbf{c}_j(t) \sin(j\beta_k t) \tag{2.60}$$

under $\lim_{j \to \infty} ||\mathbf{A}_j|| = 0$ with Eq. (1.128). The corresponding dynamical system of coefficients is

$$\begin{aligned}
\dot{\mathbf{a}}_0 &= \mathbf{F}_0(\mathbf{a}_0, \mathbf{b}, \mathbf{c}),\\
\dot{\mathbf{b}} &= -\beta_k \mathbf{k}\mathbf{c} + \mathbf{F}^c(\mathbf{a}_0, \mathbf{b}, \mathbf{c}),\\
\dot{\mathbf{c}} &= \beta_k \mathbf{k}\mathbf{b} + \mathbf{F}^s(\mathbf{a}_0, \mathbf{b}, \mathbf{c})
\end{aligned} \tag{2.61}$$

with Eqs. (1.131) and (1.132). If the equilibrium of Eq. (2.61) has a nonzero solution, i.e.,

$$(\mathbf{a}_0^*, \mathbf{b}^*, \mathbf{c}^*) \neq \mathbf{0}, \tag{2.62}$$

the new steady-state motion or equilibrium is

$$\bar{\mathbf{x}} = \mathbf{x}^* + \mathbf{y} = \mathbf{x}^* + \mathbf{a}_0^* + \sum_{j=1}^{N} \mathbf{b}_j^* \cos(j\beta_k t) + \mathbf{c}_j^* \sin(j\beta_k t). \tag{2.63}$$

1. The new solution $\bar{\mathbf{x}}$ is stable if

$$\begin{aligned}
&\lim_{t\to\infty} \|\mathbf{a}_0(t) - \mathbf{a}_0^*\| = 0,\ \lim_{t\to\infty} \|\mathbf{b}(t) - \mathbf{b}^*\| = 0,\ \lim_{t\to\infty} \|\mathbf{c}(t) - \mathbf{c}^*\| = 0\\
&(\text{or } \lim_{t\to\infty} \|\mathbf{a}_0(t) - \mathbf{a}_0^*\| = 0,\ \lim_{t\to\infty} \|\mathbf{A}(t) - \mathbf{A}^*\| = 0).
\end{aligned} \tag{2.64}$$

2. The new solution $\bar{\mathbf{x}}$ is unstable if

$$\begin{aligned}
&\lim_{t\to\infty} \|\mathbf{a}_0(t) - \mathbf{a}_0^*\| = \infty, \text{or } \lim_{t\to\infty} \|\mathbf{b}(t) - \mathbf{b}^*\| = \infty, \text{or } \lim_{t\to\infty} \|\mathbf{c}(t) - \mathbf{c}^*\| = \infty\\
&(\text{or } \lim_{t\to\infty} \|\mathbf{a}_0(t) - \mathbf{a}_0^*\| = \infty, \text{or } \lim_{t\to\infty} \|\mathbf{A}(t) - \mathbf{A}^*\| = \infty).
\end{aligned} \tag{2.65}$$

3. The new equilibrium $\bar{\mathbf{x}}^* = \mathbf{x}^* + \mathbf{a}_0^*$ is stable if

$$\lim_{t\to\infty} \|\mathbf{a}_0(t) - \mathbf{a}_0^*\| = 0, \mathbf{b}(t) = \mathbf{b}^* = \mathbf{0}, \mathbf{c}(t) = \mathbf{c}^* = \mathbf{0}. \tag{2.66}$$

4. The new equilibrium $\bar{\mathbf{x}}^*$ is unstable if

$$\lim_{t\to\infty} \|\mathbf{a}_0(t) - \mathbf{a}_0^*\| = 0, \mathbf{b}(t) = \mathbf{b}^* = \mathbf{0}, \mathbf{c}(t) = \mathbf{c}^* = \mathbf{0}. \tag{2.67}$$

Definition 2.12 Consider an n-dimensional, autonomous, nonlinear dynamical system $\dot{\mathbf{x}} = \mathbf{f}(\mathbf{x}, \mathbf{p})$ in Eq. (1.4) with an equilibrium point $\mathbf{x}^*$ and $\mathbf{f}(\mathbf{x}, \mathbf{p})$ is C^r-continuous ($r \geq 1$) in a neighborhood of the equilibrium $\mathbf{x}^*$ (i.e., $U(\mathbf{x}^*) \subset \Omega$). The corresponding solution is $\mathbf{x}(t) = \mathbf{\Phi}(\mathbf{x}_0, t - t_0, \mathbf{p})$. Suppose Eq. (1.24) holds in $U(\mathbf{x}^*) \subset \Omega$. For a linearized dynamical system in Eq. (1.19), consider a pair of complex eigenvalues $\alpha_k \pm \mathbf{i}\beta_k$ ($2k-1, 2k \in N = \{1, 2, \ldots, n\}$, $\mathbf{i} = \sqrt{-1}$) of matrix

$D\mathbf{f}(\mathbf{x}^*, \mathbf{p})$ with a pair of eigenvectors $\mathbf{v}_{2k-1} \pm \mathbf{i}\mathbf{v}_{2k}$. In the vicinity of $\mathbf{x}^*$, there exists a transformation of

$$\mathbf{x} = \mathbf{x}^* + \mathbf{y} = \mathbf{x}^* + \mathbf{a}_0(t) + \sum_{j=1}^{N} \mathbf{b}_j(t)\cos(j\beta_k t) + \mathbf{c}_j(t)\sin(j\beta_k t) \tag{2.68}$$

under $\lim_{j\to\infty} \|\mathbf{A}_j\| = 0$ with Eq. (1.128). The corresponding dynamical system of coefficients is

$$\begin{aligned} \dot{\mathbf{a}}_0 &= \mathbf{F}_0(\mathbf{a}_0, \mathbf{b}, \mathbf{c}), \\ \dot{\mathbf{b}} &= -\beta_k \mathbf{k}\mathbf{c} + \mathbf{F}^c(\mathbf{a}_0, \mathbf{b}, \mathbf{c}), \\ \dot{\mathbf{c}} &= \beta_k \mathbf{k}\mathbf{b} + \mathbf{F}^s(\mathbf{a}_0, \mathbf{b}, \mathbf{c}) \end{aligned} \tag{2.69}$$

with Eqs. (1.131) and (1.132). The equivalent equation of Eq. (2.69) is

$$\dot{\mathbf{z}} = \mathbf{g}(\mathbf{z}) \tag{2.70}$$

where

$$\mathbf{z} = (\mathbf{a}_0, \mathbf{b}, \mathbf{c})^{\mathrm{T}}, \mathbf{g} = (\mathbf{F}_0, \mathbf{F}^c, \mathbf{F}^s)^{\mathrm{T}}. \tag{2.71}$$

If the equilibrium of Eq. (2.70) has a nonzero solution, i.e.,

$$\mathbf{z}^* \neq \mathbf{0} \Rightarrow (\mathbf{a}_0^*, \mathbf{b}^*, \mathbf{c}^*) \neq \mathbf{0}, \tag{2.72}$$

and in a neighborhood of $\mathbf{z}^*$ the linearized equation with $\Delta\mathbf{z} = \mathbf{z} - \mathbf{z}^*$ is

$$\Delta\dot{\mathbf{z}} = D\mathbf{g}(\mathbf{z}^*)\Delta\mathbf{z}, \tag{2.73}$$

then the eigenvalues of $D\mathbf{g}(\mathbf{z}^*)$ determined by

$$| D\mathbf{g}(\mathbf{z}^*) - \lambda \mathbf{I}_{n(2N+1)\times n(2N+1)} | = 0 \tag{2.74}$$

yield the stability of the steady-state motion $\overline{\mathbf{x}}(t)$ and equilibrium $\overline{\mathbf{x}}^*$ of

$$(n_1 : n_2 : n_3 | n_4 : n_5 : n_6) \tag{2.75}$$

accordingly for $\Sigma_{r=1}^{6} n_r = n(2N+1)$ and $\Sigma_{r=1}^{6} n_r = n$.

The periodic motion $\overline{\mathbf{x}}(t)$ generated by the Hopf bifurcation of the equilibrium $\mathbf{x}^*$ is determined as follows:

(i) If $\text{Re}\lambda_l < 0$ ($l = 1, 2, \ldots, n(2N + 1)$), the periodic motion $\bar{\mathbf{x}}(t)$ near the equilibrium $\mathbf{x}^*$ is stable.
(ii) If $\text{Re}\lambda_l > 0$ ($l \in \{1, 2, \ldots, n(2N + 1)\}$), the periodic motion $\bar{\mathbf{x}}(t)$ near the equilibrium $\mathbf{x}^*$ is unstable.
(iii) The boundary between the stable and unstable periodic motions $\bar{\mathbf{x}}(t)$ is determined by the higher singularity.

The equilibrium $\bar{\mathbf{x}}^*$ generated by the hyperbolic bifurcation of the equilibrium $\mathbf{x}^*$ is determined as follows:

(iv) If $\text{Re}\lambda_l < 0$ ($l = 1, 2, \ldots, n$), the equilibrium $\bar{\mathbf{x}}^* = \mathbf{x}^* + \mathbf{a}_0^*$ is stable.
(v) If $\text{Re}\lambda_l > 0$ ($l \in \{1, 2, \ldots, n\}$), the equilibrium $\bar{\mathbf{x}}^* = \mathbf{x}^* + \mathbf{a}_0^*$ is unstable.
(vi) The boundary between the stable and unstable equilibrium $\bar{\mathbf{x}}^* = \mathbf{x}^* + \mathbf{a}_0^*$ is determined by the higher singularity.

2.5 Hopf Bifurcations in Second-Order Nonlinear Systems

In this section, the Hopf bifurcation of equilibrium for the second-order dynamical systems will be discussed through the finite Fourier series transformation as in Luo (2019).

Definition 2.13 Consider a $2n$-dimensional, autonomous, nonlinear dynamical system $\ddot{\mathbf{x}} = \mathbf{f}(\mathbf{x}, \dot{\mathbf{x}}, \mathbf{p})$ in Eq. (1.146) with an equilibrium point $(\mathbf{x}^*, \mathbf{0})$ and $\mathbf{f}(\mathbf{x}, \dot{\mathbf{x}}, \mathbf{p})$ is C^r-continuous ($r \geq 1$) in a neighborhood of the equilibrium $(\mathbf{x}^*, \mathbf{0})$ (i.e., $U(\mathbf{x}^*, \mathbf{0}) \subset \Omega$). The corresponding solution is $\mathbf{x}(t) = \mathbf{\Phi}(\mathbf{x}_0, t - t_0, \mathbf{p})$. Suppose Eq. (1.148) holds in $U(\mathbf{x}^*, \mathbf{0}) \subset \Omega$. For a linearized dynamical system in Eq. (1.149), consider a pair of complex eigenvalues $\alpha_k \pm \mathbf{i}\beta_k$ ($2k - 1$, $2k \in N = \{1, 2, \ldots, 2n\}$, $\mathbf{i} = \sqrt{-1}$) with a pair of eigenvectors $\mathbf{v}_{2k-1} \pm \mathbf{i}\mathbf{v}_{2k}$. In the vicinity of $(\mathbf{x}^*, \mathbf{0})$, there exists a transformation of

$$\mathbf{x} = \mathbf{x}^* + \mathbf{y} = \mathbf{x}^* + \mathbf{a}_0(t) + \sum_{j=1}^{N} \mathbf{b}_j(t)\cos(j\beta_k t) + \mathbf{c}_j(t)\sin(j\beta_k t) \tag{2.76}$$

under $\lim_{j\to\infty} \|\mathbf{A}_j\| = 0$ with Eq. (1.152). The corresponding dynamical system of coefficients is

$$\begin{aligned}
\ddot{\mathbf{a}}_0 &= \mathbf{F}_0(\mathbf{a}_0, \mathbf{b}, \mathbf{c}, \dot{\mathbf{a}}_0, \dot{\mathbf{b}}, \dot{\mathbf{c}}),\\
\ddot{\mathbf{b}} &= -2\beta_k \mathbf{k}_1 \dot{\mathbf{c}} + \beta_k^2 \mathbf{k}_2 \mathbf{b} + \mathbf{F}^c(\mathbf{a}_0, \mathbf{b}, \mathbf{c}, \dot{\mathbf{a}}_0, \dot{\mathbf{b}}, \dot{\mathbf{c}}),\\
\ddot{\mathbf{c}} &= 2\beta_k \mathbf{k}_1 \dot{\mathbf{b}} + \beta_k^2 \mathbf{k}_2 \mathbf{c} + \mathbf{F}^s(\mathbf{a}_0, \mathbf{b}, \mathbf{c}, \dot{\mathbf{a}}_0, \dot{\mathbf{b}}, \dot{\mathbf{c}})
\end{aligned} \tag{2.77}$$

with Eqs. (1.155) and (1.156). The equilibrium of Eq. (2.77) is given by

$$\begin{aligned}\mathbf{0} &= \mathbf{F}_0(\mathbf{a}_0^*, \mathbf{b}^*, \mathbf{c}^*, \mathbf{0}, \mathbf{0}, \mathbf{0}),\\ \mathbf{0} &= \beta_k^2 \mathbf{k}_2 \mathbf{b}^* + \mathbf{F}^c(\mathbf{a}_0^*, \mathbf{b}^*, \mathbf{c}^*, \mathbf{0}, \mathbf{0}, \mathbf{0}),\\ \mathbf{0} &= \beta_k^2 \mathbf{k}_2 \mathbf{c}^* + \mathbf{F}^s(\mathbf{a}_0^*, \mathbf{b}^*, \mathbf{c}^*, \mathbf{0}, \mathbf{0}, \mathbf{0}).\end{aligned} \tag{2.78}$$

If Eq. (2.78) has a nonzero solution, i.e.,

$$(\mathbf{a}_0^*, \mathbf{b}^*, \mathbf{c}^*) \neq \mathbf{0}, \tag{2.79}$$

then in the vicinity of equilibrium $\mathbf{x}^*$, there is a bifurcation solution of equilibrium $\mathbf{x}^*$.

(i) The point $(\mathbf{x}_0^*, \mathbf{p}_0)$ is called a hyperbolic bifurcation of the equilibrium $(\mathbf{x}^*, \mathbf{p})$ if

$$\begin{aligned}&\mathbf{a}_0^*(\mathbf{x}^*, \mathbf{p}) \neq \mathbf{0} \text{ and } \lim_{(\mathbf{x}^*, \mathbf{p}) \to (\mathbf{x}_0^*, \mathbf{p}_0),} \mathbf{a}_0^*(\mathbf{x}^*, \mathbf{p}) = \mathbf{0};\\ &\mathbf{A}^*(\mathbf{x}^*, \mathbf{p}) = \mathbf{0} \text{ (or } \mathbf{b}^*(\mathbf{x}^*, \mathbf{p}) = \mathbf{0}, \mathbf{c}^*(\mathbf{x}^*, \mathbf{p}) = \mathbf{0})\\ &\text{or } \beta_k = 0.\end{aligned} \tag{2.80}$$

The new equilibrium

$$\overline{\mathbf{x}}^* = \mathbf{x}^* + \mathbf{a}_0^*(\mathbf{x}^*, \mathbf{p}) \tag{2.81}$$

is called a bifurcation branch of the equilibrium $(\mathbf{x}^*, \mathbf{p})$. If there are l new branches of equilibrium points $\overline{\mathbf{x}}^*$, such a bifurcation is called an l-branch bifurcation of the equilibrium $\mathbf{x}^*$.

(i_1) For $l = 1$, the bifurcation at the point $(\mathbf{x}_0^*, \mathbf{p}_0)$ is called a transcritical bifurcation of the equilibrium $\mathbf{x}^*$ if $\overline{\mathbf{x}}^*$ is linear with parameter $\mathbf{p}$.
(i_2) For $l = 1$, the bifurcation at the point $(\mathbf{x}_0^*, \mathbf{p}_0)$ is called a saddle-node bifurcation of the equilibrium $\mathbf{x}^*$ if $\overline{\mathbf{x}}^*$ and $\mathbf{x}^*$ are parabolic with parameter $\mathbf{p}$.
(i_3) For $l = 2$, the bifurcation at the point $(\mathbf{x}_0^*, \mathbf{p}_0)$ is called a pitchfork bifurcation of the equilibrium $\mathbf{x}^*$ if two new equilibriums $\overline{\mathbf{x}}^*$ and $\mathbf{x}^*$ form a pitchfork.
(i_4) For $l = 2m - 1$, the bifurcation at the point $(\mathbf{x}_0^*, \mathbf{p}_0)$ is called the $(2m)^{\text{th}}$-order transcritical bifurcation of the equilibrium $\mathbf{x}^*$ if $\overline{\mathbf{x}}^*$ is a $(2m)^{\text{th}}$-order inflexion point at $(\mathbf{x}_0^*, \mathbf{p}_0)$ with parameter $\mathbf{p}$.
(i_5) For $l = 2m - 1$, the bifurcation at the point $(\mathbf{x}_0^*, \mathbf{p}_0)$ is called a $(2m)^{\text{th}}$-order saddle-node bifurcation of the equilibrium $\mathbf{x}^*$ if the $(2m - 1)$ branches of $\overline{\mathbf{x}}^*$ and $\mathbf{x}^*$ are parabolic-alike of the $(2m)^{\text{th}}$ order with parameter $\mathbf{p}$.
(i_6) For $l = 2m$, the bifurcation at the point $(\mathbf{x}_0^*, \mathbf{p}_0)$ is called a $(2m + 1)^{\text{th}}$-order pitchfork bifurcation of the equilibrium $\mathbf{x}^*$ if the $(2m)$ branch of $\overline{\mathbf{x}}^*$ and $\mathbf{x}^*$ forms a $(2m + 1)^{\text{th}}$-order pitchfork with $(2m + 1)$ branches.

(ii) The point $(\mathbf{x}_0^*, \mathbf{p}_0)$ is called a Hopf bifurcation of the equilibrium $(\mathbf{x}^*, \mathbf{p})$ if

$$\begin{aligned}
&\mathbf{a}_0^*(\mathbf{x}^*, \mathbf{p}) \neq \mathbf{0} \text{ and } \lim_{(\mathbf{x}^*, \mathbf{p}) \to (\mathbf{x}_0^*, \mathbf{p}_0),} \mathbf{a}_0^*(\mathbf{x}^*, \mathbf{p}) = \mathbf{0};\\
&\mathbf{A}^*(\mathbf{x}^*, \mathbf{p}) \neq \mathbf{0} \text{ and } \lim_{(\mathbf{x}^*, \mathbf{p}) \to (\mathbf{x}_0^*, \mathbf{p}_0),} \mathbf{A}^*(\mathbf{x}^*, \mathbf{p}) = \mathbf{0}.
\end{aligned} \tag{2.82}$$

The periodic motion near the equilibrium $(\mathbf{x}^*, \mathbf{p})$,

$$\bar{\mathbf{x}}(t) = \mathbf{x}^* + \mathbf{a}_0^*(\mathbf{x}^*, \mathbf{p}) + \sum_{j=1}^{N} \mathbf{b}_j^*(\mathbf{x}^*, \mathbf{p}) \cos(j\beta_k t) + \mathbf{c}_j^*(\mathbf{x}^*, \mathbf{p}) \sin(j\beta_k t), \tag{2.83}$$

is called the limit cycle (or periodic motion) near the equilibrium $(\mathbf{x}^*, \mathbf{p})$.

Definition 2.14 Consider a $2n$-dimensional, autonomous, nonlinear dynamical system $\ddot{\mathbf{x}} = \mathbf{f}(\mathbf{x}, \dot{\mathbf{x}}, \mathbf{p})$ in Eq. (1.146) with an equilibrium point $(\mathbf{x}^*, \mathbf{0})$ and $\mathbf{f}(\mathbf{x}, \dot{\mathbf{x}}, \mathbf{p})$ is C^r-continuous $(r \geq 1)$ in a neighborhood of the equilibrium $(\mathbf{x}^*, \mathbf{0})$ (i.e., $U(\mathbf{x}^*, \mathbf{0}) \subset \Omega$). The corresponding solution is $\mathbf{x}(t) = \mathbf{\Phi}(\mathbf{x}_0, t - t_0, \mathbf{p})$. Suppose Eq. (1.148) holds in $U(\mathbf{x}^*, \mathbf{0}) \subset \Omega$. For a linearized dynamical system in Eq. (1.149), consider a pair of complex eigenvalues $\alpha_k \pm \mathbf{i}\beta_k$ $(2k-1,$ $2k \in N = \{1, 2, \ldots, 2n\}$, $\mathbf{i} = \sqrt{-1})$ with a pair of eigenvectors $\mathbf{v}_{2k-1} \pm \mathbf{i}\mathbf{v}_{2k}$. In the vicinity of $(\mathbf{x}^*, \mathbf{0})$, there exists a transformation of

$$\mathbf{x} = \mathbf{x}^* + \mathbf{y} = \mathbf{x}^* + \mathbf{a}_0(t) + \sum_{j=1}^{N} \mathbf{b}_j(t) \cos(j\beta_k t) + \mathbf{c}_j(t) \sin(j\beta_k t) \tag{2.84}$$

under $\lim_{j \to \infty} \|\mathbf{A}_j\| = 0$ with Eq. (1.152). The corresponding dynamical system of coefficients is

$$\begin{aligned}
\ddot{\mathbf{a}}_0 &= \mathbf{F}_0(\mathbf{a}_0, \mathbf{b}, \mathbf{c}, \dot{\mathbf{a}}_0, \dot{\mathbf{b}}, \dot{\mathbf{c}}),\\
\ddot{\mathbf{b}} &= -2\beta_k \mathbf{k}_1 \dot{\mathbf{c}} + \beta_k^2 \mathbf{k}_2 \mathbf{b} + \mathbf{F}^c(\mathbf{a}_0, \mathbf{b}, \mathbf{c}, \dot{\mathbf{a}}_0, \dot{\mathbf{b}}, \dot{\mathbf{c}}),\\
\ddot{\mathbf{c}} &= 2\beta_k \mathbf{k}_1 \dot{\mathbf{b}} + \beta_k^2 \mathbf{k}_2 \mathbf{c} + \mathbf{F}^s(\mathbf{a}_0, \mathbf{b}, \mathbf{c}, \dot{\mathbf{a}}_0, \dot{\mathbf{b}}, \dot{\mathbf{c}})
\end{aligned} \tag{2.85}$$

with Eqs. (1.155) and (1.156). If the equilibrium of Eq. (2.85) has a nonzero solution, i.e.,

$$(\mathbf{a}_0^*, \mathbf{b}^*, \mathbf{c}^*) \neq \mathbf{0}, \tag{2.86}$$

the new steady-state motion or equilibrium is

$$\bar{\mathbf{x}} = \mathbf{x}^* + \mathbf{y} = \mathbf{x}^* + \mathbf{a}_0^* + \sum_{j=1}^{N} \mathbf{b}_j^* \cos(j\beta_k t) + \mathbf{c}_j^* \sin(j\beta_k t). \tag{2.87}$$

(i) The new solution $\bar{\mathbf{x}}$ is stable if

$$\begin{aligned}&\lim_{t\to\infty}\|\mathbf{a}_0(t)-\mathbf{a}_0^*\|=0,\ \lim_{t\to\infty}\|\mathbf{b}(t)-\mathbf{b}^*\|=0,\ \lim_{t\to\infty}\|\mathbf{c}(t)-\mathbf{c}^*\|=0\\&(\text{or }\lim_{t\to\infty}\|\mathbf{a}_0(t)-\mathbf{a}_0^*\|=0,\ \lim_{t\to\infty}\|\mathbf{A}(t)-\mathbf{A}^*\|=0).\end{aligned}\tag{2.88}$$

(ii) The new solution $\bar{\mathbf{x}}$ is unstable if

$$\begin{aligned}&\lim_{t\to\infty}\|\mathbf{a}_0(t)-\mathbf{a}_0^*\|=\infty,\text{ or }\lim_{t\to\infty}\|\mathbf{b}(t)-\mathbf{b}^*\|=\infty,\text{ or }\lim_{t\to\infty}\|\mathbf{c}(t)-\mathbf{c}^*\|=\infty\\&(\text{or }\lim_{t\to\infty}\|\mathbf{a}_0(t)-\mathbf{a}_0^*\|=\infty,\text{ or }\lim_{t\to\infty}\|\mathbf{A}(t)-\mathbf{A}^*\|=\infty).\end{aligned}\tag{2.89}$$

(iii) The new equilibrium $\bar{\mathbf{x}}^*=\mathbf{x}^*+\mathbf{a}_0^*$ is stable if

$$\lim_{t\to\infty}\|\mathbf{a}_0(t)-\mathbf{a}_0^*\|=0,\mathbf{b}(t)=\mathbf{b}^*=\mathbf{0},\mathbf{c}(t)=\mathbf{c}^*=\mathbf{0}.\tag{2.90}$$

(iv) The new equilibrium $\bar{\mathbf{x}}^*=\mathbf{x}^*+\mathbf{a}_0^*$ is unstable if

$$\lim_{t\to\infty}\|\mathbf{a}_0(t)-\mathbf{a}_0^*\|=\infty,\mathbf{b}(t)=\mathbf{b}^*=\mathbf{0},\mathbf{c}(t)=\mathbf{c}^*=\mathbf{0}.\tag{2.91}$$

Definition 2.15 Consider a $2n$-dimensional, autonomous, nonlinear dynamical system $\ddot{\mathbf{x}}=\mathbf{f}(\mathbf{x},\dot{\mathbf{x}},\mathbf{p})$ in Eq. (1.146) with an equilibrium point $(\mathbf{x}^*,\mathbf{0})$ and $\mathbf{f}(\mathbf{x},\dot{\mathbf{x}},\mathbf{p})$ is $C^r(r\geq 1)$-continuous in a neighborhood of the equilibrium $(\mathbf{x}^*,\mathbf{0})$ (i.e., $U(\mathbf{x}^*,\mathbf{0})\subset\Omega$). The corresponding solution is $\mathbf{x}(t)=\mathbf{\Phi}(\mathbf{x}_0,t-t_0,\mathbf{p})$. Suppose Eq. (1.148) holds in $U(\mathbf{x}^*,\mathbf{0})\subset\Omega$. For a linearized dynamical system in Eq. (1.149), consider a pair of complex eigenvalues $\alpha_k\pm\mathbf{i}\beta_k$ $(2k-1,$ $2k\in N=\{1,2,\ldots,2n\}$, $\mathbf{i}=\sqrt{-1})$ with a pair of eigenvectors $\mathbf{v}_{2k-1}\pm\mathbf{i}\mathbf{v}_{2k}$. In the vicinity of $(\mathbf{x}^*,\mathbf{0})$, there exists a transformation of

$$\mathbf{x}=\mathbf{x}^*+\mathbf{y}=\mathbf{x}^*+\mathbf{a}_0(t)+\sum_{j=1}^{N}\mathbf{b}_j(t)\cos(j\beta_k t)+\mathbf{c}_j(t)\sin(j\beta_k t)\tag{2.92}$$

under $\lim_{j\to\infty}\|\mathbf{A}_j\|=0$ with Eq. (1.152). The corresponding dynamical system of coefficients is

$$\begin{aligned}\ddot{\mathbf{a}}_0&=\mathbf{F}_0(\mathbf{a}_0,\mathbf{b},\mathbf{c},\dot{\mathbf{a}}_0,\dot{\mathbf{b}},\dot{\mathbf{c}}),\\\ddot{\mathbf{b}}&=-2\beta_k\mathbf{k}_1\dot{\mathbf{c}}+\beta_k^2\mathbf{k}_2\mathbf{b}+\mathbf{F}^c(\mathbf{a}_0,\mathbf{b},\mathbf{c},\dot{\mathbf{a}}_0,\dot{\mathbf{b}},\dot{\mathbf{c}}),\\\ddot{\mathbf{c}}&=2\beta_k\mathbf{k}_1\dot{\mathbf{b}}+\beta_k^2\mathbf{k}_2\mathbf{c}+\mathbf{F}^s(\mathbf{a}_0,\mathbf{b},\mathbf{c},\dot{\mathbf{a}}_0,\dot{\mathbf{b}},\dot{\mathbf{c}})\end{aligned}\tag{2.93}$$

with Eqs. (1.155) and (1.156). The equivalent equation of Eq. (2.93) is

$$\ddot{\mathbf{z}} = \mathbf{g}(\mathbf{z}, \dot{\mathbf{z}}) \tag{2.94}$$

where

$$\mathbf{z} = (\mathbf{a}_0, \mathbf{b}, \mathbf{c})^{\mathrm{T}}, \dot{\mathbf{z}} = (\dot{\mathbf{a}}_0, \dot{\mathbf{b}}, \dot{\mathbf{c}})^{\mathrm{T}}, \mathbf{g} = (\mathbf{F}_0, \mathbf{F}^c, \mathbf{F}^s)^{\mathrm{T}}. \tag{2.95}$$

The equivalent first-order equation of Eq. (2.94) is

$$\dot{\mathbf{u}} = \bar{\mathbf{f}}(\mathbf{u}) \tag{2.96}$$

where

$$\mathbf{u} = (\mathbf{z}, \dot{\mathbf{z}})^{\mathrm{T}}; \bar{\mathbf{f}} = (\dot{\mathbf{z}}, \mathbf{g})^{\mathrm{T}}. \tag{2.97}$$

If the equilibrium of Eq. (2.94) has a nonzero solution, i.e.,

$$\mathbf{z}^* \neq \mathbf{0} \Rightarrow (\mathbf{a}_0^*, \mathbf{b}^*, \mathbf{c}^*) \neq \mathbf{0}, \tag{2.98}$$

and in a neighborhood of $\mathbf{u}^*$ the linearized equation with $\Delta\mathbf{u} = \mathbf{u} - \mathbf{u}^*$ is

$$\Delta\dot{\mathbf{u}} = D\mathbf{f}(\mathbf{u}^*)\Delta\mathbf{u} \tag{2.99}$$

with

$$D\bar{\mathbf{f}}(\mathbf{u}^*) = \begin{bmatrix} \mathbf{0}_{n(2N+1)\times n(2N+1)} & \mathbf{I}_{n(2N+1)\times n(2N+1)} \\ \mathbf{G}_{n(2N+1)\times n(2N+1)} & \mathbf{H}_{n(2N+1)\times n(2N+1)} \end{bmatrix} \tag{2.100}$$

and

$$\mathbf{G} = \left.\frac{\partial \mathbf{g}}{\partial \mathbf{z}}\right|_{\mathbf{u}^*}, \mathbf{H} = \left.\frac{\partial \mathbf{g}}{\partial \dot{\mathbf{z}}}\right|_{\mathbf{u}^*} \tag{2.101}$$

then the eigenvalues of $D\bar{\mathbf{f}}(\mathbf{u}^*)$ determined by

$$|\, D\bar{\mathbf{f}}(\mathbf{u}^*) - \lambda \mathbf{I}_{2n(2N+1)\times 2n(2N+1)} \,| = 0 \tag{2.102}$$

yield the stability of the steady-state motion $\bar{\mathbf{x}}(t)$ and equilibrium $\bar{\mathbf{x}}^*$ of

$$(n_1 : n_2 : n_3 | n_4 : n_5 : n_6) \tag{2.103}$$

accordingly for $\Sigma_{r=1}^{6} n_r = 2n(2N+1)$.

The periodic motion $\bar{\mathbf{x}}(t)$ generated by the Hopf bifurcation of the equilibrium $\mathbf{x}^*$ is determined as follows:

(i) If $\text{Re}\lambda_l < 0$ ($l = 1, 2, \ldots, 2n(2N + 1)$), the periodic motion $\bar{\mathbf{x}}(t)$ near the equilibrium $\mathbf{x}^*$ is stable.
(ii) If $\text{Re}\lambda_l > 0$ ($l \in \{1, 2, \ldots, 2n(2N + 1)\}$), the periodic motion $\bar{\mathbf{x}}(t)$ near the equilibrium $\mathbf{x}^*$ is unstable.
(iii) The boundary between the stable and unstable periodic motions $\bar{\mathbf{x}}(t)$ is determined by the higher singularity.

The equilibrium $\bar{\mathbf{x}}^* = \mathbf{x}^* + \mathbf{a}_0^*$ generated by the hyperbolic bifurcation of the equilibrium $\mathbf{x}^*$ is determined as follows:

(iv) If $\text{Re}\lambda_l < 0$ ($l = 1, 2, \ldots, 2n$), the equilibrium $\bar{\mathbf{x}}^* = \mathbf{x}^* + \mathbf{a}_0^*$ is stable.
(v) If $\text{Re}\lambda_l > 0$ ($l \in \{1, 2, \ldots, 2n\}$), the equilibrium $\bar{\mathbf{x}}^* = \mathbf{x}^* + \mathbf{a}_0^*$ is unstable.
(vi) The boundary between the stable and unstable equilibrium $\bar{\mathbf{x}}^*$ is determined by the higher singularity.

Reference

Luo, A.C.J., 2019, On stability and bifurcation of equilibriums in nonlinear systems, *Journal of Vibration Testing and System Dynamics*, **3**(2), pp. 147–232.

Chapter 3
Low-Dimensional Dynamical Systems

In this chapter, low-dimensional nonlinear dynamical systems are discussed. The stability and bifurcations of the 1-dimensional systems are presented. The higher order singularity and stability for 1-dimensional nonlinear systems are developed. The stability and bifurcations of 2-dimensional systems are discussed and the Hopf bifurcation of equilibriums is presented.

3.1 1-Dimensional Nonlinear Systems

In this section, the stability and bifurcation with singularity for equilibriums will be discussed in 1-dimensional nonlinear systems and a few examples will be presented as in Luo (2012, 2019).

3.1.1 Stability and Singularity

The stability of equilibrium in 1-dimensional systems will be discussed for a better understanding of the stability and singularity theory in nonlinear dynamical systems.

Definition 3.1 Consider a *1*-dimensional, autonomous, nonlinear dynamical system

$$\dot{x} = f(x, \mathbf{p}) \tag{3.1}$$

with an equilibrium x^* of $f(x^*, \mathbf{p}) = 0$ and $f(x, \mathbf{p})$ is C^r-continuous ($r \geq 1$) in a neighborhood of x^* (i.e., $U(x^*) = (x^* - \delta, x^* + \delta) \subset \Omega$ for an arbitrary $\delta > 0$). The corresponding solution is $x(t) = \Phi(x_0, t - t_0, \mathbf{p})$. Let $y = x - x^*$.

A. C. J. Luo, *Bifurcation and Stability in Nonlinear Dynamical Systems*, Nonlinear Systems and Complexity 28, https://doi.org/10.1007/978-3-030-22910-8_3

(i) $x(t)$ at the equilibrium x^* is stable if

$$\begin{aligned} x(t+\varepsilon) - x(t) < 0 \ \text{ for } y = x(t) - x^* > 0; \\ x(t+\varepsilon) - x(t) > 0 \ \text{ for } y = x(t) - x^* < 0, \end{aligned} \tag{3.2}$$

for all $x \in U(x^*) \subset \Omega$ and all $t \in [t_0, \infty)$. The equilibrium x^* is called a sink (or stable node).

(ii) $x(t)$ at the equilibrium x^* is unstable if

$$\begin{aligned} x(t+\varepsilon) - x(t) > 0 \ \text{ for } y = x(t) - x^* > 0; \\ x(t+\varepsilon) - x(t) < 0 \ \text{ for } y = x(t) - x^* < 0, \end{aligned} \tag{3.3}$$

for all $x \in U(x^*) \subset \Omega$ and all $t \in [t_0, \infty)$. The equilibrium x^* is called a source (or unstable node).

(iii) $x(t)$ at the equilibrium x^* is increasingly unstable if

$$\begin{aligned} x(t+\varepsilon) - x(t) > 0 \ \text{ for } y = x(t) - x^* > 0; \\ x(t+\varepsilon) - x(t) > 0 \ \text{ for } y = x(t) - x^* < 0, \end{aligned} \tag{3.4}$$

for all $x \in U(x^*) \subset \Omega$ and all $t \in [t_0, \infty)$. The equilibrium x^* is called an increasing saddle (or upper saddle).

(iv) $x(t)$ at the equilibrium x^* is decreasingly unstable if

$$\begin{aligned} x(t+\varepsilon) - x(t) < 0 \ \text{ for } y = x(t) - x^* > 0; \\ x(t+\varepsilon) - x(t) < 0 \ \text{ for } y = x(t) - x^* < 0, \end{aligned} \tag{3.5}$$

for all $x \in U(x^*) \subset \Omega$ and all $t \in [t_0, \infty)$. The equilibrium x^* is called a decreasing saddle (or lower saddle).

(v) $x(t)$ at the equilibrium x^* is invariant if

$$x(t+\varepsilon) - x(t) = 0 \ \text{ for } y = x(t) - x^* \neq 0, \tag{3.6}$$

for all $x \in U(x^*) \subset \Omega$ and all $t \in [t_0, \infty)$. $x(t)$ in domain $x \in U(x^*) \subset \Omega$ is said to be static.

Theorem 3.1 *Consider a 1-dimensional, autonomous, nonlinear dynamical system $\dot{x} = f(x, \mathbf{p})$ in Eq.* (3.1) *with an equilibrium x^* and $f(x, \mathbf{p})$ is C^r-continuous ($r \geq 1$) in a neighborhood of x^** (i.e., $U(x^*) \subset \Omega$). *The corresponding solution is $x(t) = \Phi(x_0, t - t_0, \mathbf{p})$. In a neighborhood of x^*, $y = x - x^*$.*

(i) *x(t) at the equilibrium x^* is stable if and only if*

$$\begin{aligned} G(x,\mathbf{p}) &= \dot{x} = f(x,\mathbf{p}) < 0 \ \text{ for } y = x(t) - x^* > 0; \\ G(x,\mathbf{p}) &= \dot{x} = f(x,\mathbf{p}) > 0 \ \text{ for } y = x(t) - x^* < 0 \end{aligned} \tag{3.7}$$

for all $x \in U(x^*) \subset \Omega$ and all $t \in [t_0, \infty)$.

(ii) *x(t) at the equilibrium x^* is unstable if and only if*

$$\begin{aligned} G(x,\mathbf{p}) &= \dot{x} = f(x,\mathbf{p}) > 0 \ \text{ for } y = x(t) - x^* > 0; \\ G(x,\mathbf{p}) &= \dot{x} = f(x,\mathbf{p}) < 0 \ \text{ for } y = x(t) - x^* < 0 \end{aligned} \tag{3.8}$$

for all $x \in U(x^*) \subset \Omega$ and all $t \in [t_0, \infty)$.

(iii) *x(t) at the equilibrium x^* is increasingly unstable if and only if*

$$\begin{aligned} G(x,\mathbf{p}) &= \dot{x} = f(x,\mathbf{p}) > 0 \ \text{ for } y = x(t) - x^* > 0; \\ G(x,\mathbf{p}) &= \dot{x} = f(x,\mathbf{p}) > 0 \ \text{ for } y = x(t) - x^* < 0 \end{aligned} \tag{3.9}$$

for all $x \in U(x^*) \subset \Omega$ and all $t \in [t_0, \infty)$.

(iv) *x(t) at the equilibrium x^* is decreasingly unstable if and only if*

$$\begin{aligned} G(x,\mathbf{p}) &= \dot{x} = f(x,\mathbf{p}) < 0 \ \text{ for } y = x(t) - x^* > 0; \\ G(x,\mathbf{p}) &= \dot{x} = f(x,\mathbf{p}) < 0 \ \text{ for } y = x(t) - x^* < 0 \end{aligned} \tag{3.10}$$

for all $x \in U(x^*) \subset \Omega$ and all $t \in [t_0, \infty)$.

(v) *x(t) at the equilibrium x^* is invariant if*

$$G(x,\mathbf{p}) = \dot{x} = f(x,\mathbf{p}) = 0 \tag{3.11}$$

for all $\mathbf{x} \in U(\mathbf{x}^*) \subset \Omega$ and all $t \in [t_0, \infty)$.

Proof Because

$$\begin{aligned} x(t+\varepsilon) - x(t) &= x(t) + \dot{x}(t)\varepsilon + o(\varepsilon) - x(t) \\ &= \dot{x}(t)\varepsilon + o(\varepsilon), \end{aligned}$$

and $\dot{x} = f(x,\mathbf{p})$, we have

$$\begin{aligned} x(t+\varepsilon) - x(t) &= f(x,\mathbf{p})\varepsilon + o(\varepsilon) \\ &= G(x,\mathbf{p})\varepsilon + o(\varepsilon). \end{aligned}$$

(i) Due to any selection of $\varepsilon > 0$, for $y = x(t) - x^* < 0$,

$$x(t+\varepsilon) - x(t) > 0 \text{ if } G(x,\mathbf{p}) = f(x,\mathbf{p}) > 0,$$

and vice versa; and for $y = x(t) - x^* > 0$,

$$x(t+\varepsilon) - x(t) < 0 \text{ if } G(x,\mathbf{p}) = f(x,\mathbf{p}) < 0,$$

and vice versa.

(ii) For $y = x(t) - x^* > 0$,

$$x(t+\varepsilon) - x(t) > 0 \text{ if } G(x,\mathbf{p}) = f(x,\mathbf{p}) > 0,$$

and vice versa; and for $y = x(t) - x^* < 0$,

$$x(t+\varepsilon) - x(t) < 0 \text{ if } G(x,\mathbf{p}) = f(x,\mathbf{p}) < 0,$$

and vice versa.

(iii) For $y = x(t) - x^* > 0$,

$$x(t+\varepsilon) - x(t) > 0 \text{ if } G(x,\mathbf{p}) = f(x,\mathbf{p}) > 0,$$

and vice versa; and for $y = x(t) - x^* < 0$,

$$x(t+\varepsilon) - x(t) > 0 \text{ if } G(x,\mathbf{p}) = f(x,\mathbf{p}) > 0,$$

and vice versa.

(iv) For $y = x(t) - x^* > 0$,

$$x(t+\varepsilon) - x(t) < 0 \text{ if } G(x,\mathbf{p}) = f(x,\mathbf{p}) < 0,$$

and vice versa; and for $y = x(t) - x^* < 0$,

$$x(t+\varepsilon) - x(t) < 0 \text{ if } G(x,\mathbf{p}) = f(x,\mathbf{p}) < 0,$$

and vice versa.

(v) For $y = x(t) - x^* > 0$,

$$x(t+\varepsilon) - x(t) = 0 \text{ if } G(x,\mathbf{p}) = f(x,\mathbf{p}) = 0,$$

and vice versa. Similarly, for $y = x(t) - x^* < 0$,

$$x(t+\varepsilon) - x(t) = 0 \text{ if } G(x,\mathbf{p}) = f(x,\mathbf{p}) = 0,$$

and vice versa. The theorem is proved. ∎

Theorem 3.2 *Consider a 1-dimensional, autonomous, nonlinear dynamical system* $\dot{x} = f(x, \mathbf{p})$ *in Eq. (3.1) with an equilibrium* x^* *and* $f(x, \mathbf{p})$ *is* C^r*-continuous* ($r \geq 1$) *in a neighborhood of* x^* *(i.e.,* $U(x^*) \subset \Omega$*). The corresponding solution is* $x(t) = \Phi(x_0, t - t_0, \mathbf{p})$*. In a neighborhood of* x^*, $y = x - x^*$. *Under the condition in Eq. (1.24), the linearized equation of Eq. (3.1) is*

$$\dot{y} = Df(x^*, \mathbf{p})y. \tag{3.12}$$

There is a real eigenvalue $\lambda = Df(x^*, \mathbf{p}) = G^{(1)}(x^*, \mathbf{p})$ *for a specific* $\mathbf{p}$. *Suppose* $|G^{(2)}(x^*, \mathbf{p})| < \infty$ *where* $G^{(2)}(x^*, \mathbf{p}) = D^2 f(x^*, \mathbf{p})$.

(i) *$x(t)$ at the equilibrium x^* is stable if and only if*

$$G^{(1)}(x^*, \mathbf{p}) = Df(x^*, \mathbf{p}) = \lambda < 0 \tag{3.13}$$

for all $x \in U(x^*) \subset \Omega$ *and all* $t \in [t_0, \infty)$.

(ii) *$x(t)$ at the equilibrium x^* is unstable if and only if*

$$G^{(1)}(x^*, \mathbf{p}) = Df(x^*, \mathbf{p}) = \lambda > 0 \tag{3.14}$$

for all $x \in U(x^*) \subset \Omega$ *and all* $t \in [t_0, \infty)$.

(iii) *$x(t)$ at the equilibrium x^* is increasingly unstable if and only if*

$$\begin{gathered} G^{(1)}(x^*, \mathbf{p}) = Df(x^*, \mathbf{p}) = \lambda = 0 \\ \text{and } G^{(2)}(x^*, \mathbf{p}) = D^2 f(x^*, \mathbf{p}) > 0 \end{gathered} \tag{3.15}$$

for all $x \in U(x^*) \subset \Omega$ *and all* $t \in [t_0, \infty)$.

(iv) *$x(t)$ at the equilibrium x^* is decreasingly unstable if and only if*

$$\begin{gathered} G^{(1)}(x^*, \mathbf{p}) = Df(x^*, \mathbf{p}) = \lambda = 0 \\ \text{and } G^{(2)}(x^*, \mathbf{p}) = D^2 f(x^*, \mathbf{p}) < 0 \end{gathered} \tag{3.16}$$

for all $x \in U(x^*) \subset \Omega$ *and all* $t \in [t_0, \infty)$.

(v) $x(t)$ at the equilibrium x^* is invariant if and only if

$$G^{(m)}(x^*, \mathbf{p}) = D^m f(x^*, \mathbf{p}) = 0 \ \ (m = 0, 1, 2, \ldots) \tag{3.17}$$

for all $x \in U(x^*) \subset \Omega$ *and all* $t \in [t_0, \infty)$.

Proof For $x = x^*$, $y = 0$. Using the Taylor series expansion gives

$$\begin{aligned} \dot{y} &= f(x^* + y, \mathbf{p}) - f(x^*, \mathbf{p}) \\ &= Df(x^*, \mathbf{p})y + o(y) \\ &= G^{(1)}(x^*, \mathbf{p})y + o(y), \end{aligned}$$

and

$$G^{(1)}(x^*, \mathbf{p}) = Df(x^*, \mathbf{p}) = \lambda.$$

Thus, $\dot{y} = G^{(1)}(x^*, \mathbf{p})y + o(y) = \lambda y + o(y)$.

(i) For $y = x - x^* > 0$,

$$G(x, \mathbf{p}) = \dot{y} = \lambda y < 0,$$

and for $y = x - x^* < 0$,

$$G(x, \mathbf{p}) = \dot{y} = \lambda y > 0.$$

Thus, $G^{(1)}(x^*, \mathbf{p}) = Df(x^*, \mathbf{p}) = \lambda < 0$.

(ii) For $y = x - x^* > 0$,

$$G(x, \mathbf{p}) = \dot{y} = \lambda y > 0,$$

and for $y = x - x^* < 0$,

$$G(x, \mathbf{p}) = \dot{y} = \lambda y < 0,$$

Thus, $G^{(1)}(x^*, \mathbf{p}) = Df(x^*, \mathbf{p}) = \lambda > 0$.

(iii) For $y = x - x^* > 0$,

$$G(x, \mathbf{p}) = \dot{y} = \lambda y > 0,$$

and for $y = x - x^* < 0$,

$$G(x, \mathbf{p}) = \dot{y} = \lambda y > 0.$$

Thus, $G^{(1)}(x^*, \mathbf{p}) = Df(x^*, \mathbf{p}) = \lambda = 0$ and the higher order derivatives should be considered. The higher order Taylor series expansion gives

$$\begin{aligned}\dot{y} &= f(x, \mathbf{p}) - f(x^*, \mathbf{p}) \\ &= Df(x^*, \mathbf{p})y + \frac{1}{2!}D^2(x^*, \mathbf{p})y^2 + o(y^2) \\ &= \frac{1}{2!}D^2(x^*, \mathbf{p})y^2 + o(y^2) \\ &= \frac{1}{2!}G^{(2)}(\mathbf{x}^*, \mathbf{p})y^2 + o(y^2).\end{aligned}$$

For $y = x - x^* < 0$,

$$G(\mathbf{x}, \mathbf{p}) = \dot{y} = \frac{1}{2!} G^{(2)}(x^*, \mathbf{p}) y^2 > 0,$$

and for $y = x - x^* > 0$,

$$G(\mathbf{x}, \mathbf{p}) = \dot{y} = \frac{1}{2!} G^{(2)}(x^*, \mathbf{p}) y^2 > 0.$$

So we have

$$G^{(2)}(x^*, \mathbf{p}) = D^2 f(x^*, \mathbf{p}) > 0.$$

(iv) Similar to (iii), we have $G_k^{(1)}(\mathbf{x}^*, \mathbf{p}) = \lambda_k = 0$. In addition, for $y = x - x^* < 0$,

$$G(\mathbf{x}, \mathbf{p}) = \dot{y} = \frac{1}{2!} G^{(2)}(x^*, \mathbf{p}) y^2 < 0.$$

and for $y = x - x^* > 0$,

$$G(\mathbf{x}, \mathbf{p}) = \dot{y} = \frac{1}{2!} G^{(2)}(x^*, \mathbf{p}) y^2 < 0.$$

So we have

$$G^{(2)}(x^*, \mathbf{p}) = D^2 f(x^*, \mathbf{p}) < 0.$$

(v) Using the Taylor series expansion yields

$$\begin{aligned}
\dot{y} &= f(x, \mathbf{p}) - f(x^*, \mathbf{p}) \\
&= \sum\nolimits_{m=1}^{N} \frac{1}{m!} D^m f(x^*, \mathbf{p}) y^m + o(y^N) \\
&= \sum\nolimits_{m=1}^{N} \frac{1}{m!} G^{(m)}(x^*, \mathbf{p}) y^m + o(y^N) \\
&= 0 \\
&(N = 1, 2, \ldots).
\end{aligned}$$

Thus only if

$$G^{(m)}(x^*, \mathbf{p}) = D^m f(x^*, \mathbf{p}) = 0 \; (m = 1, 2, \ldots),$$

the above equation of $\dot{y} = 0$ holds, and vice versa. The theorem is proved. ∎

Definition 3.2 Consider a 1-dimensional, autonomous, nonlinear dynamical system $\dot{x} = f(x, \mathbf{p})$ in (3.1) with an equilibrium x^* and $f(x, \mathbf{p})$ is C^r-continuous ($r \geq 1$) in a neighborhood of x^* (i.e., $U(x^*) \subset \Omega$). The corresponding solution is $x(t) = \Phi(x_0, t - t_0, \mathbf{p})$. In a neighborhood of x^*, $y = x - x^*$. Under the condition in Eq. (1.24), the linearized equation of Eq. (3.1) is in Eq. (3.12). There is a real eigenvalue $\lambda = Df(x^*, \mathbf{p}) = G^{(1)}(x^*, \mathbf{p})$ for a specific $\mathbf{p}$. Suppose $|G^{(m)}(x^*, \mathbf{p})| < \infty$ where $G^{(r)}(x^*, \mathbf{p}) = D^r f(x^*, \mathbf{p})$ ($r = 1, 2, \ldots$).

(i) $x(t)$ at the equilibrium x^* is stable of the $(2m + 1)^{\text{th}}$ order if

$$\begin{aligned} &G^{(r)}(x^*, \mathbf{p}) = D^r f(x^*, \mathbf{p}) = 0, r = 0, 1, 2, \ldots, 2m; \\ &x(t + \varepsilon) - x(t) < 0 \ \text{ for } y = x(t) - x^* > 0; \\ &x(t + \varepsilon) - x(t) > 0 \ \text{ for } y = x(t) - x^* < 0 \end{aligned} \tag{3.18}$$

for all $x \in U(x^*) \subset \Omega$ and all $t \in [t_0, \infty)$. The equilibrium x^* is called the sink (or stable node) of the $(2m + 1)^{\text{th}}$ order.

(ii) $x(t)$ at the equilibrium x^* is unstable of the $(2m + 1)^{\text{th}}$ order if

$$\begin{aligned} &G^{(r)}(x^*, \mathbf{p}) = D^r f(x^*, \mathbf{p}) = 0, r = 0, 1, 2, \ldots, 2m; \\ &x(t + \varepsilon) - x(t) > 0 \ \text{ for } y = x(t) - x^* > 0; \\ &x(t + \varepsilon) - x(t) < 0 \ \text{ for } y = x(t) - x^* < 0 \end{aligned} \tag{3.19}$$

for all $x \in U(x^*) \subset \Omega$ and all $t \in [t_0, \infty)$. The equilibrium x^* is called the source (or unstable node) of the $(2m + 1)^{\text{th}}$ order.

(iii) $x(t)$ at the equilibrium x^* is increasingly unstable of the $(2m)^{\text{th}}$ order if

$$\begin{aligned} &G^{(r)}(x^*, \mathbf{p}) = D^r f(x^*, \mathbf{p}) = 0, r = 0, 1, 2, \ldots, 2m - 1; \\ &x(t + \varepsilon) - x(t) > 0 \ \text{ for } y = x(t) - x^* > 0; \\ &x(t + \varepsilon) - x(t) > 0 \ \text{ for } y = x(t) - x^* < 0 \end{aligned} \tag{3.20}$$

for all $x \in U(x^*) \subset \Omega$ and all $t \in [t_0, \infty)$. The equilibrium x^* is called the increasing saddle (or upper saddle) of the $(2m)^{\text{th}}$ order.

(iv) $x(t)$ at the equilibrium x^* is decreasingly unstable of the $(2m)^{\text{th}}$ order if

$$\begin{aligned} &G^{(r)}(x^*, \mathbf{p}) = D^r f(x^*, \mathbf{p}) = 0, r = 0, 1, 2, \ldots, 2m - 1; \\ &x(t + \varepsilon) - x(t) < 0 \ \text{ for } y = x(t) - x^* > 0; \\ &x(t + \varepsilon) - x(t) < 0 \ \text{ for } y = x(t) - x^* < 0 \end{aligned} \tag{3.21}$$

for all $x \in U(x^*) \subset \Omega$ and all $t \in [t_0, \infty)$. The equilibrium x^* is called the decreasing saddle (or lower saddle) of the $(2m)^{\text{th}}$ order.

Theorem 3.3 *Consider a* 1*-dimensional, autonomous, nonlinear dynamical system* $\dot{x} = f(x, \mathbf{p})$ *in Eq. (3.1) with an equilibrium* x^* *and* $f(x, \mathbf{p})$ *is* C^r*-continuous* ($r \geq 1$) *in a neighborhood of* x^* *(i.e.,* $U(x^*) \subset \Omega$*). The corresponding solution is* $x(t) = \Phi(x_0, t - t_0, \mathbf{p})$. *In a neighborhood of* x^*, $y = x - x^*$. *Under the condition in Eq. (1.24), the linearized equation of Eq. (3.1) is in Eq. (3.12). There is a real eigenvalue* $\lambda = Df(x^*, \mathbf{p}) = G^{(1)}(x^*, \mathbf{p})$ *for a specific* $\mathbf{p}$. *Suppose* $|G^{(r)}(x^*, \mathbf{p})| < \infty$ *where* $G^{(r)}(x^*, \mathbf{p}) = D^r f(x^*, \mathbf{p})$ ($r = 1, 2, \ldots$).

(i) $x(t)$ *at the equilibrium* x^* *is stable of the* $(2m + 1)^{\text{th}}$ *order if and only if*

$$\begin{aligned} &G^{(r)}(x^*, \mathbf{p}) = D^r f(x^*, \mathbf{p}) = 0, r = 0, 1, 2, \ldots, 2m; \\ &G^{(2m+1)}(x^*, \mathbf{p}) = D^{2m+1} f(x^*, \mathbf{p}) < 0 \end{aligned} \tag{3.22}$$

for all $x \in U(x^*) \subset \Omega$ *and all* $t \in [t_0, \infty)$.

(ii) $x(t)$ *at the equilibrium* x^* *is unstable of the* $(2m_k + 1)^{\text{th}}$ *order if and only if*

$$\begin{aligned} &G^{(r)}(x^*, \mathbf{p}) = D^r f(x^*, \mathbf{p}) = 0, r = 0, 1, 2, \ldots, 2m; \\ &G^{(2m+1)}(x^*, \mathbf{p}) = D^{2m+1} f(x^*, \mathbf{p}) > 0 \end{aligned} \tag{3.23}$$

for all $x \in U(x^*) \subset \Omega$ *and all* $t \in [t_0, \infty)$.

(iii) $x(t)$ *at the equilibrium* x^* *is increasingly unstable of the* $(2m)^{\text{th}}$ *order if and only if*

$$\begin{aligned} &G^{(r)}(x^*, \mathbf{p}) = D^r f(x^*, \mathbf{p}) = 0, r = 0, 1, 2, \ldots, 2m - 1; \\ &G^{(2m)}(x^*, \mathbf{p}) = D^{2m} f(x^*, \mathbf{p}) > 0 \end{aligned} \tag{3.24}$$

for all $x \in U(x^*) \subset \Omega$ *and all* $t \in [t_0, \infty)$.

(iv) $x(t)$ *at the equilibrium* x^* *is decreasingly unstable of the* $(2m)^{\text{th}}$ *order if and only if*

$$\begin{aligned} &G^{(r)}(x^*, \mathbf{p}) = D^r f(x^*, \mathbf{p}) = 0, r = 0, 1, 2, \ldots, 2m - 1; \\ &G^{(2m)}(x^*, \mathbf{p}) = D^{2m} f(x^*, \mathbf{p}) < 0 \end{aligned} \tag{3.25}$$

for all $x \in U(x^*) \subset \Omega$ *and all* $t \in [t_0, \infty)$.

Proof For $x = x^*$, $y = 0$. Using the Taylor series expansion gives

$$\begin{aligned} \dot{y} &= f(x, \mathbf{p}) - f(x^*, \mathbf{p}) \\ &= \sum\nolimits_{r=1}^{2m} \frac{1}{r!} D^r f(x^*, \mathbf{p}) y^r + \frac{1}{(2m+1)!} D^{2m+1} f(x^*, \mathbf{p}) y^{2m+1} + o(y^{2m+1}) \\ &= \sum\nolimits_{r=1}^{2m} \frac{1}{r!} G^{(r)}(x^*, \mathbf{p}) y^r + \frac{1}{(2m+1)!} G^{(2m+1)}(x^*, \mathbf{p}) y^{2m+1} + o(y^{2m+1}), \end{aligned}$$

and

$$G^{(r)}(x^*,\mathbf{p}) = D^r f(x^*,\mathbf{p}) = 0 \text{ for } r = 0, 1, 2, \ldots, 2m;$$

$$\begin{aligned}\dot{y} &= \frac{1}{(2m+1)!} D^{2m+1} f(x^*,\mathbf{p}) y^{2m+1} \\ &= \frac{1}{(2m+1)!} G^{(2m+1)}(x^*,\mathbf{p}) y^{2m+1}.\end{aligned}$$

(i) For $y > 0$,

$$G(\mathbf{x},\mathbf{p}) = \dot{y} = \frac{1}{(2m+1)!} G^{(2m+1)}(x^*,\mathbf{p}) y^{2m+1} < 0,$$

and for $y < 0$

$$G(\mathbf{x},\mathbf{p}) = \dot{y} = \frac{1}{(2m+1)!} G^{(2m+1)}(x^*,\mathbf{p}) y^{2m+1} > 0.$$

Thus, $G^{(2m+1)}(x^*,\mathbf{p}) = D^{2m+1} f(x^*,\mathbf{p}) < 0$.

(ii) For $y > 0$,

$$G(\mathbf{x},\mathbf{p}) = \dot{y} = \frac{1}{(2m+1)!} G^{(2m+1)}(x^*,\mathbf{p}) y^{2m+1} > 0,$$

and for $y < 0$,

$$G(\mathbf{x},\mathbf{p}) = \dot{y} = \frac{1}{(2m+1)!} G^{(2m+1)}(x^*,\mathbf{p}) y^{2m+1} < 0.$$

Thus, $G^{(2m+1)}(x^*,\mathbf{p}) = D^{2m+1} f(x^*,\mathbf{p}) > 0$.

(iii) For $\mathbf{x} = \mathbf{x}^*$, $s_k = 0$. Using the Taylor series expansion gives

$$\begin{aligned}\dot{y} &= f(x,\mathbf{p}) - f(x^*,\mathbf{p}) \\ &= \sum_{r=1}^{2m-1} \frac{1}{r!} D^r f(x^*,\mathbf{p}) y^r + \frac{1}{(2m)!} D^{2m} f(x^*,\mathbf{p}) y^{2m} + o(y^{2m}) \\ &= \sum_{r=1}^{2m} \frac{1}{r!} G^{(r)}(x^*,\mathbf{p}) y^r + \frac{1}{(2m)!} G^{(2m)}(x^*,\mathbf{p}) y^{2m} + o(y^{2m})\end{aligned}$$

and

$$G^{(r)}(x^*,\mathbf{p}) = D^r f(x^*,\mathbf{p}) = 0 \text{ for } r = 0, 1, \ldots, 2m-1,$$

$$\dot{y} = \frac{1}{(2m)!} D^{2m} f(x^*,\mathbf{p}) y^{2m} = \frac{1}{(2m)!} G^{(2m)}(x^*,\mathbf{p}) y^{2m}.$$

For $y > 0$,

$$G(\mathbf{x},\mathbf{p}) = \dot{y} = \frac{1}{(2m)!}G^{(2m)}(x^*,\mathbf{p})y^{2m} > 0,$$

and for $y < 0$,

$$G(\mathbf{x},\mathbf{p}) = \dot{y} = \frac{1}{(2m)!}G^{(2m)}(x^*,\mathbf{p})y^{2m} > 0.$$

Thus, $G^{(2m)}(x^*,\mathbf{p}) = D^{2m}f(x^*,\mathbf{p}) > 0$.

(iv) Similar to (iii), for $y > 0$,

$$G(\mathbf{x},\mathbf{p}) = \dot{y} = \frac{1}{(2m)!}G^{(2m)}(x^*,\mathbf{p})y^{2m} < 0,$$

and for $y < 0$,

$$G(\mathbf{x},\mathbf{p}) = \dot{y} = \frac{1}{(2m)!}G^{(2m)}(x^*,\mathbf{p})y^{2m} < 0.$$

Thus, $G^{(2m)}(x^*,\mathbf{p}) = D^{2m}f(x^*,\mathbf{p}) < 0$. The theorem is proved. ∎

To illustrate dynamical behaviors of 1-dimensional systems in the vicinity of equilibrium, the sink (unstable node), source (unstable node), increasing saddle (upper saddle), deceasing saddle (lower saddle), and invariant points are presented in Fig. 3.1a–e, respectively. $U_\delta(x^*) = (x^*-\delta, x^*+\delta)$ is a neighborhood of equilibrium x^*. The circular symbols are for initially selected points. The arrows are the moving directions of $x(t)$. Saddle point is the half sink and half source.

3.1.2 Bifurcations

In this section, the local bifurcation theory for 1-dimensional system is presented.

Definition 3.3 Consider a 1-dimensional, autonomous, nonlinear dynamical system $\dot{x} = f(x,\mathbf{p})$ in Eq. (3.1) with an equilibrium x^* and $f(x,\mathbf{p})$ is C^r-continuous ($r \geq 1$) in a neighborhood of x^* (i.e., $U(x^*) \subset \Omega$). The corresponding solution is $x(t) = \Phi(x_0, t-t_0,\mathbf{p})$. In a neighborhood of x^*, $y = x-x^*$. Under the condition in Eq. (1.24), the linearized equation of Eq. (3.1) is in Eq. (3.12). There is a real eigenvalue $\lambda = Df(x^*,\mathbf{p}) = G^{(1)}(x^*,\mathbf{p})$ for a specific $\mathbf{p}$. Suppose $|G^{(m)}(x^*,\mathbf{p})| < \infty$ where $G^{(r)}(x^*,\mathbf{p}) = D^{(r)}f(x^*,\mathbf{p})$ ($r = 1, 2, \ldots$). In the vicinity of point $(x_0^*,\mathbf{p}_0)$, $f(x^*,\mathbf{p})$ can be expanded as

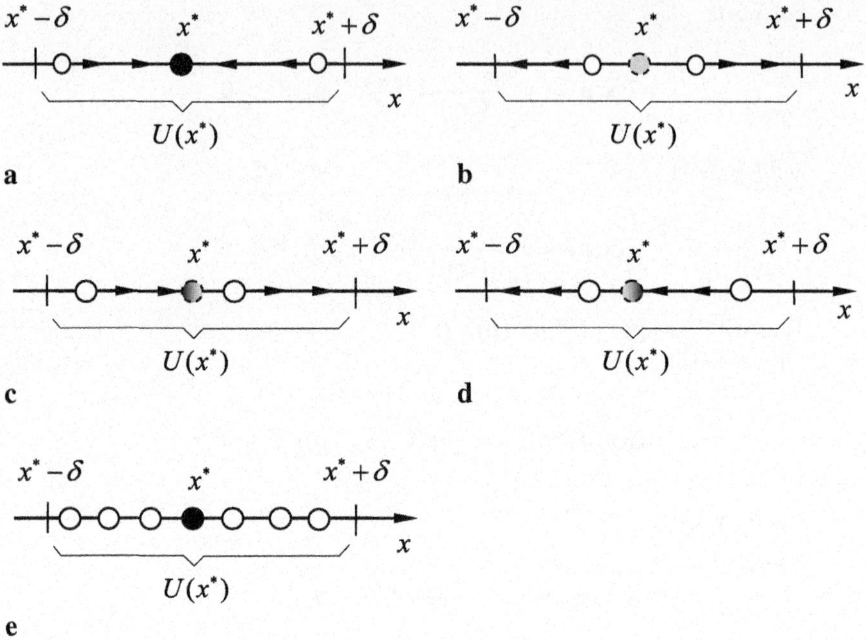

Fig. 3.1 Equilibrium x^* in a 1-dimensional nonlinear system: (**a**) stable node (sink), (**b**) unstable node (source), (**c**) increasing saddle (upper saddle), (**d**) decreasing saddle (lower saddle), (**e**) invariant point

$$\begin{aligned} f(x^*,\mathbf{p}) &= ay^* + \mathbf{b}^{\mathrm{T}}\cdot(\mathbf{p}-\mathbf{p}_0) + \sum_{q=2}^{m}\sum_{r=0}^{q}\frac{1}{q!}C_q^r\mathbf{a}^{(q-r,r)}y^{*q-r}(\mathbf{p}-\mathbf{p}_0)^r \\ &\quad + \frac{1}{(m+1)!}[y\partial_x + (\mathbf{p}-\mathbf{p}_0)\partial_{\mathbf{p}}]^{m+1}(f(x_0^*+\theta y^*,\mathbf{p}_0+\boldsymbol{\theta}\Delta\mathbf{p})), \end{aligned} \tag{3.26}$$

where

$$\begin{aligned} &y^* = x^* - x_0^*, \\ &a = \partial_x f(x,\mathbf{p})|_{(x_0^*,\mathbf{p}_0)} = Df(x,\mathbf{p})|_{(x_0^*,\mathbf{p}_0)} = \lambda_0, \\ &\mathbf{b}^T = \partial_{\mathbf{p}} f(x,\mathbf{p})|_{(x_0^*,\mathbf{p}_0)}, \\ &\mathbf{a}^{(r,s)} = \partial_x^{(r)}\partial_{\mathbf{p}}^{(s)} f(x,\mathbf{p})|_{(x_0^*,\mathbf{p}_0)}, \\ &\gamma \in (0,1), \boldsymbol{\theta} = (\theta_1,\theta_2,\ldots,\theta_m)^{\mathrm{T}}, \\ &\theta_r \in (0,1), (r = 1,2,\ldots,m). \end{aligned} \tag{3.27}$$

If $\lambda_0 = 0$ and $\mathbf{p} = \mathbf{p}_0$, the stability of current equilibrium x^* changes from stable to unstable state (or from unstable to stable state). The bifurcation branch of equilibrium is determined by

$$\left.\begin{aligned} & f(x_0^* + y^*, \mathbf{p}) = 0, \\ & \text{or } \mathbf{b}_k^{\mathrm{T}} \cdot (\mathbf{p} - \mathbf{p}_0) + \sum_{q=2}^{s} \sum_{r=0}^{q} \frac{1}{q!} C_q^r \mathbf{a}^{(q-r,r)} y^{*q-r} (\mathbf{p} - \mathbf{p}_0)^r \approx 0; \end{aligned}\right\}$$

equivalent to

$$\left.\begin{aligned} f(x_0^* + y^*, \mathbf{p}) = A_0 y^{*s} + A_1 y^{*s-1} + \cdots + A_s = 0, \\ A_i = A_i(x_0^*, \mathbf{p}) \; (i = 0, 1, 2, \ldots, s), \\ A_0 = a^{(s,0)} = D^s f(x_0^*, \mathbf{p}_0) \neq 0; \end{aligned}\right\} \tag{3.28}$$

$$\left.\begin{aligned} \lambda_0 = Df(x_0^*, \mathbf{p}_0) = 0, \\ D^r f(x_0^*, \mathbf{p}_0) = 0 \; (r = 2, 3, \ldots, s-1), \\ D^s f(x_0^*, \mathbf{p}_0) \neq 0. \end{aligned}\right\}$$

In the neighborhood of $(x_0^*, \mathbf{p}_0)$, Eq. (3.1) possesses l-equilibrium y^* $(0 < l \leq s)$ with $s = \sum_{i=1}^{l} \alpha_i$. Such l-branch solutions are called the bifurcation solutions of an equilibrium x^* in the neighborhood of $(x_0^*, \mathbf{p}_0)$. Such a bifurcation at point $(x_0^*, \mathbf{p}_0)$ is called a hyperbolic bifurcation of the sth order.

(i) Consider $f(x_0^* + y^*, \mathbf{p}) = 0$ at point $(x_0^*, \mathbf{p}_0)$ approximated by

$$\left.\begin{aligned} & f(x_0^* + y^*, \mathbf{p}) = 0 \text{ with } \lambda_0 = Df(x_0^*, \mathbf{p}_0) = 0, \\ & \text{or } \mathbf{b}_k^{\mathrm{T}} \cdot (\mathbf{p} - \mathbf{p}_0) + \sum_{q=2}^{2m+1} \sum_{r=0}^{q} \frac{1}{q!} C_q^r \mathbf{a}^{(q-r,r)} y^{*q-r} (\mathbf{p} - \mathbf{p}_0)^r \approx 0; \end{aligned}\right\}$$

equivalent to

$$\left.\begin{aligned} f(x_0^* + y^*, \mathbf{p}) = A_0 y^{*2m+1} + A_1 y^{*2m} + \cdots + A_{2m+1} = 0, \\ A_i = A_i(x_0^*, \mathbf{p}) \; (i = 0, 1, 2, \cdots, 2m+1), \\ A_0 = a^{(2m+1,0)} = D^{2m+1} f(x_0^*, \mathbf{p}_0) \neq 0. \end{aligned}\right\} \tag{3.29}$$

(i_1) If

$$\left.\begin{aligned} \lambda_0 = Df(x_0^*, \mathbf{p}_0) = 0, \\ D^r f(x_0^*, \mathbf{p}_0) = 0 \; (r = 2, 3, \ldots, 2m), \\ D^{2m+1} f(x_0^*, \mathbf{p}_0) < 0, \end{aligned}\right\} \tag{3.30}$$

the bifurcation of equilibrium at point $(x_0^*\mathbf{p}_0)$ is called the pitchfork bifurcation of the (2m+1)th order. The special bifurcation point is the (2m+1)th order sink (stable node).

(i$_2$) If

$$\left.\begin{aligned} \lambda_0 = Df(x_0^*, \mathbf{p}_0) = 0, \\ D^r f(x_0^*, \mathbf{p}_0) = 0\ (r = 2, 3, \ldots, 2m), \\ D^{2m+1} f(x_0^*, \mathbf{p}_0) > 0, \end{aligned}\right\} \tag{3.31}$$

the bifurcation of equilibrium at point $(x_0^*, \mathbf{p}_0)$ is called an unstable pitchfork bifurcation of the (2m+1)th order. The special bifurcation is a (2m+1)th-order source (unstable node).

(ii) Consider $f(x_0^* + y^*, \mathbf{p}) = 0$ at point $(x_0^*, \mathbf{p}_0)$ approximated by

$$\left.\begin{aligned} & f(x_0^* + y^*, \mathbf{p}) = 0 \text{ with } \lambda_0 = Df(x_0, \mathbf{p}_0) = 0, \\ & \text{or } \mathbf{b}_k^{\mathrm{T}} \cdot (\mathbf{p} - \mathbf{p}_0) + \sum_{q=2}^{2m} \sum_{r=0}^{q} \frac{1}{q!} C_q^r \mathbf{a}^{(q-r,r)} y^{*q-r} (\mathbf{p} - \mathbf{p}_0)^r \approx 0; \end{aligned}\right\}$$

equivalent to

$$\left.\begin{aligned} & f(x_0^* + y^*, \mathbf{p}) = A_0 (y^*)^{2m} + A_1 (y^*)^{2m-1} + \cdots + A_{2m} = 0, \\ & A_i = A_i(x_0^*, \mathbf{p})\ (i = 0, 1, 2, \ldots, 2m-1), \\ & A_0 = a^{(2m,0)} = D^{2m} f(x_0^*, \mathbf{p}_0) \neq 0. \end{aligned}\right\} \tag{3.32}$$

(ii$_1$) If

$$\left.\begin{aligned} & \lambda_0 = Df(x_0^*, \mathbf{p}_0) = 0, \\ & D^r f(x_0^*, \mathbf{p}_0) = 0\ (r = 2, 3, \ldots, 2m-1), \\ & D^{2m} f(x_0^*, \mathbf{p}_0) > 0, \end{aligned}\right\} \tag{3.33}$$

the bifurcation of equilibrium at point $(x_0^*, \mathbf{p}_0)$ is called an increasing saddle-node (or an *upper-saddle-node*) bifurcation of the (2m)th order. The special bifurcation point is a (2m)th-order increasing saddle (or upper saddle).

(ii$_2$) If

$$\left.\begin{aligned} & \lambda_0 = Df(x_0^*, \mathbf{p}_0) = 0, \\ & D^r f(x_0^*, \mathbf{p}_0) = 0\ (r = 2, 3, \ldots, 2m-1), \\ & D^{2m} f(x_0^*, \mathbf{p}_0) < 0, \end{aligned}\right\} \tag{3.34}$$

the bifurcation at point $(x_0^*, \mathbf{p}_0)$ is called a decreasing saddle-node (or a *lower-saddle-node*) bifurcation of the (2m)th order. The special bifurcation point is a (2m)th-order decreasing saddle (or *lower saddle*).

Definition 3.4 Consider a *1*-dimensional, autonomous, nonlinear dynamical system $\dot{x} = f(x, \mathbf{p})$ in Eq. (3.1) with an equilibrium x^* and $f(x, \mathbf{p})$ is C^r-continuous ($r \geq 1$) in a neighborhood of x^* (i.e., $U(x^*) \subset \Omega$). The corresponding solution is $x(t) = \Phi(x_0, t-t_0, \mathbf{p})$. In a neighborhood of x^*, $y = x-x^*$. *Under the condition in Eq. (1.24), the linearized equation of Eq. (3.1) is in Eq. (3.12). There is a real eigenvalue* $\lambda = Df(x^*, \mathbf{p}) = G^{(1)}(x^*, \mathbf{p})$ *for a specific* $\mathbf{p}$. *Suppose* $|G^{(m)}(x^*, \mathbf{p})| < \infty$ *where* $G^{(r)}(x^*, \mathbf{p}) = D^{(r)}f(x^*, \mathbf{p})$ *(*$r = 1, 2, \cdots$*).* In the vicinity of point $(x_0^*, \mathbf{p}_0)$, $f(x^*, \mathbf{p})$ can be expended in Eq. (3.26), and three special cases are defined as follows.

(i) Consider

$$\left.\begin{aligned} &\lambda_0 = Df(x_0^*, \mathbf{p}_0) = 0, f(x_0^* + y^*, \mathbf{p}_0 + \Delta\mathbf{p}) = 0 \\ &\text{or} \\ &\mathbf{b}^{\mathrm{T}} \cdot (\mathbf{p} - \mathbf{p}_0) + \frac{1}{2!}\mathbf{a}^{(0,2)} \cdot (\mathbf{p} - \mathbf{p}_0)^2 \\ &+\mathbf{a}^{(1,1)} \cdot (\mathbf{p} - \mathbf{p}_0)y^{*2} + \frac{1}{2!}a^{(2,0)}y^{*2} \approx 0; \end{aligned}\right\}$$
$$\text{equivalent to} \tag{3.35}$$
$$\dot{y} = A_0 y^2 + A_1 y + A_2 = 0 \text{ with}$$
$$\left.\begin{aligned} &A_0(x_0^*, \mathbf{p}_0) = \frac{1}{2!}a^{(2,0)}, \\ &A_1(x_0^*, \mathbf{p}_0) = \mathbf{a}^{(1,1)} \cdot (\mathbf{p} - \mathbf{p}_0) + \lambda_0, \\ &A_2(x_0^*, \mathbf{p}_0) = \mathbf{b}^{\mathrm{T}} \cdot (\mathbf{p} - \mathbf{p}_0) + \frac{1}{2!}\mathbf{a}^{(0,2)} \cdot (\mathbf{p} - \mathbf{p}_0)^2, \end{aligned}\right\}$$

with

$$\begin{aligned} a^{(2,0)} &= \partial_x^{(2)}\partial_{\mathbf{p}}^{(0)} f(x, \mathbf{p})|_{(x_0^*, \mathbf{p}_0)} \\ &= D^2 f(x, \mathbf{p})|_{(x_0^*, \mathbf{p}_0)} = G^{(2)}(x_0^*, \mathbf{p}_0) \neq 0, \\ \mathbf{a}^{(1,1)} &= \partial_x^{(1)}\partial_{\mathbf{p}}^{(1)} f(x, \mathbf{p})|_{(x_0^*, \mathbf{p}_0)}, \\ \mathbf{a}^{(0,2)} &= \partial_x^{(0)}\partial_{\mathbf{p}}^{(2)} f(x, \mathbf{p})|_{(x_0^*, \mathbf{p}_0)} = \partial_{\mathbf{p}}^{(2)} f(x, \mathbf{p})|_{(x_0^*, \mathbf{p}_0)}, \\ \mathbf{b}^{\mathrm{T}} &= \partial_{\mathbf{p}} f(x, \mathbf{p})|_{(\mathbf{x}_0^*, \mathbf{p}_0)}. \end{aligned} \tag{3.36}$$

(i_1) If the following conditions are satisfied in Eq. (3.35)

$$\begin{aligned} &A_1 = 0 \Rightarrow \mathbf{a}^{(1,1)} = \mathbf{0} \text{ and } \lambda_0 = Df(x_0^*, \mathbf{p}_0) = 0; \\ &a^{(2,0)} \times A_2 < 0, \end{aligned} \tag{3.37}$$

such a bifurcation at point $(x_0^*, \mathbf{p}_0)$ is called the *saddle-node-appearing/vanishing* bifurcation.

(i_{1a}) For

$$a^{(2,0)} = D^2 f(x, \mathbf{p})|_{(x_0^*, \mathbf{p}_0)} = G^{(2)}(x_0^*, \mathbf{p}_0) > 0, \tag{3.38}$$

the saddle-node bifurcation is called an increasing saddle-node (or an *upper-saddle-node*)-appearing/vanishing bifurcation of the second order. The bifurcation point $(x_0^*, \mathbf{p}_0)$ is an increasing saddle (or an *upper saddle*) of the second order.

(i_{1b}) For

$$a^{(2,0)} = D^2 f(x, \mathbf{p})|_{(x_0^*, \mathbf{p}_0)} = G^{(2)}(x_0^*, \mathbf{p}_0) < 0, \tag{3.39}$$

the saddle-node bifurcation is called a decreasing saddle-node (or a *lower-saddle-node*)-appearing/vanishing bifurcation. The bifurcation point $(x_0^*, \mathbf{p}_0)$ is a decreasing saddle (or a *lower saddle*) of the second order.

(i_2) If the following conditions are satisfied in Eq. (3.35)

$$\begin{aligned} A_2 = 0 &\Rightarrow \mathbf{a}^{(0,1)} = 0 \text{ and } \mathbf{a}^{(0,2)} = 0, \\ A_1 \neq 0 &\Rightarrow \mathbf{a}^{(1,1)} \neq \mathbf{0}, \end{aligned} \tag{3.40}$$

such a bifurcation at point $(x_0^*, \mathbf{p}_0)$ is called a *transcritical switching* bifurcation, which is a kind of *saddle-node*-switching bifurcation.

(i_{2a}) For

$$a^{(2,0)} = D^2 f(x, \mathbf{p})|_{(x_0^*, \mathbf{p}_0)} = G^{(2)}(x_0^*, \mathbf{p}_0) > 0, \tag{3.41}$$

the *transcritical* saddle-node bifurcation is called an increasing saddle-node (or an *upper-saddle-node*)*-switching* bifurcation of the second order. The bifurcation point $(x_0^*, \mathbf{p}_0)$ is an increasing saddle (or an *upper saddle*) of the second order.

(i_{2b}) For

$$a^{(2,0)} = D^2 f(x, \mathbf{p})|_{(x_0^*, \mathbf{p}_0)} = G^{(2)}(x_0^*, \mathbf{p}_0) < 0, \tag{3.42}$$

the *transcritical* saddle-node bifurcation is called a *decreasing saddle-node (or a lower-saddle-node)*-switching bifurcation of the second order. The bifurcation point $(x_0^*, \mathbf{p}_0)$ is a decreasing saddle (or a *lower saddle*) of the second order.

(ii) Consider

$$\left.\begin{aligned}
&\lambda_0 = Df(x_0, \mathbf{p}_0) = 0, f(x_0^* + y^*, \mathbf{p}_0 + \Delta\mathbf{p}) = 0\\
&\text{or}\\
&\mathbf{b}^{\mathrm{T}} \cdot (\mathbf{p} - \mathbf{p}_0) + \frac{1}{2!}\mathbf{a}^{(0,2)} \cdot (\mathbf{p} - \mathbf{p}_0)^2 + \mathbf{a}^{(1,1)} \cdot (\mathbf{p} - \mathbf{p}_0)y^*\\
&+\frac{1}{2!}a^{(2,0)}y^{*2} + \frac{1}{3!}\mathbf{a}^{(0,3)} \cdot (\mathbf{p} - \mathbf{p}_0)^3 + \frac{1}{2!}\mathbf{a}^{(2,1)} \cdot (\mathbf{p} - \mathbf{p}_0)y^{*2}\\
&+\frac{1}{2!}\mathbf{a}^{(1,2)} \cdot (\mathbf{p} - \mathbf{p}_0)^2 y^* + \frac{1}{3!}a^{(3,0)}y^{*3} \approx 0;
\end{aligned}\right\} \tag{3.43a}$$

equivalent to

$$\left.\begin{aligned}
&\dot{y} = A_0 y^3 + A_1 y^2 + A_2 y + A_3 = 0 \text{ with}\\
&A_0(x_0^*, \mathbf{p}_0) = \frac{1}{3!}a^{(3,0)},\\
&A_1(x_0^*, \mathbf{p}_0) = \frac{1}{2!}\mathbf{a}^{(2,1)} \cdot (\mathbf{p} - \mathbf{p}_0) + \frac{1}{2!}a^{(2,0)},\\
&A_2(x_0^*, \mathbf{p}_0) = \lambda_0 + \mathbf{a}^{(1,1)} \cdot (\mathbf{p} - \mathbf{p}_0) + \frac{1}{2!}\mathbf{a}^{(1,2)} \cdot (\mathbf{p} - \mathbf{p}_0)^2,\\
&A_3(x_0^*, \mathbf{p}_0) = \mathbf{b}^{\mathrm{T}} \cdot (\mathbf{p} - \mathbf{p}_0) + \frac{1}{2!}\mathbf{a}^{(0,2)} \cdot (\mathbf{p} - \mathbf{p}_0)^2 + \frac{1}{3!}\mathbf{a}^{(0,3)} \cdot (\mathbf{p} - \mathbf{p}_0)^3,
\end{aligned}\right\} \tag{3.43b}$$

where

$$\begin{aligned}
a^{(3,0)} &= \partial_x^{(3)}\partial_{\mathbf{p}}^{(0)} f(x, \mathbf{p})|_{(\mathbf{x}_0^*, \mathbf{p}_0)} = \partial_x^{(3)} f(x, \mathbf{p})|_{(x_0^*, \mathbf{p}_0)}\\
&= D^3 f(x, \mathbf{p})|_{(x_0^*, \mathbf{p}_0)} = G^{(3)}(x_0^*, \mathbf{p}_0) \neq 0,\\
\mathbf{a}^{(r,s)} &= \partial_x^{(r)}\partial_{\mathbf{p}}^{(s)} f(x, \mathbf{p})|_{(\mathbf{x}_0^*, \mathbf{p}_0)} (r, s = 0, 1, 2, 3; r + s = 3).
\end{aligned} \tag{3.44}$$

If the following conditions are satisfied,

$$\begin{aligned}
&A_1(x_0, \mathbf{p}_0) = 0 \Rightarrow a_k^{(2,0)} = 0 \text{ and } \mathbf{a}_k^{(2,1)} = 0,\\
&A_3(x_0, \mathbf{p}_0) = 0 \Rightarrow \mathbf{b} = 0 \text{ and } \mathbf{a}^{(0,2)} = 0\\
&A_1 y^* + A_0 y^{*3} \approx 0 \text{ with } A_1 \times A_0 < 0,
\end{aligned} \tag{3.45}$$

such a bifurcation at point $(x_0^*, \mathbf{p}_0)$ is called a *pitchfork*-switching and -appearing bifurcation of the third order.

(ii$_1$) For

$$a^{(3,0)} = D^3 f(x, \mathbf{p})|_{(x_0^*, \mathbf{p}_0)} = G^{(3)}(x_0^*, \mathbf{p}_0) > 0, \tag{3.46}$$

the *pitchfork* bifurcation is called a switching and appearing bifurcation of the third-order source. The corresponding bifurcation point is a third-order *source* (or *unstable node*).

(iii$_2$) For

$$a^{(3,0)} = D^3 f(x, \mathbf{p})|_{(x_0^*, \mathbf{p}_0)} = G^{(3)}(x_0^*, \mathbf{p}_0) < 0, \tag{3.47}$$

the *pitchfork* bifurcation is called a switching and appearing bifurcation of the third-order sink. The corresponding bifurcation point is a third-order *sink* (or *stable node*).

3.1.3 Sampled Systems

As in Luo (2012), the above three special cases can be discussed through 1-dimensional systems and intuitive illustrations are presented in Fig. 3.2. The bifurcation point is also represented by a solid circular symbol. The stable and unstable equilibrium branches are given by solid and dashed curves, respectively. The vector fields are represented by lines with arrows. If no equilibriums exist, such a region is shaded.

3.1.3a Saddle-Node-Appearing Bifurcation

Consider a saddle-node bifurcation in 1-dimensional system:

$$\dot{x} = f(x, p) \equiv p - x^2. \tag{3.48}$$

The equilibriums of the foregoing equation are $x^* = \pm\sqrt{p}$ $(p > 0)$ and no equilibriums exist for $p < 0$. From Eq. (3.48), the linearized equation in the vicinity of the equilibrium with $y = x - x^*$ is

$$\dot{y} = G^{(1)}(x^*, p)y = Df(x^*, p)y = -2x^* y. \tag{3.49}$$

For the branch of $x^* = \sqrt{p}$ $(p > 0)$, due to $Df(x^*, p) = -2\sqrt{p} < 0$, the equilibrium is a stable node because $\dot{y} > 0$ for $y < 0$ and $\dot{y} < 0$ for $y > 0$. However, for the branch of $x^* = -\sqrt{p}$ $(p > 0)$, due to $Df(x^*, p) = 2\sqrt{p} > 0$, such an equilibrium is an unstable node because $\dot{y} > 0$ for $y > 0$ and $\dot{y} < 0$ for $y < 0$.

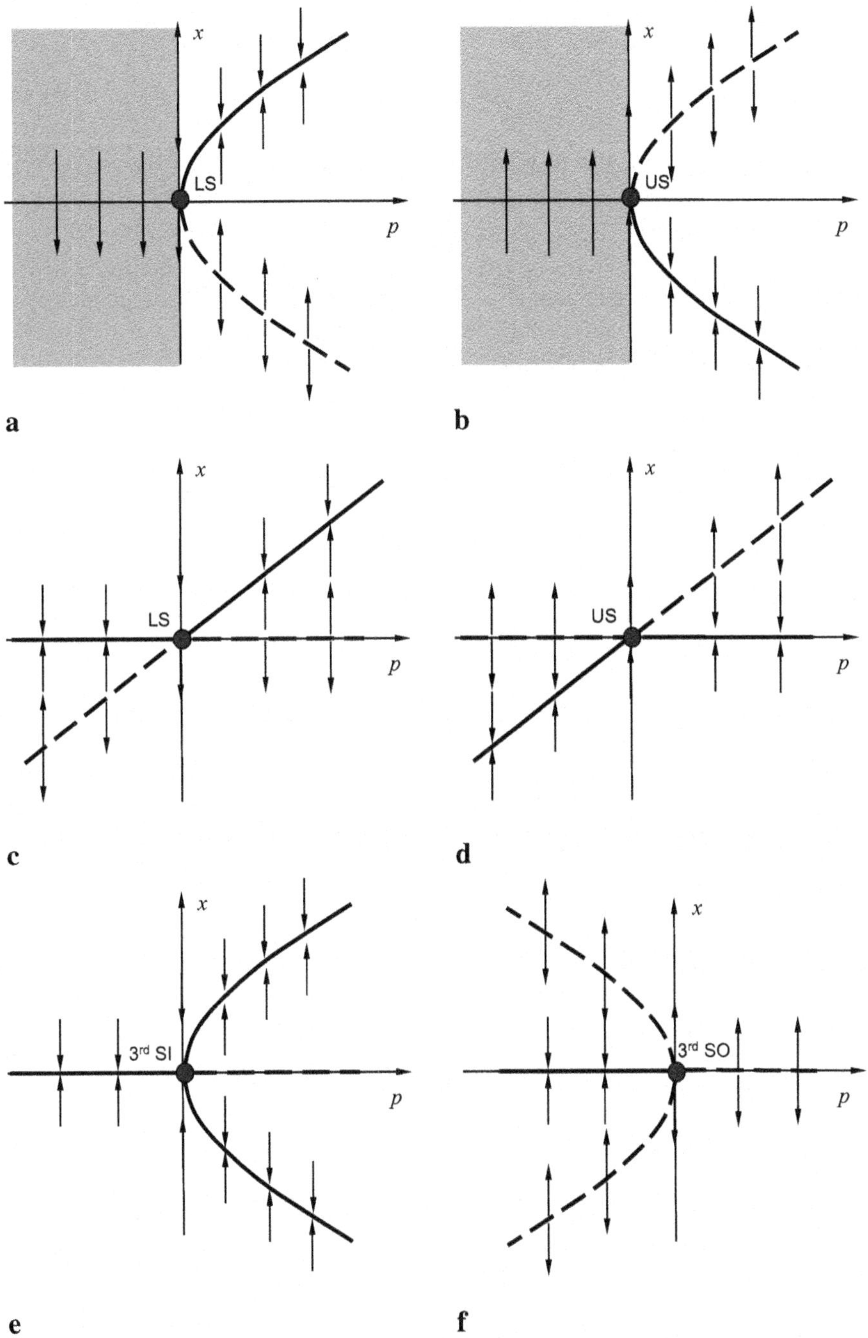

Fig. 3.2 Bifurcation diagrams: (**a**, **b**) saddle-node-appearing/vanishing bifurcation, (**c**, **d**) transcritical switching bifurcation, (**e**) pitchfork-switching/appearing bifurcation for stable symmetry, and (**f**) pitchfork bifurcation for unstable symmetry

For $p = p_0 = 0$, $x^* = x_0^* = 0$ and $Df(x_0^*, p_0) = 0$. $D^2 f(x_0^*, p_0) = -2$ is obtained. Thus

$$\dot{y} = \frac{1}{2!} D^2 f(x_0^*, p_0) y^2 = -y^2. \tag{3.50}$$

In the vicinity of $(x_0^*, p_0) = (0, 0)$, the flow vector field in Eq. (3.50) is always less than zero except for $y = 0$. The equilibrium point $(x_0^*, p_0) = (0, 0)$ is a bifurcation point, which is a *decreasing saddle* (or *lower saddle*) of the second order. For $p < 0$, the vector field of Eq. (3.48) is always less than zero without any equilibriums. The equilibrium varying with parameter p is sketched in Fig. 3.2a. On the left side of x-axis, no equilibrium exists, so only the vector field is presented.

Consider a dynamical system as

$$\dot{x} = f(x, p) \equiv -p + x^2 \tag{3.51}$$

and the corresponding equilibriums are $x^* = \pm\sqrt{p}$ with the variational equation

$$\dot{y} = G^{(1)}(x^*, p) y = Df(x^*, p) y = \pm 2\sqrt{p} y \tag{3.52}$$

where

$$G^{(1)}(x^*, p) = Df(x^*, p) = 2x^* = \pm 2\sqrt{p}. \tag{3.53}$$

For $G^{(1)}(x^*, p) = 2\sqrt{p} > 0$, $\dot{y} < 0$ for $y < 0$ and $\dot{y} > 0$ for $y > 0$. So such an equilibrium of $x^* = \sqrt{p}$ is an unstable node. If $G^{(1)}(x^*, p) = -2\sqrt{p} < 0$, $\dot{y} > 0$ for $y < 0$ and $\dot{y} < 0$ for $y > 0$. So such an equilibrium of $x^* = -\sqrt{p}$ is a stable node.

For $G^{(1)} = Df(x^*, p) = \pm 2\sqrt{p} = 0$ at $x^* = x_0^* = 0$ and $p = p_0 = 0$, we have

$$\dot{y} = \frac{1}{2!} G^{(2)}(x_0^*, p_0) y^2 = \frac{1}{2!} D^2 f(x_0^*, p_0) y^2 = y^2 \tag{3.54}$$

where

$$G^{(2)}(x_0^*, p_0) = D^2 f(x_0^*, p_0) = 2 > 0. \tag{3.55}$$

Thus, the bifurcation point at $(x_0^*, p_0) = (0, 0)$ has $\dot{y} > 0$ except for $y = 0$. The bifurcation point is the *increasing saddle (or upper saddle)* of the second order.

For $p < 0$, we have $\dot{x} > 0$ always without any equilibrium. So the flow increases always for a specific $p < 0$. The bifurcation and flow behaviors are presented in Fig. 3.2b.

3.1.3b Saddle-Node-Switching Bifurcation

Consider a saddle-node-switching bifurcation through the following system:

$$\dot{x} = f(x,p) \equiv px - x^2 \tag{3.56}$$

The equilibriums of the foregoing equation are $x^* = 0,\ p$. From Eq. (3.56), the linearized equation in the vicinity of the equilibriums with $y = x - x^*$ is

$$\dot{y} = G^{(1)}(x^*,p)y = Df(x^*,p)y = (p - 2x^*)y. \tag{3.57}$$

For the branch of $x^* = 0$ $(p > 0)$, due to $Df(x^*,p) = p > 0$ the equilibrium is an unstable node because $\dot{y} < 0$ for $y < 0$ and $\dot{y} > 0$ for $y > 0$. For the branch of $x^* = p$ $(p > 0)$, such an equilibrium is a stable node due to $Df(x^*,p) = -p < 0$. However, for the branch of $x^* = 0$ $(p < 0)$, the equilibrium is a stable node due to $Df(x^*,p) = p < 0$. For the branch of $x^* = p$ $(p < 0)$, such an equilibrium is an unstable node due to $Df(x^*,p) = -p > 0$, which causes $\dot{y} < 0$ for $y < 0$ and $\dot{y} > 0$ for $y > 0$.

For $p = p_0 = 0$, $x^* = x_0^* = 0$ and $Df(x_0^*,p_0) = p_0 = 0$. $D^2f(x_0^*,p_0) = -2$ is obtained. Thus the higher order variational equation at the equilibrium is

$$\dot{y} = \frac{1}{2!}G^{(2)}(x_0^*,p_0)y^2 = \frac{1}{2!}D^2f(x_0^*,p_0)y^2 = -y^2. \tag{3.58}$$

At $(x_0^*,p_0) = (0,0)$, the flow vector field is always less than zero except for $y = 0$. The equilibrium point $(x_0^*,p_0) = (0,0)$ is a bifurcation point. This transcritical bifurcation is a decreasing saddle (lower saddle) of the second order. The equilibrium varying with parameter p is sketched in Fig. 3.2c.

Consider another transcritical switching bifurcation of a dynamical system via

$$\dot{x} = f(x,p) \equiv -px + x^2. \tag{3.59}$$

The equilibriums of the foregoing equation are $x^* = 0,\ p$ with the variational equation as

$$\dot{y} = G^{(1)}(x^*,p)y = Df(x^*,p)y = (-p + 2x^*)y, \tag{3.60}$$

where

$$G^{(1)}(x^*,p) = Df(x^*,p) = -p + 2x^*. \tag{3.61}$$

For $G^{(1)}(x^*,p) = -p < 0$, $\dot{y} > 0$ for $y < 0$ and $\dot{y} < 0$ for $y > 0$. So such an equilibrium of $x^* = 0$ is a stable node. For $G^{(1)}(x^*,p) = -p > 0$, $\dot{y} < 0$ for $y < 0$ and $\dot{y} > 0$ for $y > 0$. So such an equilibrium of $x^* = 0$ is an unstable node. If $G^{(1)}(x^*,p) = p > 0$, $\dot{y} < 0$ for $y < 0$ and $\dot{y} > 0$ for $y > 0$. So such an equilibrium of $x^* = p$ is an unstable node. If $G^{(1)}(x^*,p) = p < 0$, $\dot{y} > 0$ for $y < 0$ and $\dot{y} < 0$ for $y > 0$. So such an equilibrium of $x^* = p$ is a stable node.

For $G^{(1)} = Df(x^*,p) = 0$ at $x^* = x_0^* = 0$ and $p = p_0 = 0$, the higher order variational equation is

$$\dot{y} = \frac{1}{2!}G^{(2)}(x_0^*, p_0)y^2 = \frac{1}{2!}D^2f(x_0^*, p_0)y^2 = y^2 \tag{3.62}$$

where

$$G^{(2)}(x_0^*, p_0) = D^2f(x_0^*, p_0) = 2 > 0. \tag{3.63}$$

Thus, the bifurcation point at $(x_0^*, p_0) = (0, 0)$ has $\dot{y} > 0$ except for $y = 0$. The transcritical bifurcation point is the *increasing saddle (or upper saddle)* of the second order. The bifurcation and flow behaviors are presented in Fig. 3.2d.

3.1.3c Pitchfork-Switching/Appearing Bifurcation

Consider the pitchfork bifurcation with stable symmetry through

$$\dot{x} = px - x^3. \tag{3.64}$$

Setting $\dot{x} = 0$ gives the equilibriums of $x^* = 0, \pm\sqrt{p}$ $(p > 0)$ and $x^* = 0$ $(p \leq 0)$. From Eq. (3.64), in the vicinity of x^* with $y = x - x^*$, the variational equation is

$$\dot{y} = G^{(1)}(x^*,p)y = Df(x^*,p)y = [p - 3(x^*)^2]y. \tag{3.64}$$

For the branch of $x^* = 0$ $(p > 0)$, because of $Df(x^*,p) = p > 0$, the equilibrium is unstable. For the branches of $x^* = \pm\sqrt{p}$ $(p > 0)$, such equilibriums are stable due to $Df(x^*,p) = 2p < 0$. However, for the branch of $x^* = 0$ $(p < 0)$, the equilibrium is stable due to $Df(x^*,p) = p < 0$.

For $p = p_0 = 0$, $x^* = x_0^* = 0$ and $Df(x_0^*, p_0) = 0$ are obtained. Because of $G^{(2)}(x_0^*, p_0) = D^2f(x_0^*, p_0) = -6x_0^* = 0$, $G^{(3)}(x_0^*, p_0) = D^3f(x_0^*, p_0) = -6 < 0$ is needed. Further, the variational equation at $(x_0^*, p_0) = (0, 0)$is

$$\dot{y} = \frac{1}{3!}G^{(3)}(x_0^*, p_0)y^3 = \frac{1}{3!}D^3f(x_0^*, p_0)y^3 = -y^3. \tag{3.65}$$

In the vicinity of $(x_0^*, p_0) = (0, 0)$, $\dot{y} < 0$ for $y > 0$ and $\dot{y} > 0$ for $y < 0$. The equilibrium point $(x_0^*, p_0) = (0, 0)$ is a bifurcation point, which is a sink of the third order. The equilibrium varying with parameter p is sketched in Fig. 3.2e.

Consider the pitchfork bifurcation for unstable symmetry as

$$\dot{x} = px + x^3. \tag{3.66}$$

Setting $\dot{x} = 0$ gives equilibriums of $x^* = 0, \pm\sqrt{-p}$ $(p < 0)$ and $x^* = 0$ $(p \geq 0)$. in the vicinity of x^*, the linearized equation with $y = x - x^*$ is

$$\dot{y} = G^{(1)}(x^*, p)y = Df(x^*, p)y = [p + 3(x^*)^2]y. \tag{3.67}$$

For the branch of $x^* = 0$ $(p < 0)$, due to $Df(x^*, p) = p < 0$, the equilibrium is stable. In the vicinity of x^*, we have $\dot{y} < 0$ for $y > 0$ and $\dot{y} > 0$ for $y < 0$. For the branches of $x^* = \pm\sqrt{-p}$ $(p < 0)$, such two equilibriums are unstable due to $Df(x^*, p) = -2p > 0$. However, for the branch of $x^* = 0$ $(p > 0)$, the equilibrium is unstable due to $Df(x^*, p) = p > 0$.

For $(x^*, p) = (x_0^*, p_0) = (0, 0)$, $Df(x_0^*, p_0) = 0$ and $D^2f(x_0^*, p_0) = 6x_0^* = 0$ are obtained. Thus, $D^3f(x_0^*, p_0) = 6 > 0$ is needed, and the variational equation at $(x_0^*, p_0) = (0, 0)$ is

$$\dot{y} = \frac{1}{3!}G^{(3)}(x_0^*, p_0)y^3 = \frac{1}{3!}D^3f(x_0^*, p_0)y^3 = y^3. \tag{3.68}$$

In the vicinity of $(x_0^*, p_0) = (0, 0)$, $\dot{y} > 0$ for $y > 0$ and $\dot{y} < 0$ for $y < 0$. The equilibrium point $(x_0^*, p_0) = (0, 0)$ is a bifurcation point, which is a source of the third order. The equilibrium varying with parameter p is sketched in Fig. 3.2f.

From the analysis, the bifurcation points possess the higher order singularity of the flow in dynamical systems. For the saddle-node bifurcation, the (2*m*)th-order singularity of the flow at the bifurcation point exists as a saddle of the (2*m*)th order. For the transcritical bifurcation, the (2*m*)th-order singularity of the flow at the bifurcation point exists as a saddle of the (2*m*)th order. However, for the stable pitchfork bifurcation, the (2*m*+1)th-order singularity of the flow at the bifurcation point exists as a sink of the (2*m*+1)th order. For the unstable pitchfork bifurcation, the (2*m*+1)th-order singularity of the flow at the bifurcation point exists as a source of the (2*m*+1)th order.

3.2 2-Dimensional Nonlinear Systems

In this section, the stability and bifurcation of equilibriums in a 2-dimensional dynamical system will be discussed for a better understanding of stability and bifurcation theory as in Luo (2019).

3.2.1 Stability and Singularity

Consider a 2-dimensional nonlinear system as

$$\dot{\mathbf{x}} = \mathbf{f}(\mathbf{x}, \mathbf{p}) \in R^2 \tag{3.70}$$

where

$$\mathbf{f} = (f_1, f_2)^{\mathrm{T}}, \mathbf{x} = (x_1, x_2)^{\mathrm{T}}, \mathbf{p} = (p_1, p_2, \ldots, p_m)^{\mathrm{T}}. \tag{3.71}$$

The equilibrium $\mathbf{x}^*$ is determined by

$$\mathbf{f}(\mathbf{x}^*, \mathbf{p}) = \mathbf{0} \Rightarrow f_1(x_1^*, x_2^*, \mathbf{p}) = 0 \text{ and } f_2(x_1^*, x_2^*, \mathbf{p}) = 0. \tag{3.72}$$

In the vicinity of $\mathbf{x}^*$, the linearized equation is

$$\dot{\mathbf{y}} = D\mathbf{f}(\mathbf{x}^*, \mathbf{p})\mathbf{y}, \tag{3.73}$$

where $\mathbf{y} = \mathbf{x} - \mathbf{x}^*$ with $\|\mathbf{y}\| \leq \delta$ for a small $\delta > 0$:

$$D\mathbf{f}(\mathbf{x}^*, \mathbf{p}) = \begin{bmatrix} \frac{\partial f_1}{\partial x_1} & \frac{\partial f_1}{\partial x_2} \\ \frac{\partial f_2}{\partial x_1} & \frac{\partial f_2}{\partial x_2} \end{bmatrix}_{\mathbf{x}^*} = \begin{bmatrix} a_{11} & a_{12} \\ a_{21} & a_{22} \end{bmatrix}. \tag{3.74}$$

The corresponding eigenvalues are determined by

$$| D\mathbf{f}(\mathbf{x}^*, \mathbf{p}) - \lambda \mathbf{I}_{2\times 2} | = 0 \Rightarrow \begin{vmatrix} a_{11} - \lambda & a_{12} \\ a_{21} & a_{22} - \lambda \end{vmatrix} = 0, \tag{3.75}$$

and

$$\lambda^2 - \mathrm{Tr}(D\mathbf{f})\lambda + \mathrm{Det}(D\mathbf{f}) = 0, \tag{3.76}$$

where

$$\mathrm{Tr}(D\mathbf{f}) = a_{11} + a_{12}, \mathrm{Det}(D\mathbf{f}) = \begin{vmatrix} a_{11} & a_{12} \\ a_{21} & a_{22} \end{vmatrix}. \tag{3.77}$$

So the eigenvalues are

$$\lambda_{1,2} = \frac{\mathrm{Tr}(D\mathbf{f}) \pm \sqrt{\Delta}}{2}, \tag{3.78}$$

where

$$\Delta = (\text{Tr}(D\mathbf{f}))^2 - 4\text{Det}(D\mathbf{f}). \tag{3.79}$$

If $\Delta > 0$, $\lambda_{1,\,2}$ are real eigenvalues.

(i) If Tr(Df) < 0 and Det(Df) > 0, then $\lambda_{1,\,2} < 0$. Thus the equilibrium $\mathbf{x}^*$ is a stable node.
(ii) If Tr(Df) < 0 and Det(Df) < 0, then $\lambda_1 < 0$ and $\lambda_2 < 0$. Thus the equilibrium $\mathbf{x}^*$ is a saddle.
(iii) If Tr(Df) > 0 and Det(Df) > 0, then $\lambda_{1,\,2} > 0$. Thus the equilibrium $\mathbf{x}^*$ is an unstable node.
(iv) If Tr(Df) > 0 and Det(Df) < 0, then $\lambda_1 < 0$ and $\lambda_2 > 0$. Thus the equilibrium $\mathbf{x}^*$ is a saddle.
(v) If Tr(Df) $= 0$, then $\lambda_1 < 0$ and $\lambda_2 > 0$. Thus the equilibrium $\mathbf{x}^*$ is a saddle.

If $\Delta = 0$, $\lambda_{1,\,2}$ are repeated real eigenvalues:

$$(\text{Tr}(D\mathbf{f}))^2 - 4\text{Det}(D\mathbf{f}) = 0 \tag{3.80}$$

A parabolic curve between Tr(Df) and Det(Df) exists.

(vi) If Tr(Df) > 0, then $\lambda_{1,\,2} = \lambda > 0$. Thus the equilibrium $\mathbf{x}^*$ is an unstable node.
(vii) If Tr(Df) < 0, then $\lambda_{1,\,2} = \lambda < 0$. Thus the equilibrium $\mathbf{x}^*$ is a stable node.
(viii) If Tr(Df) $= 0$, then $\lambda_{1,\,2} = \lambda = 0$. Thus, for $D\mathbf{f} = \mathbf{0}$, then the equilibrium $\mathbf{x}^*$ is a critical case. For $D\mathbf{f} \neq \mathbf{0}$, then the equilibrium $\mathbf{x}^*$ is an unstable node, and flow moves in one eigenvector direction.

If $\Delta < 0$, $\lambda_{1,\,2}$ are complex eigenvalues, i.e.,

$$\lambda_{1,2} = \alpha + \beta\mathbf{i}, \tag{3.81}$$

where

$$\begin{aligned} &\mathbf{i} = \sqrt{-1}, \alpha = \text{Tr}(D\mathbf{f})/2, \\ &\beta = \sqrt{-\Delta} = \sqrt{4\text{Det}(D\mathbf{f}) - (\text{Tr}(D\mathbf{f}))^2}. \end{aligned} \tag{3.82}$$

(ix) If Tr(Df) < 0, then Re$\lambda_{1,\,2} < 0$. Thus the equilibrium $\mathbf{x}^*$ is a stable focus.
(x) If Tr(Df) < 0, then Re$\lambda_{1,\,2} > 0$. Thus the equilibrium $\mathbf{x}^*$ is an unstable focus.
(xi) If Tr(Df) $= 0$, then $\lambda_{1,\,2} = \pm\beta\mathbf{i}$. Thus the equilibrium $\mathbf{x}^*$ is a center.

From the previous discussion, the equilibrium stability for stable node (sink), unstable node (source), saddle, unstable focus, stable focus, and centers is presented in Fig. 3.3.

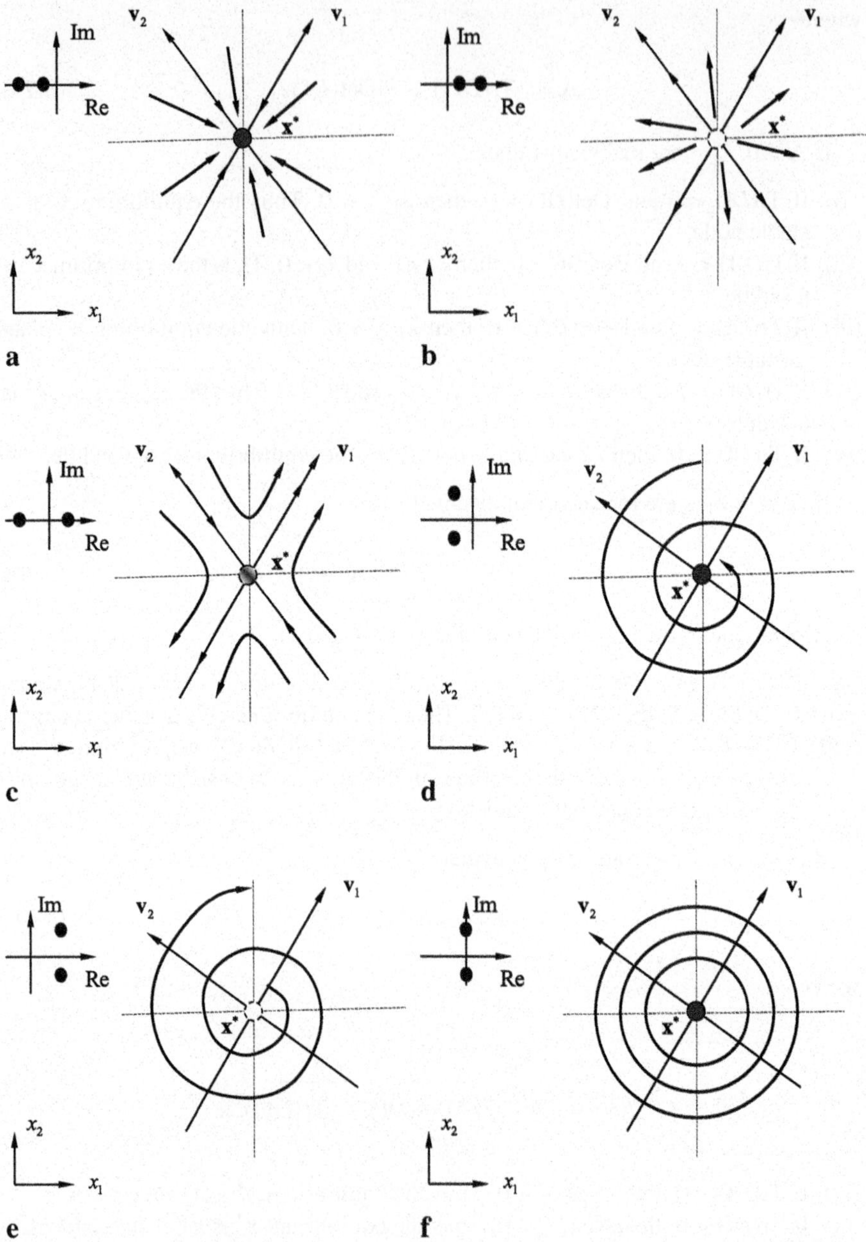

Fig. 3.3 Equilibrium stability: (**a**) stable node (sink), (**b**) unstable node (source), (**c**) saddle, (**d**) stable focus, (**e**) unstable focus, and (**f**) center. The eigenvectors may not always be orthogonal

From the eigenvalues, the eigenvectors $\mathbf{v}_j$ ($j = 1, 2$) are determined by

$$(D\mathbf{f}(\mathbf{x}^*, \mathbf{p}) - \lambda_j \mathbf{I}_{2\times 2})\mathbf{v}_j = \mathbf{0} \Rightarrow D\mathbf{f}(\mathbf{x}^*, \mathbf{p})\mathbf{v}_j = \lambda_j \mathbf{v}_j \;\; (j = 1, 2) \tag{3.83}$$

or

$$\begin{bmatrix} a_{11} - \lambda_j & a_{12} \\ a_{21} & a_{22} - \lambda_j \end{bmatrix} \begin{Bmatrix} v_{j1} \\ v_{j2} \end{Bmatrix} = \begin{Bmatrix} 0 \\ 0 \end{Bmatrix} \;\; (j = 1, 2). \tag{3.84}$$

Thus a new transformation based on the covariant eigenvectors $\mathbf{v}_j$ ($j = 1, 2$) is defined as

$$\mathbf{x} = c^j \mathbf{v}_j = c^1 \mathbf{v}_1 + c^2 \mathbf{v}_2 \;\; (\text{summation of } j = 1, 2), \tag{3.85}$$

and the matrix of the covariant eigenvectors is

$$P = (\mathbf{v}_1, \mathbf{v}_2), \tag{3.86}$$

and the contravariant eigenvectors $\mathbf{v}^j$ ($j = 1, 2$) are defined as

$$\mathbf{v}^1 = (b_{11}, b_{12})^{\mathrm{T}}, \mathbf{v}^2 = (b_{21}, b_{22})^{\mathrm{T}} \text{ and } Q = (\mathbf{v}^1, \mathbf{v}^2), \tag{3.87}$$

where

$$Q^{\mathrm{T}} = P^{-1} = (b_{ij})_{2\times 2}. \tag{3.88}$$

Therefore,

$$c^k = (\mathbf{v}^k)^{\mathrm{T}} \cdot \mathbf{x}, c^{k*} = (\mathbf{v}^k)^{\mathrm{T}} \cdot \mathbf{x}^* \;\; (k = 1, 2), \tag{3.89}$$

and

$$\dot{c}^k = (\mathbf{v}^k)^{\mathrm{T}} \cdot \dot{\mathbf{x}} = (\mathbf{v}^k)^{\mathrm{T}} \cdot \mathbf{f}(\mathbf{x}, \mathbf{p}) = G_k(\mathbf{x}, \mathbf{p}). \tag{3.90}$$

Let

$$\begin{aligned} z^k &= c^k - c^{k*} = (\mathbf{v}^k)^{\mathrm{T}} \cdot (\mathbf{x} - \mathbf{x}^*), \\ \dot{z}^k &= \dot{c}^k = (\mathbf{v}^k)^{\mathrm{T}} \cdot \dot{\mathbf{x}} = (\mathbf{v}^k)^{\mathrm{T}} \cdot \mathbf{f}(\mathbf{x}^* + \Delta\mathbf{x}, \mathbf{p}) \; (k = 1, 2). \end{aligned} \tag{3.91}$$

In the neighborhood of equilibrium $\mathbf{x}^*$, the linearized equation on the covariant eigenvector $\mathbf{v}_k$ ($k = 1, 2$) is

$$\dot{z}^k = G^{(1)}_{k(j)}(\mathbf{x}^*, \mathbf{p})z^j = \lambda_j \delta^k_j z^j = \lambda_k z^k \\ (\text{ summation on } j \text{ but not } k), \tag{3.92}$$

where

$$G_{k(j)}(\mathbf{x}^*, \mathbf{p}) = (\mathbf{v}^k)^{\mathrm{T}} \cdot D\mathbf{f}(\mathbf{x}^*, \mathbf{p})\mathbf{v}_j = (\mathbf{v}^k)^{\mathrm{T}} \cdot \lambda_j \mathbf{v}_j = \lambda_j \delta^k_j. \tag{3.93}$$

Thus

$$z^k = z^k_0 e^{\lambda_k (t - t_0)}. \tag{3.94}$$

(i) If $\lambda_k < 0$, $z^k \to 0$ as $t \to \infty$. Thus, the equilibrium $\mathbf{x}^*$ on the covariant eigenvector $\mathbf{v}_k$ ($k \in \{1, 2\}$) is a stable node.
(ii) If $\lambda_k > 0$, $z^k \to \infty$ as $t \to \infty$. Thus, the equilibrium $\mathbf{x}^*$ on the covariant eigenvector $\mathbf{v}_k$ ($k \in \{1, 2\}$) is an unstable node.
(iii) If $\lambda_k = 0$, $z^k =$ const. For linear systems, this is static. For a nonlinear system, in the vicinity of the equilibrium $\mathbf{x}^*$, the higher order variation equations on the covariant eigenvector $\mathbf{v}_k$ ($k \in \{1, 2\}$) should be considered.

In the neighborhood of equilibrium $\mathbf{x}^*$, the higher order variational equation on the covariant eigenvector $\mathbf{v}_k$ ($k = 1, 2$) is

$$\begin{gathered} G^{(r_k)}_{k(j_1 j_2 \cdots j_{r_k})} = 0, \quad r_k = 1, 2, \ldots, s_k - 1 \\ \dot{z}^k = \frac{1}{m_k!} G^{(s_k)}_{k(j_1 j_2 \cdots j_{s_k})}(\mathbf{x}^*, \mathbf{p}) z^{j_1} z^{j_1} \ldots z^{j_{s_k}} \\ (\text{summation on } j_1, j_2, \ldots, j_{s_k} = 1, 2), \end{gathered} \tag{3.95}$$

where

$$G^{(s_k)}_{k(j_1 j_2 \cdots j_{s_k})}(\mathbf{x}^*, \mathbf{p}) = (\mathbf{v}^k)^{\mathrm{T}} \cdot D^{(s_k)} \mathbf{f}(\mathbf{x}^*, \mathbf{p}) \mathbf{v}_{j_1} \mathbf{v}_{j_2} \ldots \mathbf{v}_{j_{s_k}}. \tag{3.96}$$

For $s_k = 2$,

$$\begin{aligned}
&G^{(1)}_{k(1)} = \lambda_1 \delta^k_1 = 0, G^{(1)}_{k(2)} = \lambda_2 \delta^k_2 = 0,\\
&\dot{z}^k = \frac{1}{2!} G^{(2)}_{k(j_1 j_2)}(\mathbf{x}^*, \mathbf{p}) z^{j_1} z^{j_2}\\
&= \frac{1}{2!} [G^{(2)}_{k(11)}(\mathbf{x}^*, \mathbf{p}) z^1 z^1 + G^{(2)}_{k(12)}(\mathbf{x}^*, \mathbf{p}) z^1 z^2\\
&+ G^{(2)}_{k(21)}(\mathbf{x}^*, \mathbf{p}) z^2 z^1 + G^{(2)}_{k(22)}(\mathbf{x}^*, \mathbf{p}) z^2 z^2]\\
&(k = 1, 2),
\end{aligned} \tag{3.97}$$

where

$$\begin{aligned}
&G^{(2)}_{k(j_1 j_2)}(\mathbf{x}^*, \mathbf{p}) = (\mathbf{v}^k)^{\mathrm{T}} \cdot D^2 \mathbf{f}(\mathbf{x}^*, \mathbf{p}) \mathbf{v}_{j_1} \mathbf{v}_{j_2}\\
&(j_1, j_2 = 1, 2).
\end{aligned} \tag{3.98}$$

(i) If

$$G^{(2)}_{k(j_1 j_2)}(\mathbf{x}^*, \mathbf{p}) z^{j_1} z^{j_2} > 0, \tag{3.99}$$

then the equilibrium $\mathbf{x}^*$ on the covariant eigenvector $\mathbf{v}_k$ ($k \in \{1, 2\}$) is an increasing saddle (or upper saddle) of the second order.

(ii) If

$$G^{(2)}_{k(j_1 j_2)}(\mathbf{x}^*, \mathbf{p}) z^{j_1} z^{j_2} < 0, \tag{3.100}$$

then the equilibrium $\mathbf{x}^*$ on the covariant eigenvector $\mathbf{v}_k$ ($k \in \{1, 2\}$) is a decreasing saddle (or lower saddle) of the second order.

(iii) If

$$z^k \times G^{(2)}_{k(j_1 j_2)}(\mathbf{x}^*, \mathbf{p}) z^{j_1} z^{j_2} > 0, \tag{3.101}$$

then the equilibrium $\mathbf{x}^*$ on the covariant eigenvector $\mathbf{v}_k$ ($k \in \{1, 2\}$) is a source of the second order.

(iv) If

$$z^k \times G^{(2)}_{k(j_1 j_2)}(\mathbf{x}^*, \mathbf{p}) z^{j_1} z^{j_2} < 0, \tag{3.102}$$

then the equilibrium $\mathbf{x}^*$ on the covariant eigenvector $\mathbf{v}_k$ ($k \in \{1, 2\}$) is a sink of the second order.

(v) If

$$G^{(2)}_{k(j_1j_2)}(\mathbf{x}^*,\mathbf{p}) = 0, \tag{3.103}$$

then the equilibrium $\mathbf{x}^*$ on the covariant eigenvector $\mathbf{v}_k$ ($k \in \{1,2\}$) is a critical case of the second order.

For this case, the higher order singularity should be considered for $s_k = 3$:

$$\begin{aligned}
G^{(1)}_{k(j_1)} &= \lambda_{j_1}\delta^k_{j_1} = \lambda_k = 0, G^{(2)}_{k(j_1j_2)}(\mathbf{x}^*,\mathbf{p}) = 0,\\
\dot{z}^k &= \frac{1}{3!}G^{(3)}_{k(j_1j_2j_3)}(\mathbf{x}^*,\mathbf{p})z^{j_1}z^{j_2}z^{j_3}\\
&= \frac{1}{3!}[G^{(3)}_{k(111)}(\mathbf{x}^*,\mathbf{p})z^1z^1z^1 + G^{(3)}_{k(112)}(\mathbf{x}^*,\mathbf{p})z^1z^1z^2\\
&\quad + G^{(3)}_{k(121)}(\mathbf{x}^*,\mathbf{p})z^1z^2z^1 + G^{(3)}_{k(122)}(\mathbf{x}^*,\mathbf{p})z^1z^2z^2\\
&\quad + G^{(3)}_{k(211)}(\mathbf{x}^*,\mathbf{p})z^2z^1z^1 + G^{(3)}_{k(212)}(\mathbf{x}^*,\mathbf{p})z^2z^1z^2\\
&\quad + G^{(3)}_{k(221)}(\mathbf{x}^*,\mathbf{p})z^2z^2z^1 + G^{(3)}_{k(222)}(\mathbf{x}^*,\mathbf{p})z^2z^2z^2],
\end{aligned} \tag{3.104}$$

where

$$\begin{aligned}
&G^{(3)}_{k(j_1j_2j_3)}(\mathbf{x}^*,\mathbf{p}) = (\mathbf{v}^k)^{\mathrm{T}} \cdot D^3\mathbf{f}(\mathbf{x}^*,\mathbf{p})\mathbf{v}_{j_1}\mathbf{v}_{j_2}\mathbf{v}_{j_3}\\
&(j_1,j_2,j_3 = 1,2).
\end{aligned} \tag{3.105}$$

(i) If

$$G^{(3)}_{k(j_1j_2j_3)}(\mathbf{x}^*,\mathbf{p})z^{j_1}z^{j_2}z^{j_3} > 0, \tag{3.106}$$

then the equilibrium $\mathbf{x}^*$ on the covariant eigenvector $\mathbf{v}_k$ ($k \in \{1,2\}$) is an increasing saddle of the third order (or an upper-saddle of the third-order).

(ii) If

$$G^{(3)}_{k(j_1j_2j_3)}(\mathbf{x}^*,\mathbf{p})z^{j_1}z^{j_2}z^{j_3} < 0, \tag{3.107}$$

then the equilibrium $\mathbf{x}^*$ on the covariant eigenvector $\mathbf{v}_k$ ($k \in \{1,2\}$) is a deceasing saddle of the third order (or an lower-saddle of the third-order).

(iii) If

$$z^k \times G^{(3)}_{k(j_1j_2j_3)}(\mathbf{x}^*,\mathbf{p})z^{j_1}z^{j_2}z^{j_3} > 0, \tag{3.108}$$

then the equilibrium $\mathbf{x}^*$ on the covariant eigenvector $\mathbf{v}_k$ ($k \in \{1,2\}$) is an unstable node of the third order (or a source of the third-order).

(iv) If

$$z^k \times G^{(3)}_{k(j_1 j_2 j_3)}(\mathbf{x}^*, \mathbf{p}) z^{j_1} z^{j_2} z^{j_3} < 0, \tag{3.109}$$

then the equilibrium $\mathbf{x}^*$ on the covariant eigenvector $\mathbf{v}_k$ ($k \in \{1,2\}$) is a stable node of the third order (or a sink of the third-order).

3.2.2 *Hopf Bifurcation*

Consider a dynamical system as

$$\begin{aligned} \dot{x} &= [\alpha d + a(x^2 + y^2)]x + [\beta + \alpha c + b(x^2 + y^2)]y, \\ \dot{y} &= -[\beta + \alpha c + b(x^2 + y^2)]y + [\alpha d + a(x^2 + y^2)]y. \end{aligned} \tag{3.110}$$

Setting

$$r^2 = x^2 + y^2 \text{ with } x = r\cos\theta \text{ and } y = r\sin\theta, \tag{3.111}$$

Eq. (3.110) becomes

$$\begin{aligned} \dot{r} &= f_r(r, \alpha) = [\alpha d + ar^2]r, \\ \dot{\theta} &= f_\theta(r, \alpha) = \beta_0 + \alpha c + br^2. \end{aligned} \tag{3.112}$$

The equilibrium of Eq. (3.112) is

$$\begin{aligned} r_1^* &= 0 \text{ for } \alpha \in (-\infty, +\infty), \\ r_2^* &= (-\alpha d/a)^{1/2} \text{ for } (\alpha d) \times a < 0. \end{aligned} \tag{3.113}$$

If $d \neq 0$, we have

$$Df_r(r^*, \alpha) = \alpha d + 3ar^{*2}. \tag{3.114}$$

(i) For $r_1^* = 0$, $Df_r = \alpha d$. For $d > 0$, such an equilibrium is stable as $\alpha < 0$ or unstable as $\alpha > 0$. The equilibrium is a critical point for $\alpha = 0$. However, for $d < 0$, the equilibrium is stable as $\alpha > 0$ or unstable as $\alpha < 0$.

(ii) The equilibrium of $r_2^* = (-\alpha d/a)^{1/2}$ requires $(\alpha d) \times a < 0$. For $a \times d > 0$, such an equilibrium solution exists for $\alpha < 0$. For $a \times d < 0$, the equilibrium existence condition is $\alpha > 0$. From $Df_r = -2\alpha d$, for $a \times d > 0$, the equilibrium is stable for

$(d > 0, a > 0)$ and unstable for $(d < 0, a < 0)$ because of $\alpha < 0$. For $a \times d < 0$, the equilibrium is stable for $(d < 0, a > 0)$ and unstable for $(d > 0, a < 0)$ because of $\alpha > 0$.

(iii) For $\alpha = 0$, we have $r_2^* = r_1^* \equiv r_0^* = 0$ and

$$Df_r(r_0^*, \alpha) = \alpha d = 0 \text{ and } D_\alpha Df_r(r_0^*, \alpha) = d. \tag{3.115}$$

Therefore, for $\alpha = 0$, $r_0^* = 0$ is stable for $d > 0$ and $r_0^* = 0$ is unstable for $d < 0$.

(iv) The bifurcation of equilibrium at point $(r^*, \alpha) = (0, 0)$ is the Hopf bifurcation. On the other hand, $D^2 f_r(r_0^*, \alpha) = 6ar_0^* = 0$ for $r^* = 0$. $D^3 f_r(r_0^*, \alpha) = 6a$ is obtained. Further,

$$\dot{z} = \frac{1}{3!} D^3 f_r(r_0^*, \alpha) z^3 = az^3. \tag{3.116}$$

where $z = r - r^*$.

(iv$_1$) For $a > 0$, the vector field in Eq. (3.116) is greater than zero if $y > 0$ $(r > r_0^*)$ and less than zero if $y < 0$ $(r < r_0^*)$. If $a > 0$ and $\alpha \to 0^+$, $r_2^* = (-\alpha d/a)^{1/2} \to 0^+$ is unstable with $d < 0$. If $a > 0$ and $\alpha \to 0^+$, $r_1^* \to 0^+$ is stable. If $a > 0$ and $\alpha \to 0^-$, $r_1^* \to 0^+$ is unstable. Thus, the bifurcation point possesses a source flow of the third order, and the bifurcation branch is unstable. From Eq. (3.115), we have $\alpha d < 0$ for such an unstable bifurcation because of $(\alpha d) \times a < 0$.

(iv$_2$) If $a < 0$, the vector field in Eq. (3.116) is less than zero if $y > 0$ $(r > r_0^*)$ and greater than zero if $y < 0$ $(r > r_0^*)$. If $a < 0$ and $\alpha \to 0^+$, $r_2^* = (-\alpha d/a)^{1/2} \to 0^+$ is stable with $d > 0$. If $a > 0$ and $\alpha \to 0^+$, $r_1^* \to 0^+$ is unstable. If $a > 0$ and $\alpha \to 0^-$, $r_1^* \to 0^+$ is stable. Thus, for such a case, the bifurcation point possesses a sink flow of the third order. The bifurcation branch is stable. From Eq. (3.115), we have $\alpha d > 0$ for such a stable bifurcation due to $(\alpha d) \times a < 0$.

The Hopf bifurcation with stable focus $(d > 0)$ is supercritical. The Hopf bifurcation with unstable focus $(d < 0)$ is subcritical. The supercritical and subcritical Hopf bifurcation is shown in Fig. 3.4a, b. The solid lines and curves represent stable equilibrium. The dashed lines and curves represent unstable equilibrium. Since $\dot{\theta} = \beta$ is constant and $r^* \neq 0$, one gets a periodic motion on the circle.

Consider a dynamical system in a polar coordinate frame as

$$\dot{r} = f_r(r, \alpha), \dot{\theta} = f_\theta(r, \alpha) \neq 0 \tag{3.117}$$

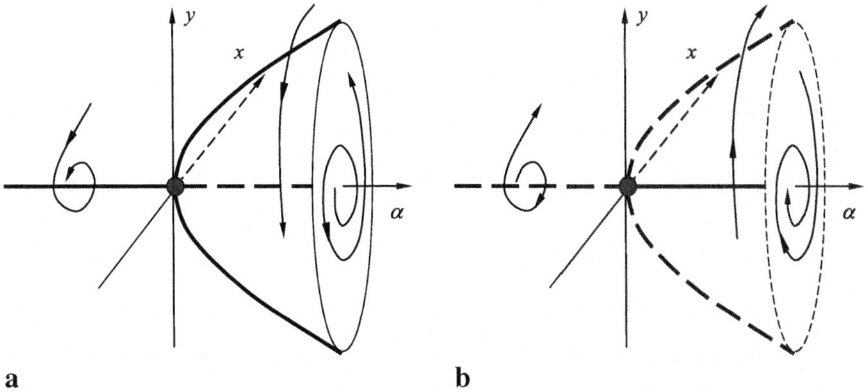

Fig. 3.4 Hopf bifurcations: (**a**) supercritical ($d > 0$, $a > 0$) and (**b**) subcritical ($d < 0$, $a > 0$)

and

$$f_r(r^*, \alpha) = 0 \Rightarrow r_l^* = \Delta_l \ \ (l = 1, 2, \ldots, l_1) \tag{3.118}$$

with

$$\begin{aligned} &\Delta \dot{r}_l = \frac{1}{m!} D^m f_r(r_l^*, \alpha)(\Delta r_l)^m \ \text{ for } \Delta r_l = r_l - r_l^*, \\ &D^{r_l} f_r(r_l^*, \alpha) = 0 \text{ for } r_l = 1, 2, \ldots, m-1, \end{aligned} \tag{3.119}$$

which can be used to determine the stability of periodic motions near the equilibrium.

At the bifurcation point r_0^*, the corresponding variational equation is

$$\dot{z} = \frac{1}{m!} D^m f_r(r_0^*, \alpha) z^m, \tag{3.120}$$

where $z = r^* - r_0^*$.

(i) For $m = 2l-1$, the periodic flow is stable of the $(2l-1)^{\text{th}}$ order if

$$G_r^{(2l-1)} = D^{2l-1} f_r(r_0^*, \alpha) < 0. \tag{3.121}$$

The periodic flow is called a periodic flow of the $(2l-1)^{\text{th}}$-order sink.

(ii) For $m = 2l-1$, the periodic flow is unstable of the $(2l-1)^{\text{th}}$ order if

$$G_r^{(2l-1)} = D^{2l-1} f_r(r_0^*, \alpha) > 0. \tag{3.122}$$

The periodic flow is called a periodic flow of the $(2l-1)^{\text{th}}$-order source.

(iii) For $m = 2l$, the periodic flow is unstable of the $(2l)^{\text{th}}$-order lower saddle if

$$G_r^{(2l)} = D^{2l} f_r(r_0^*, \alpha) < 0. \tag{3.123}$$

The periodic flow is called a periodic flow of the $(2l)^{\text{th}}$-order lower saddle.

(iv) For $m = 2l$, the periodic flow is of the $(2l)^{\text{th}}$-order upper saddle if

$$G_r^{(2l-1)} = D^{2l-1} f_r(r_0^*, \alpha) > 0. \tag{3.124}$$

The periodic flow is called a periodic flow of the $(2l)^{\text{th}}$-order upper saddle.

From the above similar analysis, the Hopf bifurcation points possess the higher order singularity of the flow in dynamical system in the corresponding radial direction, as shown in Fig. 3.5.

(i) For the stable Hopf bifurcation, the $(2l-1)$th-order singularity of the flow at the bifurcation point exists as a sink of the $(2l-1)$th order in the radial direction. Both upper and lower branches (r_1^{*+} and $r_1^{*-} = -r_1^{*+}$) are stable.
(ii) For the unstable Hopf bifurcation, the $(2l-1)$th-order singularity of the flow at the bifurcation point exists as a source of the $(2l-1)$th order in the radial direction. Both upper and lower branches (r_1^{*+} and $r_1^{*-} = -r_1^{*+}$) are unstable.
(iii) For the lower saddle Hopf bifurcation, the $(2l)$th-order singularity of the flow at the bifurcation point exists as a lower saddle of the $(2l)$th order in the radial direction. The upper branch r_1^{*+} is stable and the lower branch $r_1^{*-} = -r_1^{*+}$ is unstable.
(iv) For the upper saddle Hopf bifurcation, the $(2l)$th-order singularity of the flow at the bifurcation point exists as a source of the $(2l)$th order in the radial direction. The upper branch r_1^{*+} is unstable and the lower branch $r_1^{*-} = -r_1^{*+}$ is stable.

The stability and bifurcation of equilibrium for 2D nonlinear dynamic system are summarized with $det(D\mathbf{f}) = det(D\mathbf{f}(\mathbf{x}_0^*, \mathbf{p}_0))$ and $\text{tr}(D\mathbf{f}) = \text{tr}(D\mathbf{f}(\mathbf{x}_0^*, \mathbf{p}_0))$ in Fig. 3.6. The thick dashed lines are bifurcation lines. The stability of equilibriums is given by the eigenvalues in complex plane. The stability of equilibriums for higher dimensional systems can be identified by using a naming of stability for linear dynamical systems. The saddle-node bifurcation possesses stable saddle-node bifurcation (critical) and unstable saddle-node bifurcation (degenerate).

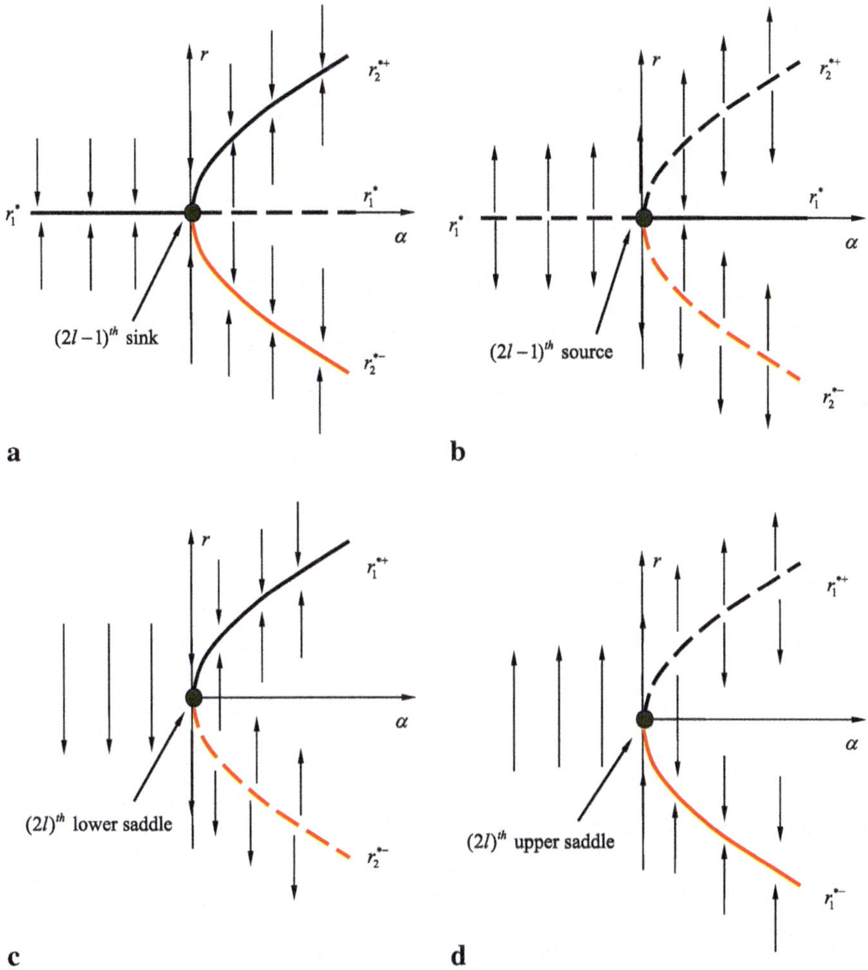

Fig. 3.5 Hopf bifurcations: (**a**) $(2l-1)^{\text{th}}$-order sink Hopf bifurcation and (**b**) $(2l-1)^{\text{th}}$-order source Hopf bifurcation, (**c**) $(2l)^{\text{th}}$-order lower-saddle-appearing/vanishing Hopf bifurcation, (**d**) $(2l)^{\text{th}}$-order upper-saddle-appearing/vanishing Hopf bifurcation. The black branch is for existence solution. The red branch is the asymmetric branch of the black branch

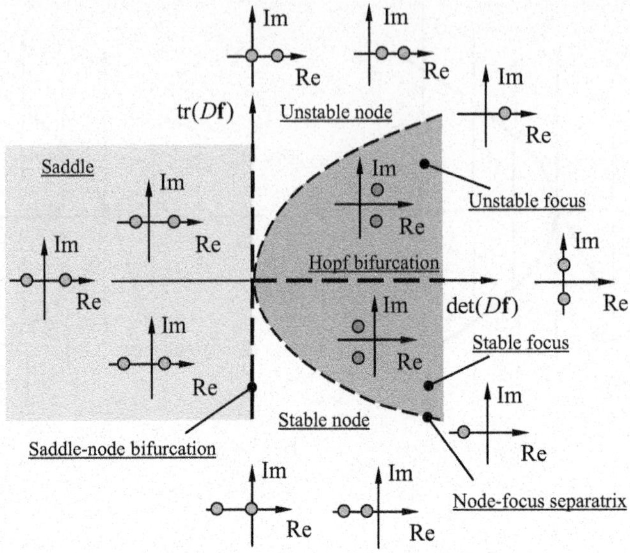

Fig. 3.6 Stability and bifurcation diagrams through the complex plane of eigenvalues for 2D dynamical systems

References

Luo, A.C.J., 2012, Continuous Dynamical Systems, HEP/L&H Scientific: Beijing/Glen Carbon.

Luo, A.C.J., 2019, On stability and bifurcation of equilibriums in nonlinear systems, *Journal of Vibration Testing and System Dynamics,* 3(2). pp. 147–232.

Chapter 4
Equilibrium Stability in 1-Dimensional Systems

In this chapter, a global analysis of equilibrium stability in 1-dimensional nonlinear dynamical systems is presented. The classification of dynamical systems is given first, and infinite-equilibrium systems are defined. The 1-dimensional dynamical systems with single equilibrium are discussed first. The 1-dimensional dynamical systems with two and three equilibriums are discussed. Simple equilibriums and higher order equilibriums in 1-dimensional dynamical systems are analyzed, and herein a higher order equilibrium is an equilibrium with higher order singularity. The separatrix flow of equilibriums in 1-dimensional systems in phase space is illustrated for a better understanding of the global stability of equilibriums in 1-dimensional dynamical systems.

4.1 System Classifications

Definition 4.1 Consider a dynamical system as

$$\dot{\mathbf{x}} = \mathbf{f}(\mathbf{x}, t, \mathbf{p}) \text{ for } t \in I, \mathbf{x} \in \Omega \text{ and } \mathbf{p} \in \Lambda \tag{4.1}$$

where

$$\mathbf{f} = (f_1, f_2, \ldots, f_n)^{\mathrm{T}}. \tag{4.2}$$

(i) The dynamical system is static if $\mathbf{f}(\mathbf{x}, t, \mathbf{p}) \equiv \mathbf{0}$ for $t \in I \subset R$ and $\mathbf{x} \in \Omega$.
(ii) The dynamical system is of the constant velocity if $\mathbf{f}(\mathbf{x}, t, \mathbf{p}) \equiv$ constant for $t \in I \subset R$ and $\mathbf{x} \in \Omega$.

A. C. J. Luo, *Bifurcation and Stability in Nonlinear Dynamical Systems*, Nonlinear Systems and Complexity 28, https://doi.org/10.1007/978-3-030-22910-8_4

(iii) The dynamical system is independent of state space if $\mathbf{f}(\mathbf{x}, t, \mathbf{p}) \equiv \mathbf{g}(t, \mathbf{p})$ for $t \in I \subset R$ and $\mathbf{x} \in \Omega$.
(iv) The dynamical system is autonomous with equilibrium if $\mathbf{f}(\mathbf{x}, t, \mathbf{p}) \equiv \mathbf{g}(\mathbf{x}, \mathbf{p})$ and

$$\mathbf{g}(\mathbf{x}, \mathbf{p})|_{\mathbf{x}=\mathbf{x}^*} = \mathbf{0} \tag{4.3}$$

for $t \in I \subset R$ and $\mathbf{x} \in \Omega$.
(v) The dynamical system is autonomous without equilibrium if $\mathbf{f}(\mathbf{x}, t, \mathbf{p}) \equiv \mathbf{g}(\mathbf{x}, \mathbf{p})$ and

$$\mathbf{g}(\mathbf{x}, \mathbf{p})|_{\mathbf{x}=\mathbf{x}^*} \neq \mathbf{0} \tag{4.4}$$

for $t \in I \subset R$ and $\mathbf{x} = \mathbf{x}^* \in \Omega$.
(vi) The dynamical system is nonautonomous if $\mathbf{f}(\mathbf{x}, t, \mathbf{p})$ is explicitly relative to time for $t \in I \subset R$ and $\mathbf{x} \in \Omega$.

Definition 4.2 Consider an autonomous dynamical system as

$$\dot{\mathbf{x}} = \mathbf{g}(\mathbf{x}, \mathbf{p}) \text{ for } t \in I, \mathbf{x} \in \Omega \text{ and } \mathbf{p} \in \Lambda \tag{4.5}$$

where

$$\mathbf{g} = (g_1, g_2, \ldots, g_n)^{\mathrm{T}}. \tag{4.6}$$

(i) Such a dynamical system is called an autonomous system without equilibrium in r-dimensional directions on domain Ω if for $\mathbf{x}^* \in \Omega$,

$$g_k(\mathbf{x}, \mathbf{p})|_{\mathbf{x}=\mathbf{x}^*} \neq 0 \text{ for } k \in \{l_1, l_2, \ldots, l_r\} \subseteq \{1, 2, \ldots, n\}. \tag{4.7}$$

(ii) Such a dynamical system is called an autonomous system without equilibrium in all directions if for $\mathbf{x}^* \in \Omega$,

$$g_k(\mathbf{x}, \mathbf{p})|_{\mathbf{x}=\mathbf{x}^*} \neq 0 \text{ for } k = 1, 2, \ldots, n. \tag{4.8}$$

(iii) Such a dynamical system is called an autonomous system with equilibrium in r-dimensional directions if for $\mathbf{x}^* \in \Omega$,

$$g_k(\mathbf{x}, \mathbf{p})|_{\mathbf{x}=\mathbf{x}^*} = 0 \text{ for } k \in \{l_1, l_2, \ldots, l_r\} \subseteq \{1, 2, \ldots, n\}. \tag{4.9}$$

(iv) Such a dynamical system is called an autonomous system with equilibrium in all directions if for $\mathbf{x}^* \in \Omega$,

$$g_k(\mathbf{x}, \mathbf{p})|_{\mathbf{x}=\mathbf{x}^*} = 0 \text{ for } k = 1, 2, \ldots, n. \tag{4.10}$$

Definition 4.3 Consider an autonomous dynamical system as

$$\begin{aligned}&\dot{\mathbf{x}} = \mathbf{g}(\mathbf{x}, \mathbf{p}) \text{ and } \mathbf{g}(\mathbf{x}, \mathbf{p})|_{\mathbf{x}=\mathbf{x}^*} = \mathbf{0}\\&\text{for } t \in I, \mathbf{x} \in \Omega \text{ and } \mathbf{p} \in \Lambda\end{aligned} \tag{4.11}$$

where

$$\mathbf{g} = (g_1, g_2, \ldots, g_n)^{\mathrm{T}}, \mathbf{A} = \frac{\partial \mathbf{g}}{\partial \mathbf{x}}\bigg|_{\mathbf{x}=\mathbf{x}^*}. \tag{4.12}$$

(i) Such a dynamical system is called an autonomous system with finite equilibriums if

$$\mathbf{g}(\mathbf{x}^*, \mathbf{p}) = \mathbf{0}, \text{and } \det(\mathbf{A}) \neq 0. \tag{4.13}$$

(ii) Such a dynamical system is called an autonomous system with infinite equilibriums if

$$\mathbf{g}(\mathbf{x}^*, \mathbf{p}) = \mathbf{0}, \text{and } \det(\mathbf{A}) = 0. \tag{4.14}$$

(iii) Such a dynamical system is called an autonomous system with at least r-sets of infinite equilibriums if

$$\mathbf{g}(\mathbf{x}^*, \mathbf{p}) = \mathbf{0}, \text{and } \det(\mathbf{A}) = 0 \,\text{with}\, \text{rank}(\mathbf{A}) = n - r. \tag{4.15}$$

Definition 4.4 Consider an autonomous dynamical system as

$$\begin{aligned}&\dot{x}_i = g_{i1}(\mathbf{x}, \mathbf{p}_{i1}) g_{i2}(\mathbf{x}, \mathbf{p}_{i2}) \ldots g_{ir_i}(\mathbf{x}, \mathbf{p}_{ir_i})\\&(i = 1, 2, \ldots, n), r_i \geq 1\end{aligned} \tag{4.16}$$

where

$$g_i(\mathbf{x}, \mathbf{p}) \equiv g_{i1}(\mathbf{x}, \mathbf{p}_{i1}) g_{i2}(\mathbf{x}, \mathbf{p}_{i2}) \cdots g_{ir_i}(\mathbf{x}, \mathbf{p}_{ir_i}). \tag{4.17}$$

(i) Such a dynamical system is called an autonomous system without equilibrium in the direction of the ith component if

$$\begin{aligned}&g_i(\mathbf{x}, \mathbf{p}) \neq 0 \text{ or } g_{ij}(\mathbf{x}, \mathbf{p}_{ij}) \neq 0\\&\text{for } i \in \{1, 2, \ldots, n\}, j = 1, 2, \ldots, r_i.\end{aligned} \tag{4.18}$$

(ii) Such a dynamical system is called an autonomous system with equilibrium in the direction of the ith component if

$$\begin{aligned} &g_i(\mathbf{x}, \mathbf{p}) = 0 \ \text{ or } \ g_{ij}(\mathbf{x}, \mathbf{p}_{ij}) = 0 \\ &\text{for } i \in \{1, 2, \ldots, n\}, j \in \{1, 2, \ldots, r_i\}. \end{aligned} \tag{4.19}$$

(iii) There is an infinite-equilibrium surface of the zeroth order in such a dynamical system if

$$\begin{aligned} &g_i(\mathbf{x}, \mathbf{p}) = 0, \ \ i = 1, 2, \ldots, n; \\ &g_{i_1 j_1}(\mathbf{x}, \mathbf{p}_{i_1 j_1}) \equiv g_{i_2 j_2}(\mathbf{x}, \mathbf{p}_{i_2 j_2}) = 0 \\ &\text{for } i_1, i_2 \in \{1, 2, \ldots, n\}, i_1 \neq i_2; \\ &j_1 \in \{1, 2, \ldots, r_{i_1}\}, j_2 \in \{1, 2, \ldots, r_{i_2}\}. \end{aligned} \tag{4.20}$$

(iv) There is an infinite-equilibrium surface of the first order in such a dynamical system as

$$\begin{aligned} &g_i(\mathbf{x}, \mathbf{p}) = 0, \ \ i = 1, 2, \ldots, n; \\ &\varphi(\mathbf{x}, \mathbf{q}) = 0 \end{aligned} \tag{4.21}$$

if

$$\begin{aligned} &\lim_{\varphi(\mathbf{x}, \mathbf{q}) \to 0} g_{i_1 j_1}(\mathbf{x}, \mathbf{p}_{i_1 j_1}) = 0, \ \ \lim_{\varphi(\mathbf{x}, \mathbf{q}) \to 0} g_{i_2 j_2}(\mathbf{x}, \mathbf{p}_{i_2 j_2}) = 0, \\ &\lim_{\varphi(\mathbf{x}, \mathbf{q}) \to 0} D_\varphi g_{i_1 j_1}(\mathbf{x}, \mathbf{p}_{i_1 j_1}) = \lim_{\varphi(\mathbf{x}, \mathbf{q}) \to 0} \frac{d}{d\varphi} g_{i_1 j_1}(\mathbf{x}, \mathbf{p}_{i_1 j_1}) \neq 0, \\ &\lim_{\varphi(\mathbf{x}, \mathbf{q}) \to 0} D_\varphi g_{i_2 j_2}(\mathbf{x}, \mathbf{p}_{i_2 j_2}) = \lim_{\varphi(\mathbf{x}, \mathbf{q}) \to 0} \frac{d}{d\varphi} g_{i_2 j_2}(\mathbf{x}, \mathbf{p}_{i_2 j_2}) \neq 0 \\ &\text{with } \lim_{\varphi(\mathbf{x}, \mathbf{q}) \to 0} \frac{g_{i_1 j_1}(\mathbf{x}, \mathbf{p}_{i_1 j_1})}{g_{i_2 j_2}(\mathbf{x}, \mathbf{p}_{i_2 j_2})} = \lim_{\varphi(\mathbf{x}, \mathbf{q}) \to 0} \frac{D_\varphi g_{i_1 j_1}(\mathbf{x}, \mathbf{p}_{i_1 j_1})}{D_\varphi g_{i_2 j_2}(\mathbf{x}, \mathbf{p}_{i_2 j_2})} = \Delta, \\ &0 \neq \mid \Delta \mid < \infty \text{ for } i_1, i_2 \in \{1, 2, \ldots, n\}, i_1 \neq i_2; j_1, j_2 \in \{1, 2, \ldots, r_i\}. \end{aligned} \tag{4.22}$$

(v) There is an infinite-equilibrium surface of the rth order in such a dynamical system as

$$\begin{aligned} &g_i(\mathbf{x}, \mathbf{p}) = 0, \ \ i = 1, 2, \ldots, n; \\ &\varphi(\mathbf{x}, \mathbf{q}) = 0 \end{aligned} \tag{4.23}$$

if

$$\lim_{\varphi(\mathbf{x},\mathbf{q})\to 0} g_{i_1j_1}(\mathbf{x},\mathbf{p}_{i_1j_1}) = 0, \quad \lim_{\varphi(\mathbf{x},\mathbf{q})\to 0} g_{i_2j_2}(\mathbf{x},\mathbf{p}_{i_2j_2}) = 0$$

$$\lim_{\varphi(\mathbf{x},\mathbf{q})\to 0} D_\varphi^k g_{i_1j_1}(\mathbf{x},\mathbf{p}_{i_1j_1}) = \lim_{\varphi(\mathbf{x},\mathbf{q})\to 0} \frac{d^k}{d\varphi^k} g_{i_1j_1}(\mathbf{x},\mathbf{p}_{i_1j_1}) = 0,$$

$$(k = 1, 2, \ldots, r-1)$$

$$\lim_{\varphi(\mathbf{x},\mathbf{q})\to 0} D_\varphi^k g_{i_2j_2}(\mathbf{x},\mathbf{p}_{i_2j_2}) = \lim_{\varphi(\mathbf{x},\mathbf{q})\to 0} \frac{d^k}{d\varphi^k} g_{i_2j_2}(\mathbf{x},\mathbf{p}_{i_2j_2}) = 0,$$

$$(k = 1, 2, \ldots, r-1)$$

$$\lim_{\varphi(\mathbf{x},\mathbf{q})\to 0} D_\varphi^r g_{i_1j_1}(\mathbf{x},\mathbf{p}_{i_1j_1}) = \lim_{\varphi(\mathbf{x},\mathbf{q})\to 0} \frac{d^r}{d\varphi^r} g_{i_1j_1}(\mathbf{x},\mathbf{p}_{i_1j_1}) \neq 0,$$

$$\lim_{\varphi(\mathbf{x},\mathbf{q})\to 0} D_\varphi^r g_{i_2j_2}(\mathbf{x},\mathbf{p}_{i_2j_2}) = \lim_{\varphi(\mathbf{x},\mathbf{q})\to 0} \frac{d^r}{d\varphi^r} g_{i_2j_2}(\mathbf{x},\mathbf{p}_{i_2j_2}) \neq 0$$

$$\text{with} \lim_{\varphi(\mathbf{x},\mathbf{q})\to 0} \frac{g_{i_1j_1}(\mathbf{x},\mathbf{p}_{i_1j_1})}{g_{i_2j_2}(\mathbf{x},\mathbf{p}_{i_2j_2})} = \lim_{\varphi(\mathbf{x},\mathbf{q})\to 0} \frac{D_\varphi^r g_{i_1j_1}(\mathbf{x},\mathbf{p}_{i_1j_1})}{D_\varphi^r g_{i_2j_2}(\mathbf{x},\mathbf{p}_{i_2j_2})} = \Delta,$$

$$|\,\Delta\,| < \infty \text{ but } \Delta \neq 0 \text{ for } i_1, i_2 \in \{1, 2, \ldots, n\}, i_1 \neq i_2; j_1, j_2 \in \{1, 2, \ldots, r_i\}. \tag{4.24}$$

4.2 Equilibrium Stability

In this section, 1-dimensional nonlinear dynamical systems will be discussed first to help one understand the stability and bifurcation of nonlinear systems as in Luo (2019).

Consider a dynamical system decomposed by

$$\dot{x} = f(x,\mathbf{p}) = g_0(x)g_1(x-a_1)g_2(x-a_2)\ldots g_{m-1}(x-a_{m-1})g_m(x-a_m) \tag{4.25}$$

where

$$\begin{aligned} &g_0(x) \neq 0 \text{ for } x \in R \\ &g_k(x-a_k) = 0 \text{ with } x_k^* = a_k \\ &(k = 1, 2, \ldots, m). \end{aligned} \tag{4.26}$$

Thus, such a system has equilibriums as

$$x_k^* = a_k \text{ for } k = 1, 2, \ldots, m. \tag{4.27}$$

Note that $g_k(x - a_k)$ can be any type of functions but not necessary to be a polynomial function.

Consider a polynomial dynamical system as

$$\dot{x} = f(x, \mathbf{p}) = g_0(x)(x - a_1)^{r_1}(x - a_2)^{r_2} \ldots (x - a_{m-1})^{r_{m-1}}(x - a_m)^{r_m} \tag{4.28}$$

where

$$\begin{aligned} &g_0(x) \neq 0 \text{ for } x \in R = (-\infty, \infty), \\ &a_k < a_{k+1} \text{with } -\infty < a_1 \text{ and } a_m < \infty, \\ &a_i \neq a_j \text{ and } r_k \in \{1, 2, \ldots\} \text{ for } i, j, k \in \{1, 2, \ldots, m\}. \end{aligned} \tag{4.29}$$

Thus, such a system has equilibriums as

$$x_k^* = a_k \text{ with the } r_k\text{th order roots for } k = 1, 2, \ldots, m. \tag{4.30}$$

From local analysis, equilibriums of Eq. (4.28) are presented in Fig. 4.1. In Fig. 4.1a, b, $g_0(x) > 0$ and $g_0(x) < 0$. The acronyms LS, US, SI, and SO represent lower saddle, upper saddle, sink (stable node), and source (unstable node), respectively. The circular symbols are equilibrium points.

For the case of $g_0(x) > 0$, there are four basic equilibriums.

(i) For an equilibrium point x_k^* $(k = 1, 2, \ldots, m)$, if

$$\Sigma_{j=k+1}^{m} r_j = 2l_{k1} \text{ and } \Sigma_{j=k}^{m} r_j = 2l_{k2} \text{ for } l_{k1}, l_{k2} \in \{1, 2, 3, \ldots\}, \tag{4.31}$$

then such an equilibrium is an upper saddle of the r_kth order because $\dot{x} > 0$ always for $x \in (x_{k-1}^*, x_k^*)$ and (x_k^*, x_{k+1}^*).

(ii) For an equilibrium point x_k^* $(k = 1, 2, \ldots, m)$, if

$$\Sigma_{j=k+1}^{m} r_j = 2l_{k1} - 1 \text{ and } \Sigma_{j=k}^{m} r_j = 2l_{k2} - 1 \text{ for } l_{k1}, l_{k2} \in \{1, 2, 3, \ldots\}, \tag{4.32}$$

then such an equilibrium is a lower saddle of the r_kth order because $\dot{x} < 0$ always for $x \in (x_{k-1}^*, x_k^*)$ and (x_k^*, x_{k+1}^*).

(iii) For an equilibrium point x_k^* $(k = 1, 2, \ldots, m)$, if

$$\Sigma_{j=k+1}^{m} r_j = 2l_{k1} \text{ and } \Sigma_{j=k}^{m} r_j = 2l_{k2} - 1 \text{ for } l_{k1}, l_{k2} \in \{1, 2, 3, \ldots\}, \tag{4.33}$$

then such an equilibrium is a source of the r_kth order because $\dot{x} > 0$ for $x \in (x_k^*, x_{k+1}^*)$ and $\dot{x} < 0$ for $x \in (x_{k-1}^*, x_k^*)$.

(iv) For an equilibrium point x_k^* $(k = 1, 2, \ldots, m)$, if

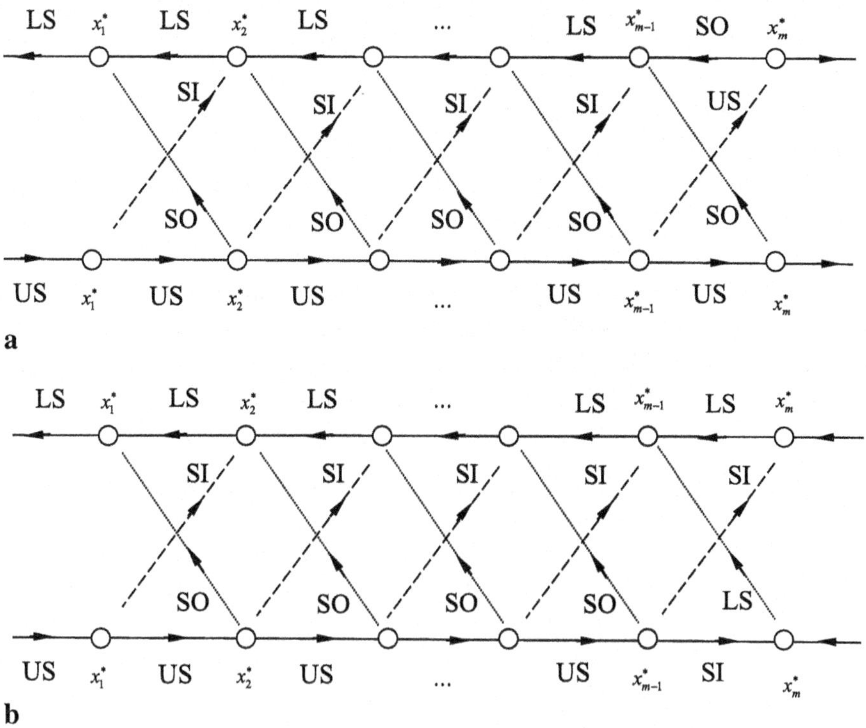

Fig. 4.1 Possible distributions of equilibriums stability of the first-order dynamical system (**a**) $g_0 > 0$ and (**b**) $g_0 < 0$. LS: lower saddle, US: upper saddle, SI: sink (stable node), SO: source (unstable node)

$$\Sigma_{j=k+1}^{m} r_j = 2l_{k1} - 1 \text{ and } \Sigma_{j=k}^{m} r_j = 2l_{k2} \text{ for } l_{k1}, l_{k2} \in \{1, 2, 3, \cdots\}, \quad (4.34)$$

then such an equilibrium is a sink of the r_kth order because $\dot{x} < 0$ for $x \in (x_k^*, x_{k+1}^*)$ and $\dot{x} > 0$ for$x \in (x_{k-1}^*, x_k^*)$.

For the case of $g_0(x) < 0$, similarly, there are four basic equilibriums.

(i) For an equilibrium point x_k^* $(k = 1, 2, \ldots, m)$, if

$$\Sigma_{j=k+1}^{m} r_j = 2l_{k1} - 1 \text{ and } \Sigma_{j=k}^{m} r_j = 2l_{k2} - 1 \text{ for } l_{k1}, l_{k2} \in \{1, 2, 3, \ldots\}, \quad (4.35)$$

then such an equilibrium is an upper saddle of the r_kth order because $\dot{x} > 0$ for $x \in (x_k^*, x_{k+1}^*)$ and $\dot{x} > 0$ for $x \in (x_{k-1}^*, x_k^*)$.

(ii) For an equilibrium point x_k^* $(k = 1, 2, \ldots, m)$, if

$$\Sigma_{j=k+1}^{m} r_j = 2l_{k1} \text{ and } \Sigma_{j=k}^{m} r_j = 2l_{k2} \text{ for } l_{k1}, l_{k2} \in \{1, 2, 3, \ldots\}, \tag{4.36}$$

then such an equilibrium is a lower saddle of the r_kth order because $\dot{x} < 0$ for $x \in (x_k^*, x_{k+1}^*)$ and $\dot{x} < 0$ for $x \in (x_{k-1}^*, x_k^*)$.

(iii) For an equilibrium point x_k^* $(k = 1, 2, \ldots, m)$, if

$$\Sigma_{j=k+1}^{m} r_j = 2l_{k1} - 1 \text{ and } \Sigma_{j=k}^{m} r_j = 2l_{k2} \text{ for } l_{k1}, l_{k2} \in \{1, 2, 3, \ldots\}, \tag{4.37}$$

then such an equilibrium is a source of the r_kth order because $\dot{x} > 0$ for $x \in (x_k^*, x_{k+1}^*)$ and $\dot{x} < 0$ for $x \in (x_{k-1}^*, x_k^*)$.

(iv) For an equilibrium point x_k^* $(k = 1, 2, \ldots, m)$, if

$$\Sigma_{j=k+1}^{m} r_j = 2l_{k1} \text{ and } \Sigma_{j=k}^{m} r_j = 2l_{k2} - 1 \text{ for } l_{k1}, l_{k2} \in \{1, 2, 3, \ldots\}, \tag{4.38}$$

then such an equilibrium is a sink of the r_kth order because $\dot{x} < 0$ for $x \in (x_k^*, x_{k+1}^*)$ and $\dot{x} > 0$ for $x \in (x_{k-1}^*, x_k^*)$.

From the above discussions, the properties of distributed equilibriums can be determined. Thus, a few examples are presented as follows.

4.3 One-Equilibrium Systems

Consider a 1-dimensional system with one equilibrium point as

$$\dot{x} = a_0(x - a_1)^r \text{ with } r = 2m - 1, 2m \text{ for } m = 1, 2, \ldots \tag{4.39}$$

where $a_0 \neq 0$. The rth-order equilibrium of $x^* = a_1$ is obtained easily and the corresponding stability is discussed. Thus, we have

$$\ddot{x} = ra_0(x - a_1)^{r-1}\dot{x} = ra_0^2(x - a_1)^{2r-1} \text{ for } x \neq x^* = a_1 \tag{4.40}$$

and

$$\begin{aligned} & f(x, \mathbf{p}) = a_0(x - a_1)^r = 0, \\ & D^{(j)}f = a_0 r(r-1)\ldots(r-j+1)(x - a_1)^{r-j} = 0, \; j = 1, 2, \ldots, r, \\ & D^{(r)}f = a_0 r! \quad \text{with } r = 2m - 1, 2m \text{ for } m = 1, 2, \ldots. \end{aligned} \tag{4.41}$$

From Eqs. (4.39) and (4.40), the signs of $\dot{x}$ and $\ddot{x}$ can be determined for $x < a_1$ and $x > a_1$, as tabulated in Table 4.1. From such signs, the flow direction in phase space of $(x, \dot{x})$ can be determined. For $a_0 > 0$, $\dot{x} = a_0(x - a_1)^r > 0$ if $x > a_1$. For $a_0 > 0$, $\dot{x} = a_0(x - a_1)^r > 0$ if $x < a_1$ for $r = 2m$, but $\dot{x} = a_0(x - a_1)^r < 0$ if $x_1 < a_1$ for $r = 2m-1$.

Table 4.1 Signs of $\dot{x}$ and $\ddot{x}$ for the 1-dimensional system in Eq. (4.39)

		$a_0 > 0$		$a_0 < 0$	
		$\dot{x}, x$	$\ddot{x}, \dot{x}$	$\dot{x}, x$	$\ddot{x}, \dot{x}$
$r = 2m - 1$	$x > a_1$	$+, \rightarrow$	$+, \uparrow$	$-, \leftarrow$	$+, \uparrow$
	$x < a_1$	$-, \leftarrow$	$-, \downarrow$	$+, \rightarrow$	$-, \downarrow$
	$x^* = a_1$	$(2m - 1)$th-order source (SO)		$(2m - 1)$th-order sink (SI)	
$r = 2m$	$x > a_1$	$+, \rightarrow$	$+, \uparrow$	$-, \leftarrow$	$+, \uparrow$
	$x < a_1$	$+, \rightarrow$	$-, \downarrow$	$-, \leftarrow$	$-, \downarrow$
	$x^* = a_1$	$(2m)$th-order upper saddle (US)		$(2m)$th-order lower saddle (LS)	

Note: $\dot{x} = a_0(x - a_1)^r$ and $\ddot{x} = ra_0^2(x - a_1)^{2r-1}$. LS: lower saddle, US: upper saddle, SI: sink (stable node), SO: source (unstable node)

For $a_0 < 0, \dot{x} = a_0(x - a_1) < 0$ if $x > a_1$. For $a_0 < 0, \dot{x} = a_0(x - a_1)^r > 0$ if $x < a_1$ for $r = 2m$ but $\dot{x} = a_0(x - a_1)^r < 0$ if $x < a_1$ for $r = 2m - 1$. For acceleration $\ddot{x}$, for $a_0 > 0$ and $a_0 < 0$, $\ddot{x} = ra_0^2(x - a_1)^{2r-1} > 0$ if $x > a_1$, but $\ddot{x} = ra_0^2(x - a_1)^{2r-1} < 0$ if $x < a_1$. Because $D^r f = a_0 r!$ with $r = 2m - 1, 2m$ $(m = 1, 2, \ldots)$, $D^r f = a_0 r! > 0$ if $a_0 > 0$ and $D^r f = a_0 r! < 0$ if $a_0 < 0$. Thus, the equilibrium should be four basic equilibriums, i.e., a source (unstable node) of the $(2m - 1)$th order, a sink (stable node) of the $(2m - 1)$th order, an upper saddle of the $(2m)$th order, and a lower saddle of the $(2m - 1)$th order. In a plane of $(x, \dot{x})$, the equilibrium and phase trajectory are presented in Fig. 4.2.

4.4 Two-Equilibrium Systems

Consider a 1-dimensional system with two equilibrium points as

$$\begin{aligned} &\dot{x} = a_0(x - a_1)^{r_1}(x - a_2)^{r_2}, \\ &a_0 \neq 0, a_i \in R \text{ with } a_1 < a_2, \\ &\text{for } r_i = 2m_i - 1, 2m_i, i \in \{1, 2\} \text{ and } m_i = 1, 2, \ldots. \end{aligned} \tag{4.42}$$

Thus

$$\begin{aligned} \ddot{x} &= a_0(r_1 + r_2)(x - a_1)^{r_1 - 1}(x - a_2)^{r_2 - 1}(x - a_{12})\dot{x} \\ &= a_0^2(r_1 + r_2)(x - a_1)^{2r_1 - 1}(x - a_{12})(x - a_2)^{2r_2 - 1}, \end{aligned} \tag{4.43}$$

where

$$a_{12} = \frac{r_1 a_2 + r_2 a_1}{r_1 + r_2} \text{ with } a_1 < a_{12} < a_2. \tag{4.44}$$

Using $\dot{x}$ and $\ddot{x}$, trajectories and equilibriums are presented in Figs. 4.3–4.6. If $a_1 = a_2 = a_{12}$, a bifurcation point exists. The two-equilibrium system becomes one

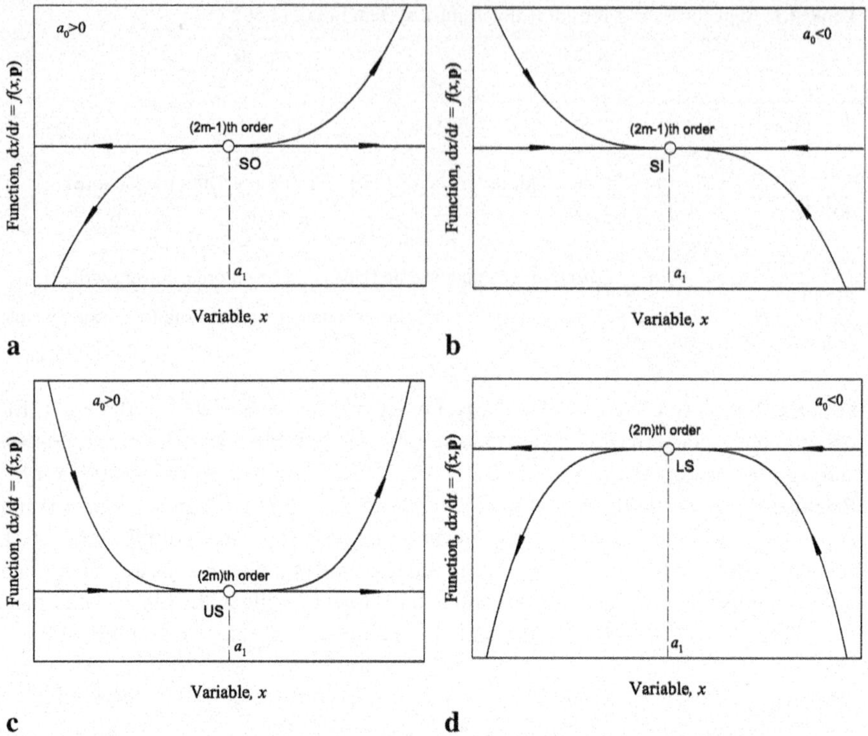

Fig. 4.2 Single equilibrium: (**a**) a source of the $(2m-1)$th order (SO) for $a_0 > 0$, (**b**) a sink of the $(2m-1)$th order (SI) for $a_0 < 0$, (**c**) an upper saddle of the $(2m)$th order (US) for $a_0 > 0$, (**d**) a lower saddle of the $(2m)$th order (LS) for $a_0 < 0$. ($m = 1, 2, \ldots$). LS: lower saddle, US: upper saddle, SI: sink (stable node), SO: source (unstable node)

equilibrium system with the $(r_1 + r_2)^{\text{th}}$-order singularity. This bifurcation generates two equilibriums with the r_1th- and r_2th-order singularity.

The (1:1)-equilibrium points (i.e., $r_1 = 1$, $r_2 = 1$) are called simple equilibriums because the two equilibriums are without any singularity. For $a_0 > 0$, using $\dot{x} = a_0(x - a_1)(x - a_2)$, $\dot{x} > 0$ for $x > a_2$, $\dot{x} < 0$ for $a_1 < x < a_2$, and $\dot{x} > 0$ for $x < a_1$. Thus, the equilibrium of $x^* = a_2$ is a source (unstable node) and the equilibrium of $x^* = a_1$ is a sink (stable node), as shown in Fig. 4.3a.

The (1:2)-equilibrium points (i.e., $r_1 = 1$, $r_2 = 2$) possess the first-order and second-order singularity for $x^* = a_1$ and $x^* = a_2$, respectively. For $a_0 > 0$, with $\dot{x} = a_0(x - a_1)(x - a_2)^2$, $\dot{x} > 0$ for $x > a_2$, $\dot{x} > 0$ for $a_1 < x < a_2$, and $\dot{x} < 0$ for $x < a_1$. Thus, the equilibrium of $x^* = a_2$ is an upper-saddle of the second-order and the equilibrium of $x^* = a_1$ is a source (unstable node), as shown in Fig. 4.3b.

The (2:1)-equilibrium points (i.e., $r_1 = 2$, $r_2 = 1$) possess the second-order and first-order singularity for $x^* = a_1$ and $x^* = a_2$, respectively. For $a_0 > 0$, with $\dot{x} = a_0(x - a_1)^2(x - a_2)$, $\dot{x} > 0$ for $x > a_2$, $\dot{x} < 0$ for $a_1 < x < a_2$, and $\dot{x} < 0$ for $x < a_1$.

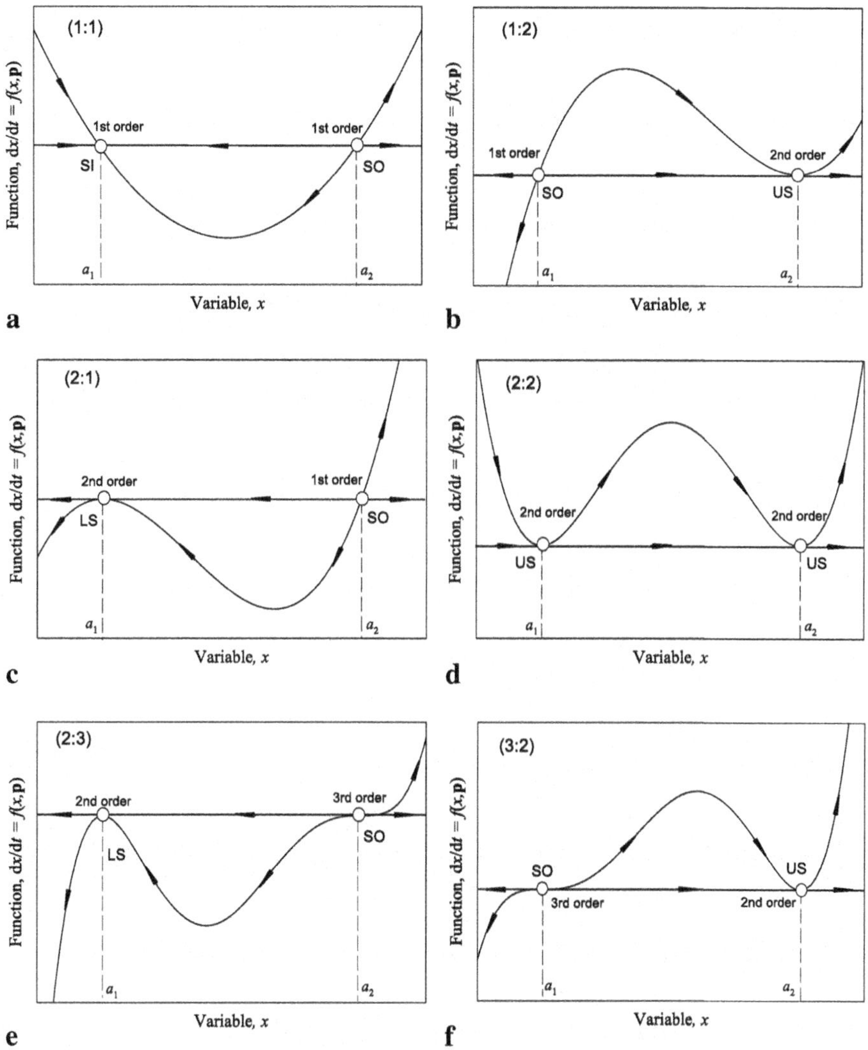

Fig. 4.3 Two equilibriums ($a_0 > 0$): (**a**) (1 : 1)-(SI:SO), (**b**) (1 : 2)-(SO:US), (**c**) (2 : 1)-(LS:SO), (**d**) (2 : 2)-(US:US), (**e**) (2 : 3)-(LS:SO), (**f**) (3 : 2)-(SO:US). LS: lower saddle, US: upper saddle, SI: sink (stable node), SO: source (unstable node)

Therefore, the equilibrium of $x^* = a_2$ is a source (unstable node) and the equilibrium of $x^* = a_1$ is a lower saddle of the second order, as shown in Fig. 4.3c.

The (2 : 2)-equilibrium points (i.e., $r_1 = 2$, $r_2 = 2$) possess the two second-order singularity for $x^* = a_1$ and $x^* = a_2$. For $a_0 > 0$, with $\dot{x} = a_0(x - a_1)^2(x - a_2)^2$, obtained are (i) $\dot{x} > 0$ for $x > a_2$, (ii) $\dot{x} > 0$ for $a_1 < x < a_2$, and (iii) $\dot{x} > 0$ for $x < a_1$. Thus, the equilibriums of $x^* = a_1$ and $x^* = a_2$ are two upper saddles of the second order (US), as shown in Fig. 4.3d.

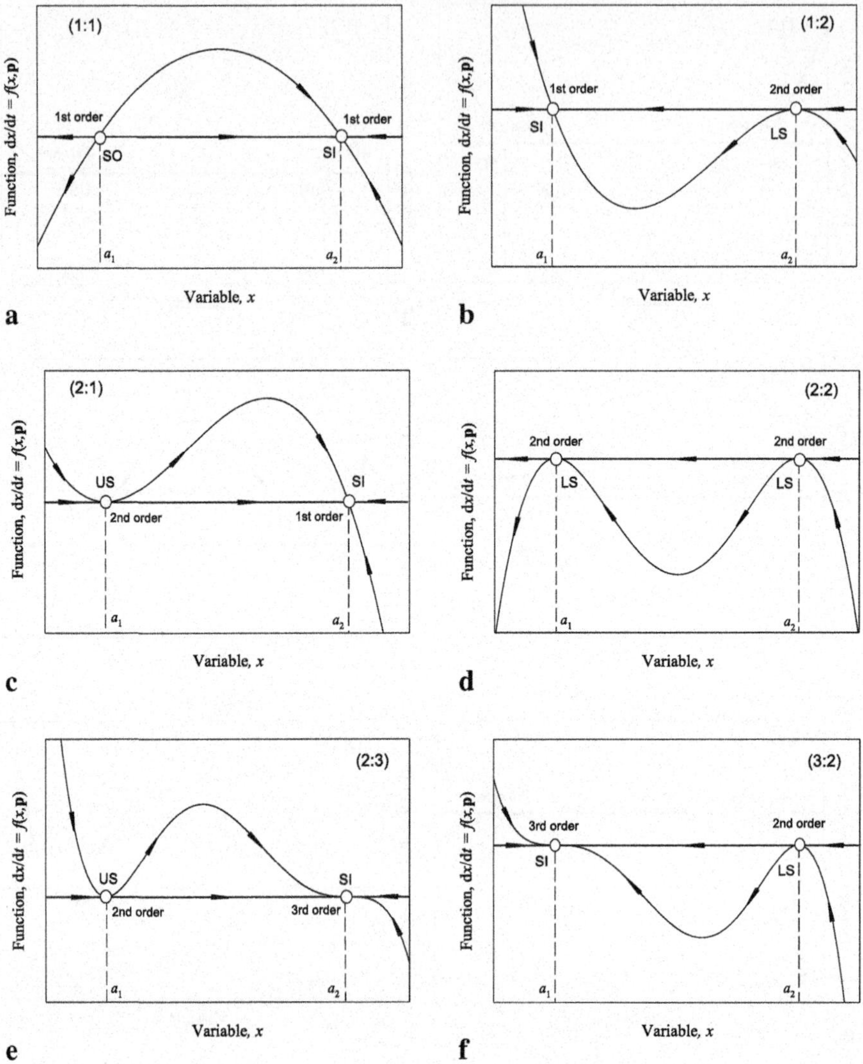

Fig. 4.4 Two equilibriums ($a_0 < 0$): (**a**) (1 : 1)-(SO:SI), (**b**) (1 : 2)-(SI:LS), (**c**) (2 : 1)-(US:SI), (**d**) (2 : 2)-(LS:LS), (**e**) (2 : 3)-(US:SI), (**f**) (3 : 2)-(SI:LS). SO: source (or unstable node), SI: sink (stable node), US: upper saddle, LS: lower saddle

The (2 : 3)-equilibrium points (i.e., $r_1 = 2$, $r_2 = 3$) possess the second-order and third-order singularity for $x^* = a_1$ and $x^* = a_2$, respectively. For $a_0 > 0$, with $\dot{x} = a_0(x - a_1)^2(x - a_2)^3$, in three intervals, $\dot{x} > 0$ for $x > a_2$, $\dot{x} < 0$ for $a_1 < x < a_2$, and $\dot{x} < 0$ for $x < a_1$ are determined. Thus, the equilibrium of $x^* = a_2$ is a source (unstable node) of the third order and the equilibrium of $x^* = a_1$ is a lower

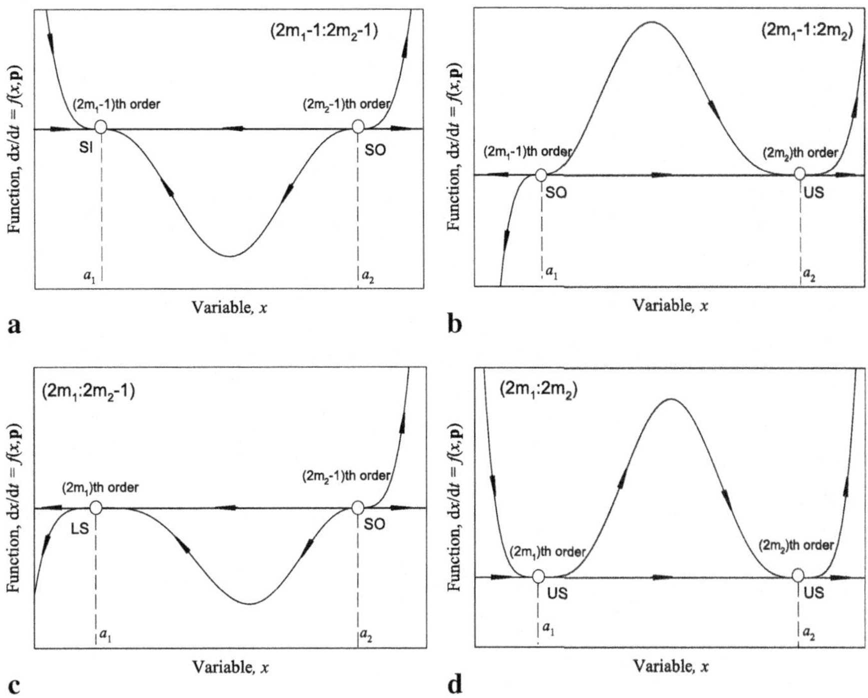

Fig. 4.5 Two equilibriums with general higher singularity: ($a_0 > 0$) (**a**) $(2m_1 - 1 : 2m_2 - 1)$-(SI: SO), (**b**) $(2m_1 - 1 : 2m_2)$-(SO:US), (**c**) $(2m_1 : 2m_2 - 1)$-(LS:SO), (**d**) $(2m_1 : 2m_2)$-(US:US). LS: lower saddle, US: upper saddle, SI: sink (stable node), SO: source (unstable node)

saddle of the second order, as shown in Fig. 4.3e. The (2 : 3)-equilibrium points are similar to the (2 : 1)-equilibrium points without singularity.

The (3 : 2)-equilibrium points (i.e., $r_1 = 3$, $r_2 = 2$) possess the third-order and second-order singularity for $x^* = a_1$ and $x^* = a_2$, respectively. For $a_0 > 0$, with $\dot{x} = a_0(x - a_1)^3(x - a_2)^2$, in three intervals, $\dot{x} > 0$ for $x > a_2$, $\dot{x} > 0$ for $a_1 < x < a_2$, and $\dot{x} < 0$ for $x < a_1$ are obtained. Thus, the equilibrium of $x^* = a_2$ is an upper saddle (unstable node) of the second order and the equilibrium of $x^* = a_1$ is a source of the third order, as shown in Fig. 4.3f. The (3 : 2)-equilibrium points are similar to the (1 : 2)-equilibrium points without singularity.

The stability and singularity of two equilibriums of $\dot{x} = a_0(x - a_1)^{r_1}(x - a_2)^{r_2}$ for $a_0 > 0$ are summarized in Table 4.2. Such analysis can help one understand the stability and singularity of equilibrium, which will be used to discuss the bifurcations of equilibriums later.

For $a_0 < 0$, the (1 : 1)-simple equilibrium points (i.e., $r_1 = 1$, $r_2 = 1$) are different from those for $a_0 > 0$. Using $\dot{x} = a_0(x - a_1)(x - a_2)$, for $a_0 < 0$, obtained are $\dot{x} < 0$ for $x > a_2$, $\dot{x} > 0$ for $a_1 < x < a_2$, and $\dot{x} < 0$ for $x < a_1$. Thus, the equilibrium of $x^* = a_2$

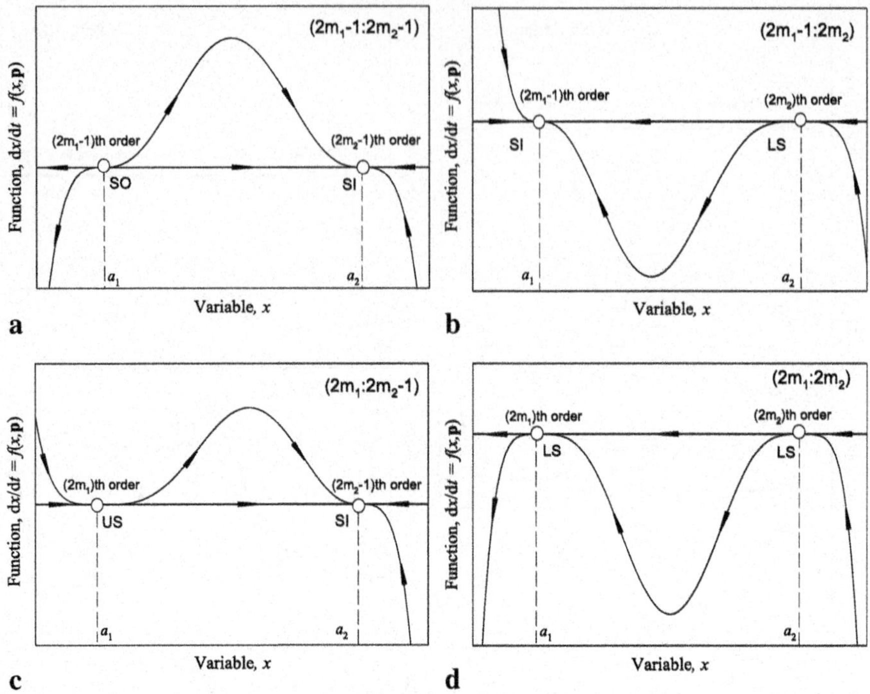

Fig. 4.6 Two equilibriums with general higher singularity: ($a_0 < 0$) (**a**) $(2m_1 - 1 : 2m_2 - 1)$-(SO: SI), (**b**) $(2m_1 - 1 : 2m_2)$-(SI:LS), (**c**) $(2m_1 : 2m_2 - 1)$-(US:SI), (**d**) $(2m_1 : 2m_2)$-(LS:LS). LS: lower saddle, US: upper saddle, SI: sink (stable node), SO: source (unstable node)

Table 4.2 Equilibrium stability and multiplicity for two-equilibrium systems ($a_0 > 0$)

	$x^* = a_1$		$x^* = a_2$	
$(r_1 : r_2)$	Stability	Repeated	Stability	Repeated
(1:1)	1st-order sink (SI)	$r_1 = 1$	1st-order source (SO)	$r_2 = 1$
(1:2)	1st-order source (SO)	$r_1 = 1$	2nd-order upper saddle (US)	$r_2 = 2$
(2:1)	2nd-order lower saddle (LS)	$r_1 = 2$	1st-order source (SO)	$r_2 = 1$
(2:2)	2nd-order upper saddle (US)	$r_1 = 2$	2nd-order upper saddle (US)	$r_2 = 2$
(2:3)	2nd-order lower saddle (LS)	$r_1 = 2$	3rd-order source (SO)	$r_2 = 3$
(3:2)	3rd-order source (SO)	$r_1 = 3$	2nd-order upper saddle (US)	$r_2 = 2$

LS: lower saddle, US: upper saddle, SI: sink (stable node), SO: source (unstable node)

is a sink (stable node) and the equilibrium of $x^* = a_1$ is a source (stable node), as shown in Fig. 4.4a.

For $a_0 < 0$, the $(1 : 2)$-equilibrium points (i.e., $r_1 = 1$, $r_2 = 2$) possess the second-order and first-order singularity for $x^* = a_1$ and $x^* = a_2$, respectively. With $\dot{x} = a_0(x - a_1)(x - a_2)^2$, $\dot{x} < 0$ for $x > a_2$, $\dot{x} < 0$ for $a_1 < x < a_2$, and $\dot{x} > 0$ for $x < a_1$ are obtained. Thus, the equilibrium of $x^* = a_2$ is a lower saddle of the second

Table 4.3 Equilibrium stability and singularity for two-equilibrium systems ($a_0 < 0$)

	$x^* = a_1$		$x^* = a_2$	
$(r_1 : r_2)$	Stability	Repeated	Stability	Repeated
(1:1)	1st-order source (SO)	$r_1 = 1$	1st-order sink (SI)	$r_2 = 1$
(1:2)	1st-order sink (SI)	$r_1 = 1$	2nd-order lower saddle (LS)	$r_2 = 2$
(2:1)	2nd-order upper saddle (US)	$r_1 = 2$	1st-order sink (SI)	$r_2 = 1$
(2:2)	2nd-order lower saddle (LS)	$r_1 = 2$	2nd-order lower saddle (LS)	$r_2 = 2$
(2:3)	2nd-order upper saddle (US)	$r_1 = 2$	3rd-order sink (SI)	$r_2 = 3$
(3:2)	3rd-order sink (SI)	$r_1 = 3$	2nd-order lower saddle (LS)	$r_2 = 2$

LS: lower saddle, US: upper saddle, SI: sink (stable node), SO: source (unstable node)

order and the equilibrium of $x^* = a_1$ is a sink of the first order (stable node) (SI), as shown in Fig. 4.4b.

For $a_0 < 0$, the $(2:1)$-equilibrium points (i.e., $r_1 = 2$, $r_2 = 1$) possess the second-order and first-order singularity for $x^* = a_1$ and $x^* = a_2$, respectively. With $\dot{x} = a_0(x - a_1)^2(x - a_2)$, $\dot{x} < 0$ for $x > a_2$, $\dot{x} > 0$ for $a_1 < x < a_2$, and $\dot{x} > 0$ for $x < a_1$. Therefore, the equilibrium of $x^* = a_2$ is a sink of the first order (stable node) (SI) and the equilibrium of $x^* = a_1$ is an upper saddle of the second order (US), as shown in Fig. 4.4c.

For $a_0 < 0$, the $(2:2)$-equilibrium points (i.e., $r_1 = 2$, $r_2 = 2$) also possess the two second-order singularity for $x^* = a_1$ and $x^* = a_2$. But the two second-order singularity is for the two lower saddles. With $\dot{x} = a_0(x - a_1)^2(x - a_2)^2$, obtained are $\dot{x} < 0$ for $x > a_2$, $\dot{x} < 0$ for $a_1 < x < a_2$, and $\dot{x} < 0$ for $x < a_1$. Thus, the equilibriums of $x^* = a_2$ and $x^* = a_1$ are two lower saddles of the second order, as shown in Fig. 4.4d.

For $a_0 < 0$, the $(2:3)$-equilibrium points (i.e., $r_1 = 2$, $r_2 = 3$) possess the second-order and third-order singularity. With $\dot{x} = a_0(x - a_1)^2(x - a_2)^3$, in three intervals, $\dot{x} < 0$ for $x > a_2$, $\dot{x} > 0$ for $a_1 < x < a_2$, and $\dot{x} > 0$ for $x < a_1$. Thus, the equilibrium of $x^* = a_2$ is a sink (stable node) of the third order and the equilibrium of $x^* = a_1$ is an upper saddle of the second order, as shown in Fig. 4.4e. The two equilibriums are also similar to the $(2:1)$-equilibrium points.

For $a_0 < 0$, the $(3:2)$-equilibrium points (i.e., $r_1 = 3$, $r_2 = 2$) possess the third-order and second-order singularity. With $\dot{x} = a_0(x - a_1)^3(x - a_2)^2$, in three intervals, $\dot{x} < 0$ for $x > a_2$, $\dot{x} < 0$ for $a_1 < x < a_2$, and $\dot{x} > 0$ for $x < a_1$ are determined. Thus, the equilibrium of $x^* = a_2$ is a lower saddle of the second order and the equilibrium of $x^* = a_1$ is a sink (stable node) of the third order, as shown in Fig. 4.4f. The two equilibriums are also similar to the $(1:2)$-equilibrium points.

The stability and singularity of two equilibriums of $\dot{x} = a_0(x - a_1)^{r_1}(x - a_2)^{r_2}$ for $a_0 < 0$ are summarized in Table 4.3, which are different from the stability and singularity for $a_0 > 0$. From the foregoing discussion, the general cases for stability and singularity for the two-equilibrium system of $\dot{x} = a_0(x - a_1)^{r_1}(x - a_2)^{r_2}$ are discussed as follows.

For $a_0 > 0$, the $(2m_1 - 1 : 2m_2 - 1)$-equilibrium points (i.e., $r_1 = 2m_1 - 1$, $r_2 = 2m_2 - 1$) possess the $(2m_1 - 1)$th-order and $(2m_2 - 1)$th-order singularity for

Table 4.4 Equilibrium stability and singularity of a two-equilibrium system in Eq. (4.42) ($a_0 > 0$)

$(r_1 : r_2)$	$x^* = a_1$	$x^* = a_2$
$(2m_1 - 1 : 2m_2 - 1)$	$(2m_1 - 1)$th-order sink (SI)	$(2m_2 - 1)$th-order source (SO)
$(2m_1 - 1 : 2m_2)$	$(2m_1 - 1)$th-order source (SO)	$(2m_2)$th-order upper saddle (US)
$(2m_1 : 2m_2 - 1)$	$(2m_1)$th-order lower saddle (LS)	$(2m_2 - 1)$th-order source (SO)
$(2m_1 : 2m_2)$	$(2m_1)$th-order upper saddle (US)	$(2m_2)$th-order upper saddle (US)

LS: lower saddle, US: upper saddle, SI: sink (stable node), SO: source (unstable node)

$x^* = a_1$ and $x^* = a_2$, respectively. With $\dot{x} = a_0(x - a_1)^{2m_1-1}(x - a_2)^{2m_2-1}$, $\dot{x} > 0$ for $x > a_2$, $\dot{x} < 0$ for $a_1 < x < a_2$, and $\dot{x} > 0$ for $x < a_1$. Thus, the equilibrium of $x^* = a_2$ is a source (unstable node) of the $(2m_1 - 1)$th order and the equilibrium of $x^* = a_1$ is a sink (stable node) of $(2m_2 - 1)$th order, as shown in Fig. 4.5a.

For $a_0 > 0$, the $(2m_1 - 1 : 2m_2)$-equilibrium points (i.e., $r_1 = 2m_1 - 1$, $r_2 = 2m_2$) possess the $(2m_1 - 1)$th-order and $(2m_2)$th-order singularity for $x^* = a_1$ and $x^* = a_2$, respectively. With $\dot{x} = a_0(x - a_1)^{2m_1-1}(x - a_2)^{2m_2}$, $\dot{x} > 0$ for $x > a_2$, $\dot{x} > 0$ for $a_1 < x < a_2$, and $\dot{x} < 0$ for $x < a_1$ are obtained. Thus, the equilibrium of $x^* = a_2$ is an upper saddle of the $(2m_2)$th order and the equilibrium of $x^* = a_1$ is a source (unstable node) of the $(2m_1 - 1)$th order, as shown in Fig. 4.5b.

For $a_0 > 0$, the $(2m_1 : 2m_2 - 1)$-equilibrium points (i.e., $r_1 = 2m_1$, $r_2 = 2m_2 - 1$) possess the $(2m_1)$th-order and $(2m_2 - 1)$th-order singularity for $x^* = a_1$ and $x^* = a_2$, respectively. With $\dot{x} = a_0(x - a_1)^{2m_1}(x - a_2)^{2m_2-1}$, obtained are $\dot{x} > 0$ for $x > a_2$, $\dot{x} < 0$ for $a_1 < x < a_2$, and $\dot{x} < 0$ for $x < a_1$. Thus, the equilibrium of $x^* = a_2$ is a source (unstable node) of the $(2m_2 - 1)$th order and the equilibrium of $x^* = a_1$ is a lower saddle of the $(2m_1)$th order (LS), as shown in Fig. 4.5c.

For $a_0 > 0$, the $(2m_1 : 2m_2)$-equilibrium points (i.e., $r_1 = 2m_1$, $r_2 = 2m_2$) possess the $(2m_1)$th-order and $(2m_2)$th-order singularity for $x^* = a_1$ and $x^* = a_2$, respectively. With $\dot{x} = a_0(x - a_1)^{2m_1}(x - a_2)^{2m_2}$, obtained are $\dot{x} > 0$ for $x > a_2$, $\dot{x} > 0$ for $a_1 < x < a_2$, and $\dot{x} > 0$ for $x < a_1$. Thus, as shown in Fig. 4.5d, the equilibriums of $x^* = a_1$ and $x^* = a_2$ are two upper saddles of the $(2m_1)$th order and $(2m_2)$th order, respectively.

The stability and higher order singularity of the two-equilibrium system for $a_0 > 0$ are summarized in Table 4.4. Similarly, for $a_0 < 0$, the stability and higher order singularity of the two-equilibrium system are discussed as follows.

The $(2m_1 - 1{:}2m_2 - 1)$-equilibrium points (i.e., $r_1 = 2m_1 - 1$, $r_2 = 2m_2 - 1$) for $a_0 < 0$ possess the $(2m_1 - 1)$th-order and $(2m_2 - 1)$th-order singularity for $x^* = a_1$ and $x^* = a_2$, respectively. With $\dot{x} = a_0(x - a_1)^{2m_1-1}(x - a_2)^{2m_2-1}$, obtained are $\dot{x} < 0$ for $x > a_2$, $\dot{x} > 0$ for $a_1 < x < a_2$, and $\dot{x} < 0$ for $x < a_1$. Thus, the equilibrium of $x^* = a_2$ is a sink (stable node) of the $(2m_2 - 1)$th order and the equilibrium of $x^* = a_1$ is a source (unstable node) of $(2m_1 - 1)$th order, as shown in Fig. 4.6a.

For $a_0 < 0$, the $(2m_1 - 1{:}2m_2)$-equilibrium points (i.e., $r_1 = 2m_1 - 1$, $r_2 = 2m_2$) possess the $(2m_1 - 1)$th-order and $(2m_2)$th-order singularity for $x^* = a_1$ and $x^* = a_2$, respectively. With $\dot{x} = a_0(x - a_1)^{2m_1-1}(x - a_2)^{2m_2}$, $\dot{x} < 0$ for $x > a_2$, $\dot{x} < 0$ for

Table 4.5 Stability and singularity of a two-equilibrium system in Eq. (4.42) ($a_0 < 0$)

$(r_1 : r_2)$	$x^* = a_1$	$x^* = a_2$
$(2m_1 - 1 : 2m_2 - 1)$	$(2m_1 - 1)$th-order source (SO)	$(2m_2 - 1)$th-order sink (SI)
$(2m_1 - 1 : 2m_2)$	$(2m_1 - 1)$th-order sink (SI)	$(2m_2)$th-order lower saddle (LS)
$(2m_1 : 2m_2 - 1)$	$(2m_1)$th-order upper saddle (US)	$(2m_2 - 1)$th-order sink (SI)
$(2m_1 : 2m_2)$	$(2m_1)$th-order lower saddle (LS)	$(2m_2)$th-order lower saddle (LS)

LS: lower saddle, US: upper saddle, SI: sink (stable node), SO: source (unstable node)

$a_1 < x < a_2$, and $\dot{x} > 0$ for $x < a_1$ are obtained. Thus, the equilibrium of $x^* = a_2$ is a lower saddle of the $(2m_2)$th order and the equilibrium of $x^* = a_1$ is a sink (stable node) of the $(2m_1 - 1)$th order, as shown in Fig. 4.6b.

For $a_0 < 0$, the $(2m_1{:}2m_2 - 1)$-equilibrium points (i.e., $r_1 = 2m_1, r_2 = 2m_2 - 1$) possess the $(2m_1)$th-order and $(2m_2 - 1)$th-order singularity for $x^* = a_1$ and $x^* = a_2$, respectively. With $\dot{x} = a_0(x - a_1)^{2m_1}(x - a_2)^{2m_2-1}$, obtained are $\dot{x} < 0$ for $x > a_2$, $\dot{x} > 0$ for $a_1 < x < a_2$, and $\dot{x} > 0$ for $x < a_1$. Thus, the equilibrium of $x^* = a_2$ is a sink (stable node) of the $(2m_2 - 1)$th order and the equilibrium of $x^* = a_1$ is an upper saddle of the $(2m_1)$th order, as shown in Fig. 4.6c.

For $a_0 < 0$, the $(2m_1{:}2m_2)$-equilibrium points (i.e., $r_1 = 2m_1, r_2 = 2m_2$) possess the $(2m_1)$th-order and $(2m_2)$th-order singularity for $x^* = a_1$ and $x^* = a_2$, respectively. With $\dot{x} = a_0(x - a_1)^{2m_1}(x - a_2)^{2m_2}$, obtained are $\dot{x} < 0$ for $x > a_2$, $\dot{x} < 0$ for $a_1 < x < a_2$, and $\dot{x} < 0$ for $x < a_1$. Thus, as shown in Fig. 4.6d, the equilibriums of $x^* = a_1$ and $x^* = a_2$ are two lower saddles of the $(2m_1)$th order and $(2m_2)$th order, respectively.

For $a_0 < 0$, the stability and higher order singularity of two equilibriums of $\dot{x} = a_0(x - a_1)^{r_1}(x - a_2)^{r_2}$ are summarized in Table 4.5. The odd-order singularity is relative to the stable and unstable nodes (i.e., sink and source). The even-order singularity is relative to the upper and lower saddles.

4.5 Three-Equilibrium Systems

Consider a 1-dimensional system with three equilibrium points as

$$\begin{aligned}&\dot{x} = a_0(x - a_1)^{r_1}(x - a_2)^{r_2}(x - a_3)^{r_3},\\ &a_0 \neq 0, a_i \in R \text{ with } a_1 < a_2 < a_3,\\ &\text{for } r_i = 2m_i - 1, 2m_i, i \in \{1, 2, 3\} \text{ and } m_i = 1, 2, \ldots.\end{aligned} \tag{4.45}$$

Thus

$$\begin{aligned}\ddot{x} &= a_0(r_1+r_2+r_3)(x-a_1)^{r_1-1}(x-a_2)^{r_2-1}(x-a_3)^{r_3-1}(x^2+a_{12}x+a_{123})\dot{x}\\ &= a_0^2(r_1+r_2+r_3)(x-a_1)^{2r_1-1}(x-a_2)^{2r_2-1}(x-a_3)^{2r_3-1}(x^2+a_{12}x+a_{123})\end{aligned} \tag{4.46}$$

where

$$\begin{aligned}a_{12} &= -\frac{r_1(a_2+a_3)+r_2(a_1+a_3)+r_3(a_1+a_2)}{r_1+r_2+r_3},\\ a_{123} &= \frac{r_1a_2a_3+r_2a_1a_3+r_3a_1a_2}{r_1+r_2+r_3}.\end{aligned} \tag{4.47}$$

Except for equilibrium points, the existence of the three-equilibrium system in Eq. (4.45) requires two points to make $\ddot{x} = 0$, i.e.,

$$x^2 + a_{12}x + a_{123} = 0. \tag{4.48}$$

Thus, the existence condition for such three-equilibrium systems is

$$\Delta = a_{12}^2 - 4a_{123} > 0, \tag{4.49}$$

with

$$x_{1,2} = \frac{a_{12}}{2} \mp \sqrt{\Delta}, a_1 < x_1 < a_2 < x_2 < a_3. \tag{4.50}$$

If $x_1 = a_1 = a_2$, the equilibrium is a bifurcation point with the $(r_1 + r_2)^{\text{th}}$-order singularity. If $x_2 = a_2 = a_3$, the equilibrium is a bifurcation point with the $(r_2 + r_3)^{\text{th}}$-order singularity. If $x_1 = x_2 = a_2$, the equilibrium of $x^* = a_2$ is a bifurcation point with the r_2th-order singularity from the three-equilibrium systems becoming a two-equilibrium system. If $x_1 = x_2 = a_1 = a_2 = a_3$, the equilibrium is a bifurcation point with the $(r_1 + r_2 + r_3)^{\text{th}}$-order singularity from the three-equilibrium systems becoming a one-equilibrium system.

Consider a few lower order singularity of three equilibriums in the three-equilibrium system to demonstrate stability and singularity of equilibriums.

First, consider a simple three-equilibrium system as

$$\dot{x} = a_0(x-a_1)(x-a_2)(x-a_3) \tag{4.51}$$

with $a_i < a_{i+1}$ $(i = 1, 2, 3)$ and $a_0 > 0$. For such a three-equilibrium system, there are three equilibriums $x^* = a_i$ $(i = 1, 2, 3)$. The equilibrium of $x^* = a_3$ is a source point. The equilibrium of $x^* = a_2$ is a sink point. The equilibrium of $x^* = a_1$ is a source point. Two equilibriums are unstable and one equilibrium is stable. The three equilibriums without any singularity are named the (1 : 1 : 1)-equilibriums, which means that three equilibriums are simple. The plane of $(x, \dot{x})$ for the (1 : 1 : 1)-equilibrium dynamical

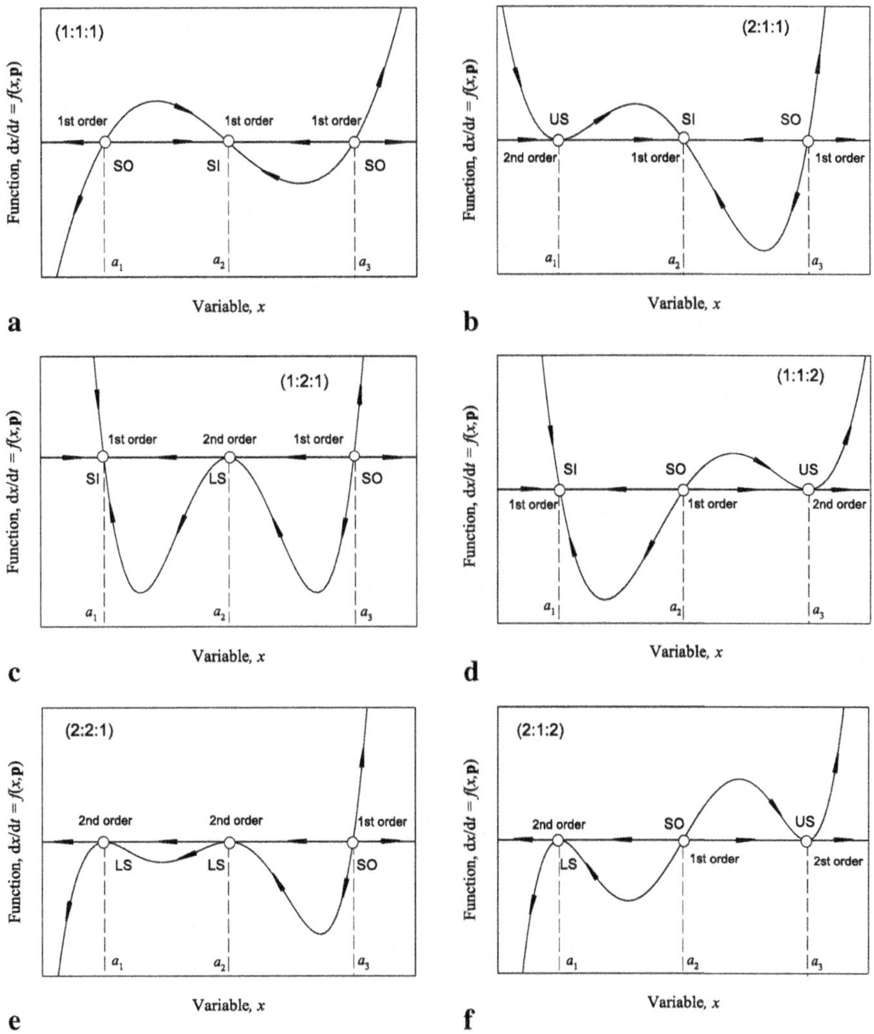

Fig. 4.7 Three-equilibrium systems ($a_0 > 0$): (**a**) (1 : 1 : 1)-(SO:SI:SO), (**b**) (2 : 1 : 1)-(US: SI:SO), (**c**) (1 : 2 : 1)-(SI: US:SO), (**d**) (1:1:2)-(SI:SO:US), (**e**) (2:2:1)-(LS:LS:SO), (**f**) (2:1:2)-(LS:SO:US), (**g**) (1:2:2)-(SO:US:US), (**h**) (3:1:1)-(SO:SI:SO), (**i**) (1:3:1)-(SO:SI:SO), (**j**) (1:1:3)-(SO:SI:SO)

system for $a_0 > 0$ is presented in Fig. 4.7a from Eq. (4.51). Such a simple three-equilibrium has a third-degree polynomial.

Consider a three-equilibrium system with a second-order singularity as

$$\dot{x} = a_0(x - a_1)^2(x - a_2)(x - a_3) \tag{4.52}$$

with $a_i < a_{i+1}$ $(i = 1, 2, 3)$ and $a_0 > 0$. For such a three-equilibrium system, the equilibrium of $x^* = a_3$ is a source point, the equilibrium of $x^* = a_2$ is a sink point,

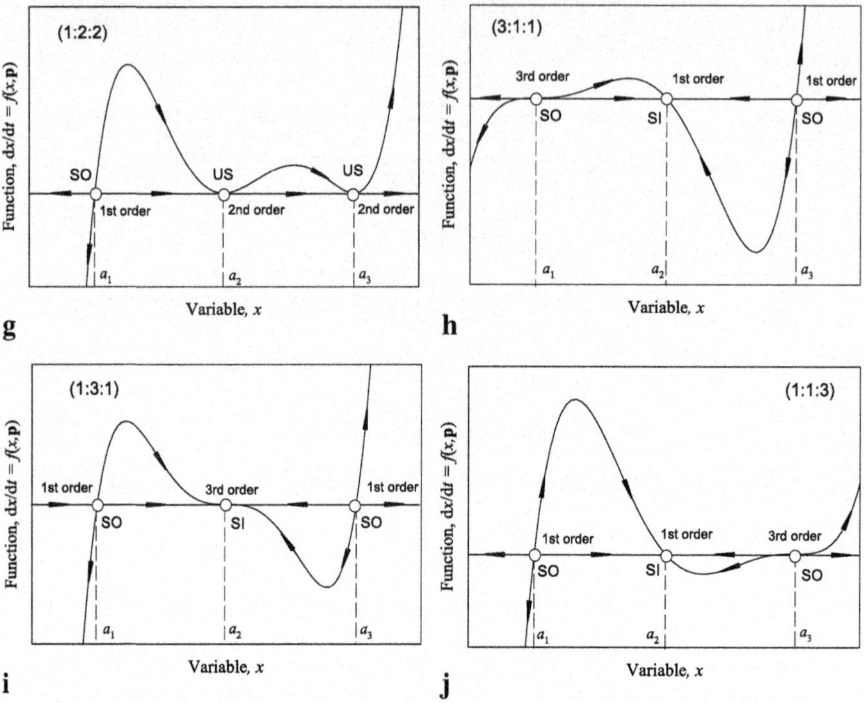

Fig. 4.7 (continued)

and the equilibrium of $x^* = a_1$ is an upper saddle point. The three-equilibrium system possesses the (2:1:1)-equilibriums. Two equilibriums are unstable and one equilibrium is stable. From Eq. (4.52), the plane of $(x, \dot{x})$ for the (2:1;1)-equilibrium dynamical system for $a_0 > 0$ is presented in Fig. 4.7b. If the double repeated equilibrium is considered as two independent equilibriums, such a system should be four equilibriums, and the corresponding equation is a fourth-degree polynomial.

Consider a three-equilibrium system with the second-order singularity on the middle of equilibrium as

$$\dot{x} = a_0(x - a_1)(x - a_2)^2(x - a_3) \tag{4.53}$$

with $a_i < a_{i+1}$ $(i = 1, 2, 3)$ and $a_0 > 0$. In fact, such a three-equilibrium system in Eq. (4.53) is the same class as in Eq. (4.52). Just the double repeated equilibrium point is different from two simple equilibriums. For such a three-equilibrium system, the equilibrium of $x^* = a_3$ is a source point, the equilibrium of $x^* = a_2$ is a lower saddle point, and the equilibrium of $x^* = a_1$ is a sink point. The three-equilibrium system possesses the (1:2;1)-equilibriums. Two equilibriums are unstable and one equilibrium is stable. From Eq. (4.53), the plane of $(x, \dot{x})$ for the (1:2:1)-equilibrium dynamical system for $a_0 > 0$ is presented in Fig. 4.7c. The three-equilibrium system also has four equilibriums with a fourth-degree polynomial.

Consider a three-equilibrium system with the second-order singularity of $x^* = a_3$ as

$$\dot{x} = a_0(x - a_1)(x - a_2)(x - a_3)^2 \tag{4.54}$$

with $a_i < a_{i+1}$ $(i = 1, 2, 3)$ and $a_0 > 0$. The three-equilibrium system in Eq. (4.54) is a class of the dynamical systems in Eqs. (4.52 and 4.53). The three systems possess one double-equilibrium point with a fourth-degree polynomial. However, the three-equilibrium systems with the second-order singularity demonstrate the different dynamical behaviors. For such a three-equilibrium system, the equilibrium of $x^* = a_3$ is an upper saddle point, the equilibrium of $x^* = a_2$ is a source point, and the equilibrium of $x^* = a_1$ is a sink point. The three-equilibrium system possesses the $(1 : 1 : 2)$-equilibriums. Two equilibriums are unstable and one equilibrium is stable. From Eq. (4.54), the plane of $(x, \dot{x})$ for the $(1 : 1 : 2)$-equilibrium dynamical system for $a_0 > 0$ is presented in Fig. 4.7d. From the aforementioned three plots, the three three-equilibrium dynamical systems have different dynamical characteristics.

Consider a three-equilibrium system with two second-order singularities as

$$\dot{x} = a_0(x - a_1)^2(x - a_2)^2(x - a_3) \tag{4.55}$$

with $a_i < a_{i+1}$ $(i = 1, 2, 3)$ and $a_0 > 0$. For such a three-equilibrium system, the equilibrium of $x^* = a_3$ is a source point, the equilibrium of $x^* = a_2$ is a lower saddle point, and the equilibrium of $x^* = a_1$ is a lower saddle point. The three-equilibrium system possesses the $(2 : 2 : 1)$-equilibriums. The three equilibriums are unstable. From Eq. (4.55), the plane of $(x, \dot{x})$ for the $(2 : 2 : 1)$-equilibrium dynamical system is presented in Fig. 4.7e for $a_0 > 0$. If a double repeated equilibrium is considered as two independent equilibriums, such a three-equilibrium system has five equilibriums and the corresponding equation should be a fifth-degree polynomial.

Consider a three-equilibrium system with two second-order singularities as

$$\dot{x} = a_0(x - a_1)^2(x - a_2)(x - a_3)^2 \tag{4.56}$$

with $a_i < a_{i+1}$ $(i = 1, 2, 3)$ and $a_0 > 0$. For such a three-equilibrium system, the equilibrium of $x^* = a_3$ is an upper saddle point, the equilibrium of $x^* = a_2$ is a source point, and the equilibrium of $x^* = a_1$ is a lower saddle point. The three-equilibrium system possesses the (2:1:2)-equilibriums. Three equilibriums are unstable. From Eq. (4.56), the plane of $(x, \dot{x})$ for the (2:1:2)-equilibrium dynamical system is presented in Fig. 4.7f for $a_0 > 0$. Such a three-equilibrium system also has five equilibriums with a fifth-degree polynomial.

Consider a three-equilibrium system with two second-order singularities as

$$\dot{x} = a_0(x - a_1)(x - a_2)^2(x - a_3)^2 \tag{4.57}$$

with $a_i < a_{i+1}$ $(i = 1,2,3)$ and $a_0 > 0$. For such a three-equilibrium system, the equilibrium of $x^* = a_3$ is an upper saddle point, the equilibrium of $x^* = a_2$ is an upper point, and the equilibrium of $x^* = a_1$ is a source point. The three-equilibrium system possesses the (1 : 2 : 2)-equilibriums. Three equilibriums are also unstable. From Eq. (4.57), the plane of $(x, \dot{x})$ for the (1 : 2 : 2)-equilibrium dynamical system is presented in Fig. 4.7g for $a_0 > 0$. Such a three-equilibrium system also has five equilibriums with a fifth-degree polynomial.

For a three-equilibrium system with a fifth-degree polynomial, consider the following three cases.

Case I A three-equilibrium dynamical system with the third-order singularity is

$$\dot{x} = a_0(x - a_1)^3(x - a_2)(x - a_3) \tag{4.58}$$

with $a_i < a_{i+1}$ $(i = 1,2,3)$ and $a_0 > 0$. For such a three-equilibrium system, the equilibrium of $x^* = a_3$ is a source point, and the equilibrium of $x^* = a_2$ is a sink point, and the equilibrium of $x^* = a_1$ is a source point of the third order. The three-equilibrium system possesses the (3 : 1 : 1)-equilibriums. Two equilibriums are unstable and one equilibrium is stable. From Eq. (4.58), the plane of $(x, \dot{x})$ for the (3 : 1 : 1)-equilibrium dynamical system is presented in Fig. 4.7h for $a_0 > 0$. Such a three-equilibrium system has five equilibriums with a fifth-degree polynomial.

Case II A three-equilibrium dynamical system with the third-order singularity is

$$\dot{x} = a_0(x - a_1)(x - a_2)^3(x - a_3) \tag{4.59}$$

with $a_i < a_{i+1}$ $(i = 1,2,3)$ and $a_0 > 0$. For such a three-equilibrium system, the equilibrium of $x^* = a_3$ is a source point, and the equilibrium of $x^* = a_2$ is a sink point of the third order, and the equilibrium of $x^* = a_1$ is a source point. The three-equilibrium system possesses the (1 : 3 : 1)-equilibriums. Two equilibriums are also unstable and one equilibrium is also stable. From Eq. (4.59), the plane of $(x, \dot{x})$ for the (1 : 3 : 1)-equilibrium dynamical system is presented in Fig. 4.7i for $a_0 > 0$. Such a three-equilibrium system also has five equilibriums with a fifth-degree polynomial.

Case III A three-equilibrium dynamical system with the third-order singularity is

$$\dot{x} = a_0(x - a_1)(x - a_2)(x - a_3)^3 \tag{4.60}$$

with $a_i < a_{i+1}$ $(i = 1,2,3)$ and $a_0 > 0$. For the three-equilibrium system, the equilibrium behaviors are changed. The equilibrium of $x^* = a_3$ is a source point of the third order, and the equilibrium of $x^* = a_2$ is a sink point, and the equilibrium of $x^* = a_1$ is a source point. The three-equilibrium system possesses the (1:1:3)-equilibriums. Two equilibriums are also unstable and one equilibrium is also stable. From Eq. (4.60), the plane of $(x, \dot{x})$ for the (1:1:3)-equilibrium dynamical system is presented in Fig. 4.7j for $a_0 > 0$. Such a three-equilibrium system also has five equilibriums with a fifth-degree polynomial.

Table 4.6 Equilibrium stability and singularity of a three-equilibrium system in Eq. (4.45) ($a_0 > 0$)

	$x^* = a_1$	$x^* = a_2$	$x^* = a_3$
(1:1:1)	Source (SO)	Sink (SI)	Source (SO)
(2:1:1)	Upper saddle (2nd)	Sink (SI)	Source (SO)
(1:2:1)	Sink (SI)	Lower saddle (2nd)	Source (SO)
(1:1:2)	Sink (SI)	Source (SO)	Upper saddle (2nd)
(2:2:1)	Lower saddle (2nd)	Lower saddle (2nd)	Source (SO)
(2:1:2)	Lower saddle (2nd)	Source (SO)	Upper saddle (2nd)
(1:2:2)	Source (SO)	Upper saddle (2nd)	Upper saddle (2nd)
(3:1:1)	Source (3rd)	Sink (SI)	Source (SO)
(1:3:1)	Source (SO)	Sink (3rd)	Source (SO)
(1:1:3)	Source (SO)	Sink (SI)	Source (3rd)

LS: lower saddle, US: upper saddle, SI: sink (stable node), SO: source (unstable node)

The stability and singularity of equilibriums for the three-equilibrium systems are summarized in Table 4.6 for $a_0 > 0$. Similarly, equilibrium and trajectories in the plane of $(x, \dot{x})$ for the three-equilibrium systems for $a_0 < 0$ are presented in Fig. 4.8. Only the stability and singularity of three equilibriums for $a_0 < 0$ are summarized in Table 4.7.

With singularity, stability for the other three-equilibrium systems can be discussed as for two-equilibrium dynamical systems. The readers can do the similar discussion as an exercise, which can help readers better understand stability and singularity of equilibriums. The equilibrium stability and singularity for the general cases of the three-equilibrium systems are tabulated in Tables 4.8 and 4.9 for $a_0 > 0$ and $a_0 < 0$, respectively. The odd-number-order singularity is for sink (stable node) and source (unstable node). The even-number-order singularity is for lower saddles and upper saddles.

For an n-equilibrium dynamical system with a $(r_1 : r_2 : \ldots : r_n)$ type, there are 2^n-potential singularities of $2m_i - 1$ and $2m_j$ number combinations for given m_i and m_j $(i,j = 1, 2, \ldots, n)$.

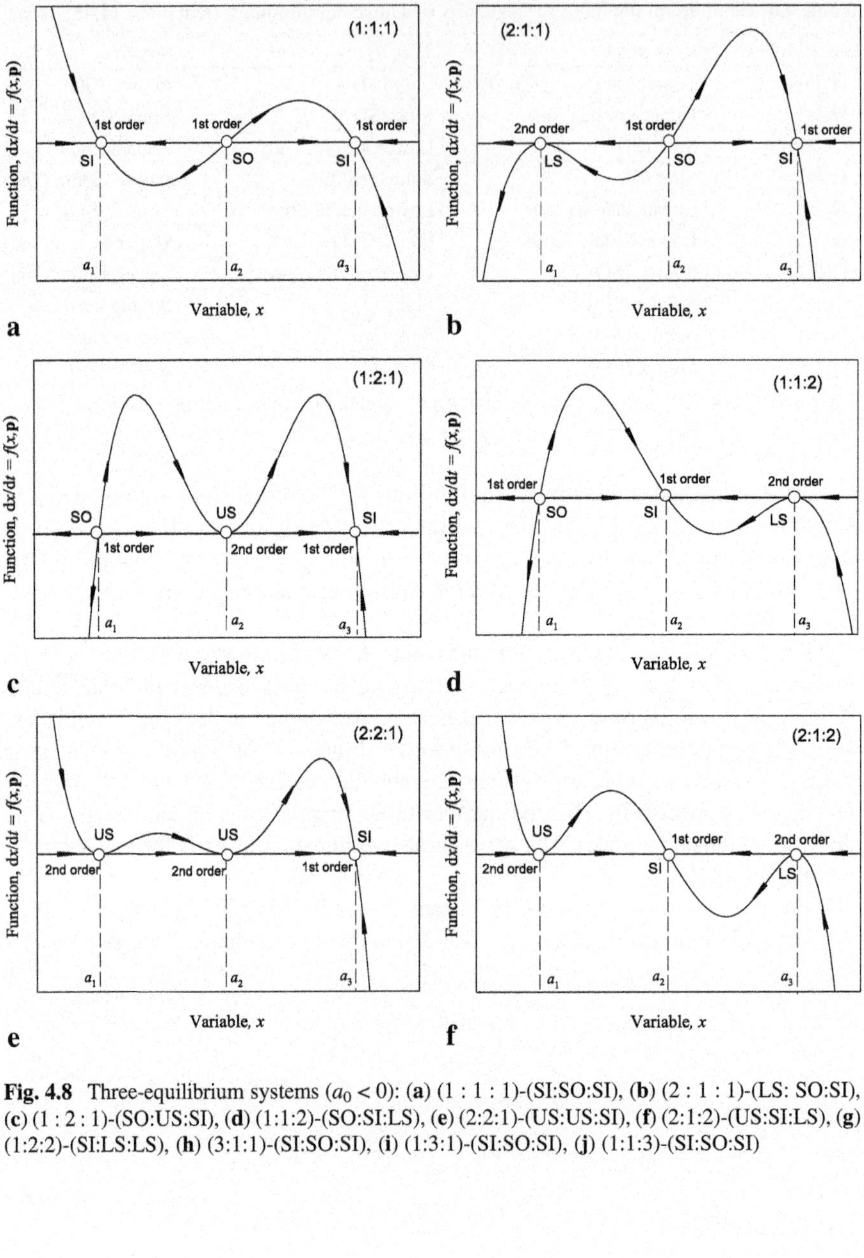

Fig. 4.8 Three-equilibrium systems ($a_0 < 0$): (**a**) (1 : 1 : 1)-(SI:SO:SI), (**b**) (2 : 1 : 1)-(LS: SO:SI), (**c**) (1 : 2 : 1)-(SO:US:SI), (**d**) (1:1:2)-(SO:SI:LS), (**e**) (2:2:1)-(US:US:SI), (**f**) (2:1:2)-(US:SI:LS), (**g**) (1:2:2)-(SI:LS:LS), (**h**) (3:1:1)-(SI:SO:SI), (**i**) (1:3:1)-(SI:SO:SI), (**j**) (1:1:3)-(SI:SO:SI)

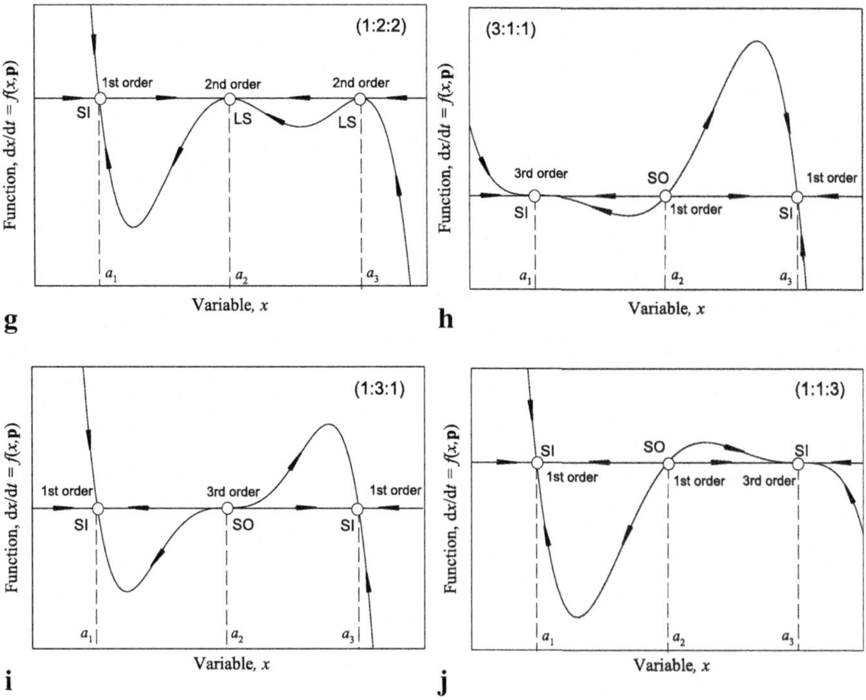

Fig. 4.8 (continued)

Table 4.7 Equilibrium stability and singularity of a three-equilibrium system in Eq. (4.45) ($a_0 < 0$)

$(r_1 : r_2 : r_3)$	$x^* = a_1$	$x^* = a_2$	$x^* = a_3$
(1:1:1)	Sink (SI)	Source (SO)	Sink (SI)
(2:1:1)	Lower saddle (2nd)	Source (SO)	Sink (SI)
(1:2:1)	Source (SO)	Upper saddle (2nd)	Sink (SI)
(1:1:2)	Source (SO)	Sink (SI)	Lower saddle (2nd)
(2:2:1)	Upper saddle (2nd)	Upper saddle (2nd)	Sink (SI)
(2:1:2)	Upper saddle (2nd)	Sink (SI)	Lower saddle (2nd)
(1:2:2)	Sink (SI)	Lower saddle (2nd)	Lower saddle (2nd)
(3:1:1)	Sink (3rd)	Source (SO)	Sink (SI)
(1:3:1)	Sink (SI)	Source (3rd)	Sink (SI)
(1:1:3)	Sink (SI)	Source (SO)	Sink (3rd)

LS: lower saddle, US: upper saddle, SI: sink (stable node), SO: source (unstable node)

Table 4.8 Equilibrium stability and singularity of a three-equilibrium system in Eq. (4.45) ($a_0 > 0$)

$(r_1 : r_2 : r_3)$	$x^* = a_1$	$x^* = a_2$	$x^* = a_3$
$(2m_1 - 1 : 2m_2 - 1 : 2m_3 - 1)$	Source	Sink	Source
$(2m_1 : 2m_2 - 1 : 2m_3 - 1)$	Upper saddle	Sink	Source
$(2m_1 - 1 : 2m_2 : 2m_3 - 1)$	Sink	Lower saddle	Source
$(2m_1 - 1 : 2m_2 - 1 : 2m_3)$	Sink	Source	Upper saddle
$(2m_1 : 2m_2 : 2m_3 - 1)$	Lower saddle	Lower saddle	Source
$(2m_1 : 2m_2 - 1 : 2m_3)$	Lower saddle	Source	Upper saddle
$(2m_1 - 1 : 2m_2 : 2m_3)$	Source	Upper saddle	Upper saddle
$(2m_1 : 2m_2 : 2m_3)$	Upper saddle	Upper saddle	Upper saddle

LS: lower saddle, US: upper saddle, SI: sink (stable node), SO: source (unstable node)

Table 4.9 Equilibrium stability and singularity of a three-equilibrium system in Eq. (4.45) ($a_0 < 0$)

$(r_1 : r_2 : r_3)$	$x^* = a_1$	$x^* = a_2$	$x^* = a_3$
$(2m_1 - 1 : 2m_2 - 1 : 2m_3 - 1)$	Sink	Source	Sink
$(2m_1 : 2m_2 - 1 : 2m_3 - 1)$	Lower saddle	Source	Sink
$(2m_1 - 1 : 2m_2 : 2m_3 - 1)$	Source	Upper saddle	Sink
$(2m_1 - 1 : 2m_2 - 1 : 2m_3)$	Source	Sink	Lower saddle
$(2m_1 : 2m_2 : 2m_3 - 1)$	Upper saddle	Upper saddle	Sink
$(2m_1 : 2m_2 - 1 : 2m_3)$	Upper saddle	Sink	Lower saddle
$(2m_1 - 1 : 2m_2 : 2m_3)$	Sink	Lower saddle	Lower saddle
$(2m_1 : 2m_2 : 2m_3)$	Lower saddle	Lower saddle	Lower saddle

LS: lower saddle, US: upper saddle, SI: sink (stable node), SO: source (unstable node)

Reference

Luo, A.C.J., 2019, The global analysis of equilibrium stability in 1-dimensional systems, *Journal of Vibration Testing and System Dynamics,* 3(3), pp. 329–346.

Chapter 5
Low-Degree Polynomial Systems

In this chapter, the global stability and bifurcation of equilibriums in low-degree polynomial systems are discussed. Appearing and switching bifurcations of simple and higher order equilibriums are discussed, and such bifurcations of equilibriums are not only for simple equilibriums but also for higher order equilibriums. The third-order sink and source bifurcations for simple equilibriums are presented. The third-order sink and source switching bifurcations for *saddle* and nodes are discovered, and the fourth-order *upper saddle* and *lower saddle* switching and appearing bifurcations are obtained for two second-order *upper saddles* and two second-order *lower saddles*, respectively. Graphical illustrations of global stability and bifurcations of equilibriums are presented.

5.1 Linear Systems

In this section, the stability and stability switching of equilibriums in linear systems are discussed. The sink and source equilibriums are discussed.

Definition 5.1 Consider a 1-dimensional linear dynamical system:

$$\dot{x} = A(\mathbf{p})x + B(\mathbf{p}) \tag{5.1}$$

where two scalar constants $A(\mathbf{p})$ and $B(\mathbf{p})$ are determined by a vector parameter

$$\mathbf{p} = (p_1, p_2, \ldots, p_m)^{\mathrm{T}}. \tag{5.2}$$

(i) If $A(\mathbf{p}) \neq 0$, there is an equilibrium point of

A. C. J. Luo, *Bifurcation and Stability in Nonlinear Dynamical Systems*, Nonlinear Systems and Complexity 28, https://doi.org/10.1007/978-3-030-22910-8_5

$$x^* = a_1(\mathbf{p}) = -\frac{B(\mathbf{p})}{A(\mathbf{p})}, \text{ with } a_0(\mathbf{p}) = A(\mathbf{p}) \tag{5.3}$$

and the corresponding dynamical system becomes

$$\dot{x} = a_0(x - a_1). \tag{5.4}$$

(ii) If $A(\mathbf{p}) = 0$, equation (5.1) becomes

$$\dot{x} = B(\mathbf{p}). \tag{5.5}$$

For $B(\mathbf{p}) \neq 0$, the 1-dimensional system is called a constant velocity system. For $B(\mathbf{p}) = 0$, the 1-dimensional system is called a permanent static system with zero velocity.

(iii) For $\|\mathbf{p}\| \to \|\mathbf{p}_0\| = \beta$, if the following relations hold

$$A(\mathbf{p}) = a_0 = \varepsilon \to 0, B(\mathbf{p}) = \varepsilon a_1(\mathbf{p}) \to 0, \tag{5.6}$$

then there is an instant equilibrium to the vector parameter $\mathbf{p}$:

$$x^* = a_1(\mathbf{p}). \tag{5.7}$$

Theorem 5.1 *Under assumption (5.6), a standard form of the 1-dimensional dynamical system in Eq. (5.1) is*

$$\dot{x} = f(x) = a_0(x - a_1) \tag{5.8}$$

(i) *If $a_0(\mathbf{p}) < 0$ (or $df/dx|_{x=a_1} < 0$), then equilibrium $x^* = a_1(\mathbf{p})$ is stable. Such a stable equilibrium is called a sink or a stable node.*
(ii) *If $a_0(\mathbf{p}) > 0$ (or $df/dx|_{x=a_1} > 0$), then equilibrium $x^* = a_1(\mathbf{p})$ is unstable. Such an unstable equilibrium is called a source or an unstable node.*
(iii) *If $a_0(\mathbf{p}) = 0$, then the flow in the neighborhood of equilibrium $x^* = a_1(\mathbf{p})$ is static (critical). Such a static equilibrium is called a critical case.*

Proof Let $y = x - a_1$ and $\dot{x} = \dot{y}$. Thus, equation (5.8) becomes

$$\dot{y} = a_0 y.$$

The corresponding solution is

$$y = y_0 e^{a_0(t - t_0)}$$

where $y_0 = y(t_0) = x_0 - a$ is an initial condition.

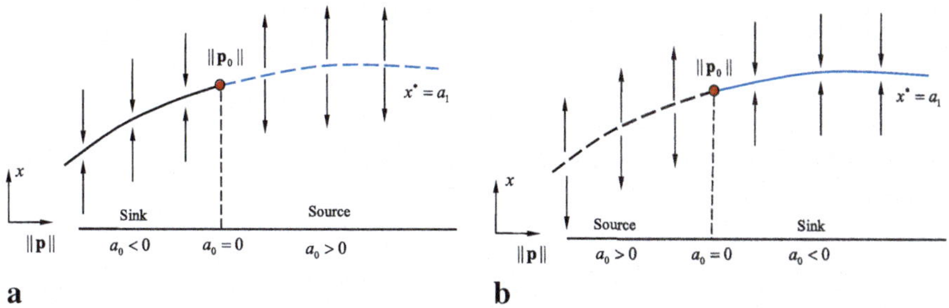

Fig. 5.1 Stability of single equilibrium in the 1-dimensional linear dynamical system: (**a**) left stable equilibrium and (**b**) right stable equilibrium. Stable and unstable equilibriums are represented by solid and dashed curves, respectively. The stability switching is labeled by a circular symbol

(i) If $a_0(\mathbf{p}) < 0$, we have

$$\lim_{t\to\infty}(x - a_1) = \lim_{t\to\infty} y = \lim_{t\to\infty} y_0 e^{a_0(t-t_0)} = 0 \Rightarrow \lim_{t\to\infty} x(t) = a_1$$

So equilibrium $x^* = a_1(\mathbf{p})$ is stable.

(ii) If $a_0(\mathbf{p}) > 0$, we have

$$\lim_{t\to\infty}(x - a_1) = \lim_{t\to\infty} y = \lim_{t\to\infty} y_0 e^{a_0(t-t_0)} = \infty \Rightarrow \lim_{t\to\infty} x(t) = \infty.$$

So equilibrium $x^* = a_1(\mathbf{p})$ is unstable.

(iii) If $a_0(\mathbf{p}) = 0$, we have

$$\lim_{t\to\infty}(x - a_1) = \lim_{t\to\infty} y = \lim_{t\to\infty} y_0 e^{a_0(t-t_0)} = y_0 \Rightarrow \lim_{t\to\infty} x(t) = x_0.$$

So the flow in the neighborhood of equilibrium $x^* = a_1(\mathbf{p})$ is static.

The theorem is proved. ∎

To illustrate the stability of equilibrium, one equilibrium point of $x^* = a_1(\mathbf{p})$ changes with a vector parameter **p**. The stability of such an equilibrium is determined by the constant $a_0(\mathbf{p})$. The stability switching at the boundary $\mathbf{p}_0 \in \partial\Omega_{12}$ with $a_0 = 0$. The stable equilibrium on the left and right sides is presented in Fig. 5.1a, b, respectively. The stable and unstable portions of the equilibrium are presented by the solid and dash curves, respectively.

5.2 Quadratic Nonlinear Systems

In this section, the stability of equilibriums in 1-dimensional quadratic nonlinear systems is discussed. The *upper-saddle-node* and *lower-saddle-node* appearing and switching bifurcations are discussed as in Luo (2019).

Definition 5.2 Consider a 1-dimensional quadratic nonlinear dynamical system:

$$\dot{x} = A(\mathbf{p})x^2 + B(\mathbf{p})x + C(\mathbf{p}) \tag{5.9}$$

where three scalar constants $A(\mathbf{p}) \neq 0$, $B(\mathbf{p})$, and $C(\mathbf{p})$ are determined by a vector parameter

$$\mathbf{p} = (p_1, p_2, \ldots, p_m)^{\mathrm{T}}. \tag{5.10}$$

(i) If

$$\Delta = B^2 - 4AC < 0 \text{ for } \mathbf{p} \in \Omega_1 \subset \mathbf{R}^m, \tag{5.11}$$

then the quadratic nonlinear system does not have any equilibriums. The flow without equilibrium is called a nonequilibrium flow.

(i_1) If $a_0(\mathbf{p}) = A(\mathbf{p}) > 0$, the nonequilibrium flow is called a positive flow.
(i_2) If $a_0(\mathbf{p}) = A(\mathbf{p}) < 0$, the nonequilibrium flow is called a negative flow.

(ii) If

$$\Delta = B^2 - 4AC > 0 \text{ for } \mathbf{p} \in \Omega_2 \subset \mathbf{R}^m, \tag{5.12}$$

then the quadratic nonlinear system has two different simple equilibriums as

$$x^* = a_1 \text{ and } x^* = a_2, \tag{5.13}$$

and the corresponding standard form is

$$\dot{x} = a_0(x - a_1)(x - a_2), \tag{5.14}$$

where

$$a_0 = A(\mathbf{p}), a_{1,2} = \frac{-B(\mathbf{p}) \pm \sqrt{\Delta}}{2A(\mathbf{p})} \text{ with } a_1 < a_2. \tag{5.15}$$

(iii) If

$$\Delta = B^2 - 4AC = 0 \text{ for } \mathbf{p} = \mathbf{p}_0 \in \partial\Omega_{12} \subset \mathbf{R}^{m-1}, \tag{5.16}$$

then the 1-dimensional dynamical system has a double-repeated equilibrium, i.e.,

$$x^* = a_1 \text{ and } x^* = a_1, \tag{5.17}$$

with the corresponding standard form of

$$\dot{x} = a_0(x - a_1)^2, \tag{5.18}$$

where

$$a_0 = A(\mathbf{p}_0), \text{and } a_1 = a_2 = -\frac{B(\mathbf{p}_0)}{2A(\mathbf{p}_0)}. \tag{5.19}$$

Such a flow with the equilibrium of $x^* = a_1$ is called a *saddle* flow of the second order.

(iii_1) If $a_0(\mathbf{p}) > 0$, then the equilibrium $x^* = a_1(\mathbf{p})$ is an *upper saddle*.
(iii_2) If $a_0(\mathbf{p}) < 0$, then the equilibrium $x^* = a_1(\mathbf{p})$ is a *lower saddle*.

(iv) The equilibrium of $x^* = a_1$ for two equilibriums vanishing or appearance is called a *saddle-node* bifurcation point of equilibrium at a point $\mathbf{p} = \mathbf{p}_0 \in \partial\Omega_{12}$, and the bifurcation condition is

$$\Delta = B^2 - 4AC = 0. \tag{5.20}$$

(iv_1) If $a_0(\mathbf{p}) > 0$, the bifurcation at $x^* = a_1(\mathbf{p})$ for two equilibriums appearing or vanishing is called an *upper-saddle-node* bifurcation.
(iv_2) If $a_0(\mathbf{p}) < 0$, the bifurcation at $x^* = a_1(\mathbf{p})$ for two equilibriums appearing or vanishing is called a *lower-saddle-node* bifurcation.

Theorem 5.2 (i) *Under a condition of*

$$\Delta = B^2 - 4AC < 0, \tag{5.21}$$

a standard form of the 1-dimensional dynamical system in Eq. (5.9) is

$$\dot{x} = a_0\left[(x - \frac{1}{2}\frac{B}{A})^2 + \frac{1}{4A^2}(-\Delta)\right] \tag{5.22}$$

with $a_0 = A(\mathbf{p})$, *which has a nonequilibrium flow.*

(i_1) If $a_0(\mathbf{p}) > 0$, *the nonequilibrium flow is called a positive flow.*
(i_2) If $a_0(\mathbf{p}) > 0$, *the non-equilibrium flow is called a negative flow.*

(ii) *Under a condition of*

$$\Delta = B^2 - 4AC > 0, \tag{5.23}$$

a standard form of the 1-dimensional dynamical system in equation (5.9) is

$$\dot{x} = f(x, \mathbf{p}) = a_0(x - a_1)(x - a_2). \tag{5.24}$$

(ii_1) If $a_0(\mathbf{p}) > 0$, then equilibrium $x^* = a_1(\mathbf{p})$ is stable (sink) with $df/dx|_{x=a_1} < 0$, and equilibrium $x^* = a_2(\mathbf{p})$ is unstable (source) with $df/dx|_{x=a_2} > 0$.

(ii_2) If $a_0(\mathbf{p}) < 0$, then equilibrium $x^* = a_1(\mathbf{p})$ is unstable (source) with $df/dx|_{x=a_1} > 0$, and equilibrium $x^* = a_2(\mathbf{p})$ is stable (sink) with $df/dx|_{x=a_2} < 0$.

(iii) *Under a condition of*

$$\Delta = B^2 - 4AC = 0, \tag{5.25}$$

a standard form of the 1-dimensional dynamical system in Eq. (5.9) is

$$\dot{x} = f(x, \mathbf{p}) = a_0(x - a_1)^2. \tag{5.26}$$

(iii_1) If $a_0(\mathbf{p}) > 0$, then the equilibrium $x^* = a_1(\mathbf{p})$ is an upper saddle *of the second order with $d^2f/dx^2|_{x=a_1} > 0$. The bifurcation at $x^* = a_1(\mathbf{p})$ for two equilibriums appearing or vanishing is called an* upper-*saddle-node bifurcation.*

(iii_2) If $a_0(\mathbf{p}) < 0$, then the equilibrium $x^* = a_1(\mathbf{p})$ is a lower saddle *of the second order with $d^2f/dx^2|_{x=a_1} < 0$. The bifurcation at $x^* = a_1(\mathbf{p})$ for two equilibriums appearing or vanishing is called a* lower-saddle-node *bifurcation.*

Proof (i) For

$$\Delta = B^2 - 4AC < 0,$$

(i_1) If $a_0 > 0$, we have

$$\dot{x} = a_0[(x - \frac{1}{2}\frac{B}{A})^2 + \frac{1}{4A^2}(-\Delta)] > 0.$$

Thus such a nonequilibrium flow is called a positive flow.

(i_2) If $a_0 < 0$, we have

$$\dot{x} = a_0[(x - \frac{1}{2}\frac{B}{A})^2 + \frac{1}{4A^2}(-\Delta)] < 0.$$

Thus such a nonequilibrium flow is called a negative flow.

Let $\Delta x_i = x - a_i$ ($i = 1, 2$) and $\dot{x} = \Delta \dot{x}_i$. Thus, equation (5.24) becomes

$$\Delta \dot{x}_i = a_0(a_i - a_j)\Delta x_i + a_0 \Delta x_i^2 \ (i,j \in \{1,2\}, j \neq i).$$

Because Δx_i is arbitrarily small, we have

$$\Delta \dot{x}_i \approx \lambda_i \Delta x_i \text{ for } \lambda_i \equiv a_0(a_i - a_j).$$

The corresponding solution is

$$\Delta x_i = \Delta x_{i0} e^{\lambda_i (t - t_0)}$$

where $\Delta x_{i0} = x_0 - a_i$ is an initial condition.

(ii) If $\lambda_i < 0$, we have

$$\lim_{t\to\infty}(x - a_i) = \lim_{t\to\infty}\Delta x_i = \lim_{t\to\infty}\Delta x_{i0} e^{\lambda_i (t-t_0)} = 0 \Rightarrow \lim_{t\to\infty} x(t) = a_i$$

So equilibrium $x^* = a_i(\mathbf{p})$ is stable.

(ii$_1$) For $\lambda_1 = a_0(a_1 - a_2) < 0$, due to $a_1 - a_2 < 0$, such an equilibrium of $x^* = a_1(\mathbf{p})$ is stable (sink) for $a_0 > 0$.

(ii$_2$) For $\lambda_2 = a_0(a_2 - a_1) < 0$, due to $a_2 - a_1 > 0$, such an equilibrium of $x^* = a_2(\mathbf{p})$ is stable (sink) for $a_0 < 0$.

(iii) If $\lambda_i > 0$, we have

$$\lim_{t\to\infty}(x - a_i) = \lim_{t\to\infty}\Delta x_i = \lim_{t\to\infty}\Delta x_{i0} e^{\lambda_i (t-t_0)} = \infty \Rightarrow \lim_{t\to\infty} x(t) = \infty$$

So equilibrium $x^* = a_i(\mathbf{p})$ is unstable.

(iii$_1$) For $\lambda_1 = a_0(a_1 - a_2) > 0$, due to $a_1 - a_2 < 0$, such an equilibrium of $x^* = a_1(\mathbf{p})$ is unstable (source) for $a_0 < 0$.

(iii$_2$) For $\lambda_2 = a_0(a_2 - a_1) > 0$, due to $a_2 - a_1 > 0$, such an equilibrium of $x^* = a_2(\mathbf{p})$ is unstable (source) for $a_0 > 0$.

(iv) If $a_1(\mathbf{p}) = a_2(\mathbf{p})$, we have

$$\Delta \dot{x}_i = a_0 \Delta x_i^2 \ (i,j \in \{1,2\}, j \neq i)$$

For $a_0 > 0$, $\Delta \dot{x}_i > 0$ exists. So a flow of x reaches to $x^* = a_i$ from the initial point of $x_0 < a_i$ and it goes to the positive infinity from $x_0 > a_i$. Such an equilibrium is unstable of the second order, which is called an *upper saddle*. Similarly, for $a_0 < 0$, $\Delta \dot{x}_i < 0$ exists. So a flow of x reaches $x^* = a_i$ from the initial point of

$x_0 > a_i$ and it goes to the negative infinity from $x_0 < a_i$. Such an equilibrium is unstable of the second order, which is called a *lower saddle*.

The theorem is proved. ∎

The stability and bifurcation of equilibriums for the 1-dimensional system in Eq. (5.9) are illustrated in Fig. 5.2. The stable and unstable equilibriums varying with the vector parameter are depicted by the solid and dashed curves, respectively. The bifurcation point of equilibriums occurs at the double-repeated equilibrium at the boundary of $\mathbf{p}_0 \in \partial\Omega_{12}$. In Fig. 5.2a, for $a_0 > 0$, the equilibriums of $x^* = a_2$ and $x^* = a_1$ for $\Delta > 0$ are unstable and stable, respectively. The bifurcation of equilibriums also occurs at $\Delta = 0$. The flow of $x(t)$ is a forward *upper* flow for $a_0 > 0$, and the equilibrium point $x^* = -\frac{B(\mathbf{p}_0)}{2A(\mathbf{p}_0)}$ at $\Delta = 0$ is termed an *upper saddle*. Such a bifurcation is termed an *upper-saddle-node* bifurcation. For $\Delta < 0$, no any equilibrium exists. Such a 1-dimensional system is termed the nonequilibrium system. For $\Delta < 0$ and $a_0 > 0$, the flow of $x(t)$ is always toward the positive direction because of $\dot{x} = a_0[(x + \frac{B}{2A})^2 + (-\frac{\Delta}{4A^2})] > 0$. The corresponding-phase portrait is presented in Fig. 5.2b. In Fig. 5.2c, the equilibriums of $x^* = a_2$ and $x^* = a_1$ for $a_0 < 0$ are stable and unstable, respectively. The bifurcation of equilibriums also occurs at $\Delta = 0$. The

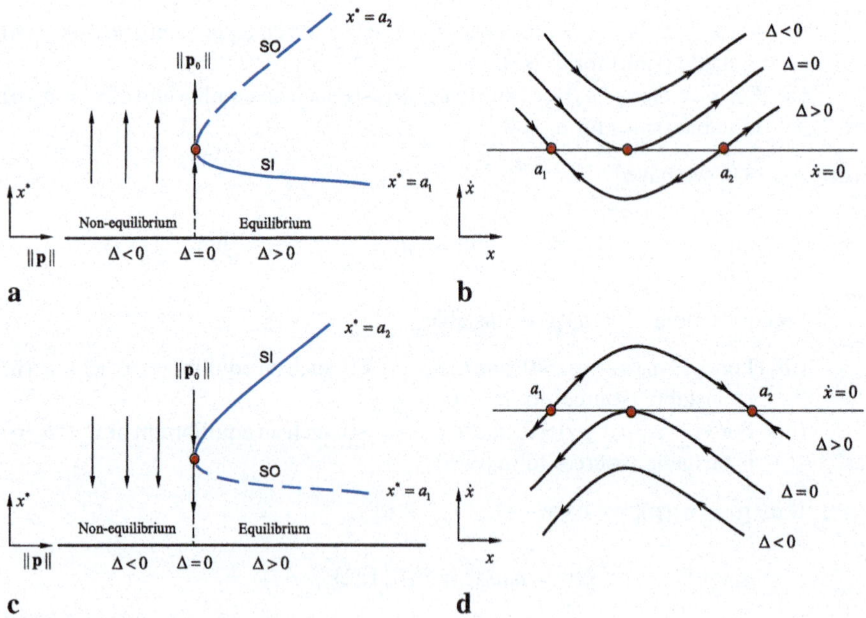

Fig. 5.2 Stability and bifurcation of two equilibriums in the quadratic dynamical system: (**a**) an *upper-saddle-node* bifurcation and (**b**) phase portrait ($a_0 > 0$), (**c**) a *lower-saddle-node* bifurcation and (**d**) phase portrait ($a_0 < 0$). Stable and unstable equilibriums are represented by solid and dashed curves, respectively

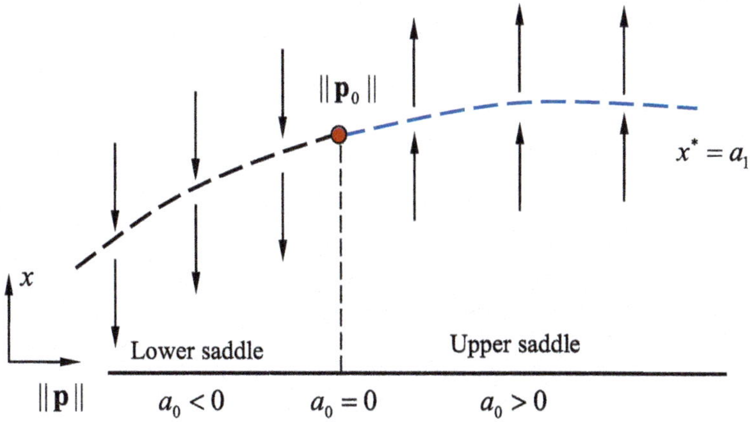

Fig. 5.3 Stability and bifurcation of a double equilibrium of the second order in the quadratic dynamical system. Unstable equilibriums are represented by a dashed curve. The stability switching from the lower saddle to upper saddle is labeled by a circular symbol

flow of $x(t)$ is a forward *lower* flow for $a_0 < 0$, and the equilibrium point of $x^* = -\frac{B(\mathbf{p}_0)}{2A(\mathbf{p}_0)}$ at $\Delta = 0$ is termed a *lower saddle*. Such a bifurcation of equilibrium is termed a *lower-saddle-node* bifurcation. For $\Delta < 0$ and $a_0 < 0$, the flow of $x(t)$ is always toward the negative direction because of $\dot{x} = a_0[(x + \frac{B}{2A})^2 + (-\frac{\Delta}{4A^2})] < 0$. The corresponding-phase portrait is presented in Fig. 5.2d.

To illustrate the stability and bifurcation of equilibrium with singularity in a 1-dimensional, quadratic nonlinear system, the equilibrium of $\dot{x} = a_0(x - a_1)^2$ is presented in Fig. 5.3. The *upper saddle* and *lower saddle* of equilibrium of $x^* = a_1$ with the second-order multiplicity are unstable, which are depicted by dashed curves. At $a_0 = 0$, the *upper saddle* and *lower saddle* equilibriums will be switched, which is marked by a circular symbol.

Definition 5.3 Consider a 1-dimensional dynamical system in Eq. (5.9) as

$$\begin{aligned}\dot{x} &= A(\mathbf{p})x^2 + B(\mathbf{p})x + C(\mathbf{p}) \\ &= a_0(\mathbf{p})(x - a(\mathbf{p}))(x - b(\mathbf{p})).\end{aligned} \tag{5.27}$$

(i) For $a < b$, the corresponding standard form is

$$\dot{x} = a_0(x - a)(x - b) \tag{5.28}$$

with two equilibriums

$$\begin{aligned}&x^* = a_1 = a \text{ and } x^* = a_2 = b \\ &\text{with } \Delta = a_0^2(a - b)^2 > 0.\end{aligned} \tag{5.29}$$

(ii) For $a > b$, the corresponding standard form is

$$\dot{x} = a_0(x - b)(x - a) \tag{5.30}$$

with two equilibriums of

$$\begin{aligned} &x^* = a_1 = b \text{ and } x^* = a_2 = a \\ &\text{with } \Delta = a_0^2(a - b)^2 > 0. \end{aligned} \tag{5.31}$$

(iii) For $a = b$, the corresponding standard form is

$$\dot{x} = a_0(x - a)^2 \tag{5.32}$$

with a double-repeated equilibrium of $x^* = a$. Such an equilibrium point is called a *saddle* of the second order.

(iii_1) If $a_0 > 0$, the equilibrium is an *upper saddle* of the second order.
(iii_2) If $a_0 < 0$, the equilibrium is a *lower saddle* of the second order.

(iv) The equilibrium of $x^* = a$ for two equilibriums switching is called a *saddle-node* bifurcation point of equilibrium at a point $\mathbf{p} = \mathbf{p}_0 \in \partial\Omega_{12}$, and the bifurcation condition is

$$\Delta = a_0^2(a - b)^2 = 0 \text{ or } a = b. \tag{5.33}$$

Theorem 5.3 (i) *Under a condition of*

$$a < b \text{ and } \Delta = a_0^2(a - b)^2 > 0 \tag{5.34}$$

a standard form of the 1-dimensional dynamical system in Eq. (5.27) is

$$\dot{x} = f(x, \mathbf{p}) = a_0(x - a)(x - b). \tag{5.35}$$

(i_1) *If $a_0(\mathbf{p}) > 0$, then equilibrium $x^* = a$ is stable with $df/dx|_{x^*=a} < 0$ and equilibrium $x^* = b$ is unstable with $df/dx|_{x^*=b} > 0$.*
(i_2) *If $a_0(\mathbf{p}) < 0$, then equilibrium $x^* = a$ is unstable with $df/dx|_{x^*=a} > 0$ and equilibrium $x^* = b$ is stable with $df/dx|_{x^*=b} < 0$.*

(ii) *Under a condition of*

$$a > b \text{ and } \Delta = a_0^2(a - b)^2 > 0 \tag{5.36}$$

a standard form of the 1-dimensional dynamical system in Eq. (5.9) is

$$\dot{x} = a_0(x-b)(x-a). \tag{5.37}$$

(ii$_1$) *If* $a_0(\mathbf{p}) > 0$, *then equilibrium* $x^* = b$ *is unstable with* $df/dx|_{x^*=a} > 0$ *and equilibrium* $x^* = b < 0$ *is stable* $df/dx|_{x^*=b} < 0$.
(ii$_2$) *If* $a_0(\mathbf{p}) < 0$, *then equilibrium* $x^* = a$ *is stable with* $df/dx|_{x^*=a} < 0$ *and equilibrium* $x^* = b$ *is unstable* $df/dx|_{x^*=b} > 0$.

(iii) *For* $a = b$, *the corresponding standard form with* $\Delta = 0$ *is*

$$\dot{x} = f(x, \mathbf{p}) = a_0(x-a)^2 \tag{5.38}$$

(iii$_1$) *If* $a_0(\mathbf{p}) > 0$, *then the equilibrium* $x^* = a$ *is an* upper saddle *of the second order with* $d^2f/dx^2|_{x^*=a} > 0$. *The equilibrium* $x^* = a$ *for two equilibriums switching is an* upper-saddle-node *bifurcation of the second order.*
(iii$_2$) *If* $a_0(\mathbf{p}) < 0$, *then the equilibrium* $x^* = a$ *is a* lower saddle *of the second order with* $d^2f/dx^2|_{x^*=a} < 0$. *The equilibrium* $x^* = a$ *for two equilibriums switching is a* lower-saddle-node *bifurcation of the second order.*

Proof The theorem can be proved as for Theorem 5.2. ∎

Definition 5.4 If $C(\mathbf{p}) = 0$ in Eq. (5.9), a 1-dimensional quadratic system is

$$\dot{x} = A(\mathbf{p})x^2 + B(\mathbf{p})x. \tag{5.39}$$

(i) For $A(\mathbf{p}) \times B(\mathbf{p}) < 0$, the corresponding standard form is

$$\dot{x} = a_0 x(x-a) \tag{5.40}$$

with two equilibriums

$$\begin{aligned} &x^* = a_1 = 0 \text{ and } x^* = a_2 = a > 0 \\ &\text{with } a_0 = A(\mathbf{p}) \text{ and } a = -\frac{B(\mathbf{p})}{A(\mathbf{p})}. \end{aligned} \tag{5.41}$$

(ii) For $A(\mathbf{p}) \times B(\mathbf{p}) > 0$, the corresponding standard form is

$$\dot{x} = a_0(x-a)x \tag{5.42}$$

with two equilibriums of

$$x^* = a_1 = a < 0 \text{ and } x^* = a_2 = 0. \tag{5.43}$$

(iii) For $B(\mathbf{p}) = 0$, the corresponding standard form is

$$\dot{x} = a_0 x^2 \tag{5.44}$$

with a double-repeated equilibrium of $x^* = 0$. Such an equilibrium is called a *saddle* of the second order. If $a_0 > 0$, the equilibrium is an *upper saddle* of the second order. If $a_0 < 0$, the equilibrium is a *lower saddle* of the second order.

(iv) The bifurcation of $x^* = 0$ for two equilibriums switching is called a *saddle-node* bifurcation at a point $\mathbf{p} = \mathbf{p}_0 \in \partial\Omega_{12}$, and the bifurcation condition is

$$B(\mathbf{p}_0) = 0. \tag{5.45}$$

Theorem 5.4 (i) *Under a condition of*

$$A(\mathbf{p}) \times B(\mathbf{p}) < 0, \tag{5.46}$$

a standard form of the 1-dimensional dynamical system in Eq. (5.39) is

$$\dot{x} = f(x, \mathbf{p}) = a_0 x(x - a). \tag{5.47}$$

(i_1) *If $a_0(\mathbf{p}) > 0$, then equilibrium $x^* = 0$ is stable with $df/dx|_{x^*=0} < 0$ and equilibrium $x^* = a > 0$ is unstable with $df/dx|_{x^*=a} > 0$.*

(i_2) *If $a_0(\mathbf{p}) < 0$, then equilibrium $x^* = 0$ is unstable with $df/dx|_{x^*=0} > 0$ and equilibrium $x^* = a > 0$ is stable with $df/dx|_{x^*=a} < 0$.*

(ii) *Under a condition of*

$$A(\mathbf{p}) \times B(\mathbf{p}) > 0, \tag{5.48}$$

a standard form of the 1-dimensional quadratic system in Eq. (5.9) is

$$\dot{x} = a_0(x - a)x. \tag{5.49}$$

(ii_1) *If $a_0(\mathbf{p}) > 0$, then equilibrium $x^* = 0$ is unstable with $df/dx|_{x^*=0} > 0$ and equilibrium $x^* = a < 0$ is stable with $df/dx|_{x^*=a} < 0$.*

(ii_2) *If $a_0(\mathbf{p}) < 0$, then equilibrium $x^* = 0$ is stable with $df/dx|_{x^*=a} < 0$ and equilibrium $x^* = a < 0$ is unstable with $df/dx|_{x^*=a} > 0$.*

(iii) *For $B(\mathbf{p}) = 0$, the corresponding standard form with $\Delta = 0$ is*

$$\dot{x} = f(x, \mathbf{p}) = a_0 x^2. \tag{5.50}$$

(iii$_1$) *If* $a_0(\mathbf{p}) > 0$, *then the equilibrium* $x^* = 0$ *is an* upper saddle *of the second order with* $d^2f/dx^2|_{x^*=0} > 0$. *The equilibrium* $x^* = 0$ *for two equilibriums switching is an* upper-saddle-node *bifurcation of the second order.*

(iii$_2$) *If* $a_0(\mathbf{p}) < 0$, *then the equilibrium* $x^* = 0$ *is a* lower saddle *of the second order with* $d^2f/dx^2|_{x^*=0} < 0$. *The equilibrium* $x^* = 0$ *for two equilibriums switching is a* lower-saddle-node *bifurcation of the second order.*

Proof The theorem can be proved as for Theorem 5.2. ∎

The stability and bifurcation of two equilibriums for the 1-dimensional system in Eq. (5.27) with $\Delta = B^2 - 4AC = a_0^2(a-b)^2 \geq 0$ are presented in Fig. 5.4. The stable and unstable equilibriums varying with the vector parameter are depicted by solid and dashed curves, respectively. The bifurcation point of equilibriums occurs at the double-repeated equilibrium at the boundary of $\mathbf{p}_0 \in \partial\Omega_{12}$. With varying parameters, the two equilibriums of $x^* = a, b$ equal each other (i.e., $x^* = a = b$). Such an equilibrium is a bifurcation point at $x^* = a = b$ for $\Delta = 0$. The equilibriums of $x^* = a, b$ with $\Delta \geq 0$ are presented in Fig. 5.4a, b for $a_0 > 0$ and $a_0 < 0$, respectively. The dynamical system in Eq. (5.39) is a special case of the dynamical system in Eq. (5.9) with $C(\mathbf{p}) = 0$. Thus $\Delta = B^2 - 4AC = B^2 \geq 0$. The equilibriums exist in the entire domain. In Fig. 5.4c, for $a_0 > 0$ and $B < 0$, the equilibriums of $x^* = 0$ and $x^* = a$ are unstable and stable, respectively. However, for $a_0 > 0$ and $B > 0$, the equilibriums of $x^* = 0$ and $x^* = a$ are stable and unstable, respectively. The

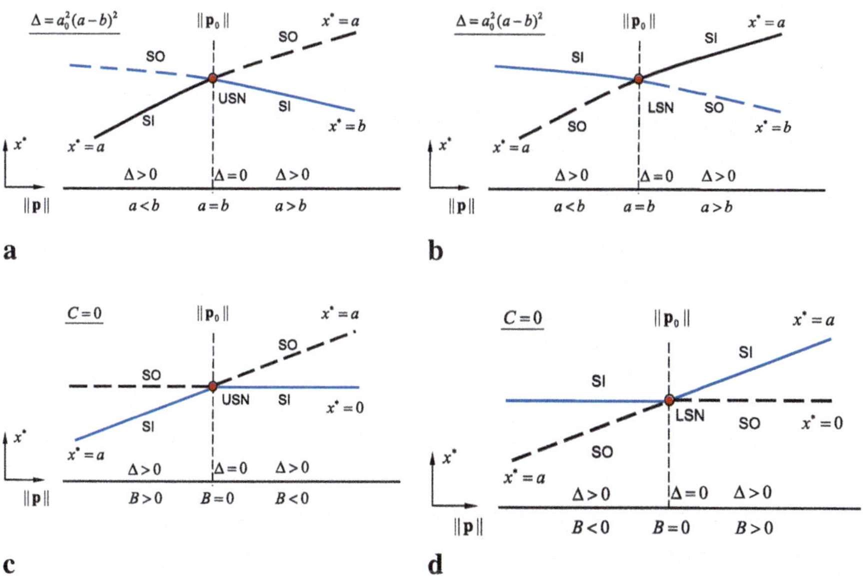

Fig. 5.4 Stability and bifurcation of two equilibriums in the quadratic dynamical system: (**a**) an *upper-saddle*-node bifurcation ($a_0 > 0$), (**b**) a *lower-saddle-node* bifurcation ($a_0 < 0$), (**c**) an upper-saddle-node bifurcation ($a_0 > 0$), (**d**) a *lower-saddle-node* bifurcation ($a_0 < 0$). Stable and unstable equilibriums are represented by solid and dashed curves, respectively

bifurcation of equilibriums occurs at $B = 0$. The flow of $x(t)$ is a forward *upper* flow for $a_0 > 0$, and the equilibrium point $x^* = 0$ at $B = 0$ is termed an *upper saddle*. Such a bifurcation is termed an *upper-saddle-node* bifurcation. In Fig. 5.4d, for $a_0 < 0$ and $B < 0$, the equilibriums of $x^* = 0$ and $x^* = a$ are stable and unstable, respectively. However, for $a_0 < 0$ and $B > 0$, the equilibriums of $x^* = 0$ and $x^* = a$ are unstable and stable, respectively. The bifurcation of equilibriums also occurs at $B = 0$. The flow of $x(t)$ is a forward *lower* flow for $a_0 < 0$, and the equilibrium point of $x^* = 0$ at $B = 0$ is termed a *lower saddle*. Such a bifurcation is termed a *lower-saddle-node* bifurcation.

Definition 5.5 If $B(\mathbf{p}) = 0$ in Eq. (5.9), a 1-dimensional quadratic system is

$$\dot{x} = A(\mathbf{p})x^2 + C(\mathbf{p}). \tag{5.51}$$

(i) For $A(\mathbf{p}) \times C(\mathbf{p}) > 0$, the system does not have any equilibriums. The nonequilibrium flow of the system is called a positive flow if $A(\mathbf{p}) > 0$. The nonequilibrium flow is called a negative flow if $A(\mathbf{p}) < 0$.

(ii) For $A(\mathbf{p}) \times C(\mathbf{p}) < 0$, the corresponding standard form is

$$\dot{x} = a_0(x + a)(x - a) \tag{5.52}$$

with two symmetric equilibriums

$$x^* = -a \text{ and } x^* = a,$$
$$\text{with } a_0 = A(\mathbf{p}_0) \text{ and } a = \sqrt{-\frac{C(\mathbf{p}_0)}{A(\mathbf{p}_0)}}. \tag{5.53}$$

(iii) For $C(\mathbf{p}_0) = 0$, the corresponding standard form with $\Delta = 0$ is

$$\dot{x} = a_0 x^2 \tag{5.54}$$

with two equilibriums of

$$x^* = -a = -0 \text{ and } x^* = a = +0. \tag{5.55}$$

Such an equilibrium point of $x^* = 0$ is called a *saddle* of the second order. If $a_0 > 0$, the equilibrium is an *upper saddle* of the second order. If $a_0 < 0$, the equilibrium is a lower saddle of the second order.

(iv) The equilibrium of $x^* = 0$ for two equilibriums appearing or vanishing is called a *saddle-node* bifurcation point of equilibrium at a point $\mathbf{p} = \mathbf{p}_0 \in \partial\Omega_{12}$, and the bifurcation condition is

$$C(\mathbf{p}_0) = 0. \tag{5.56}$$

Theorem 5.5 (i) *Under a condition of*

$$A(\mathbf{p}) \times C(\mathbf{p}) < 0, \tag{5.57}$$

a standard form of the 1-dimensional dynamical system in Eq. (5.51) is

$$\dot{x} = f(x, \mathbf{p}) = a_0(x + a)(x - a). \tag{5.58}$$

(i_1) *If* $a_0(\mathbf{p}) > 0$, *then equilibrium* $x^* = -a$ *is stable with* $df/dx|_{x^*=-a} < 0$ *and equilibrium* $x^* = a$ *is unstable with* $df/dx|_{x^*=a} > 0$.

(i_2) *If* $a_0(\mathbf{p}) < 0$, *then equilibrium* $x^* = -a$ *is unstable with* $df/dx|_{x^*=-a} > 0$ *and equilibrium* $x^* = a$ *is stable with* $df/dx|_{x^*=a} < 0$.

(ii) *Under a condition of*

$$C(\mathbf{p}) = 0, \tag{5.59}$$

a standard form of the 1-dimensional dynamical system in Eq. (5.38) is

$$\dot{x} = f(x, \mathbf{p}) = a_0 x^2. \tag{5.60}$$

(ii_1) *If* $a_0(\mathbf{p}) > 0$, *then the equilibrium of* $x^* = 0$ *is an* upper saddle *of the second order with* $d^2f/dx^2|_{x^*=0} > 0$. *Such a bifurcation for two equilibriums appearing or vanishing is an* upper-saddle-node *bifurcation of the second order.*

(ii_2) *If* $a_0(\mathbf{p}) < 0$, *then the equilibrium* $x^* = 0$ *is a* lower saddle *of the second order with* $d^2f/dx^2|_{x^*=0} < 0$. *Such a bifurcation for two equilibriums appearing or vanishing is a* lower-saddle-node *bifurcation of the second order.*

Proof The proof is similar to Theorem 5.2. The theorem is proved. ∎

The stability and bifurcation of equilibriums for the quadratic nonlinear system in Eq. (5.51) are illustrated in Fig. 5.5 as a special case of the dynamical system in Eq. (5.9) with $B(\mathbf{p}) = 0$. The stable and unstable equilibriums varying with the vector parameter are depicted by solid and dashed curves, respectively. The bifurcation point of equilibrium occurs at the double equilibrium at the boundary of $\mathbf{p}_0 \in \partial\Omega_{12}$. In Fig. 5.5a, for $\Delta = -4AC > 0$ and $a_0 = A > 0$, the equilibriums of $x^* = -a < 0$ and $x^* = a > 0$ for $C < 0$ are stable and unstable, respectively. The bifurcation of equilibrium also occurs at $C = 0$. The flow of *x*(*t*) is a forward *upper* flow for $a_0 > 0$, and the equilibrium point of $x^* = 0$ at $C = 0$ is termed the *upper saddle*. Such a bifurcation is termed the *upper-saddle-node* bifurcation. For $\Delta = -4AC < 0$ and $a_0 = A > 0$, we have $C > 0$. Thus, no equilibrium exists because of $\dot{x} = Ax^2 + C > 0$. Such a 1-dimensional system is termed a nonequilibrium system. For $a_0 = A > 0$ and

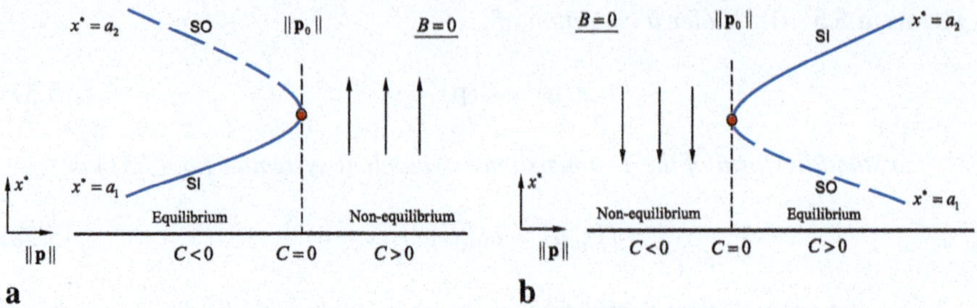

Fig. 5.5 Stability and bifurcation of two equilibriums in the quadratic dynamical system: (**a**) an *upper-saddle*-node bifurcation ($a_0 > 0$), (**b**) a *lower-saddle-node* bifurcation ($a_0 < 0$). Stable and unstable equilibriums are represented by solid and dashed curves, respectively. SO: source, SI: sink

$C > 0$, the flow of $x(t)$ is always toward the positive direction. In Fig. 5.5b, for $\Delta = -4AC > 0$ and $a_0 = A < 0$, the equilibriums of $x^* = -a$ and $x^* = a$ for $C > 0$ are unstable and stable, respectively. The bifurcation of equilibrium also occurs at $C = 0$. The flow of $x(t)$ for the bifurcation point is a forward *lower* flow for $a_0 = A < 0$, and the equilibrium bifurcation point of $x^* = 0$ at $C = 0$ is termed a *lower saddle*. Such a bifurcation is termed a *lower-saddle-node* bifurcation. For $\Delta = -4AC < 0$ and $a_0 = A < 0$, we have $C < 0$. For $a_0 = A < 0$ and $C < 0$, the flow of $x(t)$ is always toward the negative direction without any equilibrium because of $\dot{x} = Ax^2 + C < 0$.

5.3 Cubic Nonlinear Systems

In this section, the stability and stability switching of equilibriums in cubic polynomial systems are discussed. As in Luo (2019), the *upper-saddle-node* and *lower-saddle-node* appearing and switching bifurcations are discussed and the third-order sink and source switching bifurcations are discussed as well.

Definition 5.6 Consider a cubic nonlinear dynamical system

$$\begin{aligned} \dot{x} &= A(\mathbf{p})x^3 + B(\mathbf{p})x^2 + C(\mathbf{p})x + D(\mathbf{p}) \\ &\equiv a_0(\mathbf{p})(x - a(\mathbf{p}))[x^2 + B_1(\mathbf{p})x + C_1(\mathbf{p})] \end{aligned} \tag{5.61}$$

where four scalar constants $A(\mathbf{p}) \neq 0$, $B(\mathbf{p})$, $C(\mathbf{p})$, and $D(\mathbf{p})$ are determined by

$$\begin{aligned} &A = a_0, B = (-a + B_1)a_0, C = (-aB_1 + C_1)a_0, D = -aa_0C_1, \\ &\mathbf{p} = (p_1, p_2, \ldots, p_m)^{\mathrm{T}}. \end{aligned} \tag{5.62}$$

(i) If

$$\Delta_1 = B_1^2 - 4C_1 < 0 \,\text{for}\, \mathbf{p} \in \Omega_1 \subset \mathbf{R}^m \tag{5.63}$$

then the cubic nonlinear system has a simple equilibrium only as

$$x^* = a \,\text{for}\, \mathbf{p} \in \Omega_1 \subset \mathbf{R}^m \tag{5.64}$$

and the standard form of such a 1-dimensional system is

$$\dot{x} = a_0(x-a)[(x - \frac{1}{2}B_1)^2 + \frac{1}{4}(-\Delta_1)]. \tag{5.65}$$

(ii) If

$$\Delta_1 = B_1^2 - 4C_1 > 0 \,\text{for}\, \mathbf{p} \in \Omega_2 \subset \mathbf{R}^m \tag{5.66}$$

then there are three equilibriums with

$$\begin{aligned}
&a_0 = A(\mathbf{p}), b_{1,2} = -\frac{1}{2}(B_1(\mathbf{p}) \mp \sqrt{\Delta_1}) \text{ with } b_1 > b_2;\\
&a_1 = \min\{a, b_1, b_2\}, a_3 = \max\{a, b_1, b_2\},\\
&a_2 \in \{a, b_1, b_2\} \neq \{a_1, a_3\},\\
&\Delta_{ij} = (a_i - a_j)^2 > 0 \text{ for } i,j \in \{1,2,3\} \text{ but } i \neq j.
\end{aligned} \tag{5.67}$$

(ii_1) If

$$a_i \neq a_j \text{ with } \Delta_{ij} = (a_i - a_j)^2 > 0 \text{ for } i,j \in \{1,2,3\} \text{ but } i \neq j. \tag{5.68}$$

the cubic nonlinear system has three different, simple equilibriums as

$$x^* = a_1, x^* = a_2, \text{and } x^* = a_3 \tag{5.69}$$

and the corresponding standard form is

$$\dot{x} = a_0(x-a_1)(x-a_2)(x-a_3). \tag{5.70}$$

(ii_2) If at $\mathbf{p} = \mathbf{p}_1$

$$\begin{aligned}
&a_1 = b_2, a_2 = a, a_3 = b_1;\\
&\Delta_{12} = (a_1 - a_2) = (a - b_2)^2 = 0,
\end{aligned} \tag{5.71}$$

the cubic nonlinear system has a double-repeated equilibrium and a simple equilibrium as

$$x^* = a_1, x^* = a_1 \text{ and } x^* = a_2 \tag{5.72}$$

and the corresponding standard form is

$$\dot{x} = a_0(x - a_1)^2(x - a_2). \tag{5.73}$$

Such a flow at the equilibrium of $x^* = a_1$ is called a *saddle* flow of the second order.The equilibrium of $x^* = a_1$ for two different equilibriums switching is called a bifurcation point of equilibrium at a point $\mathbf{p} = \mathbf{p}_1$ with the second-order multiplicity, and the bifurcation condition is

$$a = b_1 = \min\{-\frac{1}{2}(B_1(\mathbf{p}) + \sqrt{\Delta_1}), -\frac{1}{2}(B_1(\mathbf{p}) - \sqrt{\Delta_1})\}. \tag{5.74}$$

If at $\mathbf{p} = \mathbf{p}_2$,

$$\begin{aligned} &a_2 = b_1, a_3 = a, a_1 = b_2, \\ &\Delta_{23} = (a_2 - a_3) = (a - b_1)^2 = 0, \end{aligned} \tag{5.75}$$

the cubic nonlinear system has three equilibriums as

$$x^* = a_1, x^* = a_2, \text{and } x^* = a_2 \tag{5.76}$$

and the corresponding standard form is

$$\dot{x} = a_0(x - a_1)(x - a_2)^2. \tag{5.77}$$

Such a flow at the equilibrium of $x^* = a_2$ is called a *saddle* flow of the second order.The equilibrium of $x^* = a_2$ for two different equilibriums switching is called a bifurcation point of equilibrium at a point $\mathbf{p} = \mathbf{p}_1$ with the second-order multiplicity, and the bifurcation condition is

$$a = b_2 = \max\{-\frac{1}{2}(B_1(\mathbf{p}) + \sqrt{\Delta_1}), -\frac{1}{2}(B_1(\mathbf{p}) - \sqrt{\Delta_1})\}. \tag{5.78}$$

If at $\mathbf{p} = \mathbf{p}_3$,

$$\begin{aligned} &a_1 = b_2, a_2 = a, a_3 = b_1, \\ &\Delta_{12} = (a_1 - a_2)^2 = (a - b_2)^2 = 0, \\ &\Delta_{23} = (a_2 - a_3)^2 = (a - b_1)^2 = 0, \\ &\Delta_{13} = (a_1 - a_3)^2 = (b_2 - b_1)^2 = 0, \end{aligned} \tag{5.79}$$

the cubic nonlinear system has three repeated equilibriums as

$$x^* = a_1 = a, x^* = a_2 = a \text{ and } x^* = a_3 = a \tag{5.80}$$

and the corresponding standard form is

$$\dot{x} = a_0(x-a)^3. \tag{5.81}$$

Such a flow at the equilibrium of $x^* = a$ is called a sink or source flow of the third order.

The equilibrium of $x^* = a$ at a point $\mathbf{p} = \mathbf{p}_3$ for three different equilibriums switching is called a bifurcation point of equilibrium with the third-order multiplicity, and the bifurcation condition is

$$a = b = -\frac{1}{2}B_1(\mathbf{p}). \tag{5.82}$$

(iii) If

$$\Delta_1 = B_1^2 - 4A_1C_1 = 0 \text{ for } \mathbf{p} = \mathbf{p}_0 \in \partial\Omega_{12} \subset \mathbf{R}^{m-1}, \tag{5.83}$$

then there exist

$$a_0 = A(\mathbf{p}_0), \text{and } b_1 = b_2 = b = -\frac{1}{2}B_1(\mathbf{p}_0). \tag{5.84}$$

(iii$_1$) For $a < b$, the cubic nonlinear system has a double-repeated equilibrium plus a *lower* simple equilibrium:

$$x^* = a_1 = a, \ \ x^* = a_2 = b \text{ and } x^* = a_2 = b \tag{5.85}$$

with the corresponding standard form of

$$\dot{x} = a_0(x-a_1)(x-a_2)^2. \tag{5.86}$$

Such a flow at the equilibrium of $x^* = a_2$ is called a *saddle* flow of the second order.

The equilibrium of $x^* = a_2$ for two different equilibrium switching is called a bifurcation point of equilibrium at a point $\mathbf{p} = \mathbf{p}_0 \in \partial\Omega_{12}$ with the second-order multiplicity, and the bifurcation condition is

$$\Delta_1 = B_1^2 - 4C_1 = 0 \text{ with } a < b. \tag{5.87}$$

(iii$_2$) For $a > b$, the cubic nonlinear system has a *lower* double-repeated equilibrium plus an *upper* simple equilibrium:

$$x^* = a_1 = b \text{ and } x^* = a_1 = b, x^* = a_2 = a \tag{5.88}$$

with the corresponding standard form of

$$\dot{x} = a_0(x - a_1)^2(x - a_2). \tag{5.89}$$

Such a flow at the equilibrium of $x^* = a_1$ is called a *saddle* flow of the second order.

The equilibrium of $x^* = a_1 = b$ for two different equilibriums switching is called a bifurcation point of equilibrium at a point $\mathbf{p} = \mathbf{p}_0 \in \partial\Omega_{12}$ with the second-order multiplicity, and the bifurcation condition is also

$$\Delta_1 = B_1^2 - 4C_1 = 0 \text{ with } a > b. \tag{5.90}$$

(iii$_3$) For $a = b$, the cubic nonlinear system has a triple-repeated equilibrium as

$$x^* = a_1 = a \text{ and } x^* = a_1 = a, x^* = a_2 = a \tag{5.91}$$

with the corresponding standard form of

$$\dot{x} = a_0(x - a_1)^3. \tag{5.92}$$

Such a flow at the equilibrium of $x^* = a_1$ is called a source or sink flow of the third order.

The equilibrium of $x^* = a_1 = a$ for three equilibriums switching or two equilibriums switching is called a bifurcation point of equilibrium at a point $\mathbf{p} = \mathbf{p}_0 \in \partial\Omega_{12}$ with the third-order multiplicity, and the bifurcation condition is

$$\Delta_1 = B_1^2 - 4C_1 = 0 \text{ with } a = b. \tag{5.93}$$

Theorem 5.6

(i) *Under a condition of*

$$\Delta_1 = B_1^2 - 4C_1 < 0 \tag{5.94}$$

a standard form of the 1-dimensional dynamical system in Eq. (5.61) is

$$\dot{x} = f(x, \mathbf{p}) = a_0(x - a_1)[(x + \frac{1}{2}B_1)^2 + \frac{1}{4}(-\Delta_1)]. \tag{5.95}$$

(i$_1$) *If* $a_0(\mathbf{p}) > 0$, *then equilibrium* $x^* = a_1$ *is unstable* $(df/dx|_{x^*=a_1} > 0)$.
(i$_2$) *If* $a_0(\mathbf{p}) < 0$, *then equilibrium* $x^* = a_1$ *is stable* $(df/dx|_{x^*=a_1} < 0)$.

(ii) *Under the conditions of*

$$\begin{aligned}&\Delta_1 = B_1^2 - 4C_1 > 0,\\&a_1, a_2, a_3 = \mathrm{sort}\{b_2, a, b_1\}, a_i \neq a_j, a_i < a_{i+1};\\&\Delta_{ij} = (a_i - a_j)^2 \neq 0 \text{ for } i,j \in \{1,2,3\},\end{aligned} \tag{5.96}$$

a standard form of the 1-dimensional dynamical system in Eq. (5.61) is

$$\dot{x} = f(x, \mathbf{p}) = a_0(x - a_1)(x - a_2)(x - a_3). \tag{5.97}$$

(ii$_1$) *If $a_0(\mathbf{p}) > 0$, then the equilibriums of $x^* = a_1$, $x^* = a_2$, and $x^* = a_3$ are unstable ($df/dx|_{x^*=a_1} > 0$), stable ($df/dx|_{x^*=a_2} < 0$), and unstable ($df/dx|_{x^*=a_3} > 0$), respectively.*

(ii$_2$) *If $a_0(\mathbf{p}) < 0$, then the equilibriums of $x^* = a_1$, $x^* = a_2$, and $x^* = a_3$ are stable ($df/dx|_{x^*=a_1} < 0$), unstable ($df/dx|_{x^*=a_2} > 0$), and stable ($df/dx|_{x^*=a_3} < 0$), respectively.*

(iii) *Under a condition of*

$$\begin{aligned}&\Delta_1 = B_1^2 - 4C_1 > 0,\\&a_1, a_2, a_3 = \mathrm{sort}\{b_2, \mathrm{a}, b_1\}, a_i \neq a_j, a_i \leq a_{i+1}\\&\Delta_{12} = (a_1 - a_2)^2 = 0, \text{for } i,j \in \{1,2,3\}\end{aligned} \tag{5.98}$$

a standard form of the 1-dimensional dynamical system in Eq. (5.61) is

$$\dot{x} = f(x, \mathbf{p}) = a_0(x - a_1)^2(x - a_3). \tag{5.99}$$

(iii$_1$) *If $a_0(\mathbf{p}) > 0$, then the equilibriums of $x^* = a_1$ and $x^* = a_3$ are unstable* (lower saddle, $d^2f/dx^2|_{x^*=a_1} < 0$) and *unstable (source, $df/dx|_{x^*=a_3} > 0$), respectively. The bifurcation of equilibrium at $x^* = a_1$ for the two different equilibriums switching is a* lower-saddle-node *bifurcation of the second order at a point $\mathbf{p} = \mathbf{p}_1$.*

(iii$_2$) *If $a_0(\mathbf{p}) < 0$, then the equilibriums of $x^* = a_1$ and $x^* = a_3$ are unstable* (upper saddle, $d^2f/dx^2|_{x^*=a_1} > 0$) *and stable (sink, $df/dx|_{x^*=a_3} < 0$), respectively. The bifurcation of equilibrium at $x^* = a_1$ for the two different equilibriums switching is an upper saddle of the second order at a point $\mathbf{p} = \mathbf{p}_1$.*

(iv) *For*

$$
\begin{aligned}
&\Delta_1 = B_1^2 - 4C_1 > 0, \\
&a_1, a_2, a_3 = \text{sort}\{b_2, \text{a}, b_1\}, a_i \neq a_j, a_i \leq a_{i+1} \\
&\Delta_{23} = (a_2 - a_3)^2 = 0, \text{for } i, j \in \{1, 2, 3\}
\end{aligned}
\tag{5.100}
$$

a standard form of the 1-dimensional dynamical system in Eq. (5.61) is

$$
\dot{x} = f(x, \mathbf{p}) = a_0(x - a_1)(x - a_2)^2. \tag{5.101}
$$

(iv$_1$) *If* $a_0(\mathbf{p}) > 0$, *then the equilibriums of* $x^* = a_1$ *and* $x^* = a_2$ *are unstable (source,* $df/dx|_{x^*=a_1} > 0$*) and unstable (upper saddle,* $d^2f/dx^2|_{x^*=a_2} > 0$*), respectively. The bifurcation of equilibrium at* $x^* = a_2$ *for two different equilibriums switching is an upper-saddle-node bifurcation of the second order at a point* $\mathbf{p} = \mathbf{p}_1$.

(iv$_2$) *If* $a_0(\mathbf{p}) < 0$, *then the equilibriums of* $x^* = a_1$ *and* $x^* = a_2$ *are stable (sink,* $df/dx|_{x^*=a_1} < 0$*) and unstable (lower saddle,* $d^2f/dx^2|_{x^*=a_2} < 0$*), respectively. The bifurcation of equilibrium at* $x^* = a_2$ *for two equilibrium switching is a lower-saddle-node bifurcation of the second order at a point* $\mathbf{p} = \mathbf{p}_1$.

(v) *For*

$$
\begin{aligned}
&\Delta_1 = B_1^2 - 4C_1 \geq 0, b_1 = b_2 \\
&a_1, a_2, a_3 = \text{sort}\{b_2, \text{a}, b_1\}, a_i \leq a_{i+1}, \\
&\Delta_{ij} = (a_i - a_j)^2 = 0 \text{ for } i, j = 1, 2, 3 \text{ but } i \neq j,
\end{aligned}
\tag{5.102}
$$

a standard form of the 1-dimensional dynamical system in Eq. (5.61) is

$$
\dot{x} = f(x, \mathbf{p}) = a_0(x - a_1)^3. \tag{5.103}
$$

(v$_1$) *If* $a_0(\mathbf{p}) > 0$, *then the equilibrium of* $x^* = a_1$ *is unstable (third-order source,* $d^3f/dx^3|_{x^*=a_1} > 0$*). The bifurcation of equilibrium at* $x^* = a_1$ *for three different equilibriums switching is a source switching bifurcation of the third order at a point* $\mathbf{p} = \mathbf{p}_1$.

(v$_2$) *If* $a_0(\mathbf{p}) < 0$, *then the equilibrium of* $x^* = a_1$ *is stable (third-order sink,* $d^3f/dx^3|_{x^*=a_1} < 0$*). The bifurcation of equilibrium at* $x^* = a_1$ *for three different equilibrium switching is a sink switching bifurcation of the third order at a point* $\mathbf{p} = \mathbf{p}_1$.

(vi) *For*

$$\begin{aligned}\Delta_1 &= B_1^2 - 4A_1C_1 = 0, a < b\\ a_1 &= a, a_2 = b, \Delta_{12} = (a_1 - a_2)^2 \neq 0\end{aligned} \tag{5.104}$$

at $\mathbf{p} = \mathbf{p}_0 \in \partial\Omega_{12} \subset \mathbf{R}^{m-1}$, *a standard form of the 1-dimensional dynamical system is*

$$\dot{x} = f(x, \mathbf{p}) = a_0(x - a_1)(x - a_2)^2. \tag{5.105}$$

(vi$_1$) *If* $a_0(\mathbf{p}) > 0$, *then the equilibriums of* $x^* = a_1$ *and* $x^* = a_2$ *are unstable (source,* $df/dx|_{x^*=a_1} > 0$*) and unstable (upper saddle,* $d^2f/dx^2|_{x^*=a_2} > 0$*), respectively. The bifurcation of equilibrium at* $x^* = a_2$ *for two different equilibriums vanishing or appearance is an upper-saddle-node bifurcation of the second order at a point* $\mathbf{p} = \mathbf{p}_0 \in \partial\Omega_{12}$.

(vi$_2$) *If* $a_0(\mathbf{p}) < 0$, *then the equilibriums of* $x^* = a_1$ *and* $x^* = a_2$ *are stable (sink,* $df/dx|_{x^*=a_1} < 0$*) and unstable (lower saddle,* $d^2f/dx^2|_{x^*=a_2} < 0$*), respectively. The bifurcation of equilibrium at* $x^* = a_2$ *for two different equilibriums vanishing or appearance is a lower saddle of the second order at a point* $\mathbf{p} = \mathbf{p}_0 \in \partial\Omega_{12}$.

(vii) *For*

$$\begin{aligned}\Delta_1 &= B_1^2 - 4A_1C_1 = 0, a > b\\ a_1 &= b, a_2 = a, \Delta_{12} = (a_1 - a_2)^2 \neq 0\end{aligned} \tag{5.106}$$

at $\mathbf{p} = \mathbf{p}_0 \in \partial\Omega_{12} \subset \mathbf{R}^{m-1}$, *a standard form of the 1-dimensional dynamical system is*

$$\dot{x} = f(x, \mathbf{p}) = a_0(x - a_1)^2(x - a_2). \tag{5.107}$$

(vii$_1$) *If* $a_0(\mathbf{p}) > 0$, *then the equilibriums of* $x^* = a_1$ *and* $x^* = a_2$ *are unstable (lower saddle,* $d^2f/dx^2|_{x^*=a_1} < 0$*) and unstable (source,* $df/dx|_{x^*=a_2} > 0$*), respectively. The bifurcation of equilibrium at* $x^* = a_1$ *for two different equilibriums switching is a lower-saddle-node bifurcation of the second order at a point* $\mathbf{p} = \mathbf{p}_0 \in \partial\Omega_{12}$.

(vii$_2$) *If* $a_0(\mathbf{p}) < 0$, *then the equilibriums of* $x^* = a_1$ *and* $x^* = a_2$ *are unstable (upper saddle,* $d^2f/dx^2|_{x^*=a_1} > 0$*) and stable (sink,* $df/dx|_{x^*=a_2} < 0$*), respectively. The bifurcation of equilibrium at* $x^* = a_1$ *for two different equilibriums switching is an upper-saddle-node bifurcation of the second order at a point* $\mathbf{p} = \mathbf{p}_0 \in \partial\Omega_{12}$.

(viii) *For*

$$\begin{aligned}&\Delta_1 = B_1^2 - 4A_1C_1 = 0, a = b\\&a_2 = a, a_2 = a_3 = b,\\&\Delta_{12} = (a_1 - a_2)^2 = 0\end{aligned} \tag{5.108}$$

at $\mathbf{p} = \mathbf{p}_0 \in \partial\Omega_{12} \subset \mathbf{R}^{m-1}$, *a standard form of the 1-dimensional dynamical system is*

$$\dot{x} = f(x, \mathbf{p}) = a_0(x - a_1)^3. \tag{5.109}$$

(viii$_1$) *If* $a_0(\mathbf{p}) > 0$, *then the equilibrium of* $x^* = a_1$ *is unstable (third-order source,* $d^3f/dx^3|_{x^*=a_1} > 0$*). The bifurcation of equilibrium at* $x^* = a_1$ *for one equilibrium to three different equilibriums switching is a* source *bifurcation of the third order at a point* $\mathbf{p} = \mathbf{p}_0 \in \partial\Omega_{12}$.

(viii$_2$) *If* $a_0(\mathbf{p}) < 0$, *then the equilibrium of* $x^* = a_1$ *is stable (third-order sink,* $d^3f/dx^3|_{x^*=a_1} < 0$*). The bifurcation of equilibrium at* $x^* = a_1$ *for one simple equilibrium to three different simple equilibriums switching is a* sink *bifurcation of the third order at a point* $\mathbf{p} = \mathbf{p}_0 \in \partial\Omega_{12}$.

Proof The proof is similar to Theorem 5.2. ∎

The 1-dimensional cubic nonlinear system can be expressed by a factor of $(x-a)$ and a quadratic form of $a_0(x^2 + B_1x + C_1)$ as in Eq. (5.61). Three equilibriums do not have any intersections. Thus, only one bifurcation occurs at $\Delta_1 = B_1^2 - 4C_1 = 0$. The bifurcation of equilibrium occurs at the double-repeated equilibrium at the boundary of $\mathbf{p}_0 \in \partial\Omega_{12}$. For $\Delta_1 = B_1^2 - 4C_1 > 0$, $x^2 + B_1x + C_1 = 0$ gives two equilibriums of $x^* = b_1, b_2$. For $a_0 > 0$, if $a >$ max $\{b_1, b_2\}$, then the equilibrium of $x^* = a_3 = a$ is unstable, and two equilibriums of $x^* = a_2 =$ max $\{b_1, b_2\}$ and $x^* = a_1 =$ min $\{b_1, b_2\}$ are stable and unstable, respectively. For $\Delta_1 = B_1^2 - 4C_1 < 0$, $x^2 + B_1x + C_1 = 0$ does not have any real solutions. For $\Delta_1 = B_1^2 - 4C_1 = 0$, $x^2 + B_1x + C_1 = 0$ has a double-repeated equilibrium of $x^* = b = -\frac{1}{2}B_1$. The condition of $\Delta_1 = B_1^2 - 4C_1 = 0$ gives

$$B_1{}^2 = 4C_1. \tag{5.110}$$

From Eq. (5.62), one obtains

$$B_1 = a + \frac{B}{A} \text{ and } C_1 = \frac{C}{A} + a(a + \frac{B}{A}). \tag{5.111}$$

Thus, Eq. (5.110) gives

$$a = -\frac{B}{3A} \pm \frac{2}{3A}\sqrt{B^2 - 3AC}. \tag{5.112}$$

Further, the double-repeated equilibrium of $x^* = b = -\frac{1}{2}B_1$ is given by

$$x^* = b = -\frac{B}{3A} \mp \frac{1}{3A}\sqrt{B^2 - 3AC}. \tag{5.113}$$

If $B^2 > 3AC$, such a double-repeated equilibrium exists. If $B^2 < 3AC$, such a double-repeated equilibrium does not exist. From (5.111), another equilibrium is $x^* = a$, which is different from $x^* = b$. If $B^2 = 3AC$, such a double-repeated equilibrium with equilibrium of $x^* = a$ has an intersected point at $x^* = -\frac{B}{3A}$.

The bifurcation diagram for a>max$\{b_1, b_2\}$ and $a_0 > 0$ is presented in Fig. 5.6a. The stable and unstable equilibriums varying with the vector parameter are presented by solid and dashed curves, respectively. Such an equilibrium of $x^* = b$ is a *lower-saddle-node* (LSN) bifurcation. The equilibrium of $x^* = a$ is a source, which is unstable. The equilibrium of $x^* = \max\{b_1, b_2\}$ is a sink, which is stable. The equilibrium of $x^* = \min\{b_1, b_2\}$ is a source, which is unstable. However, the bifurcation diagram for a >max $\{b_1, b_2\}$ and $a_0 < 0$ is presented in Fig. 5.6b. The equilibrium of $x^* = b$ is an *upper-saddle-node* (USN) bifurcation. The equilibrium of $x^* = a$ is a sink. The equilibrium of $x^* = \max\{b_1, b_2\}$ is a source. The equilibrium of $x^* = \min\{b_1, b_2\}$ is a sink. The bifurcation diagram for $a < \min\{b_1, b_2\}$ and $a_0 > 0$ is presented in Fig. 5.6c. The equilibrium of $x^* = b$ is an *upper-saddle-node* (USN) bifurcation. The equilibrium of $x^* = \max\{b_1, b_2\}$ is a source. The equilibrium of $x^* = \min\{b_1, b_2\}$ is a sink. The equilibrium of $x^* = a$ is a source. The bifurcation diagram for $a < \min\{b_1, b_2\}$ and $a_0 < 0$ is presented in Fig. 5.6d. The equilibrium of $x^* = b$ is a *lower-saddle-node* (LSN) bifurcation. The equilibrium of $x^* = \max\{b_1, b_2\}$ is a sink. The equilibrium of $x^* = \min\{b_1, b_2\}$ is a source. The equilibrium of $x^* = a$ is a sink. The stability and bifurcations of equilibriums of the 1-dimensional cubic nonlinear system are summarized in Table 5.1.

For $\Delta_1 = B_1^2 - 4C_1 \geq 0$, the 1-dimensional cubic nonlinear system in Eq. (5.61) has three equilibriums. Three equilibriums are $x^* = a, b_1, b_2$. Assume $a_i \leq a_{i+1}$ for $i = 1, 2$ with $a_{1,2,3} = \text{sort}(a, b_1, b_2)$. With varying parameters, two of three equilibriums (i.e., $a_i = a_j$ for $i, j \in \{1, 2, 3\}$ but $i \neq j$) will be intersected each other with the corresponding discriminant of $\Delta_{ij} = (a_i - a_j)^2 = 0$, and in the vicinity of the intersection point, $\Delta_{ij} = (a_i - a_j)^2 > 0$. The two intersected points of $a = b_{1,2}$ give

$$a = -\frac{B_1}{2} \pm \frac{1}{2}\sqrt{{B_1}^2 - 4C_1}, \tag{5.114}$$

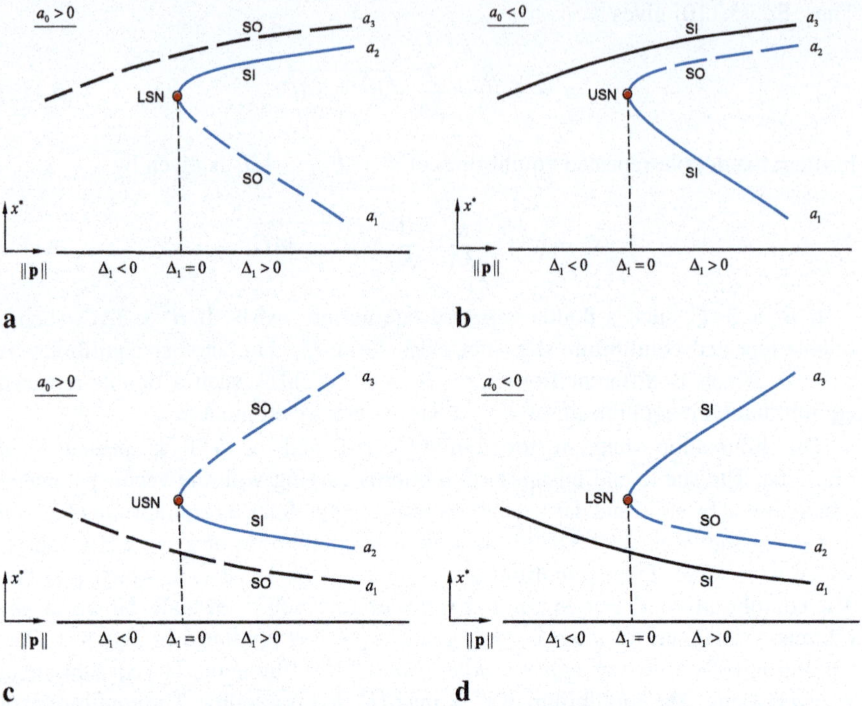

Fig. 5.6 Stability and bifurcation of three independent equilibriums in the 1-dimensional, cubic nonlinear dynamical system: For $a > \{b_1, b_2\}$: (**a**) a LSN bifurcation ($a_0 > 0$), (**b**) an USN bifurcation ($a_0 < 0$). For $a < \{b_1, b_2\}$: (**c**) an USN bifurcation ($a_0 > 0$), (**d**) a LSN bifurcation ($a_0 < 0$). LSN: *lower saddle node*, USN: *upper saddle node*. Stable and unstable equilibriums are represented by solid and dashed curves, respectively. The bifurcation points are marked by circular symbols

Table 5.1 Stability and bifurcation of a 1-dimensional cubic nonlinear system in Eq. (5.61)

Fig. 5.6		a_1	a_2	a_3	Bifurcation
(a) $a_0 > 0$	$a >$ max $\{b_1, b_2\}$	Unstable	Stable	Unstable	Second LSN
(b) $a_0 < 0$	$a >$ max $\{b_1, b_2\}$	Stable	Unstable	Stable	Second USN
(c) $a_0 < 0$	$a <$ min $\{b_1, b_2\}$	Unstable	Stable	Unstable	Second USN
(d) $a_0 < 0$	$a <$ min $\{b_1, b_2\}$	Stable	Unstable	Stable	Second LSN

Notice that $b_{1,2} = -\frac{1}{2}(B_1 \pm \sqrt{\Delta_1})$, $\Delta_1 = B_1^2 - 4C_1$. Bifurcation condition: $\Delta_1 = 0$. LSN: lower saddle node, USN: upper saddle node

or

$$a^2 + aB_1 = -C_1. \tag{5.115}$$

With Eq. (5.62) or Eq. (5.111), the foregoing equation gives

$$a = b_{1,2} = -\frac{B}{3A} \pm \frac{1}{3A}\sqrt{B^2 - 3AC}. \tag{5.116}$$

If $B^2 > 3AC$, such an intersected point of $x^* = a$ and $x^* = b_1$ or $x^* = b_2$ exists. If $B^2 < 3AC$, such an intersected point does not exist. Such an intersection point is for the two equilibriums switching, which is called the *saddle-node* bifurcation. The stability and bifurcation diagrams for $a_0 > 0$ and $a_0 < 0$ are presented in Fig. 5.7a, b, respectively. Three equilibriums are intersected at a point with $\Delta_{ij} = (a_i - a_j)^2 = 0$ and $a_1 = a_2 = a_3 = -\frac{B}{3A}$, and in the vicinity of the intersection point, $\Delta_{ij} = (a_i - a_j)^2 > 0$ for $i, j = 1, 2, 3$ but $i \neq j$. The intersection points for $a_0 > 0$ and $a_0 < 0$ are called the source and sink bifurcations of the third order, respectively. The corresponding stability and bifurcation diagrams for three equilibriums switching are presented in Fig. 5.7c, d.

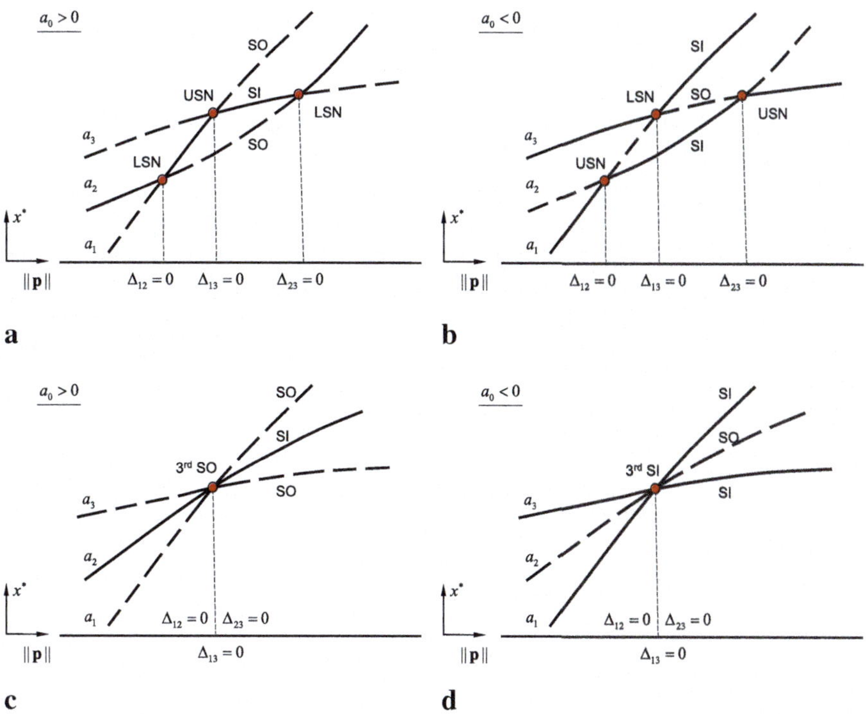

Fig. 5.7 Stability and bifurcation of equilibriums switching in the 1-dimensional, cubic nonlinear dynamical system. For two equilibriums switching: (**a**) $a_0 > 0$, (**b**) $a_0 < 0$. For three equilibriums switching: (**c**) third-order source bifurcation ($a_0 > 0$), (**d**) third-order sink bifurcation ($a_0 < 0$). LSN: *lower saddle node*, USN: *upper saddle node*. Stable and unstable equilibriums are represented by solid and dashed curves, respectively. The bifurcation points are marked by circular symbols

In the 1-dimensional cubic nonlinear system of Eq. (5.61), $x^2 + B_1 x + C_1 = 0$ gives two equilibriums of $x^* = b_1, b_2$ for $\Delta_1 = B_1^2 - 4C_1 > 0$. One of the two equilibriums has one intersection with $x^* = a$ and there are three different equilibriums for $a = a_2 \in (\min\{b_1, b_2\}, \max\{b_1, b_2\})$. For this case, the intersection point occurs at $a = \min\{b_1, b_2\}$ for $\mathbf{p}_1 \in \partial\Omega_{23}$ or $a = \max\{b_1, b_2\}$ for $\mathbf{p}_2 \in \partial\Omega_{23}$. The bifurcation point of equilibrium occurs at the double-repeated equilibrium at $\Delta_1 = B_1^2 - 4C_1 = 0$ for $\mathbf{p}_0 \in \partial\Omega_{12}$. Such a bifurcation is a *lower-* or *upper-saddle-node* bifurcation. For $a = -\frac{1}{2}B_1$ with $\Delta_1 = B_1^2 - 4C_1 = 0$, three equilibriums are repeated with three multiplicities. The intersected point of $a = -\frac{1}{2}B_1$ with Eq. (5.111) gives

$$a = -\frac{1}{2}(a + \frac{B}{A}). \tag{5.117}$$

Thus

$$a = -\frac{B}{3A}. \tag{5.118}$$

Such a bifurcation at the intersection point is also a third-order *source* or *sink* bifurcation. The bifurcation diagrams for six cases of three equilibriums with one intersection are presented in Fig. 5.8a–f and the stability and bifurcations are listed in Table 5.2.

The 1-dimensional cubic nonlinear system is expressed by a factor of $(x-a)$ and a quadratic form of $a_0(x^2 + B_1 x + C_1)$ as in Eq. (5.61). For $\Delta_1 = B_1^2 - 4C_1 > 0$, $x^2 + B_1 x + C_1 = 0$ gives two equilibriums of $x^* = b_1, b_2$. The two equilibriums *do not* have any intersections with $x^* = a$. For $\Delta_1 = B_1^2 - 4C_1 = 0$, there are two parameters of $\mathbf{p}_1 \in \partial\Omega_{12}$ and $\mathbf{p}_2 \in \partial\Omega_{12}$, and the two double-repeated equilibriums are at $x^*(\mathbf{p}_i) = -\frac{1}{2}B_1(\mathbf{p}_i)$ $(i = 1, 2)$. With the two repeated equilibriums, the two equilibriums of $x^* = b_1, b_2$ formed a closed path in the bifurcation diagram. The bifurcation points of equilibrium occur at the two double-repeated equilibriums of $\Delta_1 = B_1^2 - 4C_1 = 0$ for $\mathbf{p}_i \in \partial\Omega_{12}$ $(x^* = b_1, b_2)$. Such a bifurcation at the intersection point is also a *lower- or upper-saddle-node* bifurcation. The stable and unstable equilibriums varying with the vector parameter are also represented by solid and dashed curves, respectively. The bifurcation diagrams for four cases of three equilibriums are presented in Fig. 5.9a–d, and the stability and bifurcations are summarized in Table 5.3.

If the two repeated equilibriums have two intersections with $x^* = a(\mathbf{p}_i)$ $(i = 1, 2)$, i.e., $a(\mathbf{p}_i) = -\frac{1}{2}B_1(\mathbf{p}_i)$ then there are two triple-repeated equilibriums at $x^* = a(\mathbf{p}_i)$ $(i = 1, 2)$, which are the third-order sink or source bifurcations. The stability and bifurcation diagrams of equilibriums are formed by the equilibrium of $x^* = a(\mathbf{p})$ and the closed loop of equilibriums of $x^* = b_1, b_2$, as shown in Fig. 5.9e, f for $a_0 > 0$ and $a_0 < 0$, respectively. The stability and bifurcations are also summarized in Table 5.3.

In the 1-dimensional cubic nonlinear system in Eq. (5.61), $x^2 + B_1 x + C_1 = 0$ for $\Delta_1 = B_1^2 - 4C_1 > 0$ gives two equilibriums of $x^* = b_1, b_2$, which have an

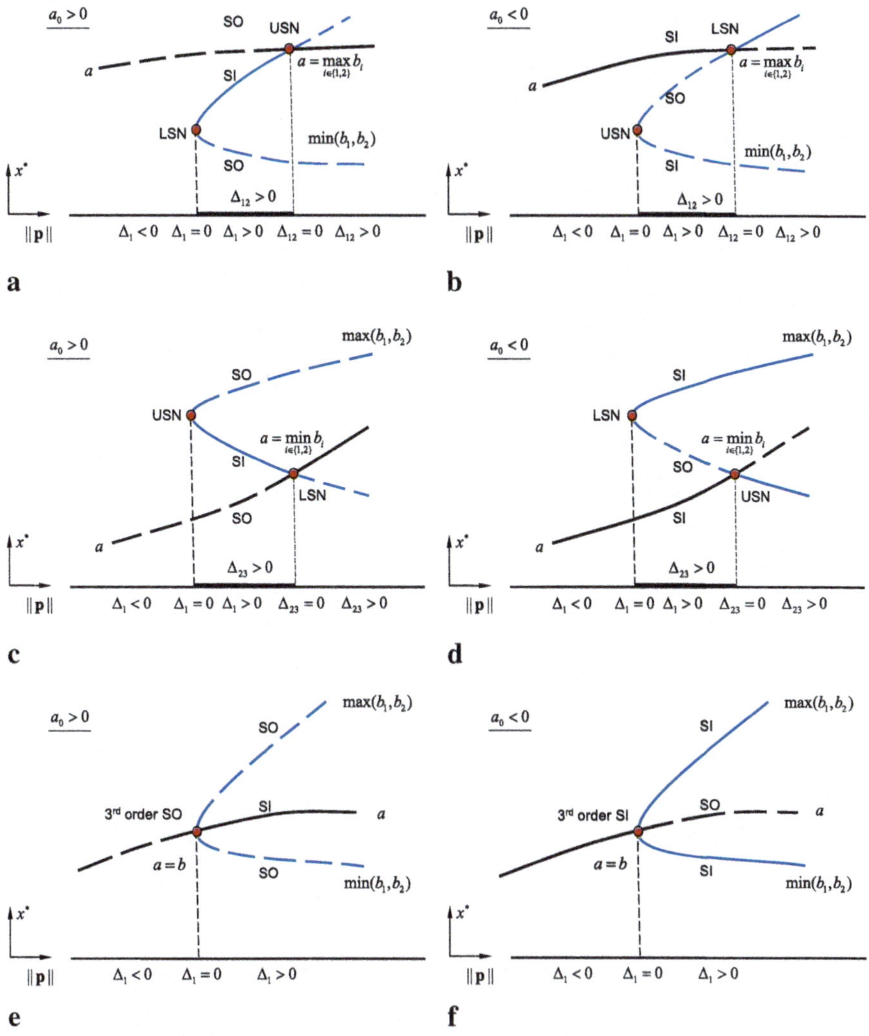

Fig. 5.8 Stability and bifurcation of equilibriums in the 1-dimensional, cubic nonlinear dynamical system: (**a**) the LSN ($\Delta_1 = 0$) and USN ($a = \max\{b_1, b_2\}$) bifurcations ($a_0 > 0$), (**b**) the USN ($\Delta_1 = 0$) and LSN ($a = \max\{b_1, b_2\}$) bifurcations ($a_0 < 0$), (**c**) the USN ($\Delta_1 = 0$) and LSN ($a = \min\{b_1, b_2\}$) bifurcations ($a_0 > 0$), (**d**) the LSN ($\Delta_1 = 0$) and USN ($a = \min\{b_1, b_2\}$) bifurcations ($a_0 < 0$), (**e**) the third-order SO bifurcation ($\Delta_1 = 0$ and $a = b$) ($a_0 > 0$), (**f**) the third-order SI bifurcation ($\Delta_1 = 0$ and $a = b$) ($a_0 < 0$). LSN: *lower saddle node*, USN: *upper saddle node*, SI: sink, SO: source. Stable and unstable equilibriums are represented by solid and dashed curves, respectively. The bifurcation points are marked by circular symbols

intersection with $x^* = a$. The intersected point is at $a = b_1$ or $a = b_2$ with Eq. (5.114). The double-repeated equilibrium requires $\Delta_1 = B_1^2 - 4C_1 = 0$ and the two equilibriums of $x^* = a, b_1$ under $\Delta_1 = B_1^2 - 4C_1 > 0$ and $x^* = b_2$ for $\Delta_1 = B_1^2 - 4C_1 < 0$. Similarly, the two equilibriums of $x^* = a, b_2$ under $\Delta_1 = B_1^2 - 4C_1 > 0$ and $x^* = b_2$

Table 5.2 Stability and bifurcation of a 1-dimensional cubic nonlinear system ($\dot{x} = a_0(x-a)[x^2 + B_1(\mathbf{p})x + C_1(\mathbf{p})], a \in (\min\{b_1, b_2\}, \max\{b_1, b_2\})$)

Fig. 5.8	a_1	a_2	a_3	B-I	B-II	B-III
(a) $a_0 > 0$	Unstable	Stable	Unstable	Second LSN	Second USN	$a = \max\{b_1, b_2\}$
(b) $a_0 < 0$	Stable	Unstable	Stable	Second USN	Second LSN	$a = \max\{b_1, b_2\}$
(c) $a_0 > 0$	Unstable	Stable	Unstable	Second USN	Second USN	$a = \min\{b_1, b_2\}$
(d) $a_0 < 0$	Stable	Unstable	Stable	Second LSN	Second USN	$a = \min\{b_1, b_2\}$
(e) $a_0 > 0$	Unstable	Stable	Unstable	$\Delta_1 = 0$	$a = -\frac{1}{2}B_1$	Third order SO
(f) $a_0 < 0$	Stable	Unstable	Stable	$\Delta_1 = 0$	$a = -\frac{1}{2}B_1$	Third order SI

Notice that $b_{1,2} = -\frac{1}{2}(B_1 \pm \sqrt{\Delta_1})$, $\Delta_1 = B_1^2 - 4C_1$. Bifurcation-I (B-I): $\Delta_1 = 0$. Bifurcation-II (B-II): $a = \max\{b_1, b_2\}$ or $a = \min\{b_1, b_2\}$. Bifurcation-III (B-III): $\Delta_1 = 0$and $a = -\frac{B_1}{2A_1}$. LSN: lower saddle node, USN: upper saddle node. SO: source, SI: sink

for $\Delta_1 = B_1^2 - 4C_1 < 0$ are required. Such a bifurcation for two equilibriums appearing and vanishing is called a *lower*- or *upper-saddle*-node bifurcation. The stable and unstable equilibriums varying with the vector parameter are also represented by solid and dashed curves, respectively. The bifurcation diagrams for four cases of three equilibriums are presented in Fig. 5.10a–f. If the double-repeated equilibrium has an intersection with $x^* = a(\mathbf{p}_0) = -\frac{1}{2}B_1 = -\frac{B}{3A}$, then there two triple-repeated equilibriums at $x^* = a(\mathbf{p}_0)$, which is the third-order sink and source bifurcations for $a_0 > 0$ and $a_0 < 0$, respectively. The stability and bifurcation diagrams of equilibriums are shown in Fig. 5.10e, f.

Consider a 1-dimensional, cubic nonlinear dynamical system with a double-repeated equilibrium and one simple equilibrium.

(i) For $b < a$, the 1-dimensional, cubic nonlinear dynamical system is

$$\dot{x} = a_0(\mathbf{p})(x - b(\mathbf{p}))^2(x - a(\mathbf{p})). \tag{5.119}$$

For such a system, if $a_0 > 0$, the double-repeated equilibrium of $x^* = b$ is a *lower saddle*, which is unstable, and the simple equilibrium of $x^* = b$ is a source, which is unstable. If $a_0 < 0$, the double-repeated equilibrium of $x^* = b$ is an *upper saddle*, which is unstable, and the simple equilibrium of $x^* = a$ is a sink, which is stable.

(ii) For $b > a$, the 1-dimensional cubic nonlinear dynamical system is

$$\dot{x} = a_0(\mathbf{p})(x - a(\mathbf{p}))(x - b(\mathbf{p}))^2. \tag{5.120}$$

For such a system, if $a_0 > 0$, the double-repeated equilibrium of $x^* = b$ is an *upper saddle*, which is unstable, and the simple equilibrium of $x^* = a$ is a source, which is unstable. If $a_0 < 0$, the double-repeated equilibrium of $x^* = b$ is a *lower saddle*, which is unstable, and the simple equilibrium of $x^* = a$ is a sink, which is stable.

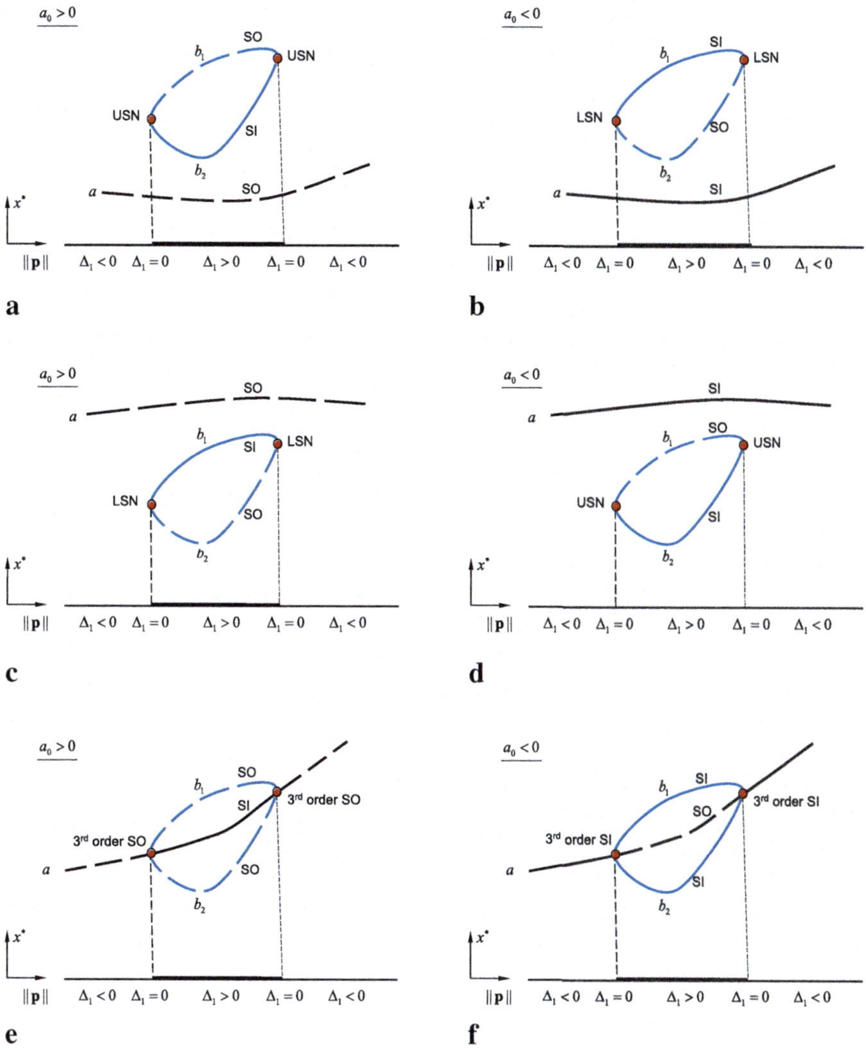

Fig. 5.9 Stability and bifurcation of three equilibriums in the 1-dimensional, cubic nonlinear dynamical system: For $a < \{b_1, b_2\}$: (**a**) two USN bifurcations ($a_0 > 0$), (**b**) two LSN bifurcations ($a_0 < 0$). For $a > \{b_1, b_2\}$: (**c**) two LSN bifurcations ($a_0 > 0$), (**d**) two USN bifurcations ($a_0 < 0$). (**e**) Two third-order SO bifurcations ($a_0 > 0$), (**f**) two third-order SI bifurcations ($a_0 < 0$). LSN: *lower saddle node,* USN: *upper saddle node*. SO: source, SI: sink. Stable and unstable equilibriums are represented by solid and dashed curves, respectively. The bifurcation points are marked by circular symbols

Table 5.3 Stability and bifurcation of a 1-dimensional cubic nonlinear system ($\dot{x} = a_0(x-a)[x^2 + B_1(\mathbf{p})x + C_1(\mathbf{p})]$,$a \in (\min\{b_1, b_2\}, \max\{b_1, b_2\})$)

Fig. 5.9	a	b_1	b_2	B-I	B-I	B-III
(a) $a_0 > 0$	Unstable	Stable	Unstable	USN	USN	–
(b) $a_0 < 0$	Stable	Unstable	Stable	LSN	LSN	–
(c) $a_0 > 0$	Unstable	Stable	Unstable	LSN	LSN	–
(d) $a_0 < 0$	Stable	Unstable	Stable	USN	USN	–
(e) $a_0 < 0$	Unstable	Stable	Unstable	–	–	Third-order SO
(f) $a_0 < 0$	Stable	Unstable	Stable	–	–	Third-order SI

Notice that $b_{1,2} = -\frac{1}{2}(B_1 \pm \sqrt{\Delta_1})$, $\Delta_1 = B_1^2 - 4C_1$. Bifurcation-I (B-I): $\Delta_1 = 0$. Bifurcation-II (B-II): $a = \max\{b_1, b_2\}$. Bifurcation-III (B-III): $a = \min\{b_1, b_2\}$. LSN: lower-saddle node, USN: upper-saddle-node

(iii) For $b = a$, the dynamical system on the boundary is

$$\dot{x} = a_0(\mathbf{p})(x - b(\mathbf{p}))^3. \tag{5.121}$$

For such a system, if $a_0 > 0$, the triple-repeated equilibrium of $x^* = b$ with the third-order multiplicity is a source bifurcation of the third-order for the (US:SO) to (SO:LS) equilibrium. If $a_0 < 0$, the triple-repeated equilibrium of $x^* = b$ with the third-order multiplicity is a sink bifurcation of the third-order for the (LS:SI) to (SO:US) equilibrium. With parameter changes, the bifurcation diagram for the cubic nonlinear system is presented in Fig. 5.11. The acronyms LS, US, SI, and SO are for *lower-saddle*, *upper-saddle*, sink, and source, respectively. Stable and unstable equilibriums are represented by solid and dashed curves, respectively. The bifurcation point is marked by a circular symbol. The third-order source bifurcation for the upper-saddle and source equilibriums to the source and lower-saddle equilibriums is presented in Fig. 5.11a. The third-order sink bifurcation for the lower-saddle and sink equilibriums to the sink and upper-saddle equilibriums is presented in Fig. 5.11b.

To illustrate the stability and bifurcation of equilibrium with singularity in a 1-dimensional, cubic nonlinear system, the equilibrium of $\dot{x} = a_0(x - a_1)^3$ is presented in Fig. 5.12. The third-order sink and source of equilibrium of $x^* = a_1$ with the third-order multiplicity are stable and unstable, respectively. The stable and unstable equilibriums are depicted by solid and dashed curves, respectively. At $a_0 = 0$, the third-order sink and source equilibriums are switched, which is marked by a circular symbol.

For the 1-dimensional, cubic nonlinear systems, the equilibrium stability and bifurcation can be described through an alternative way as follows.

Definition 5.7 Consider a 1-dimensional, cubic nonlinear dynamical system

$$\begin{aligned}\dot{x} &= A(\mathbf{p})x^3 + B(\mathbf{p})x^2 + C(\mathbf{p})x + D(\mathbf{p}) \\ &\equiv a_0(\mathbf{p})\left[(x + \frac{B}{3A})^3 + p(\mathbf{p})(x + \frac{B}{3A}) + q(\mathbf{p})\right]\end{aligned} \tag{5.122}$$

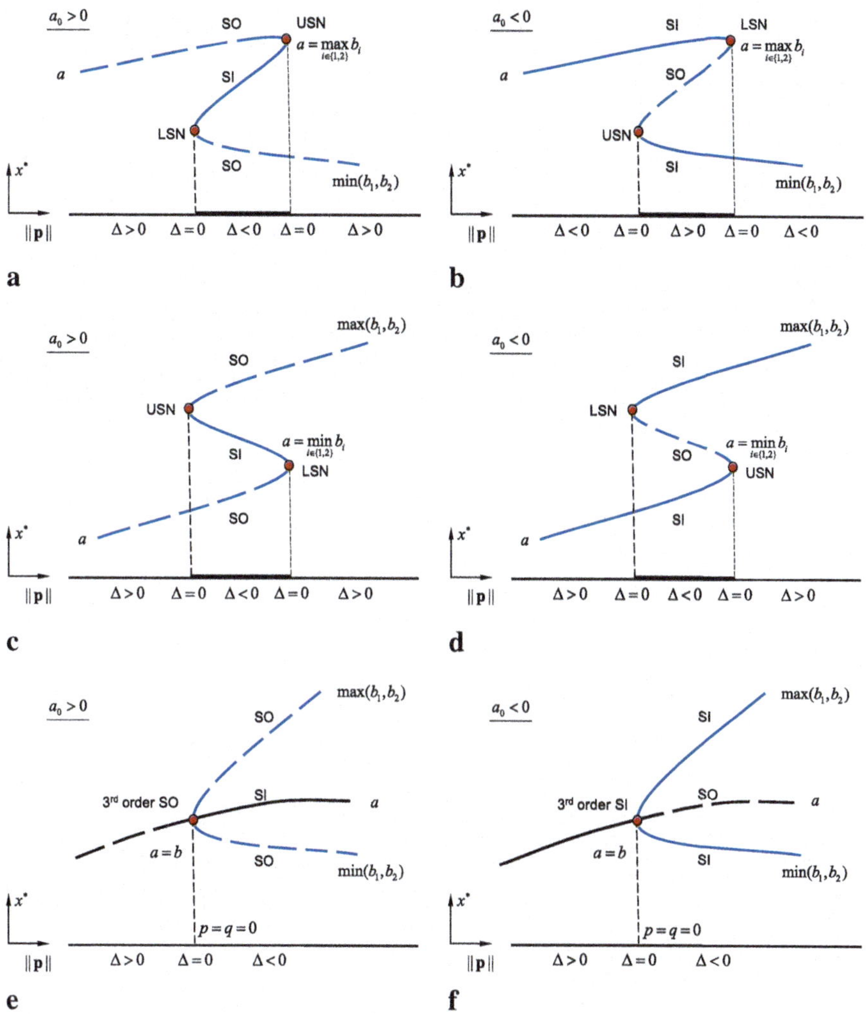

Fig. 5.10 Stability and bifurcation of equilibriums in the 1-dimensional, cubic nonlinear dynamical system: (**a**) the LSN ($\Delta_1=0$) and USN ($a=\max\{b_1,b_2\}$) bifurcations ($a_0>0$), (**b**) the USN ($\Delta_1=0$) and LSN ($a=\max\{b_1,b_2\}$) bifurcations ($a_0<0$), (**c**) the USN ($\Delta_1=0$) and LSN ($a=\min\{b_1,b_2\}$) bifurcations ($a_0>0$), (**d**) the LSN ($\Delta_1=0$) and USN ($a=\min\{b_1,b_2\}$) bifurcations ($a_0<0$), (**e**) the third-order SO bifurcation ($\Delta_1=0$ and $a=b$) ($a_0>0$), (**f**) the third-order SI bifurcation ($\Delta_1=0$ and $a=b$) ($a_0<0$). LSN: lower saddle node, USN: upper saddle node, SI: sink, SO: source. Stable and unstable equilibriums are represented by solid and dashed curves, respectively. The bifurcation points are marked by circular symbols

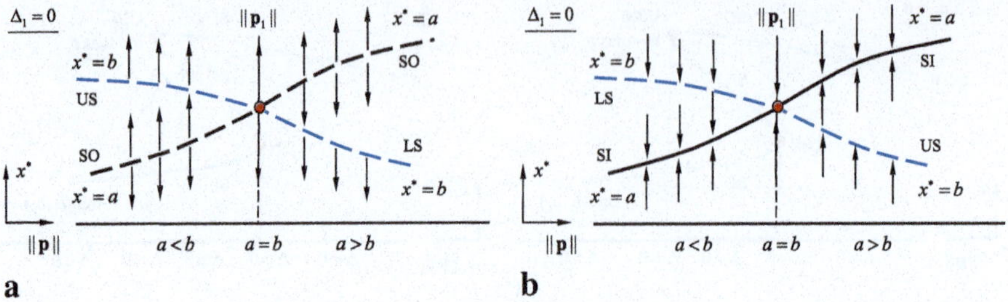

Fig. 5.11 Stability and bifurcation of a triple-repeated equilibrium with a simple equilibrium in a 1-dimensional, cubic nonlinear dynamical system: (**a**) a third-order source bifurcation for (US:SO) to (SO:LS) switching ($a_0 > 0$), (**b**) a third-order sink bifurcation ($a_0 < 0$) for (LS:SI) to (SI:US) switching. Stable and unstable equilibriums are represented by solid and dashed curves, respectively

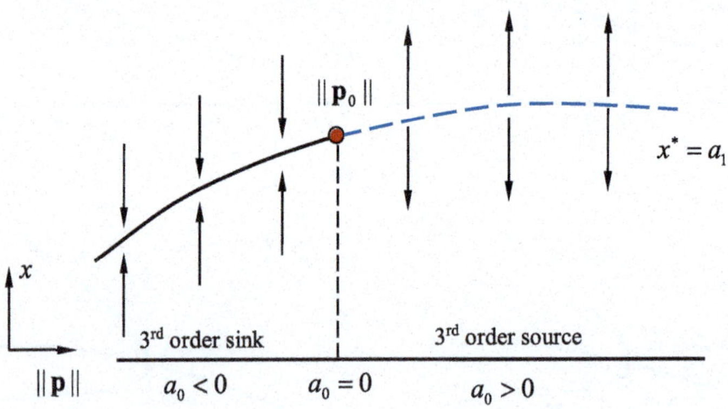

Fig. 5.12 Stability of a triple equilibrium in the 1-dimensional, cubic nonlinear dynamical system: Stable and unstable equilibriums are represented by solid and dashed curves, respectively. The stability switching is labeled by a circular symbol

where four scalar constants $A(\mathbf{p}) \neq 0$, $B(\mathbf{p})$, $C(\mathbf{p})$, and $D(\mathbf{p})$ satisfy

$$A = a_0, p = \frac{C}{A} - \frac{B^2}{3A^2}, q = \frac{D}{A} - \frac{BC}{3A^2} + \frac{2B^3}{27A^3}$$
$$\mathbf{p} = (p_1, p_2, \ldots, p_m)^{\mathrm{T}}. \tag{5.123}$$

(i) If

$$\Delta = \frac{q^2}{4} + \frac{p^3}{27} > 0 \tag{5.124}$$

the cubic nonlinear system has one equilibrium as

$$x^* = a \equiv (-\frac{q}{2} + \sqrt{\Delta})^{1/3} + (-\frac{q}{2} - \sqrt{\Delta})^{1/3} - \frac{B}{3A} \tag{5.125}$$

and the corresponding standard form is

$$\begin{aligned} \dot{x} &= a_0(x-a)[(x+\frac{1}{2}B_1)^2 + \frac{1}{4}(-\Delta_1)] \\ &= a_0(x-a)[x^2 + B_1 x + C_1)]. \end{aligned} \tag{5.126}$$

where

$$A = a_0, B = (-a + B_1)a_0, C = (-aB_1 + C_1)a_0, D = -aa_0C_1. \tag{5.127}$$

(ii) If

$$\Delta = \frac{q^2}{4} + \frac{p^3}{27} < 0 \tag{5.128}$$

the cubic nonlinear system has three equilibriums as

$$\begin{aligned} x^* &= a = (-\frac{q}{2} + \sqrt{\Delta})^{1/3} + (-\frac{q}{2} - \sqrt{\Delta})^{1/3} - \frac{B}{3A}, \\ x^* &= b_1 = \omega(-\frac{q}{2} + \sqrt{\Delta})^{1/3} + \omega^2(-\frac{q}{2} - \sqrt{\Delta})^{1/3} - \frac{B}{3A}, \\ x^* &= b_2 = \omega^2(-\frac{q}{2} + \sqrt{\Delta})^{1/3} + \omega(-\frac{q}{2} - \sqrt{\Delta})^{1/3} - \frac{B}{3A}; \\ \omega &= \frac{-1 + \mathbf{i}\sqrt{3}}{2}, \omega^2 = \frac{-1 - \mathbf{i}\sqrt{3}}{2}, \mathbf{i} = \sqrt{-1}. \end{aligned} \tag{5.129}$$

The corresponding standard form is

$$\dot{x} = a_0(x - a_1)(x - a_2)(x - a_3) \tag{5.130}$$

where

$$\begin{aligned} &a_1 = \min(b_1, b_2, a), a_3 = \max(b_1, b_2, a), a_2 \in \{b_1, b_2, a\}, \\ &a_2 \neq a_1, a_3. \end{aligned} \tag{5.131}$$

(iii) If

$$\Delta = \frac{q^2}{4} + \frac{p^3}{27} = 0, \frac{q^2}{4} = -\frac{p^3}{27} \neq 0 \tag{5.132}$$

the 1-dimensional, cubic nonlinear dynamical system has a double equilibrium with the second multiplicity plus a simple equilibrium as

$$
\begin{aligned}
& x^* = a = 2(-\tfrac{q}{2})^{1/3} - \frac{B}{3A}, \\
& x^* = b_1 = b, x^* = b_2 = b; \\
& b = \omega(-\tfrac{q}{2})^{1/3} + \omega^2(-\tfrac{q}{2})^{1/3} - \frac{B}{3A}, \\
& \omega = \frac{-1 + \mathbf{i}\sqrt{3}}{2}, \omega^2 = \frac{-1 - \mathbf{i}\sqrt{3}}{2}, \mathbf{i} = \sqrt{-1}.
\end{aligned}
\tag{5.133}
$$

(iii$_1$) The corresponding standard form for $a < b$ is

$$\dot{x} = a_0(x - a)(x - b)^2. \tag{5.134}$$

Such a flow at the equilibrium of $x^* = b$ is called a *saddle* flow of the second order. The bifurcation of equilibrium at $x^* = b$ for two different equilibriums appearing or vanishing is called a *saddle-node* bifurcation of the second order at a point $\mathbf{p} = \mathbf{p}_1 \in \partial\Omega_{12}$, and the bifurcation condition is

$$\Delta = \frac{q^2}{4} + \frac{p^3}{27} = 0, \frac{q^2}{4} = -\frac{p^3}{27} \neq 0, a < b. \tag{5.135}$$

(iii$_2$) The corresponding standard form for $a > b$ is

$$\dot{x} = a_0(x - b)^2(x - a). \tag{5.136}$$

Such a flow at the equilibrium of $x^* = b$ is called a *saddle* flow of the second order. The bifurcation of equilibrium at $x^* = b$ for two equilibriums appearing or vanishing is called a *saddle-node* bifurcation of the second order at a point $\mathbf{p} = \mathbf{p}_1 \in \partial\Omega_{12}$, and the bifurcation condition is

$$\Delta = \frac{1}{4}q^2 + \frac{1}{27}p^3 = 0, \frac{1}{4}q^2 = -\frac{1}{27}p^3 \neq 0, a > b. \tag{5.137}$$

(iv) If

$$\Delta = \frac{1}{4}q^2 + \frac{1}{27}p^3 = 0, \frac{1}{4}q^2 = -\frac{1}{27}p^3 = 0, \tag{5.138}$$

the 1-dimensional dynamical system has a triple-repeated equilibrium as

$$x^* = a = -\frac{B}{3A}, x^* = b_1 = -\frac{B}{3A}, x^* = b_2 = -\frac{B}{3A}. \tag{5.139}$$

The corresponding standard form is

$$\dot{x} = a_0(x - a)^3. \tag{5.140}$$

Such a flow at the equilibrium of $x^* = a$ is called a source or sink flow of the third order. The bifurcation of equilibrium at $x^* = -\frac{B}{3A}$ for one equilibrium to three equilibriums is called a source or think bifurcation of the third order at $\mathbf{p} = \mathbf{p}_1 \in \partial\Omega_{12}$, and the bifurcation condition is

$$\Delta = \frac{1}{4}q^2 + \frac{1}{27}p^3 = 0, \frac{1}{4}q^2 = -\frac{1}{27}p^3 = 0. \tag{5.141}$$

From the afore-described stability and bifurcation of the 1-dimensional, cubic nonlinear dynamical systems, the equilibrium stability and bifurcations of equilibrium in Eq. (5.122) are similar to Theorem 5.6.

The 1-dimensional cubic nonlinear system has the following four cases:

(i) One real solution of equilibrium of $x^* = a$ requires $\Delta = (\frac{q}{2})^2 + (\frac{p}{3})^3 > 0$ for Eq. (5.122), equivalent to $\Delta_1 = B_1^2 - 4C_1 < 0$ for Eq. (5.61):

$$Aa^3 + Ba^2 + Ca + D = 0. \tag{5.142}$$

(ii) Three different solutions of equilibriums of $x^* = a, b_1, b_2$ require $\Delta < 0$ for Eq. (5.122), equivalent to $\Delta_1 = B_1^2 - 4C_1 > 0$ for Eq. (5.61):

$$Aa^3 + Ba^2 + Ca + D = 0 \text{ and } b_{1,2} = \frac{1}{2}(B_1 \pm \sqrt{B_1^2 - 4C_1}). \tag{5.143}$$

(iii) The double-repeated equilibriums require $\Delta = 0$ and $(\frac{q}{2})^2 = -(\frac{p}{3})^3 \neq 0$ for Eq. (5.122), equivalent to $\Delta_1 = B_1^2 - 4C_1 = 0$ or $a = b_{1,2}$ for Eq. (5.61):

$$a = b_{1,2} = -\frac{B}{3A} \pm \frac{1}{3A}\sqrt{B^2 - 3AC} \text{ with } B^2 > 3AC. \tag{5.144}$$

(iv) The triple-repeated equilibriums require $\Delta = 0$ and $(\frac{q}{2})^2 = -(\frac{p}{3})^3 = 0$ for Eq. (5.122), equivalent to $\Delta_1 = B_1^2 - 4C_1 = 0$ and $a = b_{1,2}$ for Eq. (5.61):

$$a = b_{1,2} = -\frac{B}{3A} \text{ with } B^2 = 3AC. \tag{5.145}$$

5.4 Quartic Nonlinear Systems

In this section, the stability and bifurcation of the quartic nonlinear systems will be presented as in Luo (2019). The fourth-order upper-saddle and lower-saddle appearing bifurcations of two second-order upper saddles and lower saddles will be presented. The third-order switching sink and source bifurcations of lower saddle with sink and upper saddle with source will be discussed.

Definition 5.8 Consider a 1-dimensional, quartic nonlinear dynamical system

$$\begin{aligned}\dot{x} &= A(\mathbf{p})x^4 + B(\mathbf{p})x^3 + C(\mathbf{p})x^2 + D(\mathbf{p})x + E(\mathbf{p}) \\ &= a_0(\mathbf{p})[x^2 + B_1(\mathbf{p})x + C_1(\mathbf{p})][x^2 + B_2(\mathbf{p})x + C_2(\mathbf{p})]\end{aligned} \tag{5.146}$$

where $A(\mathbf{p}) \neq 0$, and

$$\mathbf{p} = (p_1, p_2, \ldots, p_m)^{\mathrm{T}}. \tag{5.147}$$

(i) If

$$\Delta_i = B_i^2 - 4C_i < 0 \text{ for } i = 1, 2 \tag{5.148}$$

the quartic nonlinear dynamical system does not have any equilibrium, and the corresponding standard form is

$$\dot{x} = a_0[(x + \frac{1}{2}B_1)^2 + \frac{1}{4}(-\Delta_1)][(x + \frac{1}{2}B_2)^2 + \frac{1}{4}(-\Delta_2)]. \tag{5.149}$$

The flow of such a system without equilibriums is called a nonequilibrium flow.

(i_1) If $a_0 > 0$, the nonequilibrium flow is called the positive flow.
(i_2) If $a_0 < 0$, the nonequilibrium flow is called the negative flow.

(ii) If

$$\Delta_i = B_i^2 - 4C_i > 0 \text{ and } \Delta_j = B_j^2 - 4C_j < 0 \text{ for } i, j \in \{1, 2\}, i \neq j \tag{5.150}$$

the quartic polynomial system has two simple equilibriums, i.e.,

$$x^* = b_1^{(i)} = -\frac{1}{2}(B_i + \sqrt{\Delta_i}), x^* = b_2^{(i)} = -\frac{1}{2}(B_i - \sqrt{\Delta_i}). \tag{5.151}$$

The corresponding standard form is

$$\dot{x} = a_0(x - a_1)(x - a_2)[(x + \frac{1}{2}B_j)^2 + \frac{1}{4}(-\Delta_j)] \tag{5.152}$$

where

$$a_1 = \min\{b_1^{(i)}, b_2^{(i)}\} \text{ and } a_2 = \max\{b_1^{(i)}, b_2^{(i)}\}. \tag{5.153}$$

Such a flow of equilibriums is called a flow of two simple equilibriums.

(iii) If

$$\Delta_i = B_i^2 - 4C_i = 0 \text{ and } \Delta_j = B_j^2 - 4C_j < 0 \text{ for } i,j \in \{1,2\}, i \neq j \tag{5.154}$$

the quartic polynomial system has a double-repeated equilibrium, i.e.,

$$x^* = b_1^{(i)} = -\frac{1}{2}B_i, x^* = b_2^{(i)} = -\frac{1}{2}B_i. \tag{5.155}$$

The corresponding standard form is

$$\dot{x} = a_0(x - a_1)^2[(x + \frac{1}{2}B_j)^2 + \frac{1}{4}(-\Delta_j)] \tag{5.156}$$

where

$$a_1 = b_1^{(i)} = b_2^{(i)}. \tag{5.157}$$

Such a flow of the equilibrium of $x^* = a_1$ is called a *saddle* flow of the second order. The equilibrium of $x^* = a_1$ for two equilibriums switching or appearing or vanishing is called a bifurcation point of equilibrium at $\mathbf{p} = \mathbf{p}_1 \in \partial\Omega_{12}$, and the bifurcation condition is

$$\Delta_i = B_i^2 - 4C_i = 0 \ (i \in \{1,2\}) \text{ and } a_1 = -\frac{1}{2}B_i. \tag{5.158}$$

(iv) If

$$\Delta_i = B_i^2 - 4C_i \geq 0 \text{ for } i = 1,2 \tag{5.159}$$

the quartic nonlinear dynamical system has four equilibriums, i.e.,

$$x^* = b_1^{(i)} = -\frac{1}{2}(B_i + \sqrt{\Delta_i}), x^* = b_2^{(i)} = -\frac{1}{2}(B_i - \sqrt{\Delta_i}) \text{ for } i = 1,2. \tag{5.160}$$

(iv$_1$) A standard form is

$$\dot{x} = a_0(x - a_1)(x - a_2)(x - a_3)(x - a_4) \tag{5.161}$$

where

$$\begin{aligned} &\Delta_i = B_i^2 - 4C_i > 0, i = 1, 2; \\ &b_k^{(1)} \neq b_l^{(2)} \text{ for } k, l \in \{1, 2\}; \\ &a_{1,2,3,4} \in \{b_1^{(1)}, b_2^{(1)}, b_1^{(2)}, b_2^{(2)}\} \text{ with } a_m < a_{m+1}. \end{aligned} \tag{5.162}$$

Such a flow of equilibriums is called a flow of four simple equilibriums.

(iv$_2$) The corresponding standard form is

$$\dot{x} = a_0(x - a_{i_1})^2(x - a_{i_2})(x - a_{i_3}) \tag{5.163}$$

where

$$\begin{aligned} &\Delta_i = B_i^2 - 4C_i > 0,\ \ \Delta_j = B_j^2 - 4C_j > 0 \text{ for } i, j = 1, 2; \\ &a_{i_1} = b_k^{(i)} = b_l^{(j)}, (i, k) \neq (j, l); i, j, k, l \in \{1, 2\} \\ &a_{i_1} \notin \{a_{i_2}, a_{i_3}\} \text{ for } i_\alpha \in \{1, 2, 3, 4\} \text{ and } \alpha \in \{1, 2, 3, 4\}. \end{aligned} \tag{5.164}$$

Such a flow of equilibrium $x^* = a_{i_1}$ is called a *saddle* flow of the second order. The equilibrium of $x^* = a_{i_1}$ for two equilibriums switching or appearing or vanishing is called a *saddle* bifurcation of equilibrium at a point $\mathbf{p} = \mathbf{p}_1 \in \partial\Omega_{12}$, and the bifurcation condition is

$$\begin{aligned} &\Delta_i = B_i^2 - 4C_i > 0\ (i \in \{1, 2\}) \text{ and} \\ &\Delta_j = B_j^2 - 4C_j > 0\ (j \in \{1, 2\}) \\ &b_k^{(i)} = b_l^{(j)}, (i, k) \neq (j, l), (i, j, k, l \in \{1, 2\}). \end{aligned} \tag{5.165}$$

(iv$_3$) The corresponding standard form is

$$\dot{x} = a_0(x - a_{i_1})^3(x - a_{i_2}) \tag{5.166}$$

where

$$\begin{aligned} &\Delta_i = B_i^2 - 4C_i > 0,\ \ \Delta_j = B_j^2 - 4C_j = 0 \text{ for } i, j = 1, 2; \\ &a_{i_1} = -\frac{1}{2}B_j = b_l^{(i)}, a_{i_2} = b_k^{(i)}, k \neq l; k, l \in \{1, 2\} \\ &a_{i_1} = a_{i_3} \text{ for } i_\alpha \in \{1, 2, 3, 4\} \text{ and } \alpha \in \{1, 2, 3, 4\}. \end{aligned} \tag{5.167}$$

Such a flow of the equilibrium of $x^* = a_{i_1}$ is called a source and sink flow of the third order. The equilibrium of $x^* = a_{i_1}$ for one equilibrium to three equilibrium is called a third-order source or sink bifurcation of equilibrium at a point $\mathbf{p} = \mathbf{p}_1 \in \partial\Omega_{12}$, and the bifurcation condition is

$$\begin{aligned} &\Delta_i = B_i^2 - 4C_i > 0\ (i \in \{1,2\}) \text{ and} \\ &\Delta_j = B_j^2 - 4C_j = 0\ (j \in \{1,2\}) \\ &b_k^{(i)} = -\frac{1}{2}B_j, b_k^{(i)} \neq b_l^{(i)}, (k \neq l, k, l \in \{1,2\}). \end{aligned} \tag{5.168}$$

(iv$_4$) The corresponding standard form is

$$\dot{x} = a_0(x - a_1)^2(x - a_2)^2 \tag{5.169}$$

where

$$\begin{aligned} &\Delta_i = B_i^2 - 4C_i = 0, i = 1,2 \\ &b_1^{(1)} = b_2^{(1)} = -\frac{1}{2}B_1, b_1^{(2)} = b_2^{(2)} = -\frac{1}{2}B_2, B_1 \neq B_2; \\ &a_1 = \min\{-\frac{1}{2}B_1, -\frac{1}{2}B_2\}, a_2 = \max\{-\frac{1}{2}B_1, -\frac{1}{2}B_2\}. \end{aligned} \tag{5.170}$$

Such a flow with the two equilibriums of $x^* = a_1$ and $x^* = a_2$ is called a (US:US) or (LS:LS) flow. The equilibriums of $x^* = a_1$ and $x^* = a_2$ for two equilibriums switching are called two bifurcations of equilibrium at a point $\mathbf{p} = \mathbf{p}_1 \in \partial\Omega_{12}$, and the bifurcation condition is

$$\begin{aligned} &\Delta_i = B_i^2 - 4C_i = 0, i = 1,2 \\ &b_1^{(1)} = b_2^{(1)} = -\frac{1}{2}B_1, b_1^{(2)} = b_2^{(2)} = -\frac{1}{2}B_2. \end{aligned} \tag{5.171}$$

(iv$_5$) The corresponding standard form is

$$\dot{x} = a_0(x - a_1)^4 \tag{5.172}$$

where

$$\begin{aligned} &\Delta_i = B_i^2 - 4C_i = 0, i = 1,2 \\ &b_1^{(i)} = b_2^{(i)} = -\frac{1}{2}B_i, B_1 = B_2. \end{aligned} \tag{5.173}$$

Such a flow at the equilibrium of $x^* = a_1$ is called a *saddle* flow of the fourth order. The equilibrium of $x^* = a_1$ for two second-order equilibriums switching or four simple equilibriums appearing or from two equilibriums to four simple equilibriums switching is called a fourth-order *saddle* bifurcation of equilibrium at a point $\mathbf{p} = \mathbf{p}_1 \in \partial\Omega_{12}$, and the bifurcation condition is

$$\Delta_i = B_i^2 - 4C_i = 0, a_1 = b_1^{(i)} = b_2^{(i)}, i = 1, 2. \tag{5.174}$$

Theorem 5.7 (i) *Under conditions of*

$$\Delta_i = B_i^2 - 4C_i < 0 \text{ for } i = 1, 2 \tag{5.175}$$

a standard form of Eq. (5.146) is

$$\dot{x} = f(x, \mathbf{p}) = a_0\left[(x + \frac{1}{2}B_1)^2 + \frac{1}{4}(-\Delta_1)\right]\left[(x + \frac{1}{2}B_2)^2 + \frac{1}{4}(-\Delta_2)\right] \tag{5.176}$$

with $a_0 = A(\mathbf{p})$, which has a nonequilibrium flow.

(i$_1$) *If $a_0(\mathbf{p}) > 0$, the nonequilibrium flow is called a positive flow.*
(i$_2$) *If $a_0(\mathbf{p}) > 0$, the nonequilibrium flow is called a negative flow.*

(ii) *Under a condition of*

$$\Delta_i = B_i^2 - 4C_i > 0 \text{ and } \Delta_j = B_j^2 - 4C_j < 0 \text{ for } i, j \in \{1, 2\}, i \neq j \tag{5.177}$$

a standard form of Eq. (5.146) is

$$\dot{x} = f(x, \mathbf{p}) = a_0(x - a_1)(x - a_2)\left[(x + \frac{1}{2}B_j)^2 + \frac{1}{4}(-\Delta_j)\right] \tag{5.178}$$

where

$$\begin{aligned} a_1 &= \min\,(b_1^{(i)}, b_2^{(i)}) \text{ and } a_2 = \max\,(b_1^{(i)}, b_2^{(i)}), \\ b_1^{(i)} &= -\frac{1}{2}(B_i + \sqrt{\Delta_i}), b_2^{(i)} = -\frac{1}{2}(B_i - \sqrt{\Delta_i}). \end{aligned} \tag{5.179}$$

(ii$_1$) *For $a_0(\mathbf{p}) > 0$, the equilibriums of $x^* = a_1$ and $x^* = a_2$ are stable (sink, $df/dx|_{x^*=a_1} < 0$) and unstable (source, $df/dx|_{x^*=a_2} > 0$), respectively.*
(ii$_2$) *For $a_0(\mathbf{p}) < 0$, the equilibriums of $x^* = a_1$ and $x^* = a_2$ are unstable (source, $df/dx|_{x^*=a_1} > 0$) and stable (sink, $df/dx|_{x^*=a_2} < 0$), respectively.*

(iii) *Under conditions of*

$$\Delta_i = B_i^2 - 4C_i = 0 \text{ and } \Delta_j = B_j^2 - 4C_j < 0 \quad \text{for } i,j \in \{1,2\}, i \neq j \tag{5.180}$$

a standard form of Eq. (5.146) is

$$\dot{x} = f(x, \mathbf{p}) = a_0(x - a_1)^2\left[(x + \frac{1}{2}B_j)^2 + \frac{1}{4}(-\Delta_j)\right] \tag{5.181}$$

where

$$a_1 = b_1^{(i)} = b_2^{(i)} = -\frac{1}{2}B_i. \tag{5.182}$$

(iii$_1$) *For* $a_0(\mathbf{p}) > 0$, *the equilibrium of* $x^* = a_1$ *is unstable (an* upper saddle, $d^2f/dx^2|_{x^*=a_1} > 0$*). Such a flow at the equilibrium of* $x^* = a_1$ *is called an upper-saddle flow of the second order. The bifurcation of equilibrium of at* $x^* = a_1$ *for two equilibriums appearing or vanishing is called an* upper-saddle-node *bifurcation of the second order at a point* $\mathbf{p} = \mathbf{p}_1 \in \partial\Omega_{12}$.

(iii$_2$) *For* $a_0(\mathbf{p}) < 0$, *the equilibrium of* $x^* = a_1$ *is unstable (a* lower saddle, $d^2f/dx^2|_{x^*=a_1} < 0$*). Such a flow at the equilibrium of* $x^* = a_1$ *is called a lower-saddle flow of the second order. The bifurcation of equilibrium of at* $x^* = a_1$ *for two equilibriums appearing or vanishing is called a lower-saddle-node bifurcation of the second order at a point* $\mathbf{p} = \mathbf{p}_1 \in \partial\Omega_{12}$.

(iv) *Under conditions of*

$$\begin{aligned} &\Delta_i = B_i^2 - 4C_i > 0, i = 1,2 \\ &b_k^{(1)} \neq b_l^{(2)} \text{ for } k,l \in \{1,2\}; \\ &b_1^{(i)} = -\frac{1}{2}(B_i + \sqrt{\Delta_i}), b_2^{(i)} = -\frac{1}{2}(B_i - \sqrt{\Delta_i}) \text{ for } i = 1,2 \end{aligned} \tag{5.183}$$

a standard form is

$$\dot{x} = f(x, \mathbf{p}) = a_0(x - a_1)(x - a_2)(x - a_3)(x - a_4) \tag{5.184}$$

where

$$a_{1,2,3,4} \in \cup_{i=1}^{2}\{b_1^{(i)}, b_2^{(i)}\} \text{ with } a_m < a_{m+1}. \tag{5.185}$$

(iv$_1$) *For* $a_0(\mathbf{p}) > 0$, *the equilibriums of* $x^* = a_1, a_2, a_3, a_4$ *are stable, unstable, stable, and unstable, respectively. The flow is called a (SI:SO:SI:SO) flow.*

(iv$_2$) *For* $a_0(\mathbf{p}) < 0$, *the equilibriums of* $x^* = a_1, a_2, a_3, a_4$ *are unstable, stable, unstable, and stable, respectively. The flow is called a (SO:SI:SO:SI) flow.*

The equilibrium of $x^* = a_i\,(i = 1, 2, 3, 4)$ *is unstable (source,* $df/dx|_{x^*=a_i} > 0$*) and stable (sink,* $df/dx|_{x^*=a_i} < 0$*).*

(v) *Under conditions of*

$$\begin{aligned} &\Delta_i = B_i^2 - 4C_i > 0\ (i \in \{1,2\}) \text{ and} \\ &\Delta_j = B_j^2 - 4C_j > 0\ (j \in \{1,2\}), \\ &b_1^{(\alpha)} = -\frac{1}{2}(B_\alpha + \sqrt{\Delta_\alpha}), b_2^{(\alpha)} = -\frac{1}{2}(B_\alpha - \sqrt{\Delta_\alpha}) \text{ for } \alpha = i,j \\ &b_k^{(i)} = b_l^{(j)}, (i,k) \neq (j,l), (i,j,k,l \in \{1,2\}), \end{aligned} \tag{5.186}$$

a standard form of Eq. (5.146) is

$$\dot{x} = f(x, \mathbf{p}) = a_0(x - a_{i_1})^2(x - a_{i_2})(x - a_{i_3}) \tag{5.187}$$

where

$$\begin{aligned} &a_{i_1} = b_k^{(i)} = b_l^{(j)} \in \cup_{i=1}^2\{b_1^{(i)}, b_1^{(i)}\}, (i,k) \neq (j,l); i,j,k,l \in \{1,2\} \\ &a_{i_1} \notin \{a_{i_2}, a_{i_3}\} \subset \cup_{i=1}^2\{b_1^{(i)}, b_1^{(i)}\} \text{ for } i_\alpha \in \{1,2,3\} \text{ and } \alpha \in \{1,2,3\}. \end{aligned} \tag{5.188}$$

(v$_1$) *The equilibriums of* $x^* = a_{i_2}, a_{i_3}$ *are unstable (source,* $df/dx|_{x^*=a_{i_2}, a_{i_3}} > 0$*) and stable (sink,* $df/dx|_{x^*=a_{i_2}, a_{i_3}} < 0$*).*

(v$_2$) *The equilibrium of* $x^* = a_{i_1}$ *is unstable (an upper saddle,* $d^2f/dx^2|_{x^*=a_{i_1}} > 0$*) and unstable (a lower saddle,* $d^2f/dx^2|_{x^*=a_{i_1}} < 0$*).*

The bifurcation of equilibrium at $x^* = a_{i_1}$ *for two equilibriums switching or vanishing is called the upper-saddle-node or lower-saddle-node bifurcation of the second order at a point* $\mathbf{p} = \mathbf{p}_1 \in \partial\Omega_{12}$.

(vi) *Under conditions of*

$$\begin{aligned} &\Delta_i = B_i^2 - 4C_i > 0\ (i \in \{1,2\}) \text{ and} \\ &\Delta_j = B_j^2 - 4C_j = 0\ (j \in \{1,2\}) \\ &b_1^{(i)} = -\frac{1}{2}(B_i + \sqrt{\Delta_i}), b_2^{(i)} = -\frac{1}{2}(B_i - \sqrt{\Delta_i}) \\ &b_{1,2}^{(j)} = -\frac{1}{2}B_j, b_k^{(i)} = b_l^{(i)}, (k \neq l, k, l \in \{1,2\}) \end{aligned} \tag{5.189}$$

a standard form of Eq. (5.146) is

$$\dot{x} = f(x, \mathbf{p}) = a_0(x - a_{i_1})^3(x - a_{i_2}) \tag{5.190}$$

where

$$\begin{aligned} &a_{i_1} = b_{1,2}^{(j)} = b_l^{(i)}, a_{i_2} = b_k^{(i)}, a_1 < a_2; \\ &\text{for } i,j,l \in \{1,2\}, i_\alpha \in \{1,2\} \text{ and } \alpha \in \{1,2\}. \end{aligned} \tag{5.191}$$

(vi$_1$) *The equilibrium of* $x^* = a_{i_2}$ *is unstable (source,* $df/dx|_{x^*=a_{i_2}} > 0$*) or stable (sink,* $df/dx|_{x^*=a_{i_2}} < 0$*).*

(vi$_2$) *The equilibrium of* $x^* = a_{i_1}$ *is unstable (the third-order source,* $d^3f/dx^3|_{x^*=a_{i_1}} > 0$*) and stable (the third-order sink,* $d^3f/dx^3|_{x^*=a_{i_1}} < 0$*). The bifurcation of equilibrium at* $x^* = a_{i_1}$ *for one equilibrium to three equilibriums is called the source or sink bifurcation of the third order at a point* $\mathbf{p} = \mathbf{p}_1 \in \partial\Omega_{12}$.

(vii) *Under conditions of*

$$\begin{aligned} &\Delta_i = B_i^2 - 4C_i = 0\ (i \in \{1,2\}) \text{ and} \\ &\Delta_j = B_j^2 - 4C_j = 0\ (j \in \{1,2\}) \\ &b_1^{(\alpha)} = b_2^{(\alpha)} = -\frac{1}{2}B_\alpha \text{ for } \alpha = i,j \\ &B_1 \neq B_2, \end{aligned} \tag{5.192}$$

a standard form of Eq. (5.146) *is*

$$\dot{x} = a_0(x - a_1)^2(x - a_2)^2 \tag{5.193}$$

where

$$a_1 = \min\{-\frac{1}{2}B_1, -\frac{1}{2}B_2\}, a_2 = \max\{-\frac{1}{2}B_1, -\frac{1}{2}B_2\}. \tag{5.194}$$

(vii$_1$) *For* $a_0(\mathbf{p}) > 0$*, the equilibriums of* $x^* = a_i\ (i = 1, 2)$ *are unstable (upper saddle,* $d^2f/dx^2|_{x^*=a_i} > 0$*). The equilibrium of* $x^* = a_i$ *for two equilibriums vanishing and appearing are called an upper-saddle-node bifurcation of the second order at a point* $\mathbf{p} = \mathbf{p}_1 \in \partial\Omega_{12}$.

(vii$_2$) *For* $a_0(\mathbf{p}) < 0$*, the equilibriums of* $x^* = a_i\ (i = 1, 2)$ *are unstable (lower saddle,* $d^2f/dx^2|_{x^*=a_i} < 0$*). The equilibrium of* $x^* = a_1$ *for two equilibriums vanishing and appearing are called a lower-saddle-node bifurcation of the second order at a point* $\mathbf{p} = \mathbf{p}_1 \in \partial\Omega_{12}$.

(viii) *Under conditions of*

$$\begin{aligned} &\Delta_i = B_i^2 - 4C_i = 0 \ (i \in \{1,2\}) \text{ and} \\ &\Delta_j = B_j^2 - 4C_j = 0 \ (j \in \{1,2\}) \\ &b_1^{(\alpha)} = b_2^{(\alpha)} = -\frac{1}{2}B_\alpha \text{ for } \alpha = i,j \\ &B_1 \neq B_2, \end{aligned} \tag{5.195}$$

the corresponding standard form is

$$\dot{x} = a_0(x - a_1)^4 \tag{5.196}$$

where

$$a_1 = -\frac{1}{2}B_1 = -\frac{1}{2}B_2. \tag{5.197}$$

(viii$_1$) *For* $a_0(\mathbf{p}) > 0$, *the equilibriums of* $x^* = a_1$ *are unstable (upper saddle,* $d^4f/dx^4|_{x^*=a_1} > 0$*). The equilibrium of* $x^* = a_1$ *for four equilibriums vanishing and appearing are called an upper-saddle-node bifurcation of the fourth order at a point* $\mathbf{p} = \mathbf{p}_1 \in \partial\Omega_{12}$.

(viii$_2$) *For* $a_0(\mathbf{p}) < 0$, *the equilibriums of* $x^* = a_1$ *are unstable (lower saddle,* $d^4f/dx^4|_{x^*=a_1} < 0$*). The equilibrium of* $x^* = a_1$ *for four equilibriums vanishing and appearing are called a lower-saddle-node bifurcation of the fourth order at a point* $\mathbf{p} = \mathbf{p}_1 \in \partial\Omega_{12}$.

As discussed before, a quartic nonlinear system is expressed by the product of two quadratic polynomials, i.e.,

$$\dot{x} = a_0(\mathbf{p})[x^2 + B_1(\mathbf{p})x + C_1(\mathbf{p})][x^2 + B_2(\mathbf{p})x + C_2(\mathbf{p})]. \tag{5.198}$$

Thus, for $\dot{x} = 0$, the equilibriums are determined by the roots of two quadratic polynomial equations, i.e.,

$$x^2 + B_1(\mathbf{p})x + C_1(\mathbf{p}) = 0 \text{ and/or } x^2 + B_2(\mathbf{p})x + C_2(\mathbf{p}) = 0. \tag{5.199}$$

The roots of such quadratic polynomial equations are determined by the corresponding discriminant of the quadratic equations, i.e.,

$$\Delta_i = B_i^2 - 4C_i \text{ for } i = 1,2. \tag{5.200}$$

If $\Delta_i < 0$, the quadratic equation of $x^2 + B_i(\mathbf{p})x + C_i(\mathbf{p}) = 0$ does not have any roots. If $\Delta_i > 0$, the quadratic equation of $x^2 + B_i(\mathbf{p})x + C_i(\mathbf{p}) = 0$ possesses two roots.

If $\Delta_i = 0$, the quadratic equation of $x^2 + B_i(\mathbf{p})x + C_i(\mathbf{p}) = 0$ has a repeated root. With parameter variation, suppose one of the two quadratic polynomial equations has one root intersected with the roots of the other quadratic polynomial equation. There are six cases for $a_0 > 0$: (a) $b_2^{(i)} = b_1^{(j)}$, (b) $b_1^{(j)} = b_1^{(i)} = b_2^{(i)} = -\frac{1}{2}B_i$, (c) $b_1^{(i)} = b_1^{(j)}$, (d) $b_2^{(i)} = b_2^{(j)}$, (e) $b_2^{(j)} = b_1^{(i)} = b_2^{(i)} = -\frac{1}{2}B_i$, and (f) $b_1^{(i)} = b_2^{(j)}$, as presented in Fig. 5.13. The intersected point for non-repeated roots is a saddle-node bifurcation for the subcritical case. The lower-saddle-node and upper-saddle-node bifurcations are shown in Fig. 5.13a, c and d, f, respectively. The bifurcation dynamics for the 1-dimensional quartic system is determined by $\dot{x} = a_0(x - a_{i_1})^2(x - a_{i_2})(x - a_{i_3})$ with i_α, $\alpha \in \{1, 2, 3\}$ for four equilibriums or $\dot{x} = a_0(x - a_i)^2[(x + \frac{1}{2}B_j)^2 - \frac{1}{4}\Delta_j]$ with $i, j \in \{1, 2\}$ for two equilibriums. If the intersected point occurs at the repeated root, the third-order source and sink bifurcations are presented in Fig. 5.13b, e, respectively. The corresponding bifurcation dynamics for the 1-dimensional quartic system is determined by $\dot{x} = a_0(x - a_{i_1})^2(x - a_{i_2})$ with i_α, $\alpha \in \{1, 2\}$. The stable and unstable equilibriums are presented by the solid and dashed curves, respectively. The intersected points are marked by circular symbols, which are the bifurcation points. Without losing generality, suppose the two roots of the quadratic polynomial equation have a relation of $b_1^{(i)} > b_2^{(i)}$ for $i = 1, 2$. The repeated roots of the two quadratic polynomial equations are also the upper- or lower-saddle-node bifurcations for two equilibriums appearing and vanishing. Similarly, the six cases of stability and bifurcation diagrams varying with parameter for $a_0 < 0$ are presented in Fig. 5.14. The stability and bifurcation conditions for $a_0 < 0$ are opposite to $a_0 > 0$.

If the roots of two quadratic equations do not have any intersections, the open loops for stability and bifurcation diagrams of equilibriums for $a_0 > 0$ and $a_0 < 0$ are presented in Fig. 5.15. There are four cases of open loops for $a_0 > 0$: (a) $B_i < B_j$, (b) $B_i > B_j$, (c) $b_2^{(j)} < -\frac{1}{2}B_i < b_1^{(j)}$, and (d) $\Delta_i = \Delta_j$, $B_i \neq B_j$ and four cases of open loops for $a_0 < 0$: (e) $B_i < B_j$, (f) $B_i > B_j$, (g) $b_2^{(j)} < -\frac{1}{2}B_i < b_1^{(j)}$, and (h) $\Delta_i = \Delta_j = 0$, $B_i \neq B_j$. The two bifurcations occur at the same time because the quadratic equations have $\Delta_i = \Delta_j = 0$, $B_i \neq B_j$. The bifurcation points are only for two equilibriums appearing or vanishing from the discriminants of the quadratic equations. The bifurcation dynamics for the 1-dimensional quartic system is determined by $\dot{x} = a_0(x - a_{i_1})^2(x - a_{i_2})(x - a_{i_3})$ with i_α, $\alpha \in \{1, 2, 3\}$.

With varying vector parameter, the open loops of stability and bifurcation diagrams will become closed loops. Thus, the closed loops of stability and bifurcation diagrams of equilibriums for $a_0 > 0$ and $a_0 < 0$ are presented in Fig. 5.16. There are six cases of closed loops: (a) $B_i < B_j$, (b) $B_i > B_j$, and (c) $b_2^{(j)} < -\frac{1}{2}B_i < b_1^{(j)}$ for $a_0 > 0$ and (d) $B_i < B_j$, (e) $B_i > B_j$, and (f) $b_2^{(j)} < -\frac{1}{2}B_i < b_1^{(j)}$ for $a_0 < 0$ For such a closed loop, the bifurcation points are the upper- or lower-saddle bifurcations of the second order at both ends. The bifurcation points are determined from the discriminants of the quadratic equations.

If $\Delta_i = \Delta_j = 0$ occurs at the same parameter, the bifurcation dynamics for the quartic dynamical system is determined by $\dot{x} = a_0(x - a_1)^4$, as shown in Fig. 5.17.

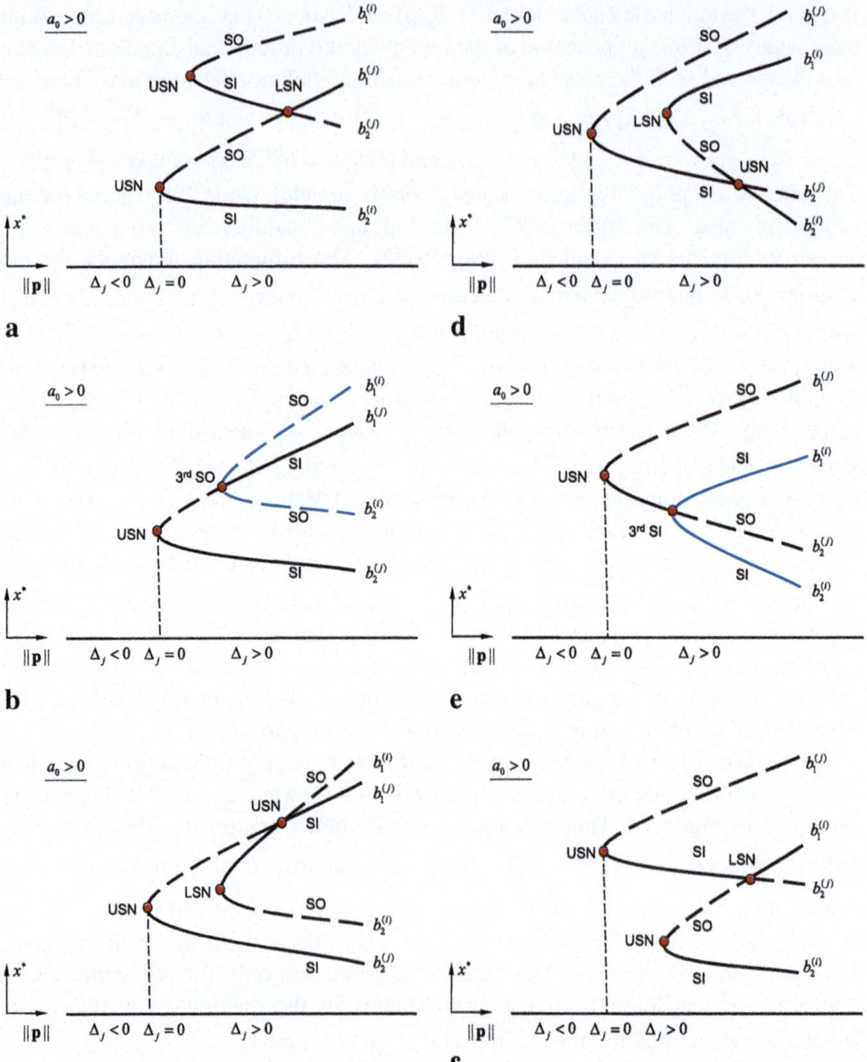

Fig. 5.13 Open loops for stability and bifurcations of equilibriums in the 1-dimensional, quartic nonlinear dynamical system ($a_0 > 0$): **(a)** $b_2^{(i)} = b_1^{(j)}$, **(b)** $b_1^{(j)} = b_1^{(i)} = b_2^{(i)} = -\frac{1}{2}B_i$, **(c)** $b_1^{(i)} = b_1^{(j)}$, **(d)** $b_2^{(i)} = b_2^{(j)}$, **(e)** $b_2^{(j)} = b_1^{(i)} = b_2^{(i)} = -\frac{1}{2}B_i$, **(f)** $b_1^{(i)} = b_2^{(j)}$. LSN: lower saddle node, USN: upper saddle node, SI: sink, SO: source. Stable and unstable equilibriums are represented by solid and dashed curves, respectively. The bifurcation points are marked by circular symbols

There are six cases: two closed loops ($\Delta_i > \Delta_j$, $B_i = B_j$): (a) $a_0 > 0$ and (b) $a_0 > 0$; two open loops ($\Delta_i > \Delta_j$, $B_i = B_j$): (c) $a_0 > 0$ and (d) $a_0 > 0$; and two open loops ($B_i \leq B_j$): (e) $a_0 > 0$ and (f) $a_0 > 0$. The bifurcation points are the upper- and lower-saddle-node bifurcations of the fourth order. In Fig. 5.17a, b, the fourth-order upper- and lower-

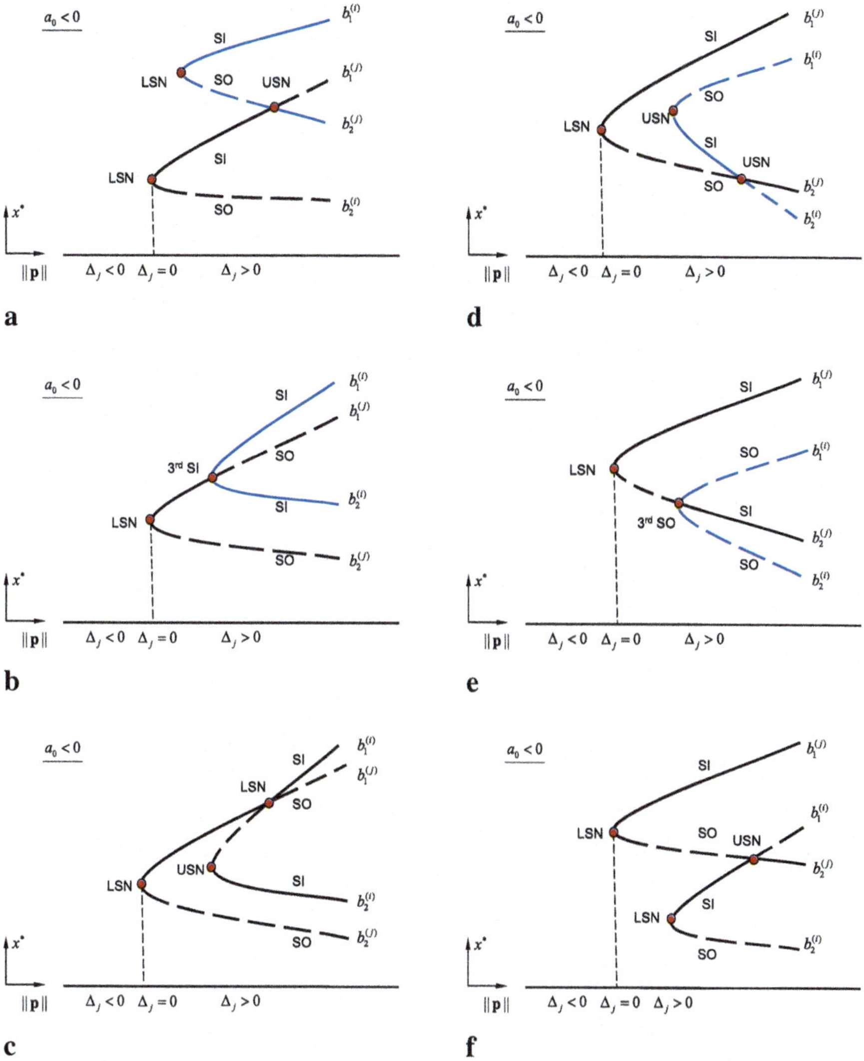

Fig. 5.14 Open loops for stability and bifurcation of equilibriums in the 1-dimensional, quartic nonlinear dynamical system ($a_0 < 0$): **(a)** $b_2^{(i)} = b_1^{(j)}$, **(b)** $b_1^{(j)} = b_1^{(i)} = b_2^{(i)} = -\frac{1}{2}B_i$, **(c)** $b_1^{(i)} = b_1^{(j)}$, **(d)** $b_2^{(i)} = b_2^{(j)}$, **(e)** $b_2^{(j)} = b_1^{(i)} = b_2^{(i)} = -\frac{1}{2}B_i$, **(f)** $b_1^{(i)} = b_2^{(j)}$. LSN: lower saddle node, USN: upper saddle node, SI: sink, SO: source. Stable and unstable equilibriums are represented by solid and dashed curves, respectively. The bifurcation points are marked by circular symbols

saddle-node bifurcations in the closed loop of bifurcation diagrams for $a_0 > 0$ and $a_0 > 0$ are presented, respectively. However, for the open loop of bifurcation diagrams, the fourth-order upper- and lower-saddle-node bifurcations for $a_0 > 0$ and $a_0 > 0$ are presented in Fig. 5.17c, d, respectively. The fourth-order saddle-node

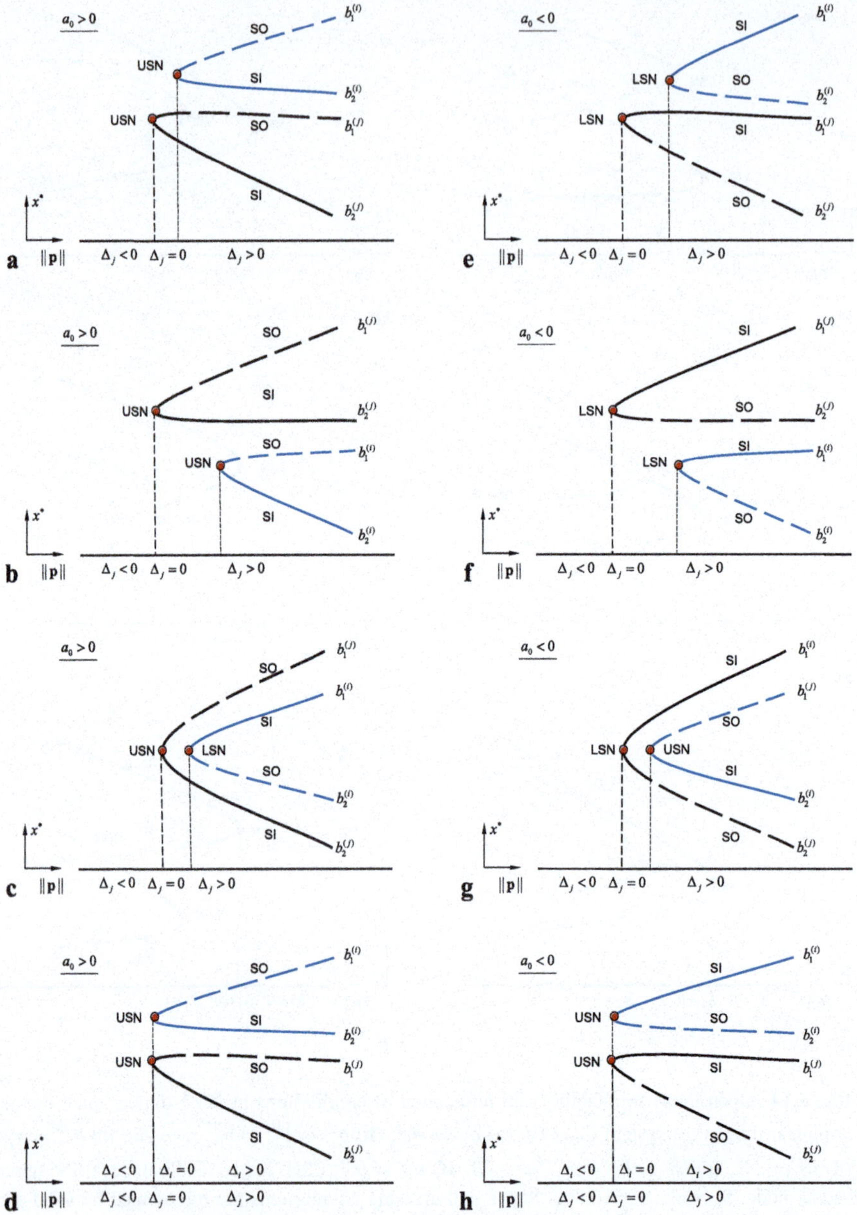

Fig. 5.15 Open loops of stability and bifurcation of equilibriums in the 1-dimensional, quartic nonlinear dynamical system. ($a_0 > 0$): **(a)** $B_i < B_j$, **(b)** $B_i > B_j$, **(c)** $b_2^{(j)} < -\frac{1}{2}B_i < b_1^{(j)}$, **(d)** $\Delta_i = \Delta_j = 0$, $B_i \neq B_j$. ($a_0 < 0$): **(e)** $B_i < B_j$, **(f)** $B_i > B_j$, **(g)** $b_2^{(j)} < -\frac{1}{2}B_i < b_1^{(j)}$, **(h)** $\Delta_i = \Delta_j = 0$, $B_i \neq B_j$. LSN: lower saddle node, USN: upper saddle node, SI: sink, SO: source. Stable and unstable equilibriums are represented by solid and dashed curves, respectively. The bifurcation points are marked by circular symbols

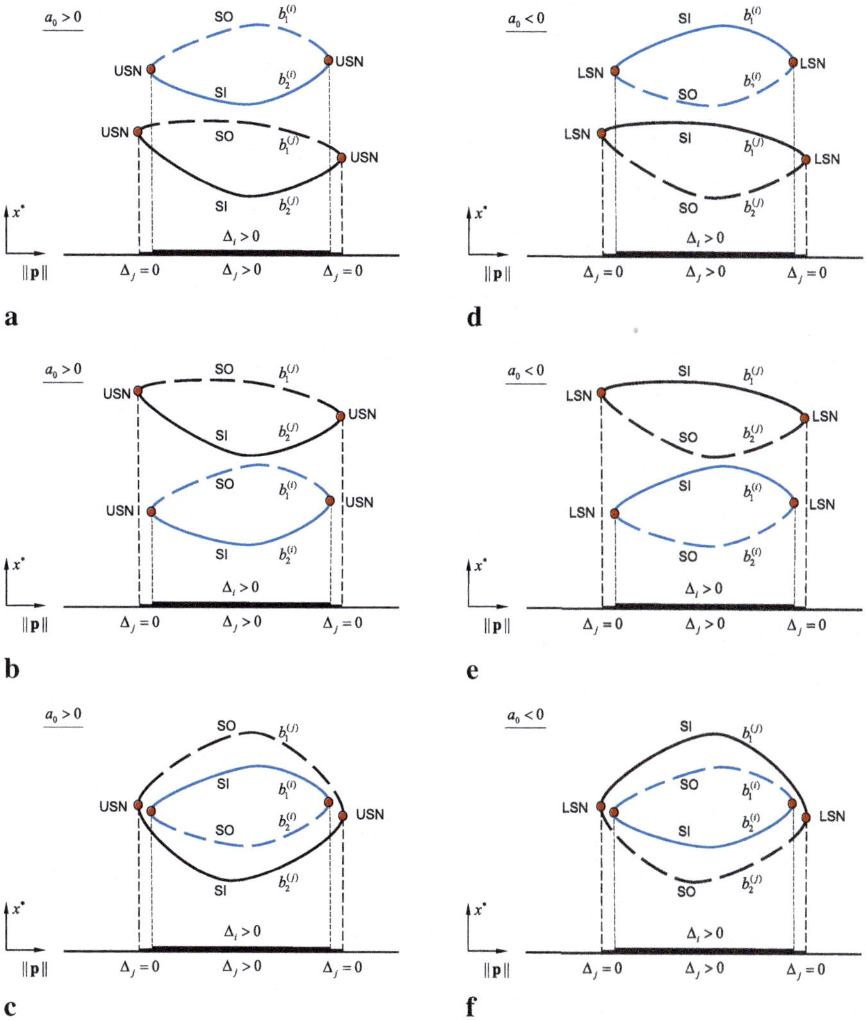

Fig. 5.16 Closed loops for stability and bifurcation of equilibriums in the 1-dimensional, quartic nonlinear dynamical system: ($a_0 > 0$): (**a**) $B_i < B_j$, (**b**) $B_i > B_j$, (**c**) $b_2^{(j)} < -\frac{1}{2}B_i < b_1^{(j)}$. ($a_0 < 0$): (**d**) $B_i < B_j$, (**e**) $B_i > B_j$, (**f**) $b_2^{(j)} < -\frac{1}{2}B_i < b_1^{(j)}$. LSN: lower saddle node, USN: upper saddle node, SI: sink, SO: source. Stable and unstable equilibriums are represented by solid and dashed curves, respectively. The bifurcation points are marked by circular symbols

bifurcation possesses four branches rather than two branches for the second-order saddle-node bifurcation. For $B_i \leq B_j$, the fourth-order upper- and lower-saddle-node bifurcations are presented in Fig. 5.17e, f. The quadratic equation gives the two equilibriums, which are of the same stability. That is, both equilibriums for the same quadratic equation are stable or unstable.

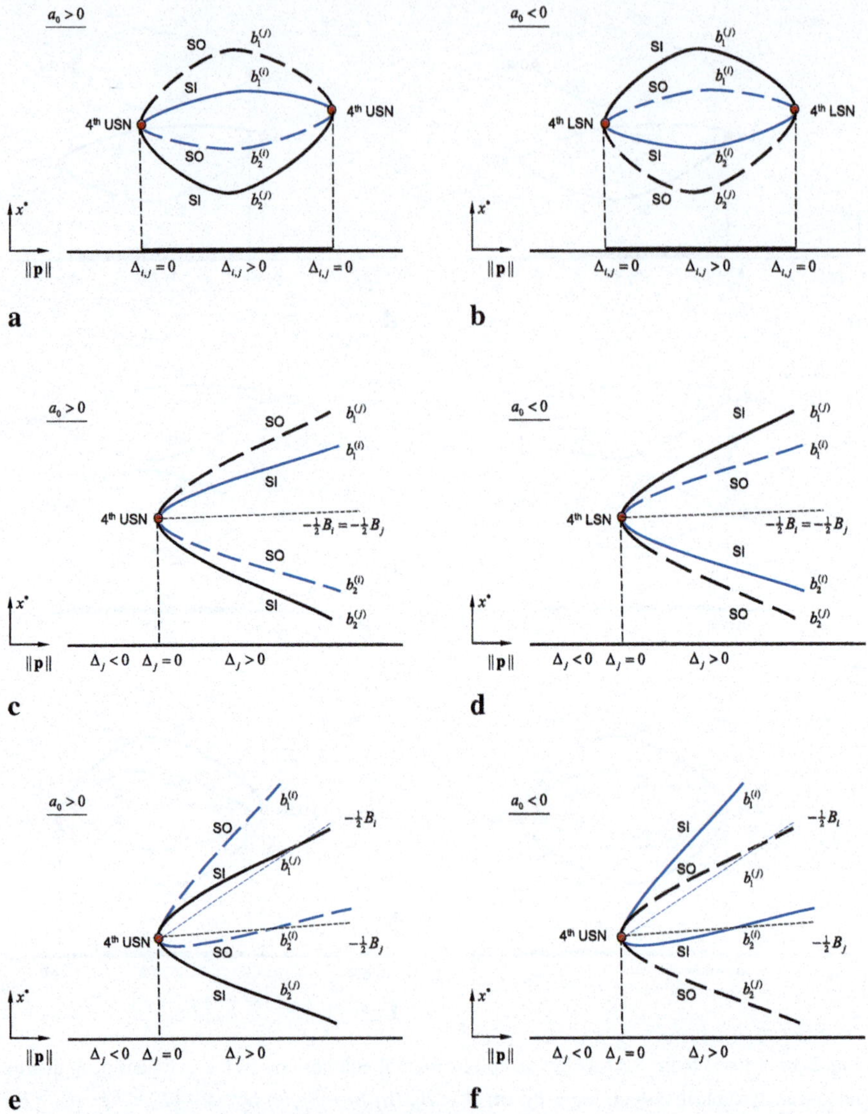

Fig. 5.17 Stability and bifurcation of equilibriums in the 1-dimensional, quartic nonlinear dynamical system. Closed loop ($\Delta_i > \Delta_j$, $B_i = B_j$) for **(a)** $a_0 > 0$ and **(b)** $a_0 > 0$; open loop ($\Delta_i > \Delta_j$, $B_i = B_j$) for **(c)** $a_0 > 0$ and **(d)** $a_0 > 0$; open loop ($B_i < B_j$) for **(e)** $a_0 > 0$ and **(f)** $a_0 > 0$. LSN: lower saddle node, USN: upper saddle node, SI: sink, SO: source. Stable and unstable equilibriums are represented by solid and dashed curves, respectively. The bifurcation points are marked by circular symbols

If $\Delta_i = 0$ ($i \in \{1, 2\}$) occurs for new equilibrium appearing or vanishing only, the open loop of the bifurcation diagrams for $a_0 > 0$ is presented in Fig. 5.18. There are four cases for $a_0 > 0$: (a) $b_2^{(i)} = b_1^{(j)}$, (b) $b_1^{(i)} = b_1^{(j)}$, (c) $b_2^{(i)} = b_2^{(j)}$, and (d) $b_1^{(i)} = b_2^{(j)}$. The bifurcation points are the upper- and lower-saddle-node bifurcations of the

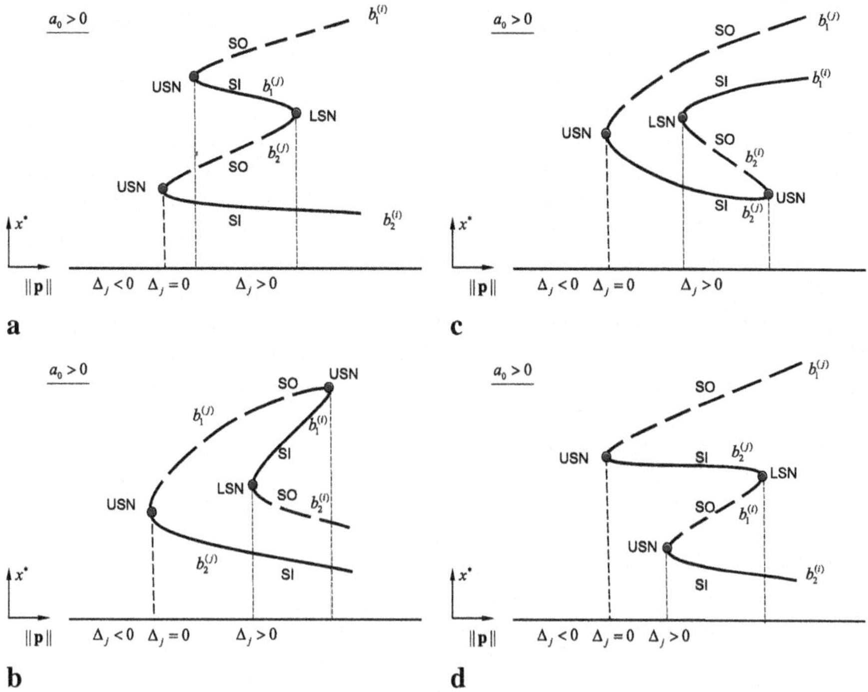

Fig. 5.18 Open loops of stability and bifurcation of equilibriums in the 1-dimensional, quartic nonlinear dynamical system ($a_0 > 0$): (**a**) $b_2^{(i)} = b_1^{(j)}$, (**b**) $b_1^{(i)} = b_1^{(j)}$, (**c**) $b_2^{(i)} = b_2^{(j)}$, (**d**) $b_1^{(i)} = b_2^{(j)}$. LSN: lower saddle node, USN: upper saddle node, SI: sink, SO: source. Stable and unstable equilibriums are represented by solid and dashed curves, respectively. The bifurcation points are marked by circular symbols

second order. In Fig. 5.19a–d, the upper- and lower-saddle-node bifurcations of the second order in the open loop of bifurcation diagrams for $a_0 < 0$ are presented. Such a diagram of stability and bifurcation possesses three saddle-node bifurcations.

If $\Delta_i = 0$ and $\Delta_j > 0$ ($i, j \in \{1, 2\}$, $i \neq j$) exist for equilibrium appearing or vanishing with the source and sink bifurcations of the third order, one closed loop of the bifurcation diagrams for $a_0 > 0$ is presented in Fig. 5.20. Four cases for $a_0 > 0$ exist with the four saddle-node bifurcations in one closed loop: (a) $b_2^{(i)} = b_1^{(j)}$ and $b_1^{(i)} = b_2^{(j)}$, (b) $b_1^{(i)} = b_1^{(j)}$ and $b_2^{(i)} = b_2^{(j)}$, (c) $b_2^{(i)} = b_2^{(j)}$ and $b_1^{(i)} = b_1^{(j)}$, (d) $b_1^{(i)} = b_2^{(j)}$ and $b_2^{(i)} = b_1^{(j)}$. Two cases for $a_0 > 0$ exist with the three saddle-node bifurcations plus the source and sink bifurcations of the third order in two closed loops: (e) $b_1^{(j)} = b_1^{(i)} = b_2^{(i)} = -\frac{1}{2}B_i$ with $b_1^{(i)} = b_1^{(j)}$ and $b_2^{(i)} = b_2^{(j)}$, and (f) $b_2^{(j)} = b_1^{(i)} = b_2^{(i)} = -\frac{1}{2}B_i$ with $b_1^{(i)} = b_1^{(j)}$ and $b_2^{(i)} = b_2^{(j)}$. However, in Fig. 5.21a–f, the upper- and lower-

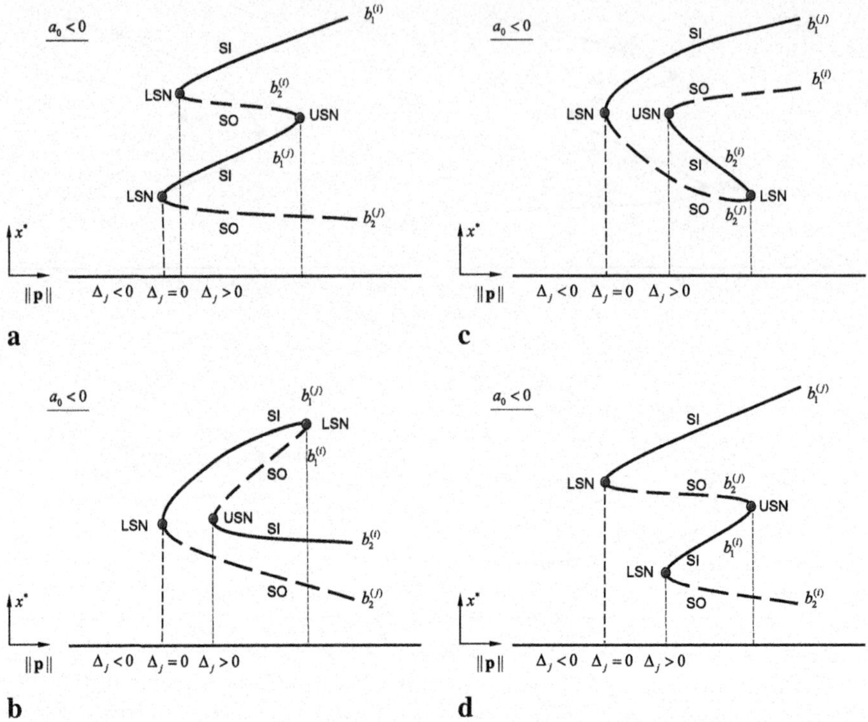

Fig. 5.19 Open loops for stability and bifurcation of equilibriums in the 1-dimensional, quartic nonlinear dynamical system ($a_0 < 0$): (**a**) $b_2^{(i)} = b_1^{(j)}$, (**b**) $b_1^{(i)} = b_1^{(j)}$, (**c**) $b_2^{(i)} = b_2^{(j)}$, (**d**) $b_1^{(i)} = b_2^{(j)}$. LSN: lower saddle-node, USN: upper-saddle-node, SI: sink, SO: source. Stable and unstable equilibriums are represented by solid and dashed curves, respectively. The bifurcation points are marked by circular symbols

saddle-node bifurcations of the second order plus the source and sink bifurcations of the third order in the closed loop of bifurcation diagrams for $a_0 < 0$ are presented.

Definition 5.9 Consider a 1-dimensional, quartic nonlinear dynamical system:

$$\begin{aligned} \dot{x} &= A(\mathbf{p})x^4 + B(\mathbf{p})x^3 + C(\mathbf{p})x^2 + D(\mathbf{p})x + E(\mathbf{p}) \\ &= a_0(\mathbf{p})(x - b_1)^2[x^2 + B_2(\mathbf{p})x + C_2(\mathbf{p})] \end{aligned} \tag{5.201}$$

where $A(\mathbf{p}) \neq 0$, and

$$\mathbf{p} = (p_1, p_2, \ldots, p_m)^{\mathrm{T}}. \tag{5.202}$$

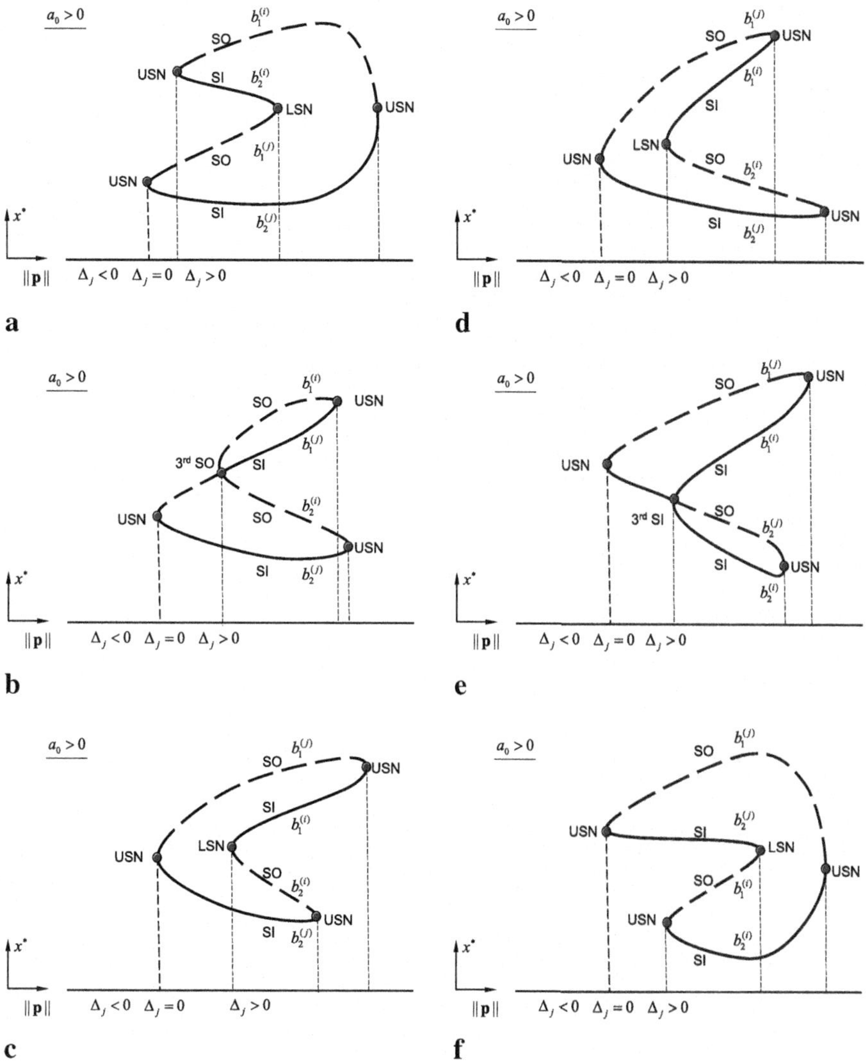

Fig. 5.20 Closed loops of stability and bifurcation of equilibriums in the 1-dimensional, quartic nonlinear dynamical system ($a_0 > 0$): **(a)** $b_2^{(i)} = b_1^{(j)}$ and $b_1^{(i)} = b_2^{(j)}$, **(b)** $b_1^{(j)} = b_1^{(i)} = b_2^{(i)} = -\frac{1}{2}B_i$ with $b_1^{(i)} = b_1^{(j)}$ and $b_2^{(i)} = b_2^{(j)}$, **(c)** $b_1^{(i)} = b_1^{(j)}$ and $b_2^{(i)} = b_2^{(j)}$, **(d)** $b_2^{(i)} = b_2^{(j)}$ and $b_1^{(i)} = b_1^{(j)}$, **(e)** $b_2^{(j)} = b_1^{(i)} = b_2^{(i)} = -\frac{1}{2}B_i$ with $b_1^{(i)} = b_1^{(j)}$ and $b_2^{(i)} = b_2^{(j)}$, **(f)** $b_1^{(i)} = b_2^{(j)}$ and $b_2^{(i)} = b_1^{(j)}$. LSN: lower saddle node, USN: upper saddle node, SI: sink, SO: source. Stable and unstable equilibriums are represented by solid and dashed curves, respectively. The bifurcation points are marked by circular symbols

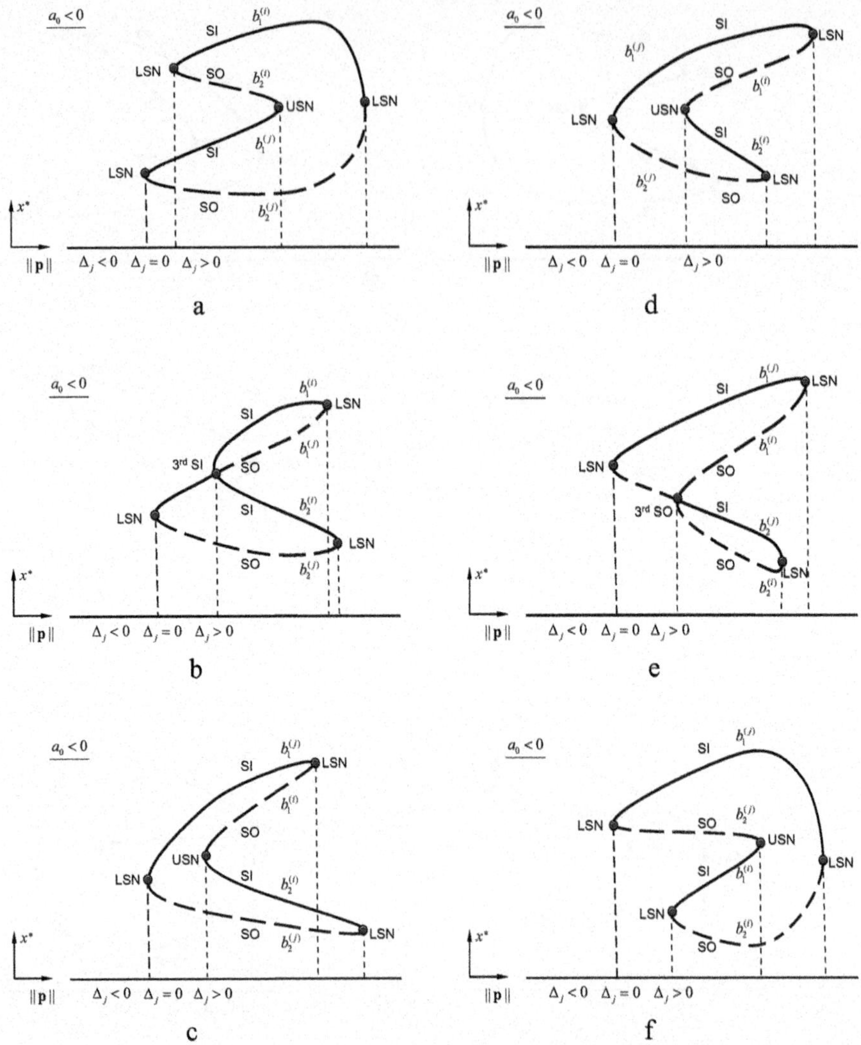

Fig. 5.21 A closed loop for stability and bifurcations of equilibriums in the 1-dimensional, quartic nonlinear dynamical system ($a_0 < 0$): ($a_0 > 0$): **(a)** $b_2^{(i)} = b_1^{(j)}$ and $b_1^{(i)} = b_2^{(j)}$, **(b)** $b_1^{(j)} = b_1^{(i)} = b_2^{(i)} = -\frac{1}{2}B_i$ with $b_1^{(i)} = b_1^{(j)}$ and $b_2^{(i)} = b_2^{(j)}$, **(c)** $b_1^{(i)} = b_1^{(j)}$ and $b_2^{(i)} = b_2^{(j)}$, **(d)** $b_2^{(i)} = b_2^{(j)}$ and $b_1^{(i)} = b_1^{(j)}$, **(e)** $b_2^{(j)} = b_1^{(i)} = b_2^{(i)} = -\frac{1}{2}B_i$ with $b_1^{(i)} = b_1^{(j)}$ and $b_2^{(i)} = b_2^{(j)}$, **(f)** $b_1^{(i)} = b_2^{(j)}$ and $b_2^{(i)} = b_1^{(j)}$. LSN: lower saddle node, USN: upper saddle node, SI: sink, SO: source. Stable and unstable equilibriums are represented by solid and dashed curves, respectively. The bifurcation points are marked by circular symbols

(i) If

$$\Delta_2 = B_2^2 - 4C_2 < 0 \tag{5.203}$$

the corresponding standard form is

$$\dot{x} = a_0(x - b_1)^2[(x + \frac{1}{2}B_2)^2 + \frac{1}{4}(-\Delta_2)] \tag{5.204}$$

with a double equilibrium:

$$x^* = a_1 = b_1. \tag{5.205}$$

(a) For $a_0 > 0$, the flow is called an upper-saddle (US) flow.
(b) For $a_0 < 0$, the flow is called a lower-saddle (LS) flow.

(ii) If

$$\Delta_2 = B_2^2 - 4C_2 > 0 \tag{5.206}$$

the quartic nonlinear dynamical system has two more equilibriums as

$$x^* = b_1^{(2)} = -\frac{1}{2}(B_2 + \sqrt{\Delta_2}), x^* = b_2^{(2)} = -\frac{1}{2}(B_2 - \sqrt{\Delta_2}). \tag{5.207}$$

(ii_1) The corresponding standard form is

$$\dot{x} = a_0(x - a_1)^2(x - a_2)(x - a_3) \tag{5.208}$$

where

$$\begin{aligned} a_1 &= b_1 < \min\{b_1^{(2)}, b_2^{(2)}\}, a_2 = \min\{b_1^{(2)}, b_2^{(2)}\}, \\ a_3 &= \max\{b_1^{(2)}, b_2^{(2)}\}. \end{aligned} \tag{5.209}$$

(a) For $a_0 > 0$, the flow is called a (US:SI:SO) flow.
(b) For $a_0 < 0$, the flow is called an (LS:SO:SI) flow.

(ii_2) The corresponding standard form is

$$\dot{x} = a_0(x - a_1)(x - a_2)^2(x - a_3) \tag{5.210}$$

where

$$\begin{aligned} a_1 &= \min\{b_1^{(2)}, b_2^{(2)}\}, a_2 = b_1 > \min\{b_1^{(2)}, b_2^{(2)}\}, \\ a_3 &= \max\{b_1^{(2)}, b_2^{(2)}\} > b_1. \end{aligned} \tag{5.211}$$

(a) For $a_0 > 0$, the flow is called an (SI:LS:SO) flow.
(b) For $a_0 < 0$, the flow is called an (SO:US:SI) flow.

(ii_3) The corresponding standard form is

$$\dot{x} = a_0(x - a_1)(x - a_2)(x - a_3)^2 \tag{5.212}$$

where

$$\begin{aligned} a_1 &= \min\{b_1^{(2)}, b_2^{(2)}\},\ a_2 = \max\{b_1^{(2)}, b_2^{(2)}\},\\ a_3 &= b_1 > \max\{b_1^{(2)}, b_2^{(2)}\}. \end{aligned} \tag{5.213}$$

(a) For $a_0 > 0$, the flow is called an (SI:SO:US) flow.
(b) For $a_0 < 0$, the flow is called an (SO:SI:LS) flow.

(ii$_4$) The corresponding standard form is

$$\dot{x} = a_0(x - a_1)^3(x - a_2) \tag{5.214}$$

where

$$a_1 = b_1 = \min\{b_1^{(2)}, b_2^{(2)}\},\ a_2 = \max\{b_1^{(2)}, b_2^{(2)}\}. \tag{5.215}$$

(a) For $a_0 > 0$, the flow is called a (3rd order-SI:SO) flow. The bifurcation of equilibrium for (US:SI:SO)-equilibriums to (SI:LS:SO)-equilibrium is called the sink bifurcation of the third order.
(b) For $a_0 < 0$, the flow is called a (3rd order-SO:SI) flow. The bifurcation of equilibrium for (LS:SO:SI)-equilibrium to (SO:US:SI)-equilibrium is called the source bifurcation of the third order.

(ii$_5$) The corresponding standard form is

$$\dot{x} = a_0(x - a_1)(x - a_2)^3 \tag{5.216}$$

where

$$a_1 = \min\{b_1^{(2)}, b_2^{(2)}\},\ a_2 = b_1 = \max\{b_1^{(2)}, b_2^{(2)}\}. \tag{5.217}$$

(a) For $a_0 > 0$, the flow is called a (SI:3rd order-SO) flow. The bifurcation of equilibrium for (SI:SO:US)-equilibriums to (SI:US:SO)-equilibrium is called the source bifurcation of the third order.
(b) For $a_0 < 0$, the flow is called a (SO:3rd order-SI) flow. The bifurcation of equilibrium for (SO:SI:LS)-equilibriums to (SO:US:SI)-equilibriums is called the sink bifurcation of the third order.

(iii) If

$$\Delta_2 = B_2^2 - 4C_2 = 0 \tag{5.218}$$

the quartic nonlinear dynamical system has two equilibriums as

$$x^* = b_1^{(2)} = b_2^{(2)} = b_2 = -\frac{1}{2}B_2. \tag{5.219}$$

(iii$_1$) The corresponding standard form for $b_1 < b_2$ is

$$\dot{x} = a_0(x - b_1)^2(x - b_2)^2. \tag{5.220}$$

(a) For $a_0 > 0$, the flow is called a (US:US) flow. The bifurcation of equilibrium for the (US:SI:SO)-equilibrium appearance is called the upper-saddle bifurcation of the second order.
(b) For $a_0 < 0$, the flow is called an (LS:LS) flow. The bifurcation of equilibrium for (LS:SO:SI)-equilibrium appearance is called the lower-saddle bifurcation of the second order.

(iii$_2$) The corresponding standard form for $b_1 > b_2$ is

$$\dot{x} = a_0(x - b_2)^2(x - b_1)^2. \tag{5.221}$$

(a) For $a_0 > 0$, the flow is called a (US:US) flow. The bifurcation of equilibrium for the (SI:SO:US)-equilibrium appearance is called the upper-saddle bifurcation of the second order.
(b) For $a_0 < 0$, the flow is called an (LS:LS) flow. The bifurcation of equilibrium for (SO:SI:LS)-equilibrium appearance is called the lower-saddle bifurcation of the second order.

(iii$_3$) The corresponding standard form with $b_1 = b_2 = a_1$ is

$$\dot{x} = a_0(x - a_1)^4. \tag{5.222}$$

(a) For $a_0 > 0$, the flow is called a fourth-order upper-saddle (US) flow. The bifurcation of equilibrium for the upper-saddle (US) equilibrium to (SO:LS:SI)-equilibriums is called the upper-saddle-node bifurcation of the fourth order.
(b) For $a_0 < 0$, the flow is called a fourth-order lower-saddle (LS) flow. The bifurcation of equilibrium for the lower-saddle (LS) equilibrium to (SO:US:SI)-equilibrium appearance is called the lower-saddle bifurcation of the fourth order.

Definition 5.10 Consider a 1-dimensional, quartic nonlinear dynamical system

$$\begin{aligned}\dot{x} &= A(\mathbf{p})x^4 + B(\mathbf{p})x^3 + C(\mathbf{p})x^2 + D(\mathbf{p})x + E(\mathbf{p}) \\ &= a_0(\mathbf{p})[x^2 + B_1(\mathbf{p})x + C_1(\mathbf{p})]^2\end{aligned} \tag{5.223}$$

where $A(\mathbf{p}) \neq 0$, and

$$\mathbf{p} = (p_1, p_2, \ldots, p_m)^{\mathrm{T}}. \tag{5.224}$$

(i) If

$$\Delta_1 = B_1^2 - 4C_1 < 0, \tag{5.225}$$

the quartic nonlinear dynamical system does not have any equilibrium.

(a) For $a_0 > 0$, the nonequilibrium flow is called a positive flow.
(b) For $a_0 < 0$, the nonequilibrium flow is called a negative flow.

(ii) If

$$\Delta_1 = B_1^2 - 4C_1 > 0 \tag{5.226}$$

the 1-dimensional quartic nonlinear dynamical system has two double-repeated equilibriums as

$$x^* = b_1^{(1)} = -\frac{1}{2}(B_1 + \sqrt{\Delta_1}), x^* = b_1^{(1)} = -\frac{1}{2}(B_1 - \sqrt{\Delta_1}). \tag{5.227}$$

The corresponding standard form is

$$\dot{x} = a_0(x - a_1)^2 (x - a_2)^2 \tag{5.228}$$

where

$$a_1 = \min\{b_1^{(1)}, b_2^{(1)}\}, a_2 = \max\{b_1^{(1)}, b_2^{(1)}\}. \tag{5.229}$$

(a) For $a_0 > 0$, the flow is called a (US:US) flow.
(b) For $a_0 < 0$, the flow is called an (LS:LS) flow.

(iii) If

$$\Delta_1 = B_1^2 - 4C_1 = 0 \tag{5.230}$$

the 1-dimensional quartic nonlinear dynamical system has two equilibriums as

$$x^* = b_1^{(2)} = b_2^{(2)} = -\frac{1}{2}B_2 = a_1. \tag{5.231}$$

The corresponding standard form is

$$\dot{x} = a_0(x - a_1)^4. \tag{5.232}$$

(a) For $a_0 > 0$, the flow is called a fourth-order US flow. The bifurcation of equilibrium for (US:US)-equilibrium appearance is called the upper-saddle bifurcation of the fourth order.
(b) For $a_0 < 0$, the flow is called a fourth-order LS flow. The bifurcation of equilibrium for the LS equilibrium to (LS:LS)-equilibrium appearance is called the lower-saddle bifurcation of the fourth order.

From a 1-dimensional, quartic nonlinear system with singularity, the saddle equilibrium with and without intersection with simple equilibriums is presented in Figs. 5.22 and 5.23. In Fig. 5.22a, d, the upper-saddle equilibrium for $a_0 > 0$ does not intersect with any branch of the simple equilibriums. In Fig. 5.23b, c, the upper-saddle equilibrium for $a_0 > 0$ intersects with one branch of the simple equilibriums, and the upper-saddle equilibrium switches to the lower-saddle equilibrium with source and sink equilibriums, which are called the source and sink bifurcations of the third order, accordingly. In Fig. 5.22e, the upper-saddle equilibrium for $a_0 > 0$ intersects with a double-repeated equilibrium with upper saddle. The intersected point is an unstable equilibrium, which is called a fourth-order upper-saddle-node bifurcation. In Fig. 5.22f, the two second-order upper-saddle equilibriums are presented. The two upper-saddle equilibriums appear at the bifurcation of the fourth-order upper-saddle bifurcation.

Similarly, the lower-saddle equilibrium for $a_0 < 0$ does not intersect with any branch of the simple equilibriums, as shown in Fig. 5.23a, d. In Fig. 5.23b, c, the lower-saddle equilibrium for $a_0 < 0$ intersects with one branch of the simple equilibriums, and the lower-saddle equilibrium switches to the upper-saddle equilibrium with source and sink equilibriums, which are called the source and sink bifurcations of the third order, accordingly. In Fig. 5.23e, the lower-saddle equilibrium for $a_0 < 0$ intersects with a repeated equilibrium with a lower saddle. The intersection point is an unstable equilibrium, which is called a fourth-order lower-saddle-node bifurcation. In Fig. 5.23f, the two second-order lower-saddle equilibriums are presented for $a_0 < 0$. The two lower-saddle equilibriums appear at the fourth-order lower-saddle bifurcation.

Consider a 1-dimensional, quartic nonlinear dynamical system with two double equilibriums.

(i) For $b \neq a$, the dynamical system is

$$\dot{x} = a_0(\mathbf{p})(x - b(\mathbf{p}))^2 (x - a(\mathbf{p}))^2. \tag{5.233}$$

For such a system, if $a_0 > 0$, two double-repeated equilibriums of $x^* = a, b$ are two upper saddles, which are unstable. If $a_0 < 0$, two double-repeated equilibriums of $x^* = a, b$ are two lower saddles, which are unstable.

(ii) For $b = a$, the dynamical system on the boundary is

$$\dot{x} = a_0(\mathbf{p})(x - b(\mathbf{p}))^4. \tag{5.234}$$

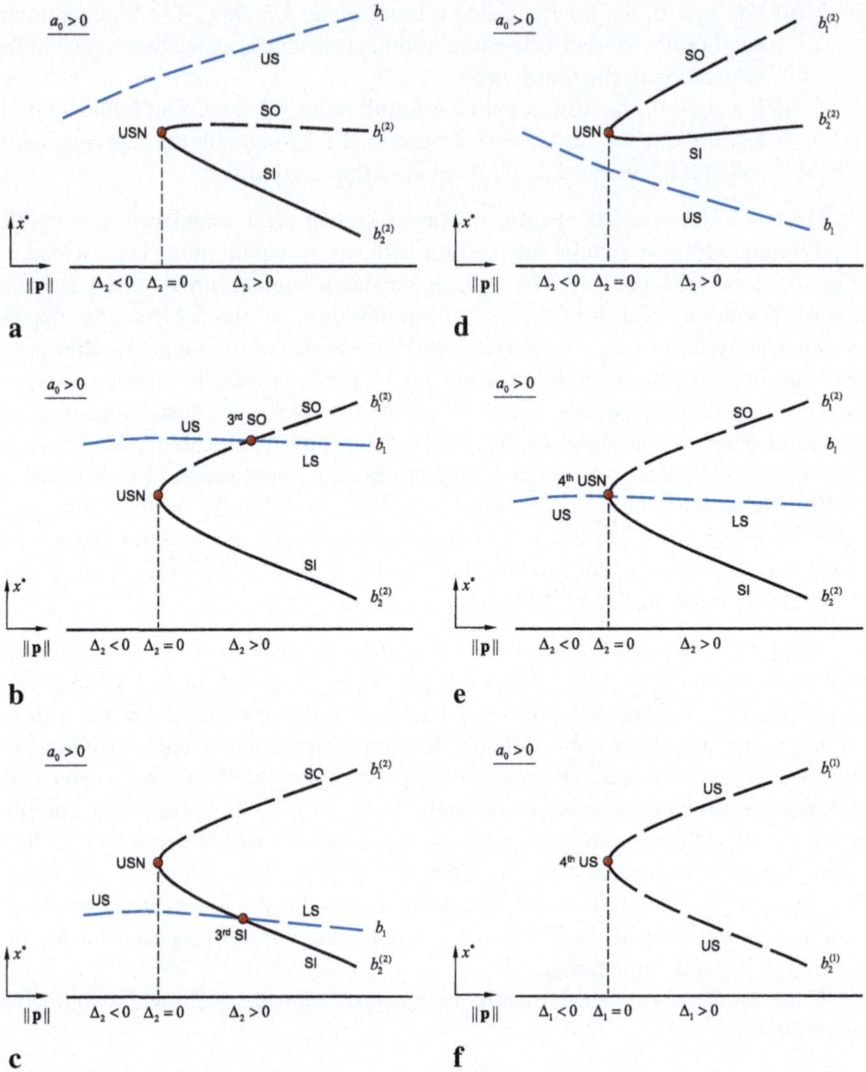

Fig. 5.22 Stability and bifurcation of three equilibriums with and without intersection in the 1-dimensional, quartic nonlinear dynamical system ($a_0 > 0$): (**a**) without intersection $b_1 > b_1^{(2)}$, (**b**) an intersection at $b_1 = b_1^{(2)}$, (**c**) an intersection at $b_1 = b_2^{(2)}$, (**d**) without intersection $b_1 < b_1^{(2)}$, (**e**) an intersection at $b_1 = -\frac{1}{2}B_i$, (**f**) $\Delta_1 = 0$. LSN: lower saddle node, USN: upper saddle node, SI: sink, SO: source. Stable and unstable equilibriums are represented by solid and dashed curves, respectively. The bifurcation points are marked by circular symbols

With parameter changes, the bifurcation diagram for the quartic nonlinear system is presented in Fig. 5.24. Stable and unstable equilibriums are represented by solid and dashed curves, respectively. The bifurcation point is marked by a circular symbol. In Fig. 5.24a, if $a_0 > 0$, two double-repeated equilibriums of $x^* = a, b$ are

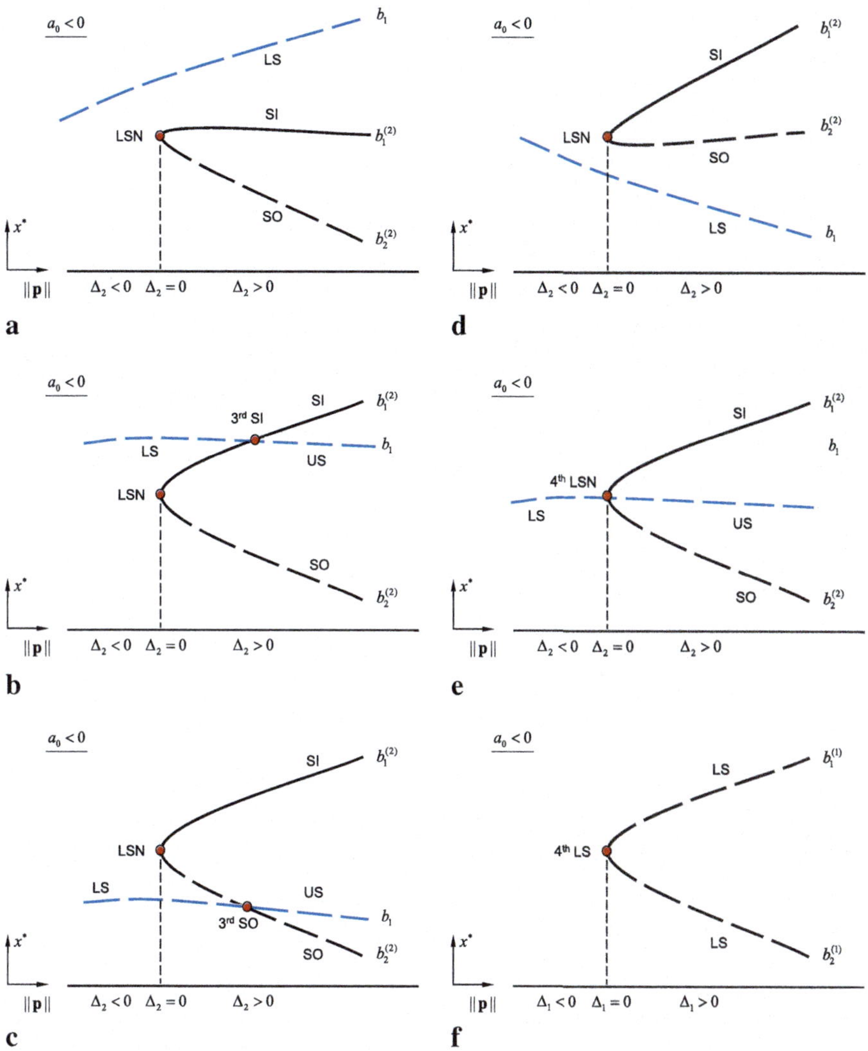

Fig. 5.23 Stability and bifurcation of four equilibriums with intersection in the 1-dimensional, quartic nonlinear dynamical system ($a_0 < 0$): (**a**) without intersection $b_1 > b_1^{(2)}$, (**b**) an intersection at $b_1 = b_1^{(2)}$, (**c**) an intersection at $b_1 = b_1^{(2)}$, (**d**) without intersection $b_1 < b_1^{(2)}$, (**e**) an intersection at $b_1 = -\frac{1}{2}B_i$, (**f**) $\Delta_1 = 0$. LSN: lower saddle node, USN: upper saddle node, SI: sink, SO: source. Stable and unstable equilibriums are represented by solid and dashed curves, respectively. The bifurcation points are marked by circular symbols

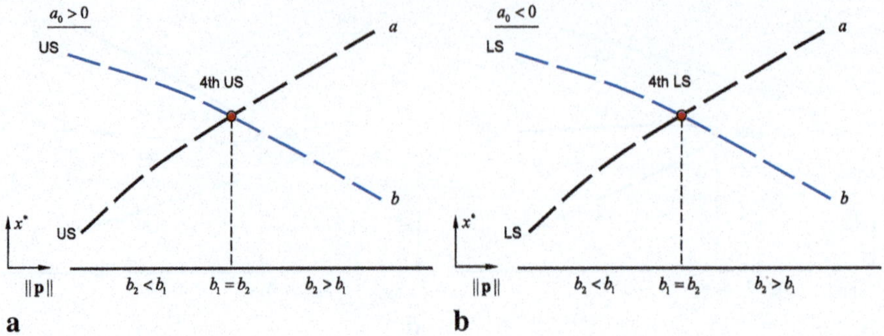

Fig. 5.24 Stability and bifurcation of two US or LS equilibriums with intersection in the 1-dimensional, quartic nonlinear dynamical system: (**a**) (US:US)-flow ($a_0 > 0$), (**b**) (LS:LS)-flow ($a_0 < 0$). Fourth LS: fourth-order lower-saddle bifurcation, fourth US: fourth-order upper-saddle bifurcation. Stable and unstable equilibriums are represented by solid and dashed curves, respectively. The bifurcation points are marked by circular symbols

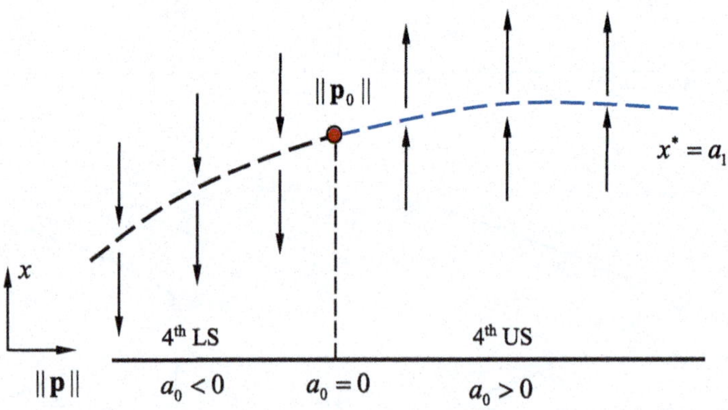

Fig. 5.25 Stability of a repeated equilibrium with the fourth multiplicity in the 1-dimensional, quartic nonlinear dynamical system: Stable and unstable equilibriums are represented by solid and dashed curves, respectively. The stability switching is labeled by a circular symbol

the upper saddles of the second order. The two upper saddles intersect at a point of $x^* = a = b$ with the fourth-order multiplicity, which is an upper-saddle bifurcation of the fourth order for the (US:US) to (US:US) equilibriums. If $a_0 < 0$, two double equilibriums of $x^* = a, b$ are the lower saddle of the second order, which are intersected at a point of $x^* = a = b$, as shown in Fig. 5.24b. Such a quartic equilibrium with the fourth-order multiplicity is called a lower-saddle bifurcation of the fourth order for the (LS:LS) to (LS:LS) equilibrium.

To illustrate the stability and bifurcation of equilibrium with singularity in a 1-dimensional, quadratic nonlinear system, the equilibrium of $\dot{x} = a_0(x - a_1)^4$ is presented in Fig. 5.25. The fourth-order upper and lower saddles of equilibrium of

$x^* = a_1$ with the fourth-order multiplicity are unstable, and the fourth-order upper- and lower-saddle equilibriums of the fourth order are invariant. At $a_0 = 0$, the lower-saddle equilibrium switches to the upper-saddle equilibrium, which is a switching point marked by a circular symbol.

For further discussion on the switching bifurcations in the quartic nonlinear system, the following definitions are presented.

Definition 5.11 Consider a 1-dimensional, quartic nonlinear dynamical system

$$\begin{aligned}\dot{x} &= A(\mathbf{p})x^4 + B(\mathbf{p})x^3 + C(\mathbf{p})x^2 + D(\mathbf{p})x + E(\mathbf{p}) \\ &= a_0(\mathbf{p})(x-a)(x-b)[x^2 + B_2(\mathbf{p})x + C_2(\mathbf{p})]\end{aligned} \tag{5.235}$$

where $A(\mathbf{p}) \neq 0$ and

$$\mathbf{p} = (p_1, p_2, \ldots, p_m)^{\mathrm{T}}. \tag{5.236}$$

(i) If

$$\begin{aligned}&\Delta_2 = B_2^2 - 4C_2 < 0, \\ &\{a_1, a_2\} = \mathrm{sort}\{a, b\}, a_1 \leq a_2,\end{aligned} \tag{5.237}$$

the quartic nonlinear dynamical system has two equilibriums. The standard form is

$$\dot{x} = a_0(\mathbf{p})(x-a_1)(x-a_2)[(x^2 + \frac{1}{2}B_2)^2 + \frac{1}{4}(-\Delta_2)]. \tag{5.238}$$

(i_1) For $a_0 > 0$, the equilibrium flow is an (SI:SO) flow. The equilibrium of $x^* = a_1$ is stable (sink, $df/dx|_{x^*=a_1} < 0$) and the equilibrium of $x^* = a_2$ is unstable (source, $df/dx|_{x^*=a_2} > 0$).

(i_2) For $a_0 < 0$, the equilibrium flow is an (SO:SI) flow. The equilibrium of $x^* = a_1$ is unstable (source, $df/dx|_{x^*=a_1} > 0$) and the equilibrium of $x^* = a_2$ is stable (sink, $df/dx|_{x^*=a_2} < 0$).

(i_3) Under

$$\Delta_{12} = (a_1 - a_2)^2 = 0 \text{ with } a_1 = a_2, \tag{5.239}$$

the quartic nonlinear dynamical system has a standard form as

$$\dot{x} = f(x, \mathbf{p}) = a_0(x-a_1)^2[(x^2 + \frac{1}{2}B_2)^2 + \frac{1}{4}(-\Delta_2)]. \tag{5.240}$$

(a) For $a_0\,(\mathbf{p}) > 0$, the equilibrium of $x^* = a_1$ is unstable (an upper saddle of second order, $d^2f/dx^2|_{x^*=a_1} > 0$). Such a flow is called an upper-saddle flow. The bifurcation of equilibrium at $x^* = a_1$ for two equilibriums switching of $x^* = a_1,\ a_2$ is called an upper-saddle-node switching bifurcation of the second order at a point $\mathbf{p} = \mathbf{p}_1$.

(b) For $a_0\,(\mathbf{p}) < 0$, the equilibrium of $x^* = a_1$ is unstable (a lower saddle of the second order, $d^2f/dx^2|_{x^*=a_1} < 0$). Such a flow is called a lower-saddle flow. The bifurcation of equilibrium at $x^* = a_1$ for two equilibriums switching of $x^* = a_1,\ a_2$ is called a lower-saddle-node switching bifurcation of the second order at a point $\mathbf{p} = \mathbf{p}_1$.

(ii) If

$$\Delta_2 = B_2^2 - 4C_2 > 0, \tag{5.241}$$

the 1-dimensional quartic nonlinear dynamical system has four equilibriums as

$$\begin{aligned} &x_k^* = b_1^{(2)} = -\frac{1}{2}(B_2 + \sqrt{\Delta_2}), x_k^* = b_2^{(2)} = -\frac{1}{2}(B_2 - \sqrt{\Delta_2}) \\ &\{a_1, a_2, a_3.a_4\} = \text{sort}\{a, b, b_1^{(2)}, b_2^{(2)}\}, a_i < a_{i+1} \end{aligned} \tag{5.242}$$

The corresponding standard form is

$$\dot{x} = a_0(x - a_1)(x - a_2)(x - a_3)(x - a_4). \tag{5.243}$$

(ii$_1$) For $a_0 > 0$, the flow is called an (SI:SO:SI:SO) flow.
(ii$_2$) For $a_0 < 0$, the flow is called an (SO:SI:SO:SO) flow.
(ii$_3$) Under

$$\Delta_{i_1 i_2} = (a_{i_1} - a_{i_2})^2 = 0, a_{i_1} = a_{i_2}, i_1, i_2 \in \{1, 2, 3, 4\}, i_1 \neq i_2, \tag{5.244}$$

the standard form is

$$\begin{aligned} &\dot{x} = f(x, \mathbf{p}) = a_0(x - a_{i_1})^2(x - a_{i_3})(x - a_{i_4}) \\ &i_\alpha \in \{1, 2, 3, 4\}, \alpha = 1, 3, 4 \end{aligned} \tag{5.245}$$

(a) The equilibrium of $x^* = a_{i_1}$ is unstable (an upper saddle of the second order, $d^2f/dx^2|_{x^*=a_{i_1}} > 0$). Such a flow is called an upper-saddle flow at $x^* = a_{i_1}$. The bifurcation of equilibrium at $x^* = a_{i_1}$ for two equilibriums switching of $x^* = a_{i_1}, a_{i_2}$ is called an upper-saddle-node switching bifurcation of the second order at a point $\mathbf{p} = \mathbf{p}_1$.

(b) The equilibrium of $x^* = a_{i_1}$ is unstable (a lower saddle of the second order, $d^2f/dx^2|_{x^*=a_{i_1}} < 0$). Such a flow is called a lower-saddle flow at $x^* = a_{i_1}$. The bifurcation of equilibrium at $x^* = a_{i_1}$ for two equilibriums switching of $x^* = a_{i_1}, a_{i_2}$ is called a lower-saddle-node switching bifurcation of the second order at a point $\mathbf{p} = \mathbf{p}_1$.

(ii$_4$) Under

$$\begin{aligned} &\Delta_{i_1 i_2} = (a_{i_1} - a_{i_2})^2 = 0, \Delta_{i_2 i_3} = (a_{i_2} - a_{i_3})^2 = 0 \\ &a_{i_1} = a_{i_2} = a_3, i_1, i_2, i_3 \in \{1,2,3,4\}, i_1 \neq i_2 \neq i_3, \end{aligned} \tag{5.246}$$

the standard form is

$$\begin{aligned} &\dot{x} = f(x, \mathbf{p}) = a_0 (x - a_{i_1})^3 (x - a_{i_4}) \\ &i_\alpha \in \{1,2,3,4\}, \alpha = 1,4. \end{aligned} \tag{5.247}$$

(a) The equilibrium of $x^* = a_{i_1}$ is unstable (a source of the third order, $d^3f/dx^3|_{x^*=a_{i_1}} > 0$). Such a flow is called a third-order source flow at $x^* = a_{i_1}$. The bifurcation of equilibrium at $x^* = a_{i_1}$ for three simple equilibrium bundle-switching of $x^* = a_{i_1}, a_{i_2}, a_{i_3}$ is called a third-order source bundle-switching bifurcation at a point $\mathbf{p} = \mathbf{p}_1$.
(b) The equilibrium of $x^* = a_{i_1}$ is stable (a sink of the third order, $d^3f/dx^3|_{x^*=a_{i_1}} < 0$). Such a flow is called a third-order sink flow at $x^* = a_{i_1}$. The bifurcation of equilibrium at $x^* = a_{i_1}$ for three simple equilibrium bundle-switching of $x^* = a_{i_1}, a_{i_2}, a_{i_3}$ is called a third-sink bundle-switching bifurcation at a point $\mathbf{p} = \mathbf{p}_1$.

(ii$_5$) Under

$$\begin{aligned} &\Delta_{i_1 i_2} = (a_{i_1} - a_{i_2})^2 = 0, \Delta_{i_2 i_3} = (a_{i_2} - a_{i_3})^2 = 0, \\ &\Delta_{i_3 i_4} = (a_{i_4} - a_{i_4})^2 = 0, a_{i_1} = a_{i_2} = a_{i_3} = a_{i_4}, \\ &i_1, i_2, i_3, i_4 \in \{1,2,3,4\}, i_1 \neq i_2 \neq i_3 \neq i_4, \end{aligned} \tag{5.248}$$

the standard form is

$$\dot{x} = f(x, \mathbf{p}) = a_0 (x - a_{i_1})^4. \tag{5.249}$$

(a) The equilibrium of $x^* = a_{i_1}$ is unstable (an upper saddle of the fourth order, $d^4f/dx^4|_{x^*=a_{i_1}} > 0$). Such a flow is called a fourth-order upper-saddle flow at $x^* = a_{i_1}$. The bifurcation of equilibrium at $x^* = a_{i_1}$ for four simple equilibriums bundle-switching of $x^* = a_{1,2,3,4}$ is called a fourth-order upper-saddle-node bundle-switching bifurcation at a point $\mathbf{p} = \mathbf{p}_1$.

(b) The equilibrium of $x^* = a_{i_1}$ is unstable (a lower-saddle of the fourth order, $d^4f/dx^4|_{x^*=a_{i_1}} < 0$). Such a flow is called a fourth-order lower-saddle flow at $x^* = a_{i_1}$. The bifurcation of equilibrium at $x^* = a_{i_1}$ for four simple equilibriums bundle-switching of $x^* = a_{1,2,3,4}$ is called a fourth-order lower-saddle bundle-switching bifurcation at a point $\mathbf{p} = \mathbf{p}_1$.

(iii) If

$$\Delta_2 = B_2^2 - 4C_2 = 0, \tag{5.250}$$

the 1-dimensional quartic nonlinear dynamical system has three equilibrium as

$$\begin{aligned}
&x_k^* = b_1^{(2)} = b_2^{(2)} = -\frac{1}{2}B_2;\\
&\{a_1, a_2, a_3\} = \text{sort}\{a, b, b_1^{(2)} = b_2^{(2)}\}, a_i < a_{i+1};\\
&a_{i_1, i_2} = b_1^{(2)} = b_2^{(2)}, a_{i_3} = a, a_{i_4} = b;\\
&i_\alpha \in \{1,2,3\}, \alpha \in \{1,2,3,4\},
\end{aligned} \tag{5.251}$$

and the corresponding standard form is

$$\dot{x} = a_0(x - a_{i_1})^2(x - a_{i_2})(x - a_{i_3}). \tag{5.252}$$

(iii$_1$) The equilibrium of $x^* = a_{i_1}$ is unstable (an upper saddle, $d^2f/dx^2|_{x^*=a_1} > 0$), the flow is an upper-saddle flow at $x^* = a_{i_1}$. The bifurcation of equilibrium at $x^* = a_{i_1}$ for the appearing or vanishing of two simple equilibriums is called the upper-saddle-node bifurcation of the second order.

(iii$_2$) The equilibrium of $x^* = a_{i_1}$ is unstable (a lower saddle, $d^2f/dx^2|_{x^*=a_1} < 0$), the flow is a lower-saddle flow at $x^* = a_{i_1}$. The bifurcation of equilibrium at $x^* = a_{i_1}$ for the appearing or vanishing of two simple equilibriums is called the lower-saddle-node bifurcation of the second order.

(iii$_3$) Under

$$\begin{aligned}
&\Delta_{i_3 i_4} = (a_{i_3} - a_{i_4})^2 = 0, a_{i_3} = a_{i_4}, a_{i_1} \neq a_{i_3},\\
&i_\alpha \in \{1,2,3\}, \alpha \in \{1,2,3,4\};
\end{aligned} \tag{5.253}$$

the corresponding standard form is

$$\begin{aligned}
&\dot{x} = f(x, \mathbf{p}) = a_0(x - a_{i_1})^2(x - a_{i_3})^2\\
&i_\alpha \in \{1,2\}, \alpha = 1,3
\end{aligned} \tag{5.254}$$

The equilibriums of $x^* = a_{i_1}, a_{i_3}$ are unstable (an upper saddle of the second order, $d^2f/dx^2|_{x^*=a_{i_1}, a_{i_3}} > 0$) and unstable (a lower saddle of the second order, $d^2f/dx^2|_{x^*=a_{i_1}, a_{i_3}} < 0$). Such a flow is called an (US:US) or a (LS:LS) flow. The bifurcation of equilibrium at $x^* = a_{i_1}$ for two simple equilibrium onset of $x^* = a_{i_1}, a_{i_2}$ and at $x^* = a_{i_3}$ for two equilibrium switching of $x^* = a_{i_3}, a_{i_4}$ is called an (US:US) or a (LS:LS) bifurcation at a point $\mathbf{p} = \mathbf{p}_1$.

(iii$_4$) Under

$$\begin{aligned} &\Delta_{i_1 i_3} = (a_{i_1} - a_{i_3})^2 = 0, a_{i_1} = a_{i_2}, a_{i_1} = a_{i_3} a_{i_1} \neq a_{i_4}, \\ &i_\alpha \in \{1,2,3\}, \alpha \in \{1,2,3,4\}; \end{aligned} \tag{5.255}$$

the standard form is

$$\begin{aligned} &\dot{x} = f(x, \mathbf{p}) = a_0(x - a_{i_1})^3(x - a_{i_4}) \\ &i_\alpha \in \{1,2\}, \alpha = 1,4. \end{aligned} \tag{5.256}$$

(a) The equilibrium of $x^* = a_{i_1}$ is unstable (a third-order source, $d^3f/dx^3|_{x^*=a_{i_1}} > 0$). Such a flow is called a third-order source flow. The bifurcation of equilibrium at $x^* = a_{i_1}$ for one simple equilibrium of $x^* = a_{i_3}$ switching to three simple equilibriums of $x^* = a_{i_1, i_2, i_3}$ is called a third-order source pitchfork bifurcation at a point $\mathbf{p} = \mathbf{p}_1$.

(b) The equilibrium of $x^* = a_{i_1}$ is stable (a third-order sink, $d^3f/dx^3|_{x^*=a_{i_1}} < 0$). Such a flow is called a third-order sink flow. The bifurcation of equilibrium at $x^* = a_{i_1}$ for one simple equilibrium of $x^* = a_{i_1}$ switching to three simple equilibriums of $x^* = a_{i_1, i_2, i_3}$ is called a third-order sink pitchfork-switching bifurcation at a point $\mathbf{p} = \mathbf{p}_1$.

(iii$_5$) Under

$$\begin{aligned} &\Delta_{i_1 i_3} = (a_{i_1} - a_{i_3})^2 = 0, \Delta_{i_3 i_4} = (a_{i_3} - a_{i_4})^2 = 0; \\ &a_{i_1} = a_{i_2}, a_{i_1} = a_{i_3} a_{i_1} = a_{i_4}, \\ &i_\alpha \in \{1,2,3\}, \alpha \in \{1,2,3,4\}; \end{aligned} \tag{5.257}$$

the standard form is

$$\begin{aligned} &\dot{x} = f(x, \mathbf{p}) = a_0(x - a_{i_1})^4 \\ &i_\alpha \in \{1,2\}, \alpha = 1,3 \end{aligned} \tag{5.258}$$

(a) For $a_0 > 0$, the equilibrium of $x^* = a_{i_1}$ is unstable (a fourth-order upper saddle, $d^4f/dx^4|_{x^*=a_{i_1}} > 0$). Such a flow is called a fourth-order upper-saddle flow. The bifurcation of equilibrium at $x^* = a_{i_1}$ for two simple equilibriums switching to four simple equilibriums is called a fourth-order upper-saddle-node flower-bundle-switching bifurcation at a point $\mathbf{p} = \mathbf{p}_1$.

(b) For $a_0 > 0$, the equilibrium of $x^* = a_{i_1}$ is unstable (a fourth-order lower saddle, $d^4f/dx^4|_{x^*=a_{i_1}} < 0$). Such a flow is called a fourth-order lower-saddle flow. The bifurcation of equilibrium at $x^* = a_{i_1}$ for two simple equilibriums switching to four simple equilibriums is called a fourth-order lower-saddle-node flower-bundle-switching bifurcation at a point $\mathbf{p} = \mathbf{p}_1$.

Based on the previous definition, the stability and bifurcations of equilibriums in the 1-dimensional, quartic nonlinear dynamical system ($a_0 > 0$) are presented in Fig. 5.26. In Fig. 5.26a–c, upper-saddle-node (USN) and lower-saddle-node (LSN) switching bifurcations are at two locations for two simple equilibriums, and one upper-saddle-node (USN) appearing bifurcation is for two simple equilibriums. In Fig. 5.26d, a third-order sink (3rd SI) pitchfork-switching bifurcation for a switching of one sink equilibrium to three simple equilibriums is presented, and one upper-saddle-node (USN) switching bifurcation for two simple equilibriums switching is also presented. In Fig. 5.26e, a third-order source (3rd SO) bundle-switching bifurcation for three equilibrium bundle-switching is presented, and an upper-saddle-node (USN) appearing bifurcation for two equilibrium onset is also presented. In Fig. 5.26f, a fourth-order upper-saddle (4th US) flower-bundle-switching bifurcation for four simple equilibriums is presented.

Similarly, the stability and bifurcations of equilibriums in the 1-dimensional, quartic nonlinear dynamical system ($a_0 < 0$) are presented in Fig. 5.27. In Fig. 5.27 (a–c), lower-saddle-node (LSN) and upper-saddle-node (USN) switching bifurcations are at two locations for two simple equilibriums, and one lower-saddle-node (LSN) appearing bifurcation is for two simple equilibriums appearing. In Fig. 5.27 (d), a third-order source (3rd SO) pitchfork-switching bifurcation for a switching of one source equilibrium to three simple equilibriums is presented, and one lower-saddle-node (LSN) switching bifurcation for two simple equilibriums switching is also presented. In Fig. 5.27(e), a third-order sink (3rd SI) bundle-switching bifurcation for three equilibrium bundle-switching is presented, and a lower-saddle-node (LSN) appearing bifurcation for two equilibrium onset is also presented. In Fig. 5.27 (f), a fourth-order lower-saddle (4th LS) flower-bundle-switching bifurcation for four simple equilibriums is presented.

For further discussion on the switching bifurcation, the following definition is given for the 1-dimensional, quartic nonlinear dynamical system.

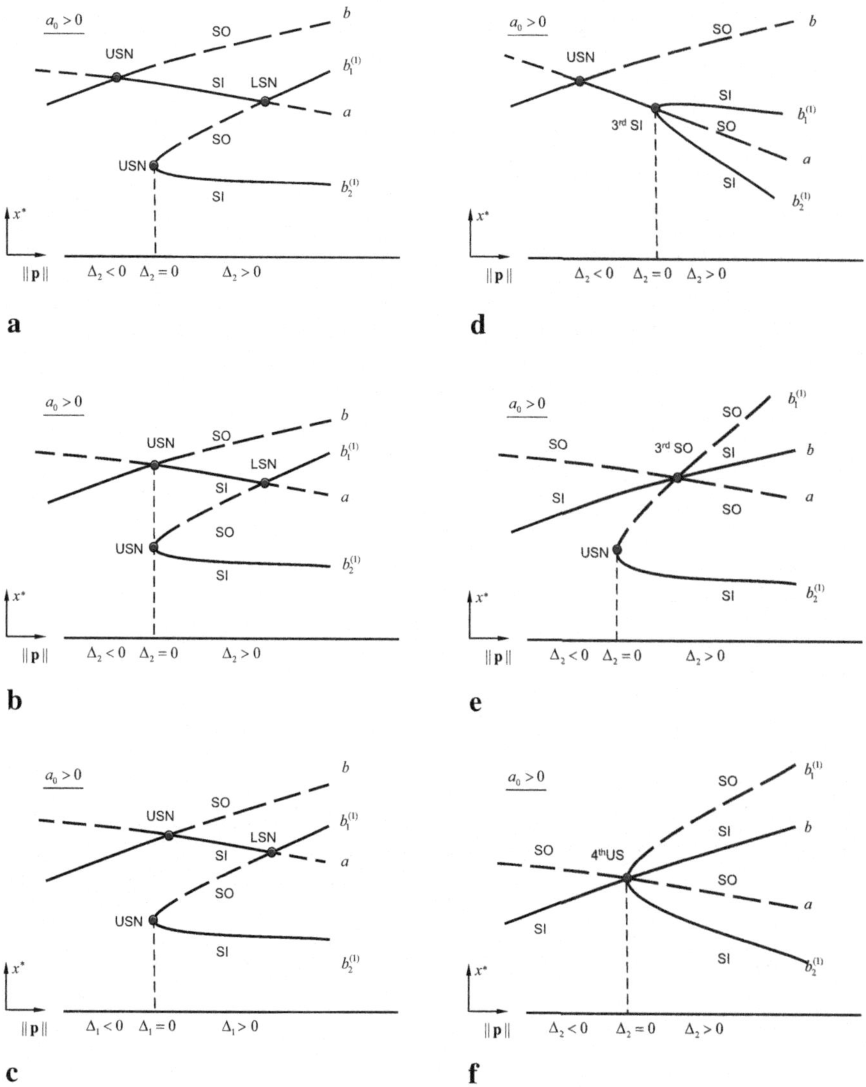

Fig. 5.26 Stability and bifurcations of equilibriums in the 1-dimensional, quartic nonlinear dynamical system ($a_0 > 0$): (**a**–**c**) Two (USN and LSN) switching and one (USN) appearing bifurcations, (**d**) 3rd SI pitchfork-switching bifurcation, (**e**) 3rd SO bundle-switching bifurcation, (**f**) 4th US flower-bundle-switching bifurcation. LSN: lower-saddle-node, USN: upper-saddle-node, SI: sink, SO: source. Stable and unstable equilibriums are represented by solid and dashed curves, respectively. The bifurcation points are marked by circular symbols

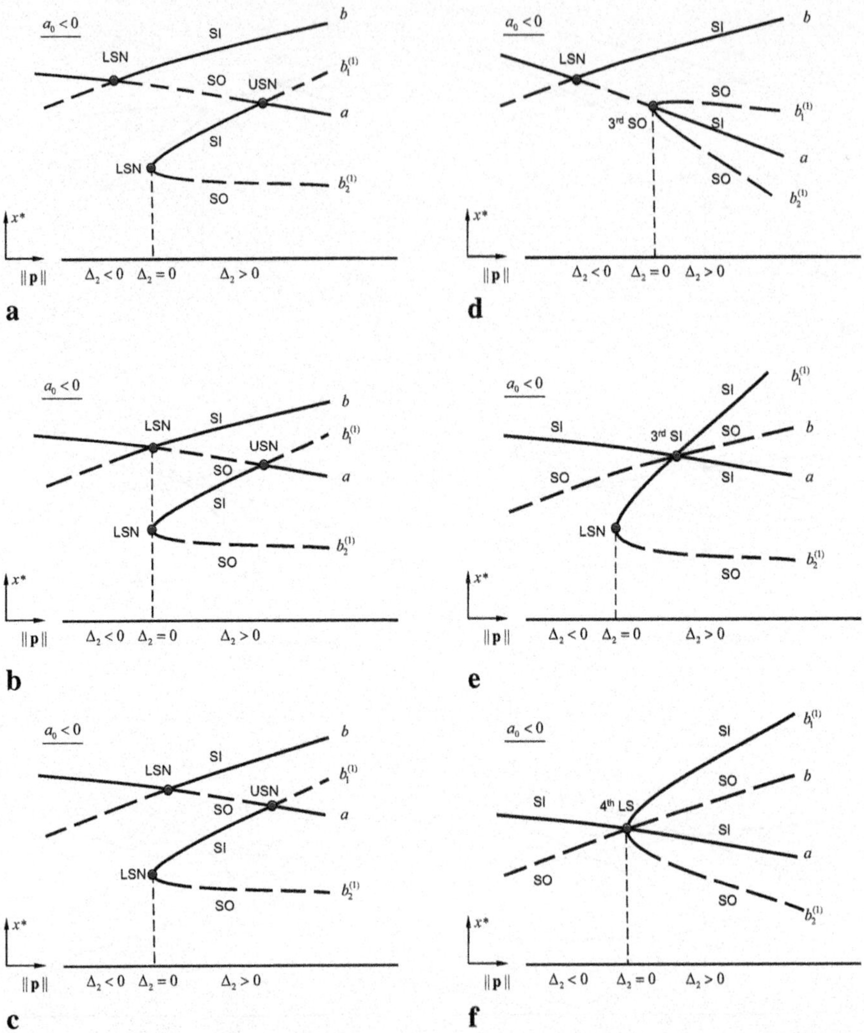

Fig. 5.27 Stability and bifurcations of equilibriums in the 1-dimensional, quartic nonlinear dynamical system ($a_0 < 0$): (**a–c**) Two (LSN and USN) switching and one LSN appearing bifurcations, (**d**) 3rd SO pitchfork-switching bifurcation, (**e**) 3rd SI bundle-switching bifurcation, (**f**) 4th LSN flower-bundle-switching bifurcation. LSN: lower-saddle-node, USN: upper-saddle-node, SI: sink, SO: source. Stable and unstable equilibriums are represented by solid and dashed curves, respectively. The bifurcation points are marked by circular symbols

Definition 5.12 Consider a 1-dimensional, quartic nonlinear dynamical system

$$\begin{aligned}\dot{x} &= A(\mathbf{p})x^4 + B(\mathbf{p})x^3 + C(\mathbf{p})x^2 + D(\mathbf{p})x + E(\mathbf{p}) \\ &= a_0(\mathbf{p})(x-a)(x-b)(x-c)(x-d)\end{aligned} \tag{5.259}$$

where $A(\mathbf{p}) \neq 0$, and

$$\mathbf{p} = (p_1, p_2, \ldots, p_m)^{\mathrm{T}}. \tag{5.260}$$

(i) If

$$\{a_1, a_2, a_3, a_4\} = \mathrm{sort}\{a, b, c, d\}, a_i \leq a_{i+1}, \tag{5.261}$$

the quartic nonlinear dynamical system has four simple equilibriums, and the standard form is

$$\dot{x} = a_0(\mathbf{p})(x-a_1)(x-a_2)(x-a_3)(x-a_4). \tag{5.262}$$

(i_1) For $a_0 > 0$, the equilibrium flow is an (SI:SO:SI:SO) flow. The equilibrium of $x^* = a_{1,3}$ is stable (sink, $df/dx|_{x^*=a_{1,3}} < 0$) and the equilibrium of $x^* = a_{2,4}$ is unstable (source, $df/dx|_{x^*=a_{2,4}} > 0$).

(i_2) For $a_0 < 0$, the equilibrium flow is an (SO:SI:SO:SI) flow. The equilibrium of $x^* = a_1$ is unstable (source, $df/dx|_{x^*=a_1} > 0$) and the equilibrium of $x^* = a_2$ is stable (sink, $df/dx|_{x^*=a_2} < 0$).

(ii) If

$$\Delta_{i_1 i_2} = (a_{i_1} - a_{i_2})^2 = 0 \text{ with } a_{i_1} = a_{i_2}; i_1, i_2 \in \{1, 2, 3, 4\}, \tag{5.263}$$

the quartic nonlinear dynamical system has a standard form as

$$\dot{x} = f(x, \mathbf{p}) = a_0(x - a_{i_1})^2(x - a_{i_3})(x - a_{i_4}). \tag{5.264}$$

(ii_1) The equilibrium of $x^* = a_{i_1}$ is unstable (an upper saddle of the second order, $d^2f/dx^2|_{x^*=a_{i_1}} > 0$). Such a flow is called an upper-saddle flow at $x^* = a_{i_1}$. The bifurcation of equilibrium at $x^* = a_{i_1}$ for two equilibriums switching of $x^* = a_{i_1}, a_{i_2}$ is called an upper-saddle-node bifurcation of the second order at a point $\mathbf{p} = \mathbf{p}_1$.

(ii_2) The equilibrium of $x^* = a_{i_1}$ is unstable (a lower saddle of the second order, $d^2f/dx^2|_{x^*=a_{i_1}} < 0$). Such a flow is called a lower-saddle flow at $x^* = a_{i_1}$. The bifurcation of equilibrium at $x^* = a_{i_1}$ for two equilibriums switching of $x^{*}=a_{i_1}, a_{i_2}$ is called a lower-saddle-node bifurcation of the second order at a point $\mathbf{p}=\mathbf{p}_1$.

(iii) If

$$\begin{aligned} &\Delta_{i_1 i_2}=(a_{i_1}-a_{i_2})^2=0, \Delta_{i_2 i_3}=(a_{i_2}-a_{i_3})^2=0 \\ &a_{i_1}=a_{i_2}=a_{i_3};\ i_1, i_2, i_3 \in \{1,2,3,4\}, i_1 \neq i_2 \neq i_3, \end{aligned} \tag{5.265}$$

the corresponding standard form is

$$\begin{aligned} &\dot{x}=f(x, \mathbf{p})=a_0(x-a_{i_1})^3(x-a_{i_4}) \\ &i_\alpha \in \{1,2,3,4\},\ \alpha=1,4. \end{aligned} \tag{5.266}$$

(iii$_1$) The equilibrium of $x^{*}=a_{i_1}$ is unstable with $d^3 f/dx^3|_{x^{*}=a_{i_1}}>0$ (a third-order source). Such a flow is called a third-order source flow at $x^{*}=a_{i_1}$. The bifurcation of equilibrium at $x^{*}=a_{i_1}$ for a bundle switching of three simple equilibriums of $x^{*}=a_{i_1}, a_{i_2}, a_{i_3}$ is called a third-order source bundle-switching bifurcation at a point $\mathbf{p}=\mathbf{p}_1$.

(iii$_2$) The equilibrium of $x^{*}=a_{i_1}$ is stable with $d^3 f/dx^3|_{x^{*}=a_{i_1}}<0$ (a third-order sink). Such a flow is called a third-order sink flow at $x^{*}=a_{i_1}$. The bifurcation of equilibrium at $x^{*}=a_{i_1}$ for a bundle switching of three simple equilibriums of $x^{*}=a_{i_1}, a_{i_2}, a_{i_3}$ is called a third-order sink bundle-switching bifurcation at a point $\mathbf{p}=\mathbf{p}_1$.

(iv) If

$$\begin{aligned} &\Delta_{i_1 i_2}=(a_{i_1}-a_{i_2})^2=0, \Delta_{i_2 i_3}=(a_{i_2}-a_{i_3})^2=0, \\ &\Delta_{i_3 i_4}=(a_{i_4}-a_{i_4})^2=0, a_{i_1}=a_{i_2}=a_{i_3}=a_{i_4}, \\ &i_1, i_2, i_3, i_4 \in \{1,2,3,4\}, i_1 \neq i_2 \neq i_3 \neq i_4 \end{aligned} \tag{5.267}$$

the corresponding standard form is

$$\dot{x}=f(x, \mathbf{p})=a_0(x-a_{i_1})^4 \tag{5.268}$$

(iv$_1$) The equilibrium of $x^{*}=a_{i_1}$ is unstable (a fourth-order upper saddle, $d^4 f/dx^4|_{x^{*}=a_{i_1}}>0$). Such a flow is called a fourth-order upper-saddle flow at $x^{*}=a_{i_1}$. The bifurcation of equilibrium at $x^{*}=a_{i_1}$ for a bundle switching of four simple equilibriums of $x^{*}=a_{1,2,3,4}$ is called a fourth-order upper-saddle-node bundle-switching bifurcation at a point $\mathbf{p}=\mathbf{p}_1$.

(iv$_2$) The equilibrium of $x^{*}=a_{i_1}$ is unstable (a fourth-order lower saddle, $d^4 f/dx^4|_{x^{*}=a_{i_1}}<0$). Such a flow is called a fourth-order lower-saddle flow at $x^{*}=a_{i_1}$. The bifurcation of equilibrium at $x^{*}=a_{i_1}$ for a bundle switching of four simple equilibriums of $x^* = a_{1,2,3,4}$ is called a fourth-order lower-saddle bundle-switching bifurcation at a point $\mathbf{p} = \mathbf{p}_1$.

From the previous definition, stability and bifurcations of equilibriums in the 1-dimensional, quartic nonlinear dynamical system are presented in Fig. 5.28. For $a_0 > 0$, the bifurcations and stability of equilibriums are presented in Fig. 5.28a–c. In Fig. 5.28a, four upper-saddle-node (USN) and two lower-saddle-

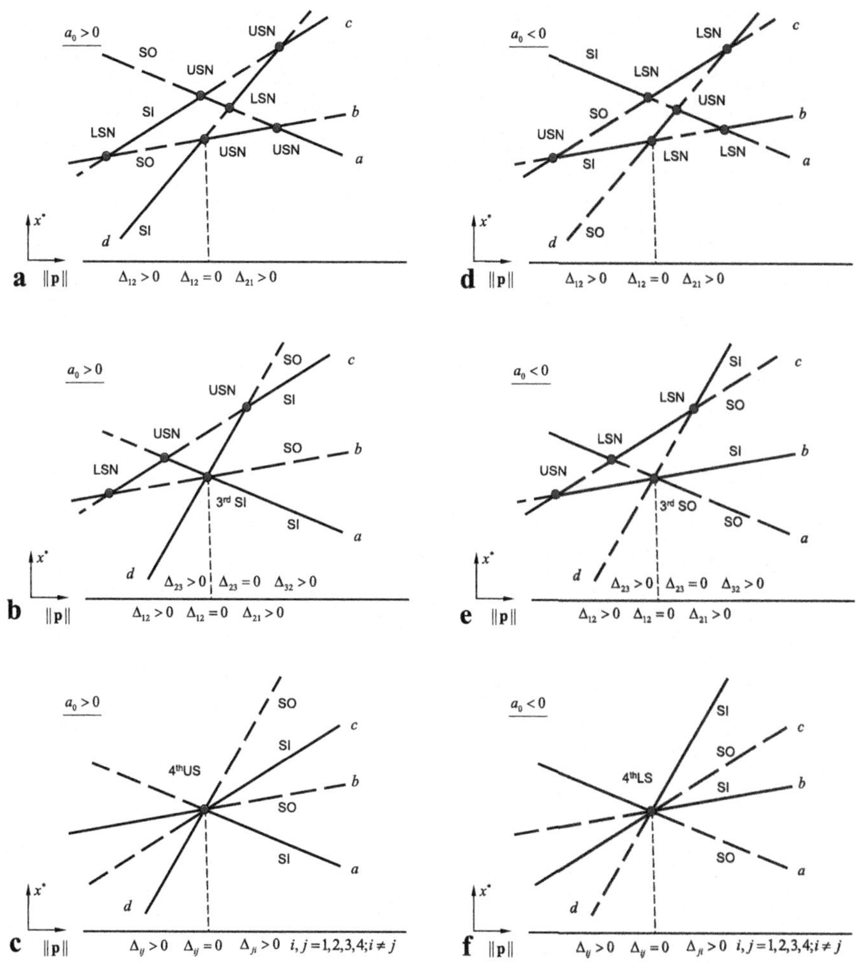

Fig. 5.28 Stability and bifurcations of equilibriums in the 1-dimensional, quartic nonlinear dynamical system ($a_0 > 0$): (**a**) Four USN and two LSN switching bifurcation network, (**b**) third-order SI bundle-switching bifurcation, (**c**) fourth-order US bundle-switching bifurcation ($a_0 < 0$): (**d**) Four LSN and two USN switching bifurcation network, (**e**) third-order SO bundle-switching bifurcation, (**f**) fourth-order LS bundle-switching bifurcation. LSN: lower-saddle-node, USN: upper-saddle-node, SI: sink, SO: source. Stable and unstable equilibriums are represented by solid and dashed curves, respectively. The bifurcation points are marked by circular symbols

node (LSN) switching bifurcation network are presented for all possible switching bifurcation between two simple equilibriums. In Fig. 5.28b, a third-order sink (3rd SI) bundle-switching bifurcation for three simple equilibrium is presented, and there are three possible upper-saddle-node (USN) and lower-saddle-node (LSN) switching bifurcations for two simple equilibriums. In Fig. 5.28c a fourth-order upper-saddle (4th US) bundle-switching bifurcation for four simple equilibriums is presented. Similarly, For $a_0 < 0$, the bifurcations and stability of equilibriums are presented in Fig. 5.28d–f. In Fig. 5.28d, four lower-saddle-node (LSN) and two upper-saddle-node (USN) switching bifurcation network are presented for all possible switching bifurcation between two simple equilibriums. In Fig. 5.28e, a third-order source (3rd SO) bundle-switching bifurcation for three simple equilibrium is presented, and there are three possible lower-saddle-node (LSN) and upper-saddle-node (USN) switching bifurcations for two simple equilibriums. In Fig. 5.28f a fourth-order lower-saddle (4th LS) bundle-switching bifurcation for four simple equilibriums is presented.

For the switching bifurcation between the second-order and simple equilibriums, the following definition is given for the 1-dimensional, quartic nonlinear dynamical system.

Definition 5.13 Consider a 1-dimensional, quartic nonlinear dynamical system

$$\begin{aligned}\dot{x} &= A(\mathbf{p})x^4 + B(\mathbf{p})x^3 + C(\mathbf{p})x^2 + D(\mathbf{p})x + E(\mathbf{p}) \\ &= a_0(\mathbf{p})(x-a)^2(x-b)(x-c)\end{aligned} \tag{5.269}$$

where $A(\mathbf{p}) \neq 0$, and

$$\mathbf{p} = (p_1, p_2, \ldots, p_m)^{\mathrm{T}}. \tag{5.270}$$

(i) If

$$\begin{aligned}&\{a_1, a_2, a_3\} = \mathrm{sort}\{a, b, c\}, a_i < a_{i+1} \\ &i_1, i_2, i_3 \in \{1, 2, 3\},\end{aligned} \tag{5.271}$$

the quartic nonlinear dynamical system has a standard form as

$$\dot{x} = f(x, \mathbf{p}) = a_0(x - a_{i_1})^2(x - a_{i_2})(x - a_{i_3}). \tag{5.272}$$

(i_1) The equilibrium of $x^* = a_{i_1}$ is unstable (an upper saddle of the second order, $d^2f/dx^2|_{x^*=a_{i_1}} > 0$). Such a flow is called an upper-saddle flow at $x^* = a_{i_1}$.

(i_2) The equilibrium of $x^* = a_{i_1}$ is unstable (a lower saddle of the second order, $d^2f/dx^2|_{x^*=a_{i_1}} < 0$). Such a flow is called a lower-saddle flow at $x^* = a_{i_1}$.

(ii) If

$$\begin{aligned}&a = a_{i_1}, b = a_{i_2}, c = a_{i_3};\\&\Delta_{i_2 i_3} = (a_{i_2} - a_{i_3})^2 = 0, a_{i_2} = a_{i_3};\\&i_1, i_2, i_3 \in \{1,2\}, i_1 \neq i_2 \neq i_3;\end{aligned} \quad (5.273)$$

the corresponding standard form is

$$\begin{aligned}&\dot{x} = f(x, \mathbf{p}) = a_0(x - a_{i_1})^2(x - a_{i_2})^2\\&i_\alpha \in \{1,2\}, \alpha = 1,2.\end{aligned} \quad (5.274)$$

The equilibriums of $x^* = a_{i_1}, a_{i_2}$ are unstable (an upper saddle of the second order, $d^2f/dx^2|_{x^*=a_{i_1},\, a_{i_2}} > 0$ or a lower saddle of the second order, $d^2f/dx^2|_{x^*=a_{i_2},\, a_{i_1}} < 0$). Such a flow is called an (US:US) or a (LS:LS) flow. The bifurcation of equilibrium at $x^* = a_{i_2}$ for two simple equilibriums switching of $x^* = b$, c is called an upper-saddle or lower-saddle switching bifurcation at a point $\mathbf{p} = \mathbf{p}_1$.

(iii) If

$$\begin{aligned}&a = a_{i_1}; a_{i_3}, a_{i_2} \in \{b, c\};\\&\Delta_{i_1 i_3} = (a_{i_1} - a_{i_3})^2 = 0, a_{i_1} = a_{i_3}, a_{i_3} \neq a_{i_2};\\&i_1, i_2, i_3 \in \{1,2,3\}, i_1 \neq i_2 \neq i_3,\end{aligned} \quad (5.275)$$

the corresponding standard form is

$$\begin{aligned}&\dot{x} = f(x, \mathbf{p}) = a_0(x - a_{i_1})^3(x - a_{i_2})\\&i_\alpha \in \{1,2\}, \alpha = 1,2.\end{aligned} \quad (5.276)$$

(iii$_1$) The equilibrium of $x^* = a_{i_1}$ is unstable (a third-order source, $d^3f/dx^3|_{x^*=a_{i_1}} > 0$). Such a flow is called a third-order source flow at $x^* = a_{i_1}$. The bifurcation of equilibrium at $x^* = a_{i_1}$ for a switching of one second-order and one simple equilibriums of $x^* = a_{i_1}, a_{i_2}$ is called a third-order source switching bifurcation at a point $\mathbf{p} = \mathbf{p}_1$.

(iii$_2$) The equilibrium of $x^* = a_{i_1}$ is stable (a third-order sink, $d^3f/dx^3|_{x^*=a_{i_1}} < 0$). Such a flow is called a third-order sink flow at $x^* = a_{i_1}$. The bifurcation of equilibrium at $x^* = a_{i_1}$ for a switching of one second-order and one simple equilibriums of $x^* = a_{i_1}, a_{i_2}$ is called a third-order sink switching bifurcation at a point $\mathbf{p} = \mathbf{p}_1$.

(iv) If

$$\Delta_{i_1 i_2} = (a_{i_1} - a_{i_2})^2 = 0, \Delta_{i_2 i_3} = (a_{i_2} - a_{i_3})^2 = 0,\\ i_1, i_2, i_3 \in \{1,2,3\}, i_1 \neq i_2 \neq i_3, \tag{5.277}$$

the corresponding standard form is

$$\dot{x} = f(x, \mathbf{p}) = a_0(x - a_{i_1})^4. \tag{5.278}$$

(iv_1) The equilibrium of $x^* = a_{i_1}$ is unstable (a fourth-order upper saddle, $d^4f/dx^4|_{x^*=a_{i_1}} > 0$). Such a flow is called a fourth-order upper-saddle flow at $x^* = a_{i_1}$. The bifurcation of equilibrium at $x^* = a_{i_1}$ for a bundle switching of one second-order and two simple equilibriums of $x^* = a_{1,2,3}$ is called a fourth-order upper-saddle bundle-switching bifurcation at a point $\mathbf{p} = \mathbf{p}_1$.

(iv_2) The equilibrium of $x^* = a_{i_1}$ is unstable (a fourth-order lower saddle, $d^4f/dx^4|_{x^*=a_{i_1}} < 0$). Such a flow is called a fourth-order lower-saddle flow at $x^* = a_{i_1}$. The bifurcation of equilibrium at $x^* = a_{i_1}$ for a bundle switching of one second-order and two simple equilibriums of $x^* = a_{1,2,3}$ is called a fourth-order lower-saddle bundle-switching bifurcation at a point $\mathbf{p} = \mathbf{p}_1$.

From Definition 5.13, stability and bifurcations of equilibriums in the 1-dimensional, quartic nonlinear dynamical system are presented in Fig. 5.29. For $a_0 > 0$, the bifurcations and stability of equilibriums are presented in Fig. 5.29a–c. In Fig. 5.29a, there is a switching bifurcation network with two third-order source switching bifurcations and one upper-saddle-node bifurcation. The two third-order source (3rd SO) switching bifurcations are for (LS:SO) switching to (US:SO) equilibriums and for (SO:US) switching to (SO:LS)-equilibriums. The upper-saddle-node (USN) switching bifurcation is for two simple equilibriums. In Fig. 5.29b, a fourth-order upper-saddle (4th US) bundle-switching bifurcation for (SI:LS:SO) switching to (SO:LS:SI) equilibriums. In Fig. 5.29c, a fourth-order upper-saddle (4th US) bundle-switching bifurcation for (SI:SO:US) switching to (SO:SI:US) equilibriums. Similarly, for $a_0 < 0$, the bifurcations and stability of equilibriums are presented in Fig. 5.29d–f. In Fig. 5.29d, the switching bifurcation network consists of two third-order sink switching bifurcations and one lower-saddle-node bifurcation. The two third-order sink (3rd SI) switching bifurcations are for the (LS:SO) switching to (US:SO)-equilibriums and for the (SO:US) switching to (SO:LS)-equilibriums. The upper-saddle-node (USN) switching bifurcation is for two simple equilibriums. In Fig. 5.29e, a fourth-order lower-saddle (4th LS) bundle-switching bifurcation for (SO:US:SI) switching to (SI:US:SO) equilibriums. In Fig. 5.29f, a fourth-order lower-saddle (4th LS) bundle-switching bifurcation for (SO:SI: LS) switching to (SI:SO:LS) equilibriums.

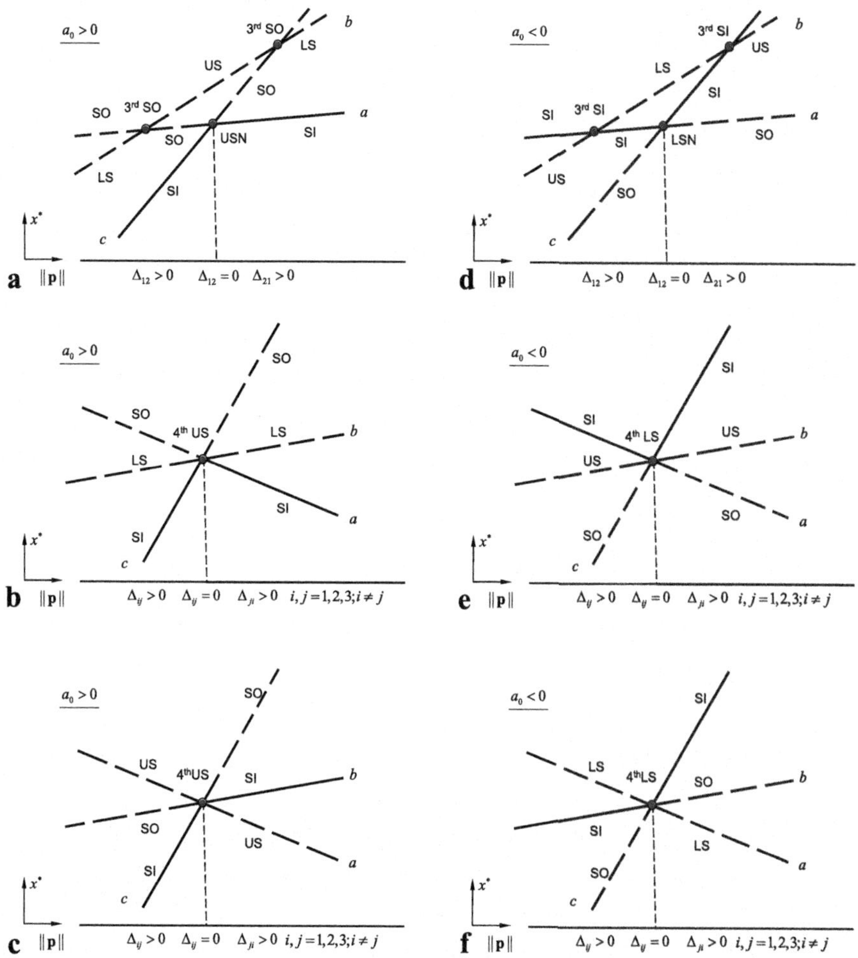

Fig. 5.29 Stability and bifurcations of equilibriums in the 1-dimensional, quartic nonlinear dynamical system ($a_0 > 0$): (**a**) two third-order SO and USN switching bifurcation network, (**b**) fourth-order bundle-switching bifurcation, (**c**) fourth-order US bundle-switching bifurcation; ($a_0 < 0$): (**d**) two third-order SI and LSN switching bifurcation network, (**e**) fourth-order LS bundle-switching bifurcation, (**f**) fourth-order LS bundle switching. LSN: lower-saddle-node, USN: upper-saddle-node, SI: sink, SO: source. Stable and unstable equilibriums are represented by solid and dashed curves, respectively. The bifurcation points are marked by circular symbols

For the switching bifurcation between the third-order and simple equilibriums, the following definition is given for the 1-dimensional, quartic nonlinear dynamical system.

Definition 5.14 Consider a 1-dimensional, quartic nonlinear dynamical system

$$\begin{aligned}\dot{x} &= A(\mathbf{p})x^4 + B(\mathbf{p})x^3 + C(\mathbf{p})x^2 + D(\mathbf{p})x + E(\mathbf{p})\\ &= a_0(\mathbf{p})(x-a)^3(x-b)\end{aligned} \tag{5.279}$$

where $A(\mathbf{p}) \neq 0$, and

$$\mathbf{p} = (p_1, p_2, \ldots, p_m)^{\mathrm{T}}. \tag{5.280}$$

(i) If

$$\begin{aligned}&\{a_1, a_2\} = \mathrm{sort}\{a, b\}, a_i < a_{i+1}\\ &i_1, i_2 \in \{1, 2\},\end{aligned} \tag{5.281}$$

the quartic nonlinear dynamical system has a standard form as

$$\dot{x} = f(x, \mathbf{p}) = a_0(x - a_{i_1})^3(x - a_{i_2}). \tag{5.282}$$

(i_1) The equilibrium of $x^* = a_{i_1}$ is unstable (a third-order source, $d^3f/dx^3|_{x^*=a_{i_1}} > 0$). The equilibrium of $x^* = a_{i_2}$ is stable (a sink, $df/dx|_{x^*=a_{i_2}} < 0$). Such a flow is called a (3rd SO:SI) or an (SI:3rd SO) flow.

(i_2) The equilibrium of $x^* = a_{i_1}$ is stable (a third-order sink, $d^3f/dx^3|_{x^*=a_{i_1}} < 0$). The equilibrium of $x^* = a_{i_2}$ is unstable (a source, $df/dx|_{x^*=a_{i_2}} > 0$). Such a flow is called a (3rd SI:SO) or an (SO:3rd SI) flow.

(ii) If

$$\begin{aligned}&a = a_{i_1}, b = a_{i_2};\\ &\Delta_{i_1 i_2} = (a_{i_1} - a_{i_2})^2 = 0, a_{i_1} = a_{i_2};\\ &i_1, i_2 \in \{1, 2\}, i_1 \neq i_2;\end{aligned} \tag{5.283}$$

the corresponding standard form is

$$\dot{x} = f(x, \mathbf{p}) = a_0(x - a_{i_1})^4 \tag{5.284}$$

(ii_1) The equilibrium of $x^* = a_{i_1}$ is unstable (a fourth-order upper saddle, $d^4f/dx^4|_{x^*=a_{i_1}} > 0$). Such a flow is called a fourth-order upper-saddle flow. The bifurcation of equilibrium at $x^* = a_{i_1}$ for a switching of one

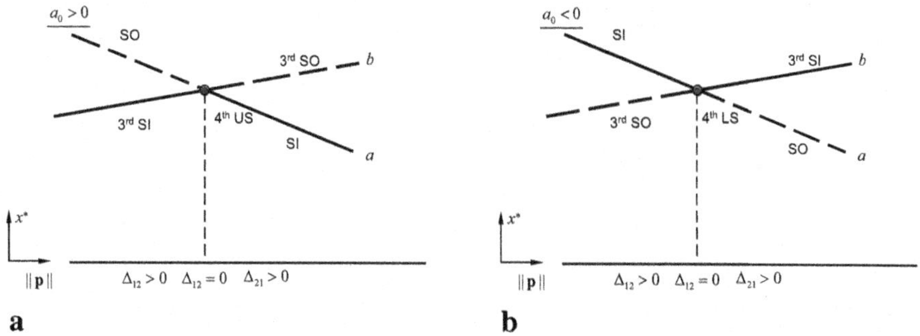

Fig. 5.30 Stability and bifurcations of equilibriums in the 1-dimensional, quartic nonlinear dynamical system ($a_0 > 0$): (**a**) fourth-order US switching bifurcation of (3rd SI:SO) to (SI:3rd SO), ($a_0 < 0$): (**b**) fourth-order LS switching bifurcation of (3rd SO:SI) to (SO:3rd SI). LS: lower-saddle, US: upper-saddle, SI: sink, SO: source. Stable and unstable equilibriums are represented by solid and dashed curves, respectively. The bifurcation points are marked by circular symbols

third-order and one simple equilibriums of $x^* = a_{1,2}$ is called a fourth-order upper-saddle switching bifurcation at a point $\mathbf{p} = \mathbf{p}_1$.

(ii$_2$) The equilibrium of $x^* = a_{i_1}$ is unstable (a fourth-order lower saddle, $d^4f/dx^4|_{x^*=a_{i_1}} < 0$). Such a flow is called a fourth-order lower-saddle flow. The bifurcation of equilibrium at $x^* = a_{i_1}$ for a switching of one third-order and one simple equilibriums of $x^* = a_{1,2}$ is called a fourth-order lower-saddle switching bifurcation at a point $\mathbf{p} = \mathbf{p}_1$.

From Definition 5.14, the stability and bifurcations of equilibriums in the 1-dimensional, quartic nonlinear dynamical system are presented in Fig. 5.30. In Fig. 5.30(a), the fourth-order upper-saddle (4th US) switching bifurcation for $a_0 > 0$ is presented for the third-order sink (3rd SI) with simple source (SO) equilibriums (i.e., (3rd SI:SO)) to the third-order source (3rd SO) with a simple sink (SI) equilibriums (i.e., (SI:3rd SO)). Similarly, in Fig. 5.30(b), the fourth-order lower-saddle (4th LS) switching bifurcation for $a_0 < 0$ is presented for the third-order source (3rd SO) with simple sink (SI) equilibriums (i.e., (3rd SO:SI)) to the 3rd order sink (3rd SI) with a simple source (SO) equilibriums (i.e., (SO:3rd SI)).

For the switching bifurcation between the two second-order equilibriums, the following definition was presented in Definition 5.10, and the corresponding illustrations are presented in Fig. 5.24.

Reference

Luo, A.C.J., 2019, The stability and bifurcation of equilibriums in low-degree polynomial systems, *Journal of Vibration Testing and System Dynamics*, 3(4), pp. 403–451.

Chapter 6
$(2m)^{\text{th}}$-Degree Polynomial Systems

In this chapter, the global stability and bifurcations of equilibriums in the $(2m)^{\text{th}}$-degree polynomial system are discussed for a better understanding of the complexity of bifurcations and stability of equilibriums. The appearing and switching bifurcations for simple equilibriums are presented, and the appearing and switching bifurcations for higher order equilibriums are discussed as well. The parallel *appearing* bifurcations, *spraying-appearing* bifurcations, and *sprinkler-spraying-appearing* bifurcations for simple and higher order equilibriums are presented. The *antenna-switching* bifurcations for simple and higher order equilibriums are discussed and the parallel *straw-bundle-switching bifurcations* and *flower-bundle-switching bifurcations* for simple and higher order equilibriums are presented as well.

6.1 Global Stability and Bifurcations

In a similar fashion of low-degree polynomial systems in Chap. 5, the global stability and bifurcation of equilibriums in the $(2m)^{\text{th}}$-degree polynomial nonlinear systems are discussed as in Luo (2020). The stability and bifurcation of each individual equilibrium are analyzed from the local analysis in Chaps. 1–3.

Definition 6.1 Consider a $(2m)^{\text{th}}$-degree polynomial nonlinear system:

$$\begin{aligned}\dot{x} &= A_0(\mathbf{p})x^{2m} + A_1(\mathbf{p})x^{2m-1} + \cdots + A_{2m-2}(\mathbf{p})x^2 + A_{2m-1}x + A_{2m}(\mathbf{p}) \\ &= a_0(\mathbf{p})[x^2 + B_1(\mathbf{p})x + C_1(\mathbf{p})]\cdots[x^2 + B_m(\mathbf{p})x + C_m(\mathbf{p})]\end{aligned} \tag{6.1}$$

where $A_0(\mathbf{p}) \neq 0$, and

A. C. J. Luo, *Bifurcation and Stability in Nonlinear Dynamical Systems*, Nonlinear Systems and Complexity 28, https://doi.org/10.1007/978-3-030-22910-8_6

$$\mathbf{p} = (p_1, p_2, \ldots, p_{m_1})^{\mathrm{T}}. \tag{6.2}$$

(i) If

$$\Delta_i = B_i^2 - 4C_i < 0 \text{ for } i = 1, 2, \ldots, m, \tag{6.3}$$

the 1-dimensional nonlinear dynamical system with a $(2m)^{\text{th}}$-degree polynomial does not have any equilibrium, and the corresponding standard form is

$$\dot{x} = a_0[(x + \frac{1}{2}B_1)^2 + \frac{1}{4}(-\Delta_1)] \cdots [(x + \frac{1}{2}B_m)^2 + \frac{1}{4}(-\Delta_m)]. \tag{6.4}$$

The flow of such a system without equilibriums is called a nonequilibrium flow.

(a) If $a_0 > 0$, the nonequilibrium flow is called a positive flow.
(b) If $a_0 < 0$, the nonequilibrium flow is called a negative flow.

(ii) If

$$\begin{aligned}
&\Delta_i = B_i^2 - 4C_i > 0, i = i_1, i_2, \ldots, i_l \in \{1, 2, \ldots, m\}, \\
&\Delta_j = B_j^2 - 4C_j < 0, j = i_{l+1}, i_{l+2}, \ldots, i_m \in \{1, 2, \ldots, m\} \\
&\text{with } l \in \{0, 1, \ldots, m\},
\end{aligned} \tag{6.5}$$

the 1-dimensional, $(2m)^{\text{th}}$-degree polynomial system has $2l$-equilibriums as

$$\begin{aligned}
&x^* = b_1^{(i)} = -\frac{1}{2}(B_i + \sqrt{\Delta_i}), x^* = b_2^{(i)} = -\frac{1}{2}(B_i - \sqrt{\Delta_i}) \\
&i \in \{i_1, i_2, \ldots, i_l\} \subseteq \{1, 2, \ldots, m\}.
\end{aligned} \tag{6.6}$$

(ii_1) If

$$\begin{aligned}
&b_r^{(i)} \neq b_s^{(j)} \text{ for } r,s \in \{1,2\}; i,j = 1,2, \ldots, l \\
&\{a_1, a_2, \ldots, a_{2l}\} = \text{sort}\{b_1^{(1)}, b_2^{(1)}, \ldots, b_1^{(l)}, b_2^{(l)}\}, a_s < a_{s+1},
\end{aligned} \tag{6.7}$$

then the corresponding standard form is

$$\dot{x} = a_0 \prod_{i=1}^{l} (x - a_{2i-1})(x - a_{2i}) \prod_{k=l+1}^{m} [(x + \frac{1}{2}B_{i_k})^2 + \frac{1}{4}(-\Delta_{i_k})]. \tag{6.8}$$

(a) If $a_0 > 0$, the simple equilibrium separatrix flow is called a (SI : SO : $\cdots$: SI : SO : $\cdots$SI : SO) -flow.

(b) If $a_0 < 0$, the simple equilibrium separatrix flow is called a (SO : SI : $\cdots$: SO : SI : $\cdots$SO : SI) -flow.

(ii$_2$) If

$$\begin{aligned}
&\{a_1, a_2, \ldots, a_{2l}\} = \text{sort}\{b_1^{(1)}, b_2^{(1)}, \ldots, b_1^{(l)}, b_2^{(l)}\},\\
&a_{i_1} \equiv a_1 = \cdots = a_{l_1},\\
&a_{i_2} \equiv a_{l_1+1} = \cdots = a_{l_1+l_2},\\
&\vdots\\
&a_{i_r} \equiv a_{\Sigma_{i=1}^{r-1} l_i+1} = \cdots = a_{\Sigma_{i=1}^{r-1} l_i+l_r} = a_{2l}\\
&\text{with } \Sigma_{s=1}^{r} l_s = 2l,
\end{aligned} \tag{6.9}$$

then the corresponding standard form is

$$\dot{x} = a_0 \prod_{s=1}^{r} (x - a_{i_s})^{l_s} \prod_{k=l+1}^{m} [(x + \frac{1}{2} B_{i_k})^2 + \frac{1}{4}(-\Delta_{i_k})]. \tag{6.10}$$

The equilibrium separatrix flow is called an (l_1th XX : l_2th XX : $\cdots$: l_rth XX)-flow.

(a) For $a_0 > 0$ and $p = 1, 2, \ldots, r$,

$$l_p\text{th XX} = \begin{cases} (2r_p - 1)^{\text{th}}\text{order source}, & \text{for } \alpha_p = 2M_p - 1, l_p = 2r_p - 1;\\ (2r_p - 1)^{\text{th}}\text{order sink}, & \text{for } \alpha_p = 2M_p, l_p = 2r_p - 1;\\ (2r_p)^{\text{th}}\text{order lower-saddle}, & \text{for } \alpha_p = 2M_p - 1, l_p = 2r_p;\\ (2r_p)^{\text{th}}\text{order upper-saddle}, & \text{for } \alpha_p = 2M_p, l_p = 2r_p; \end{cases}, \tag{6.11}$$

where

$$\alpha_p = \sum\nolimits_{s=p}^{r} l_s. \tag{6.12}$$

(b) For $a_0 < 0$ and $p = 1, 2, \ldots, r$,

$$l_p\text{th XX} = \begin{cases} (2r_p - 1)^{\text{th}}\text{order sink}, & \text{for } \alpha_p = 2M_p - 1, l_p = 2r_p - 1;\\ (2r_p - 1)^{\text{th}}\text{order source}, & \text{for } \alpha_p = 2M_p, l_p = 2r_p - 1;\\ (2r_p)^{\text{th}}\text{order upper-saddle}, & \text{for } \alpha_p = 2M_p - 1, l_p = 2r_p;\\ (2r_p)^{\text{th}}\text{order lower-saddle}, & \text{for } \alpha_p = 2M_p, l_p = 2r_p. \end{cases} \tag{6.13}$$

(c) The equilibrium of $x^* = a_{i_p}$ for $(l_p > 1)$-repeated equilibriums switching is called an l_pth XX bifurcation of $(l_{p_1}\text{th XX} : l_{p_2}\text{th XX} : \cdots : l_{p_\beta}\text{th XX})$ equilibrium switching at a point $\mathbf{p} = \mathbf{p}_1 \in \partial\Omega_{12}$, and the bifurcation condition is

$$\begin{aligned} &a_{i_p} \equiv a_{\Sigma_{i=1}^{p-1} l_i+1} = \cdots = a_{\Sigma_{i=1}^{p-1} l_i+l_p}, \\ &a^{\pm}_{\Sigma_{i=1}^{p-1} l_i+1} \neq \cdots \neq a^{\pm}_{\Sigma_{i=1}^{p-1} l_i+l_p}; l_p = \sum\nolimits_{i=1}^{\beta} l_{p_i}. \end{aligned} \tag{6.14}$$

(iii) If

$$\begin{aligned} &\Delta_i = B_i^2 - 4C_i = 0, i \in \{i_{11}, i_{12}, \ldots, i_{1s}\} \subseteq \{i_1, i_2, \ldots, i_l\} \subseteq \{1, 2, \ldots, m\}, \\ &\Delta_k = B_k^2 - 4C_k > 0, k \in \{i_{21}, i_{22}, \ldots, i_{2r}\} \subseteq \{i_1, i_2, \ldots, i_l\} \subseteq \{1, 2, \ldots, m\}, \\ &\Delta_j = B_j^2 - 4C_j < 0,\ j \in \{i_{l+1}, i_{l+2}, \ldots, i_m\} \subseteq \{1, 2, \ldots, m\}, \end{aligned} \tag{6.15}$$

the 1-dimensional, $(2m)^{\text{th}}$-degree polynomial system has $2l$-equilibriums as

$$\begin{aligned} &x^* = b_1^{(i)} = -\frac{1}{2}B_i, x^* = b_2^{(i)} = -\frac{1}{2}B_i \text{ for } i \in \{i_{11}, i_{12}, \ldots, i_{1s}\}, \\ &x^* = b_1^{(k)} = -\frac{1}{2}(B_k + \sqrt{\Delta_k}), x^* = b_2^{(k)} = -\frac{1}{2}(B_k - \sqrt{\Delta_k}) \\ &\text{for } k \in \{i_{21}, i_{22}, \ldots, i_{2r}\}. \end{aligned} \tag{6.16}$$

If

$$\begin{aligned} &\{a_1, a_2, \ldots, a_{2l}\} = \text{sort}\{b_1^{(1)}, b_2^{(1)}, \ldots, b_1^{(l)}, b_2^{(l)}\}, \\ &a_{i_1} \equiv a_1 = \cdots = a_{l_1}, \\ &a_{i_2} \equiv a_{l_1+1} = \cdots = a_{l_1+l_2}, \\ &\vdots \\ &a_{i_r} \equiv a_{\Sigma_{i=1}^{r-1} l_i+1} = \cdots = a_{\Sigma_{i=1}^{r-1} l_i+l_r} = a_{2l} \\ &\text{with } \Sigma_{s=1}^{r} l_s = 2l, \end{aligned} \tag{6.17}$$

then the corresponding standard form is

$$\dot{x} = a_0 \prod_{s=1}^{r} (x - a_{i_s})^{l_s} \prod_{k=l+1}^{m} [(x + \frac{1}{2}B_{i_k})^2 + \frac{1}{4}(-\Delta_{i_k})]. \tag{6.18}$$

The equilibrium separatrix flow is called an ($l_1$thXX : $l_2$thXX : $\cdots$: l_rth XX)-flow.

(a) The equilibrium of $x^* = a_{i_p}$ for ($l_p > 1$)-repeated equilibriums appearing (or vanishing) is called an $l_p$thXX bifurcation of equilibrium at a point $\mathbf{p} = \mathbf{p}_1 \in \partial\Omega_{12}$, and the bifurcation condition is

$$\begin{aligned} & a_{i_p} \equiv a_{\Sigma_{i=1}^{p-1} l_i+1} = \cdots = a_{\Sigma_{i=1}^{p-1} l_i+l_p} = -\frac{1}{2}B_{i_q}, \\ & \text{with } \Delta_{i_q} = B_{i_q}^2 - 4C_{i_q} = 0 \ \big(i_q \in \{i_1, i_2, \cdots, i_l\}\big), \\ & a^+_{\Sigma_{i=1}^{p-1} l_i+1} \neq \cdots \neq a^+_{\Sigma_{i=1}^{p-1} l_i+l_p} \text{ or } a^-_{\Sigma_{i=1}^{p-1} l_i+1} \neq \cdots \neq a^-_{\Sigma_{i=1}^{p-1} l_i+l_p}. \end{aligned} \tag{6.19}$$

(b) The equilibrium of $x^* = a_{i_q}$ for ($l_q > 1$)-repeated equilibriums switching is called an l_qth XX bifurcation of (l_{q_1}th XX : l_{q_2}th XX : $\cdots$: l_{q_β}th XX) equilibrium switching at a point $\mathbf{p} = \mathbf{p}_1 \in \partial\Omega_{12}$, and the bifurcation condition is

$$\begin{aligned} & a_{i_q} \equiv a_{\Sigma_{i=1}^{q-1} l_i+1} = \cdots = a_{\Sigma_{i=1}^{q-1} l_i+l_p}, \\ & a^{\pm}_{\Sigma_{i=1}^{q-1} l_i+1} \neq \cdots \neq a^{\pm}_{\Sigma_{i=1}^{q-1} l_i+l_q}; \ l_q = \sum\nolimits_{i=1}^{\beta} l_{q_i}. \end{aligned} \tag{6.20}$$

(c) The equilibrium of $x^* = a_{i_p}$ for $\big(l_{p_1} \geq 1\big)$-repeated equilibriums appearing and $\big(l_{p_2} \geq 2\big)$-repeated equilibriums switching of $\big(l_{p_{21}}$th XX : $l_{p_{22}}$th XX : $\cdots$: $l_{p_{2\beta}}$th XX$\big)$ is called an l_pth XX bifurcation of equilibrium at a point $\mathbf{p} = \mathbf{p}_1 \in \partial\Omega_{12}$, and the bifurcation condition is

$$\begin{aligned} & a_{i_p} \equiv a_{\Sigma_{i=1}^{p-1} l_i+1} = \cdots = a_{\Sigma_{i=1}^{p-1} l_i+l_p} \\ & \text{with } \Delta_{i_q} = B_{i_q}^2 - 4C_{i_q} = 0 \ (i_q \in \{i_1, i_2, \ldots, i_l\}) \\ & a^+_{\Sigma_{i=1}^{p-1} l_i+j_1} \neq \cdots \neq a^+_{\Sigma_{i=1}^{p-1} l_i+j_{p_1}} \text{ or } a^-_{\Sigma_{i=1}^{p_1-1} l_i+j_1} \neq \cdots \neq a^-_{\Sigma_{i=1}^{p_1-1} l_i+j_{p_1}}, \\ & \text{for } \{j_1, j_2, \ldots, j_{p_1}\} \subseteq \{1, 2, \ldots, l_p\}, \\ & a^{\pm}_{\Sigma_{i=1}^{p-1} l_i+k_1} \neq \cdots \neq a^{\pm}_{\Sigma_{i=1}^{p-1} l_i+k_{p_2}} \\ & \text{for } \{k_1, k_2, \ldots, k_{p_2}\} \subseteq \{1, 2, \ldots, l_p\}, \\ & \text{with } l_{p_1} + l_{p_2} = l_p; \ l_{p_2} = \textstyle\sum_{i=1}^{\beta} l_{p_{2i}}. \end{aligned} \tag{6.21}$$

(iv) If

$$\Delta_i = B_i^2 - 4C_i > 0 \text{ for } i = 1, 2, \ldots, m \tag{6.22}$$

the 1-dimensional, $(2m)^{\text{th}}$-degree polynomial system has $(2m)$ equilibriums as

$$x^* = b_1^{(i)} = -\frac{1}{2}(B_i + \sqrt{\Delta_i}), x^* = b_2^{(i)} = -\frac{1}{2}(B_i - \sqrt{\Delta_i})$$
$$\text{for } i = 1, 2, \ldots, m. \tag{6.23}$$

(iv_1) If

$$\begin{array}{l} b_r^{(i)} \neq b_s^{(j)} \text{ for } r,s \in \{1,2\}; i,j = 1,2, \ldots, m \\ \{a_1, a_2, \ldots, a_{2m}\} = \text{sort}\{b_1^{(1)}, b_2^{(1)}, \ldots, b_1^{(m)}, b_2^{(m)}\}, a_s < a_{s+1}, \end{array} \tag{6.24}$$

the corresponding standard form is

$$\dot{x} = a_0(x - a_1)(x - a_2)(x - a_3)(x - a_4)\cdots(x - a_{2m-1})(x - a_{2m}). \tag{6.25}$$

Such a flow is formed with all the simple equilibriums.

(a) If $a_0 > 0$, the simple equilibrium separatrix flow is called a (SO : SI : $\cdots$: SO : SI : $\cdots$SO : SI) -flow.
(b) If $a_0 < 0$, the simple equilibrium separatrix flow is called a (SI : SO : $\cdots$: SI : SO : $\cdots$SI : SO) -flow.

(iv_2) If

$$\begin{array}{l} \{a_1, a_2, \ldots, a_{2m}\} = \text{sort}\{b_1^{(1)}, b_2^{(1)}, \ldots, b_1^{(m)}, b_2^{(m)}\}, \\ a_{i_1} \equiv a_1 = \cdots = a_{l_1}, \\ a_{i_2} \equiv a_{l_1+1} = \cdots = a_{l_1+l_2}, \\ \vdots \\ a_{i_r} \equiv a_{\Sigma_{i=1}^{r-1} l_i + 1} = \cdots = a_{\Sigma_{i=1}^{r-1} l_i + l_r} = a_{2m} \\ \text{with } \Sigma_{s=1}^{r} l_s = 2m, \end{array} \tag{6.26}$$

then the corresponding standard form is

$$\dot{x} = a_0 \prod_{s=1}^{r} (x - a_{i_s})^{l_s}. \tag{6.27}$$

The equilibrium separatrix flow is called an (l_1th XX : l_2th XX :$\cdots$: l_rth XX)-flow. The equilibrium of $x^* = a_{i_p}$ for l_p-repeated equilibriums switching is called an l_pth XX bifurcation of $(l_{p_1}\text{th XX} : l_{p_2}\text{th XX} : \cdots : l_{p_\beta}\text{th XX})$ equilibrium switching at a point $\mathbf{p} = \mathbf{p}_1 \in \partial\Omega_{12}$, and the bifurcation condition is

$$\begin{aligned} &a_{i_p} \equiv a_{\Sigma_{i=1}^{p-1} l_i+1} = \cdots = a_{\Sigma_{i=1}^{p-1} l_i+l_p}, \\ &a^{\pm}_{\Sigma_{i=1}^{p-1} l_i+1} \neq \cdots \neq a^{\pm}_{\Sigma_{i=1}^{p-1} l_i+l_p};\ l_p = \sum\nolimits_{i=1}^{\beta} l_{p_i}. \end{aligned} \tag{6.28}$$

Definition 6.2 Consider a 1-dimensional, $(2m)^{\text{th}}$-degree polynomial nonlinear dynamical system as

$$\begin{aligned} \dot{x} &= A_0(\mathbf{p})x^{2m} + A_1(\mathbf{p})x^{2m-1} + \cdots + A_{2m-2}(\mathbf{p})x^2 + A_{2m-1}x + A_{2m}(\mathbf{p}) \\ &= a_0(\mathbf{p})\prod_{i=1}^{n}[x^2 + B_i(\mathbf{p})x + C_i(\mathbf{p})]^{q_i} \end{aligned} \tag{6.29}$$

where $A_0(\mathbf{p}) \neq 0$, and

$$\mathbf{p} = (p_1, p_2, \ldots, p_{m_1})^{\mathrm{T}}, m = \sum\nolimits_{i=1}^{n} q_i. \tag{6.30}$$

(i) If

$$\Delta_i = B_i^2 - 4C_i < 0 \text{ for } i = 1, 2, \ldots, n \tag{6.31}$$

the 1-dimensional nonlinear dynamical system with a $(2m)^{\text{th}}$-degree polynomial does not have any equilibriums, and the corresponding standard form is

$$\dot{x} = a_0 \prod_{i=1}^{n} [(x + \frac{1}{2}B_i)^2 + \frac{1}{4}(-\Delta_i)]^{q_i}. \tag{6.32}$$

The flow of such a system without equilibriums is called a nonequilibrium flow.

(a) If $a_0 > 0$, the nonequilibrium flow is called the positive flow.
(b) If $a_0 < 0$, the nonequilibrium flow is called the negative flow.

(ii) If

$$\begin{aligned} &\Delta_i = B_i^2 - 4C_i > 0, i \in \{i_1, i_2, \ldots, i_l\} \subseteq \{1, 2, \ldots, n\}, \\ &\Delta_j = B_j^2 - 4C_j < 0, j \in \{i_{l+1}, i_{l+2}, \ldots, i_n\} \subseteq \{1, 2, \ldots, n\} \end{aligned} \tag{6.33}$$

the 1-dimensional, $(2m)^{\text{th}}$-degree polynomial system has $2l$-equilibriums as

$$\begin{aligned} & x^* = b_1^{(i)} = -\frac{1}{2}(B_i + \sqrt{\Delta_i}), x^* = b_2^{(i)} = -\frac{1}{2}(B_i - \sqrt{\Delta_i}) \\ & i \in \{i_1, i_2, \ldots, i_l\} \subseteq \{1, 2, \ldots, n\}. \end{aligned} \tag{6.34}$$

(ii$_1$) If

$$\begin{aligned} & b_r^{(i)} \neq b_s^{(j)} \text{ for } r,s \in \{1,2\}; i,j = 1,2, \ldots, l \\ & \{a_1, a_2, \ldots, a_{2l}\} = \text{sort}\{b_1^{(1)}, b_2^{(1)}, \ldots, b_1^{(l)}, b_2^{(l)}\}, a_s < a_{s+1}, \end{aligned} \tag{6.35}$$

then the corresponding standard form is

$$\begin{aligned} & \dot{x} = a_0 \prod_{s=1}^{2l} (x - a_s)^{l_s} \prod_{k=l+1}^{n} [(x + \frac{1}{2}B_{i_k})^2 + \frac{1}{4}(-\Delta_{i_k})]^{q_{i_k}} \\ & \text{with } l_s \in \{q_{i_1}, q_{i_2}, \ldots, q_{i_l}\}. \end{aligned} \tag{6.36}$$

The equilibrium separatrix flow is called an (l_1th XX : l_2th XX : $\cdots$: l_{2l}th XX) -flow.

(a) For $a_0 > 0$ and $p = 1, 2, \ldots, 2l$,

$$l_p\text{th XX} = \begin{cases} (2r_p - 1)^{\text{th}}\text{order source}, & \text{for } \alpha_p = 2M_p - 1, l_p = 2r_p - 1; \\ (2r_p - 1)^{\text{th}}\text{order sink}, & \text{for } \alpha_p = 2M_p, l_p = 2r_p - 1; \\ (2r_p)^{\text{th}}\text{order lower-saddle}, & \text{for } \alpha_p = 2M_p - 1, l_p = 2r_p; \\ (2r_p)^{\text{th}}\text{order upper-saddle}, & \text{for } \alpha_p = 2M_p, l_p = 2r_p; \end{cases}, \tag{6.37}$$

where

$$\alpha_p = \sum\nolimits_{s=p}^{2l} l_s. \tag{6.38}$$

(b) For $a_0 < 0$ and $p = 1, 2, \ldots, 2l$,

$$l_p\text{th XX} = \begin{cases} (2r_p - 1)^{\text{th}}\text{order sink}, & \text{for } \alpha_p = 2M_p - 1, l_p = 2r_p - 1; \\ (2r_p - 1)^{\text{th}}\text{order source}, & \text{for } \alpha_p = 2M_p, l_p = 2r_p - 1; \\ (2r_p)^{\text{th}}\text{order upper-saddle}, & \text{for } \alpha_p = 2M_p - 1, l_p = 2r_p; \\ (2r_p)^{\text{th}}\text{order lower-saddle}, & \text{for } \alpha_p = 2M_p, l_p = 2r_p. \end{cases} \tag{6.39}$$

(ii$_2$) If

$$\begin{aligned}&\{a_1, a_2, \ldots, a_{2l}\} = \text{sort}\{b_1^{(1)}, b_2^{(1)}, \ldots, b_1^{(l)}, b_2^{(l)}\},\\ &a_{i_1} \equiv a_1 = \cdots = a_{l_1},\\ &a_{i_2} \equiv a_{l_1+1} = \cdots = a_{l_1+l_2},\\ &\vdots\\ &a_{i_r} \equiv a_{\Sigma_{i=1}^{r-1} l_i+1} = \cdots = a_{\Sigma_{i=1}^{r-1} l_i + l_r} = a_{2l}\\ &\text{with } \Sigma_{s=1}^{r} l_s = 2l,\end{aligned} \tag{6.40}$$

then the corresponding standard form is

$$\dot{x} = a_0 \prod_{s=1}^{r} (x - a_{i_s})^{l_s} \prod_{k=l+1}^{n} [(x + \frac{1}{2}B_{i_k})^2 + \frac{1}{4}(-\Delta_{i_k})]^{q_i}. \tag{6.41}$$

The equilibrium separatrix flow is called an (l_1th XX: l_2th XX: $\cdots$: l_rth XX)-flow.

(a) For $a_0 > 0$ and $s = 1, 2, \ldots, r$,

$$l_p\text{th XX} = \begin{cases} (2r_p - 1)^{\text{th}}\text{order source}, & \text{for } \alpha_p = 2M_p - 1, l_p = 2r_p - 1;\\ (2r_p - 1)^{\text{th}}\text{order sink}, & \text{for } \alpha_p = 2M_p, l_p = 2r_p - 1;\\ (2r_p)^{\text{th}}\text{order lower-saddle}, & \text{for } \alpha_p = 2M_p - 1, l_p = 2r_p;\\ (2r_p)^{\text{th}}\text{order upper-saddle}, & \text{for } \alpha_p = 2M_p, l_p = 2r_p; \end{cases} \tag{6.42}$$

where

$$\alpha_p = \sum\nolimits_{s=p}^{r} l_s. \tag{6.43}$$

(b) For $a_0 < 0$ and $p = 1, 2, \ldots, r$,

$$l_p\text{th XX} = \begin{cases} (2r_p - 1)^{\text{th}}\text{order sink}, & \text{for } \alpha_p = 2M_p - 1, l_p = 2r_p - 1;\\ (2r_p - 1)^{\text{th}}\text{order source}, & \text{for } \alpha_p = 2M_p, l_p = 2r_p - 1;\\ (2r_p)^{\text{th}}\text{order upper-saddle}, & \text{for } \alpha_p = 2M_p - 1, l_p = 2r_p;\\ (2r_p)^{\text{th}}\text{order lower-saddle}, & \text{for } \alpha_p = 2M_p, l_p = 2r_p. \end{cases} \tag{6.44}$$

(c) The equilibrium of $x^* = a_{i_p}$ for ($l_p > 1$)-repeated equilibriums switching is called an l_pth XX bifurcation of (l_{p_1}th XX : l_{p_2}th XX : $\cdots$: l_{p_β}th XX) equilibrium switching at a point $\mathbf{p} = \mathbf{p}_1 \in \partial\Omega_{12}$, and the bifurcation condition is

$$\begin{aligned} & a_{i_p} \equiv a_{\Sigma_{i=1}^{p-1} l_i+1} = \cdots = a_{\Sigma_{i=1}^{p-1} l_i+l_p}, \\ & a^{\pm}_{\Sigma_{i=1}^{p-1} l_i+1} \neq \cdots \neq a^{\pm}_{\Sigma_{i=1}^{p-1} l_i+l_p};\ l_p = \sum\nolimits_{i=1}^{\beta} l_{p_i}. \end{aligned} \tag{6.45}$$

(iii) If

$$\begin{aligned} & \Delta_i = B_i^2 - 4C_i = 0, i \in \{i_{11}, i_{12}, \ldots, i_{1s}\} \subseteq \{i_1, i_2, \ldots, i_l\} \subseteq \{1, 2, \ldots, n\}, \\ & \Delta_k = B_k^2 - 4C_k > 0, k \in \{i_{21}, i_{22}, \ldots, i_{2r}\} \subseteq \{i_1, i_2, \ldots, i_l\} \subseteq \{1, 2, \ldots, n\}, \\ & \Delta_j = B_j^2 - 4C_j < 0,\ j \in \{i_{l+1}, i_{l+2}, \ldots, i_n\} \subseteq \{1, 2, \ldots, n\} \text{ with } i \neq j \neq k, \end{aligned} \tag{6.46}$$

the 1-dimensional, $(2m)^{\text{th}}$-degree polynomial system has $2l$-equilibriums as

$$\begin{aligned} & x^* = b_1^{(i)} = -\frac{1}{2}B_i, x^* = b_2^{(i)} = -\frac{1}{2}B_i \text{ for } i \in \{i_{11}, i_{12}, \ldots, i_{1s}\}, \\ & x^* = b_1^{(k)} = -\frac{1}{2}(B_k + \sqrt{\Delta_k}), x^* = b_2^{(k)} = -\frac{1}{2}(B_k - \sqrt{\Delta_k}) \\ & \text{for } k \in \{i_{21}, i_{22}, \ldots, i_{2r}\}. \end{aligned} \tag{6.47}$$

If

$$\begin{aligned} & \{a_1, a_2, \ldots, a_{2l}\} = \text{sort}\{b_1^{(1)}, b_2^{(1)}, \ldots, b_1^{(l)}, b_2^{(l)}\}, \\ & a_{i_1} \equiv a_1 = \cdots = a_{l_1}, \\ & a_{i_2} \equiv a_{l_1+1} = \cdots = a_{l_1+l_2}, \\ & \vdots \\ & a_{i_r} \equiv a_{\Sigma_{i=1}^{r-1} l_i+1} = \cdots = a_{\Sigma_{i=1}^{r-1} l_i+l_r} = a_{2l} \\ & \text{with } \Sigma_{s=1}^{r} l_s = 2l, \end{aligned} \tag{6.48}$$

then the corresponding standard form is

$$\dot{x} = a_0 \prod_{s=1}^{r} (x - a_{i_s})^{l_s} \prod_{k=l+1}^{n} [(x + \frac{1}{2}B_{i_k})^2 + \frac{1}{4}(-\Delta_{i_k})]^{q_i}. \tag{6.49}$$

The equilibrium separatrix flow is called an (l_1th XX : l_2th XX : $\cdots$: l_rth XX) -flow.

(a) The equilibrium of $x^* = a_{i_p}$ for ($l_p > 1$)-repeated equilibriums appearing (or vanishing) is called an l_pth XX bifurcation of equilibrium at a point $\mathbf{p} = \mathbf{p}_1 \in \partial\Omega_{12}$, and the bifurcation condition is

$$\begin{aligned}
&a_{i_p} \equiv a_{\Sigma_{i=1}^{p-1} l_i+1} = \cdots = a_{\Sigma_{i=1}^{p-1} l_i+l_p} = -\frac{1}{2}B_{i_q} \\
&\text{with } \Delta_{i_q} = B_{i_q}^2 - 4C_{i_q} = 0 \ (i_q \in \{i_1, i_2, \cdots, i_l\}), \\
&a^{+}_{\Sigma_{i=1}^{p-1} l_i+1} \neq \cdots \neq a^{+}_{\Sigma_{i=1}^{p-1} l_i+l_p} \text{ or } a^{-}_{\Sigma_{i=1}^{p-1} l_i+1} \neq \cdots \neq a^{-}_{\Sigma_{i=1}^{p-1} l_i+l_p}.
\end{aligned} \tag{6.50}$$

(b) The equilibrium of $x^* = a_{i_q}$ for ($l_q > 1$)-repeated equilibriums switching is called an l_qth XX bifurcation of (l_{q_1}th XX : l_{q_2}th XX : $\cdots$: l_{q_β}th XX)-equilibrium switching at a point $\mathbf{p} = \mathbf{p}_1 \in \partial\Omega_{12}$, and the bifurcation condition is

$$\begin{aligned}
&a_{i_q} \equiv a_{\Sigma_{i=1}^{q-1} l_i+1} = \cdots = a_{\Sigma_{i=1}^{q-1} l_i+l_q}, \\
&a^{\pm}_{\Sigma_{i=1}^{q-1} l_i+1} \neq \cdots \neq a^{\pm}_{\Sigma_{i=1}^{q-1} l_i+l_q};\ l_q = \sum\nolimits_{i=1}^{\beta} l_{q_i}.
\end{aligned} \tag{6.51}$$

(iv) If

$$\Delta_i = B_i^2 - 4C_i > 0 \ \text{ for } i = 1, 2, \ldots, n \tag{6.52}$$

the 1-dimensional, $(2m)^{\text{th}}$-degree polynomial system has $(2n)$-equilibriums as

$$\begin{aligned}
&x^* = b_1^{(i)} = -\frac{1}{2}(B_i + \sqrt{\Delta_i}), x^* = b_2^{(i)} = -\frac{1}{2}(B_i - \sqrt{\Delta_i}) \\
&\text{for } i = 1, 2, \ldots, n.
\end{aligned} \tag{6.53}$$

(iv_1) If

$$\begin{aligned}
&b_r^{(i)} \neq b_s^{(j)} \text{ for } r,s \in \{1,2\}; i,j = 1,2, \ldots, n \\
&\{a_1, a_2, \ldots, a_{2n}\} = \text{sort}\{b_1^{(1)}, b_2^{(1)}, \ldots, b_1^{(n)}, b_2^{(n)}\}, a_s < a_{s+1},
\end{aligned} \tag{6.54}$$

the corresponding standard form is

$$\dot{x} = a_0 \prod_{s=1}^{2n} (x - a_s)^{l_s} \text{ with } l_s \in \{q_{i_1}, q_{i_2}, \ldots, q_{i_n}\}. \tag{6.55}$$

The equilibrium separatrix flow is called an (l_1th XX : l_2th XX : $\cdots$: l_{2n}th XX)-flow.

(a) For $a_0 > 0$ and $p = 1, 2, \ldots, 2n$,

$$l_p\text{th XX} = \begin{cases} (2r_p - 1)^{\text{th}}\text{order source, for } \alpha_p = 2M_p - 1, l_p = 2r_p - 1; \\ (2r_p - 1)^{\text{th}}\text{order sink, for } \alpha_p = 2M_p, l_p = 2r_p - 1; \\ (2r_p)^{\text{th}}\text{order lower-saddle, for } \alpha_p = 2M_p - 1, l_p = 2r_p; \\ (2r_p)^{\text{th}}\text{order upper-saddle, for } \alpha_p = 2M_p, l_p = 2r_p; \end{cases} \tag{6.56}$$

where

$$\alpha_p = \sum\nolimits_{s=p}^{2n} l_s. \tag{6.57}$$

(b) For $a_0 < 0$ and $p = 1, 2, \ldots, 2n$,

$$l_p\text{th XX} = \begin{cases} (2r_p - 1)^{\text{th}}\text{order sink, for } \alpha_p = 2M_p - 1, l_p = 2r_p - 1; \\ (2r_p - 1)^{\text{th}}\text{order source, for } \alpha_p = 2M_p, l_p = 2r_p - 1; \\ (2r_p)^{\text{th}}\text{order upper-saddle, for } \alpha_p = 2M_p - 1, l_p = 2r_p; \\ (2r_p)^{\text{th}}\text{order lower-saddle, for } \alpha_p = 2M_p, l_p = 2r_p. \end{cases} \tag{6.58}$$

(iv_2) If

$$\begin{aligned} &\{a_1, a_2, \ldots, a_{2n}\} = \text{sort}\{b_1^{(1)}, b_2^{(1)}, \ldots, b_1^{(n)}, b_2^{(n)}\}, \\ &a_{i_1} \equiv a_1 = \cdots = a_{l_1}, \\ &a_{i_2} \equiv a_{l_1+1} = \cdots = a_{l_1+l_2}, \\ &\vdots \\ &a_{i_r} \equiv a_{\Sigma_{i=1}^{r-1} l_i+1} = \cdots = a_{\Sigma_{i=1}^{r-1} l_i+l_r} = a_{2n}, \\ &\text{with } \Sigma_{s=1}^{r} l_s = 2n, \end{aligned} \tag{6.59}$$

then the corresponding standard form is

$$\dot{x} = a_0 \prod_{s=1}^{r} (x - a_{i_s})^{l_s}. \tag{6.60}$$

The equilibrium separatrix flow is called an (l_1th XX: l_2th XX:$\cdots$: l_rth XX)-flow. The equilibrium of $x^* = a_{i_p}$ for l_p-repeated equilibriums switching is called a l_pth XX bifurcation of (l_{p_1}th XX : l_{p_2}th XX : $\cdots$: l_{p_β}th XX) equilibrium switching at a point $\mathbf{p} = \mathbf{p}_1 \in \partial\Omega_{12}$, and the bifurcation condition is

$$\begin{aligned} &a_{i_p} \equiv a_{\Sigma_{i=1}^{p-1} l_i+1} = \cdots = a_{\Sigma_{i=1}^{p-1} l_i+l_p}, \\ &a^{\pm}_{\Sigma_{i=1}^{p-1} l_i+1} \neq \cdots \neq a^{\pm}_{\Sigma_{i=1}^{p-1} l_i+l_p};\ l_p = \sum\nolimits_{i=1}^{\beta} l_{p_i}. \end{aligned} \tag{6.61}$$

Definition 6.3 Consider a 1-dimensional, $(2m)^{\text{th}}$-degree polynomial nonlinear dynamical system

$$\begin{aligned} \dot{x} &= A_0(\mathbf{p})x^{2m} + A_1(\mathbf{p})x^{2m-1} + \cdots + A_{2m-2}(\mathbf{p})x^2 + A_{2m-1}x + A_{2m}(\mathbf{p}) \\ &= a_0(\mathbf{p}) \prod_{s=1}^{r} (x - c_{i_s}(\mathbf{p}))^{l_s} \prod_{i=r+1}^{n} \left[x^2 + B_i(\mathbf{p})x + C_i(\mathbf{p})\right]^{q_i} \end{aligned} \tag{6.62}$$

where $A_0(\mathbf{p}) \neq 0$, and

$$\sum\nolimits_{s=1}^{r} l_s = 2l, \sum\nolimits_{i=r+1}^{n} q_i = (m-l), \mathbf{p} = (p_1, p_2, \ldots, p_{m_1})^{\mathrm{T}}. \tag{6.63}$$

(i) If

$$\begin{aligned} &\Delta_i = B_i^2 - 4C_i < 0 \ \text{ for } i = r+1, r+2, \ldots, n, \\ &\{a_1, a_2, \ldots, a_r\} = \text{sort}\{c_1, c_2, \ldots, c_r\} \text{ with } a_i < a_{i+1} \end{aligned} \tag{6.64}$$

the 1-dimensional nonlinear dynamical system with a $(2m)^{\text{th}}$-degree polynomial has equilibriums of $x^* = a_{i_s}(\mathbf{p})$ ($s = 1, 2, \ldots, r$), and the corresponding standard form is

$$\dot{x} = a_0(\mathbf{p}) \prod_{s=1}^{r} (x - a_{i_s})^{l_s} \prod_{i=r+1}^{n} [(x + \frac{1}{2}B_i)^2 + \frac{1}{4}(-\Delta_i)]^{l_i}. \tag{6.65}$$

The equilibrium separatrix flow is called an (l_1th XX : l_2th XX : $\cdots$: l_rth XX)-flow.

(a) For $a_0 > 0$ and $s = 1, 2, \ldots, r$,

$$l_p\text{th XX} = \begin{cases} (2r_p - 1)^{\text{th}}\text{order source, for } \alpha_p = 2M_p - 1, l_p = 2r_p - 1; \\ (2r_p - 1)^{\text{th}}\text{order sink, for } \alpha_p = 2M_p, l_p = 2r_p - 1; \\ (2r_p)^{\text{th}}\text{order lower-saddle, for } \alpha_p = 2M_p - 1, l_p = 2r_p; \\ (2r_p)^{\text{th}}\text{order upper-saddle, for } \alpha_p = 2M_p, l_p = 2r_p; \end{cases} \tag{6.66}$$

where

$$\alpha_p = \sum\nolimits_{s=p}^{r} l_s. \tag{6.67}$$

(b) For $a_0 < 0$ and $p = 1, 2, \ldots, r$,

$$l_p\text{th XX} = \begin{cases} (2r_p - 1)^{\text{th}}\text{order sink, for } \alpha_p = 2M_p - 1, l_p = 2r_p - 1; \\ (2r_p - 1)^{\text{th}}\text{order source, for } \alpha_p = 2M_p, l_p = 2r_p - 1; \\ (2r_p)^{\text{th}}\text{order upper-saddle, for } \alpha_p = 2M_p - 1, l_p = 2r_p; \\ (2r_p)^{\text{th}}\text{order lower-saddle, for } \alpha_p = 2M_p, l_p = 2r_p. \end{cases} \tag{6.68}$$

(ii) If

$$\begin{aligned} &\Delta_i = B_i^2 - 4C_i > 0, i = j_1, j_2, \ldots, j_s \in \{l+1, l+2, \ldots, n\}, \\ &\Delta_j = B_j^2 - 4C_j < 0, j = j_{s+1}, j_{s+2}, \ldots, j_n \in \{l+1, l+2, \ldots, n\} \\ &\text{with } s \in \{1, \ldots, n-l\}, \end{aligned} \tag{6.69}$$

the 1-dimensional, $(2m)^{\text{th}}$-degree polynomial system has $2n_2$-equilibriums as

$$\begin{aligned} &x^* = b_1^{(i)} = -\frac{1}{2}(B_i + \sqrt{\Delta_i}), x^* = b_2^{(i)} = -\frac{1}{2}(B_i - \sqrt{\Delta_i}) \\ &i \in \{j_1, j_2, \cdots, j_{n_1}\} \subseteq \{l+1, l+2, \ldots, n\}. \end{aligned} \tag{6.70}$$

If

$$\{a_1, a_2, \ldots, a_{2n_2}\} = \text{sort}\{c_1, c_2, \ldots, c_{2l}, \underbrace{b_1^{(r+1)}, b_2^{(r+1)}}_{q_{r+1}\text{ sets}}, \ldots, \underbrace{b_1^{(n_1)}, b_2^{(n_1)}}_{q_{n_1}\text{ sets}}\},$$

$$\begin{aligned} &a_{i_1} \equiv a_1 = \cdots = a_{l_1}, \\ &a_{i_2} \equiv a_{l_1+1} = \cdots = a_{l_1+l_2}, \\ &\vdots \\ &a_{i_{n_1}} \equiv a_{\Sigma_{i=1}^{n_1-1} l_i+1} = \cdots = a_{\Sigma_{i=1}^{n_1-1} l_i + l_{n_1}} = a_{2n_2} \\ &\text{with } \Sigma_{s=1}^{n_1} l_s = 2n_2, \end{aligned} \tag{6.71}$$

then the corresponding standard form is

$$\dot{x} = a_0 \prod_{s=1}^{n_1} (x - a_{i_s})^{l_s} \prod_{i=n_2+1}^{n} [(x + \frac{1}{2}B_i)^2 + \frac{1}{4}(-\Delta_i)]^{q_i}. \tag{6.72}$$

The equilibrium separatrix flow is called an $(l_1\text{th XX} : l_2\text{thXX} : \cdots : l_{n_1}\text{th XX})$-flow.

(a) For $a_0 > 0$ and $p = 1, 2, \ldots, r, r+1, \ldots, n_1$,

$$l_p\text{th XX} = \begin{cases} (2r_p - 1)^{\text{th}}\text{order source, for } \alpha_p = 2M_p - 1, l_p = 2r_p - 1; \\ (2r_p - 1)^{\text{th}}\text{order sink, for } \alpha_p = 2M_p, l_p = 2r_p - 1; \\ (2r_p)^{\text{th}}\text{order lower-saddle, for } \alpha_p = 2M_p - 1, l_p = 2r_p; \\ (2r_p)^{\text{th}}\text{order upper-saddle, for } \alpha_p = 2M_p, l_p = 2r_p; \end{cases} \tag{6.73}$$

where

$$\alpha_p = \sum\nolimits_{s=p}^{n_1} l_s. \tag{6.74}$$

(b) For $a_0 < 0$ and $p = 1, 2, \ldots, r, r+1, \ldots, n_1$,

$$l_p\text{th XX} = \begin{cases} (2r_p - 1)^{\text{th}}\text{order sink, for } \alpha_p = 2M_p - 1, l_p = 2r_p - 1; \\ (2r_p - 1)^{\text{th}}\text{order source, for } \alpha_p = 2M_p, l_p = 2r_p - 1; \\ (2r_p)^{\text{th}}\text{order upper-saddle, for } \alpha_p = 2M_p - 1, l_p = 2r_p; \\ (2r_p)^{\text{th}}\text{order lower-saddle, for } \alpha_p = 2M_p, l_p = 2r_p. \end{cases} \tag{6.75}$$

(c) The equilibrium of $x^* = a_{i_p}$ for $(l_p > 1)$-repeated equilibriums switching is called an l_pth XX switching bifurcation of $(l_{p_1}\text{th XX} : l_{p_2}\text{th XX} : \cdots : l_{p_\beta}\text{th XX})$ equilibrium at a point $\mathbf{p} = \mathbf{p}_1 \in \partial\Omega_{12}$, and the bifurcation condition is

$$a_{i_p} \equiv a_{\Sigma_{i=1}^{p-1} l_i + 1} = \cdots = a_{\Sigma_{i=1}^{p-1} l_i + l_p}, a^{\pm}_{\Sigma_{i=1}^{p-1} l_i + 1} \neq \cdots \neq a^{\pm}_{\Sigma_{i=1}^{p-1} l_i + l_p}. \tag{6.76}$$

(iii) If

$$\begin{aligned} &\Delta_i = B_i^2 - 4C_i = 0, \\ &\text{for } i \in \{i_{11}, i_{12}, \ldots, i_{1s}\} \subseteq \{i_{l+1}, i_{l+2}, \ldots, i_{n_2}\} \subseteq \{l+1, l+2, \ldots, n\}, \\ &\Delta_k = B_k^2 - 4C_k > 0, \\ &\text{for } k \in \{i_{21}, i_{22}, \ldots, i_{2r}\} \subseteq \{i_{l+1}, i_{l+2}, \ldots, i_{n_2}\} \subseteq \{l+1, l+2, \ldots, n\}, \\ &\Delta_j = B_j^2 - 4C_j < 0, \\ &\text{for } j \in \{i_{n_2+1}, i_{n_2+2}, \ldots, i_n\} \subseteq \{l+1, l+2, \ldots, n\}, \end{aligned} \tag{6.77}$$

the 1-dimensional, $(2m)^{\text{th}}$-degree polynomial system has $2n_2$-equilibriums as

$$
\begin{aligned}
&x^* = b_1^{(i)} = -\frac{1}{2}B_i, x^* = b_2^{(i)} = -\frac{1}{2}B_i \text{ for } i \in \{i_{11}, i_{12}, \ldots, i_{1s}\},\\
&x^* = b_1^{(k)} = -\frac{1}{2}(B_k + \sqrt{\Delta_k}), x^* = b_2^{(k)} = -\frac{1}{2}(B_k - \sqrt{\Delta_k})\\
&\text{for } i \in \{i_{21}, i_{22}, \ldots, i_{2r}\}.
\end{aligned}
\tag{6.78}
$$

If

$$
\begin{aligned}
&\{a_1, a_2, \ldots, a_{2n_2}\} = \text{sort}\{c_1, c_2, \ldots, c_{2l}, \underbrace{b_1^{(r+1)}, b_2^{(r+1)}}_{q_{r+1}\text{ sets}}, \ldots, \underbrace{b_1^{(n_1)}, b_2^{(n_1)}}_{q_{n_1}\text{ sets}}\},\\
&a_{i_1} \equiv a_1 = \cdots = a_{l_1},\\
&a_{i_2} \equiv a_{l_1+1} = \cdots = a_{l_1+l_2},\\
&\vdots\\
&a_{i_{n_1}} \equiv a_{\Sigma_{i=1}^{n_1-1} l_i+1} = \cdots = a_{\Sigma_{i=1}^{n_1-1} l_i+l_{n_1}} = a_{2n_2}\\
&\text{with } \Sigma_{s=1}^{n_1} l_s = 2n_2,
\end{aligned}
\tag{6.79}
$$

then the corresponding standard form is

$$
\dot{x} = a_0 \prod_{s=1}^{n_1} (x - a_{i_s})^{l_s} \prod_{i=n_2+1}^{n} [(x + \frac{1}{2}B_i)^2 + \frac{1}{4}(-\Delta_i)]^{q_i}. \tag{6.80}
$$

The equilibrium separatrix flow is called an l_1th XX : l_2th XX : $\cdots$: l_{n_1}th XX)-flow.

(a) The equilibrium of $x^* = a_{i_p}$ for $(q_p > 1)$-repeated equilibriums appearing or vanishing is called an l_pth XX bifurcation of equilibrium at a point $\mathbf{p} = \mathbf{p}_1 \in \partial\Omega_{12}$, and the bifurcation condition is

$$
\begin{aligned}
&a_{i_p} \equiv a_{\Sigma_{i=1}^{p-1} l_i+1} = \cdots = a_{\Sigma_{i=1}^{p-1} l_i+l_p} = -\frac{1}{2}B_{i_q}\\
&\text{with } \Delta_{i_q} = B_{i_q}^2 - 4C_{i_q} = 0 \; (i_q \in \{i_1, i_2, \ldots, i_l\})\\
&a^+_{\Sigma_{i=1}^{p-1} q_i+1} \neq \cdots \neq a^+_{\Sigma_{i=1}^{p-1} q_i+q_p} \text{ or } a^-_{\Sigma_{i=1}^{p-1} q_i+1} \neq \cdots \neq a^-_{\Sigma_{i=1}^{p-1} q_i+q_p}.
\end{aligned}
\tag{6.81}
$$

(b) The equilibrium of $x^* = a_{i_q}$ for $(l_p > 1)$-repeated equilibriums switching is called an l_pth XX bifurcation of $(l_{p_1}$th XX : l_{p_2}th XX : $\cdots$: l_{p_β}th XX) equilibrium switching at a point $\mathbf{p} = \mathbf{p}_1 \in \partial\Omega_{12}$, and the bifurcation condition is

$$a_{i_q} \equiv a_{\Sigma_{i=1}^{q-1} l_i+1} = \cdots = a_{\Sigma_{i=1}^{q-1} l_i+l_p}, a^{\pm}_{\Sigma_{i=1}^{q-1} l_i+1} \neq \cdots \neq a^{\pm}_{\Sigma_{i=1}^{q-1} l_i+l_q}, l_p = \sum\nolimits_{i=1}^{\beta} l_{p_i}. \quad (6.82)$$

(c) The equilibrium of $x^* = a_{i_p}$ for $(l_{p_1} \geq 1)$-repeated equilibriums appearing (or vanishing) and $(l_{p_2} \geq 2)$-repeated equilibriums switching of ($l_{p_{21}}$th XX : $l_{p_{22}}$th XX : $\cdots$: $l_{p_{2\beta}}$th XX) is called an q_pth XX bifurcation of equilibrium at a point $\mathbf{p} = \mathbf{p}_1 \in \partial\Omega_{12}$, and the bifurcation condition is

$$\begin{aligned}
&a_{i_p} \equiv a_{\Sigma_{i=1}^{p-1} q_i+1} = \cdots = a_{\Sigma_{i=1}^{p-1} q_i+q_p} \\
&\text{with } \Delta_{i_q} = B_{i_q}^2 - 4C_{i_q} = 0 \ (i_q \in \{i_1, i_2, \cdots, i_l\}) \\
&a^{+}_{\Sigma_{i=1}^{p-1} q_i+j_1} \neq \cdots \neq a^{+}_{\Sigma_{i=1}^{p-1} q_i+j_{p_1}} \text{ or } a^{-}_{\Sigma_{i=1}^{p_1-1} q_i+j_1} \neq \cdots \neq a^{-}_{\Sigma_{i=1}^{p_1-1} q_i+j_{p_1}}, \\
&\text{for } \{j_1, j_2, \ldots, j_{p_1}\} \subseteq \{1, 2, \ldots, q_p\}, \\
&a^{\pm}_{\Sigma_{i=1}^{p-1} q_i+k_1} \neq \cdots \neq a^{\pm}_{\Sigma_{i=1}^{p-1} q_i+k_{p_2}} \\
&\text{for } \{k_1, k_2, \ldots, k_{p_2}\} \subseteq \{1, 2, \ldots, q_p\}, \\
&\text{with } l_{p_1} + l_{p_2} = l_p.
\end{aligned} \quad (6.83)$$

(iv) If

$$\Delta_i = B_i^2 - 4C_i > 0 \ \text{ for } i = l+1, l+2, \ldots, n \quad (6.84)$$

the 1-dimensional, $(2m)^{\text{th}}$-degree polynomial system has $(2m)$-equilibriums as

$$\begin{gathered}
x^* = b_1^{(i)} = -\frac{1}{2}(B_i + \sqrt{\Delta_i}), x^* = b_2^{(i)} = -\frac{1}{2}(B_i - \sqrt{\Delta_i}) \\
\text{for } i = l+1, l+2, \ldots, n.
\end{gathered} \quad (6.85)$$

If

$$\begin{aligned}
&\{a_1, a_2, \ldots, a_{2m}\} = \text{sort}\{c_1, c_2, \ldots, c_{2l}, \underbrace{b_1^{(r+1)}, b_2^{(r+1)}}_{q_{r+1} \text{ sets}}, \ldots, \underbrace{b_1^{(n)}, b_2^{(n)}}_{q_n \text{sets}}\}, \\
&a_{i_1} \equiv a_1 = \cdots = a_{l_1}, \\
&a_{i_2} \equiv a_{l_1+1} = \cdots = a_{l_1+l_2}, \\
&\vdots \\
&a_{i_n} \equiv a_{\Sigma_{i=1}^{n-1} l_i+1} = \cdots = a_{\Sigma_{i=1}^{n-1} l_i+l_r} = a_{2m} \\
&\text{with } \Sigma_{s=1}^{n} l_s = 2m,
\end{aligned} \quad (6.86)$$

then the corresponding standard form is

$$\dot{x} = a_0 \prod_{s=1}^{n} (x - a_{i_s})^{l_s}. \tag{6.87}$$

The equilibrium separatrix flow is called an (l_1th XX : l_2th XX : $\cdots$: l_nth XX)-flow. The equilibrium of $x^* = a_{i_p}$ for l_p- repeated equilibriums switching is called an l_pth XX switching bifurcation of (l_{p_1}th XX : l_{p_2}thXX : $\cdots$: l_{p_β}th XX) equilibrium at a point $\mathbf{p} = \mathbf{p}_1 \in \partial\Omega_{12}$, and the bifurcation condition is

$$\begin{aligned} &a_{i_p} \equiv a_{\Sigma_{i=1}^{p-1} l_i+1} = \cdots = a_{\Sigma_{i=1}^{p-1} l_i+l_p}, \\ &a^{\pm}_{\Sigma_{i=1}^{p-1} l_i+1} \neq \cdots \neq a^{\pm}_{\Sigma_{i=1}^{p-1} l_i+l_p};\ l_p = \sum\nolimits_{i=1}^{\beta} l_{p_i}. \end{aligned} \tag{6.88}$$

6.2 Simple Equilibrium Bifurcations

From the global analysis in Chap. 4, in this section, the bifurcations of simple equilibriums in the $(2m)^{\text{th}}$-degree polynomial systems are discussed, which include appearing/vanishing bifurcations, switching bifurcations, and switching and appearing bifurcations as in Luo (2020).

6.2.1 *Appearing Bifurcation*

Consider a $(2m)^{\text{th}}$-degree polynomial system in a form of

$$\dot{x} = a_0 Q(x) \prod_{i=1}^{n} \left(x^2 + B_i x + C_i\right). \tag{6.89}$$

Without loss of generality, a function of $Q(x) > 0$ is either a polynomial function or a non-polynomial function. The roots of $x^2 + B_i x + C_i = 0$ are

$$\begin{aligned} &b_{1,2}^{(i)} = -\frac{1}{2}B_i \pm \frac{1}{2}\sqrt{\Delta_i}, \Delta_i = B_i^2 - 4C_i \geq 0\ (i = 1, 2, \cdots, n); \\ &\{a_1, a_2, \cdots, a_{2l}\} \subset \text{sort}\{b_1^{(1)}, b_2^{(1)}, b_1^{(2)}, b_2^{(2)}, \cdots, b_1^{(n)}, b_2^{(n)}\}, a_s \leq a_{s+1}; \\ &\left.\begin{aligned} &B_i \neq B_j\ (i, j = 1, 2, \cdots, n; i \neq j) \\ &\Delta_i = 0\ (i = 1, 2, \cdots, n) \end{aligned}\right\} \text{at bifurcation.} \end{aligned} \tag{6.90}$$

The second-order singularity bifurcation is for the birth of a pair of simple sink and source. There are two *appearing* bifurcations for $i \in \{1, 2, \cdots, n\}$:

$$2^{\text{nd}}\text{order US} \xrightarrow[\text{appearing bifurcation}]{i^{\text{th}}\text{quadratic factor}} \begin{cases} \text{SO}, & \text{for } x^* = a_{2i}, \\ \text{SI}, & \text{for } x^* = a_{2i-1}. \end{cases} \tag{6.91}$$

$$2^{\text{nd}}\text{order LS} \xrightarrow[\text{appearing bifurcation}]{i^{\text{th}}\text{quadratic factor}} \begin{cases} \text{SI}, & \text{for } x^* = a_{2i}, \\ \text{SO}, & \text{for } x^* = a_{2i-1}. \end{cases} \tag{6.92}$$

If $Q(x) = 1$ and $n = m$, a set of paralleled different simple upper-saddle *appearing* bifurcations in the $(2m)^{\text{th}}$-degree polynomial nonlinear system is called the m-upper-saddle-node (m-USN) *parallel appearing* bifurcation. Such a bifurcation is also called the m-upper-saddle-node (m-USN) *teeth comb appearing* bifurcation. At the *appearing* bifurcation point, $\Delta_i = 0$ ($i = 1, 2, \cdots, m$), and the m-USN *teeth comb appearing* bifurcation structure is

$$m\text{-USN}\begin{cases} \text{US} \xrightarrow[\textit{appearing}]{m^{\text{th}}\textit{bifurcation}} \begin{cases} \text{SO}, & \text{for } x^* = a_{2m}, \\ \text{SI}, & \text{for } x^* = a_{2m-1}; \end{cases} \\ \vdots \\ \text{US} \xrightarrow[\textit{appearing}]{i^{\text{th}}\textit{bifurcation}} \begin{cases} \text{SO}, & \text{for } x^* = a_{2i}, \\ \text{SI}, & \text{for } x^* = a_{2i-1}; \end{cases} \\ \vdots \\ \text{US} \xrightarrow[\textit{appearing}]{1^{\text{st}}\textit{bifurcation}} \begin{cases} \text{SO}, & \text{for } x^* = a_{2}, \\ \text{SI}, & \text{for } x^* = a_{1}. \end{cases} \end{cases} \tag{6.93}$$

Similarly, a set of paralleled different simple lower-saddle *appearing* bifurcations is called the m-lower-saddle-node (m-LSN) parallel *appearing* bifurcation for the $(2m)^{\text{th}}$-degree polynomial nonlinear system. The lower-saddle-node bifurcation is called the m-lower-saddle-node (m-LSN) *teeth comb appearing* bifurcation. At the bifurcation point, $\Delta_i = 0$ ($i = 1, 2, \ldots, m$), and the m-LSN *appearing* bifurcation structure is

$$m\text{-LSN}\begin{cases} \text{LS} \xrightarrow[\textit{appearing}]{m^{\text{th}}\textit{bifurcation}} \begin{cases} \text{SI}, & \text{for } x^* = a_{2m}, \\ \text{SO}, & \text{for } x^* = a_{2m-1}; \end{cases} \\ \vdots \\ \text{LS} \xrightarrow[\textit{appearing}]{i^{\text{th}}\textit{bifurcation}} \begin{cases} \text{SI}, & \text{for } x^* = a_{2i}, \\ \text{SO}, & \text{for } x^* = a_{2i-1}; \end{cases} \\ \vdots \\ \text{LS} \xrightarrow[\textit{appearing}]{1^{st}\textit{bifurcation}} \begin{cases} \text{SI}, & \text{for } x^* = a_{2}, \\ \text{SO}, & \text{for } x^* = a_{1}. \end{cases} \end{cases} \tag{6.94}$$

Consider an appearing bifurcation for a cluster of sink and source equilibriums with the following conditions:

$$\left.\begin{array}{l} B_i = B_j\ (i,j \in \{1,2,\ldots,n\}; i \neq j) \\ \Delta_i = 0\ (i = 1,2,\ldots,n) \end{array}\right\} \text{at bifurcation.} \tag{6.95}$$

Thus, the $(2l)^{\text{th}}$-order appearing bifurcation is for a cluster of simple sinks and sources. Two $(2l)^{\text{th}}$-order *appearing* bifurcations for $l \in \{1,2,\ldots,s\}$ are

$$(2l)^{\text{th}}\text{order US} \xrightarrow[\text{appearing bifurcation}]{\text{cluster of } l\text{-quadratics}} \begin{cases} \text{SO}, & \text{for } x^* = a_{2s_l}, \\ \text{SI}, & \text{for } x^* = a_{2s_l-1}, \\ \vdots & \\ \text{SO}, & \text{for } x^* = a_{2s_1}, \\ \text{SI}, & \text{for } x^* = a_{2s_1-1}. \end{cases} \tag{6.96}$$

$$(2l)^{\text{th}}\text{order LS} \xrightarrow[\text{appearing bifurcation}]{\text{cluster of } l\text{-quadratics}} \begin{cases} \text{SI}, & \text{for } x^* = a_{2s_l}, \\ \text{SO}, & \text{for } x^* = a_{2s_l-1}, \\ \vdots & \\ \text{SI}, & \text{for } x^* = a_{2s_1}, \\ \text{SO}, & \text{for } x^* = a_{2s_1-1}. \end{cases} \tag{6.97}$$

A set of paralleled, different, higher order upper-saddle-node bifurcations with multiplicity is the $((2l_1)^{\text{th}}\text{US} : (2l_2)^{\text{th}}\text{US} : \cdots : (2l_s)^{\text{th}}\text{US})$ *parallel appearing* bifurcation in the $(2m)^{\text{th}}$-degree polynomial system. $(2l_i)^{\text{th}}$US for $(i = 1, 2, \ldots, s)$ with sources and sinks is the $(2l_i)^{\text{th}}$-order upper saddle with l_i-pairs of simple source and sink equilibriums. With different orders of l_i-pairs of simple sources and sinks, the $(2l_i)^{\text{th}}$USN bifurcation possesses different *spraying-appearing clusters* of sinks and sources. $\sum_{i=1}^{s} l_i = n \le m$ where s, $l_i \in \{0,1,2,\cdots,m\}$. If $l_i = 1$ for $i = 1, 2, \cdots, m$ with $n = m$, the simple upper-saddle-node parallel bifurcation or the upper-saddle-node *teeth comb appearing* bifurcation is recovered. Introduce

$$((2l_1)^{\text{th}}\text{US} : (2l_2)^{\text{th}}\text{US} : \cdots : (2l_s)^{\text{th}}\text{US}) \equiv (2l_1 : 2l_2 : \cdots : 2l_s)^{\text{th}}\text{US}. \tag{6.98}$$

At the *sprinkler-spraying-appearing* bifurcation, $\Delta_i = 0$ $(i = 1, 2, \cdots, s)$ and $B_i = B_j$ $(i, j \in \{1,2,\cdots,s\};\ i \neq j)$. The *sprinkler-spraying* USN *appearing* bifurcation is

$$(2l_1 : 2l_2 : \cdots : 2l_s)^{\text{th}}\text{US} = \begin{cases} (2l_s)^{\text{th}} \text{ order US}, \\ \vdots \\ (2l_2)^{\text{th}} \text{ order US}, \\ (2l_1)^{\text{th}} \text{ order US}. \end{cases} \tag{6.99}$$

Thus, the $(2l_1 : 2l_2 : \cdots : 2l_s)^{\text{th}}$ US appearing (or vanishing) bifurcation is called the $(2l_1 : 2l_2 : \cdots : 2l_s)^{\text{th}}$ USN *sprinkler-spraying-appearing (or vanishing)* bifurcation.

Similarly, a set of paralleled different lower-saddle *appearing* bifurcations with multiplicity is the $((2l_1)^{\text{th}}\text{LS} : (2l_2)^{\text{th}}\text{LS} : \cdots : (2l_s)^{\text{th}}\text{LS})$ *appearing* bifurcation in the $(2m)^{\text{th}}$-degree polynomial system. Thus, the $(2l_1 : 2l_2 : \cdots : 2l_s)^{\text{th}}$ LS *appearing (or vanishing)* bifurcation is also called the $(2l_1 : 2l_2 : \cdots : 2l_s)^{\text{th}}$ LS *sprinkler-spraying-appearing (or vanishing)* bifurcation. Again, at the LS *sprinkler-spraying* bifurcation, $\Delta_i = 0\ (i = 1, 2, \ldots, n)$ and $B_i = B_j\ (i, j \in \{1, 2, \ldots, n\}; i \neq j)$. Thus, the *sprinkler-spraying* LSN *appearing* bifurcation is

$$(2l_1 : 2l_2 : \cdots : 2l_s)^{\text{th}}\text{LS} = \begin{cases} (2l_s)^{\text{th}} \text{ order LS}, \\ \vdots \\ (2l_2)^{\text{th}} \text{ order LS}, \\ (2l_1)^{\text{th}} \text{ order LS}. \end{cases} \tag{6.100}$$

Two m-USN and m-LSN *teeth comb appearing* bifurcations are presented in Fig. 6.1a, b for $a_0 > 0$ and $a_0 < 0$, respectively. The set of paralleled $(4^{\text{th}}\text{US} : \cdots : (2r)^{th}\text{US} : \cdots : 4^{\text{th}}\text{US} : 6^{\text{th}}\text{US})$ appearing bifurcations for simple sinks and sources is presented in Fig. 6.1c for $a_0 > 0$, where $l_1 = 2, \ldots, l_i = r, \ldots,$ and $l_{s-1} = 2, l_s = 3$ with $\sum_{i=1}^{s} l_i = m$. The $(4 : \cdots : 2r : \cdots : 4 : 6)^{\text{th}}$-USN *appearing* bifurcation is a USN *sprinkler-spraying-appearing* bifurcation. However, for $a_0 < 0$, the $(4^{\text{th}}\text{LS} : \cdots : (2r)^{th}\text{LS} : \cdots : 4^{\text{th}}\text{LS} : 6^{\text{th}}\text{LS})$ appearing bifurcations for simple sources and sinks are presented in Fig. 6.1d. The $(4 : \cdots : 2r : \cdots : 4 : 6)^{\text{th}}$-LSN *appearing* bifurcation is a LSN *sprinkler-spraying-appearing* bifurcation.

For a cluster of m-quadratics, $B_i = B_j\ (i, j \in \{1, 2, \ldots, m\};\ i \neq j)$ and $\Delta_i = 0\ (i = 1, 2, \ldots, m)$. The $(2m)^{\text{th}}$-order upper-saddle-node appearing bifurcation for m-pairs of sink and source equilibriums is

$$(2m)^{\text{th}}\text{order US} \xrightarrow[\text{appearing bifurcation}]{\text{cluster of } m\text{-quadratics}} \begin{cases} \text{SO}, & \text{for } x^* = a_{2m}, \\ \text{SI}, & \text{for } x^* = a_{2m-1}, \\ \vdots \\ \text{SO}, & \text{for } x^* = a_2, \\ \text{SI}, & \text{for } x^* = a_1. \end{cases} \tag{6.101}$$

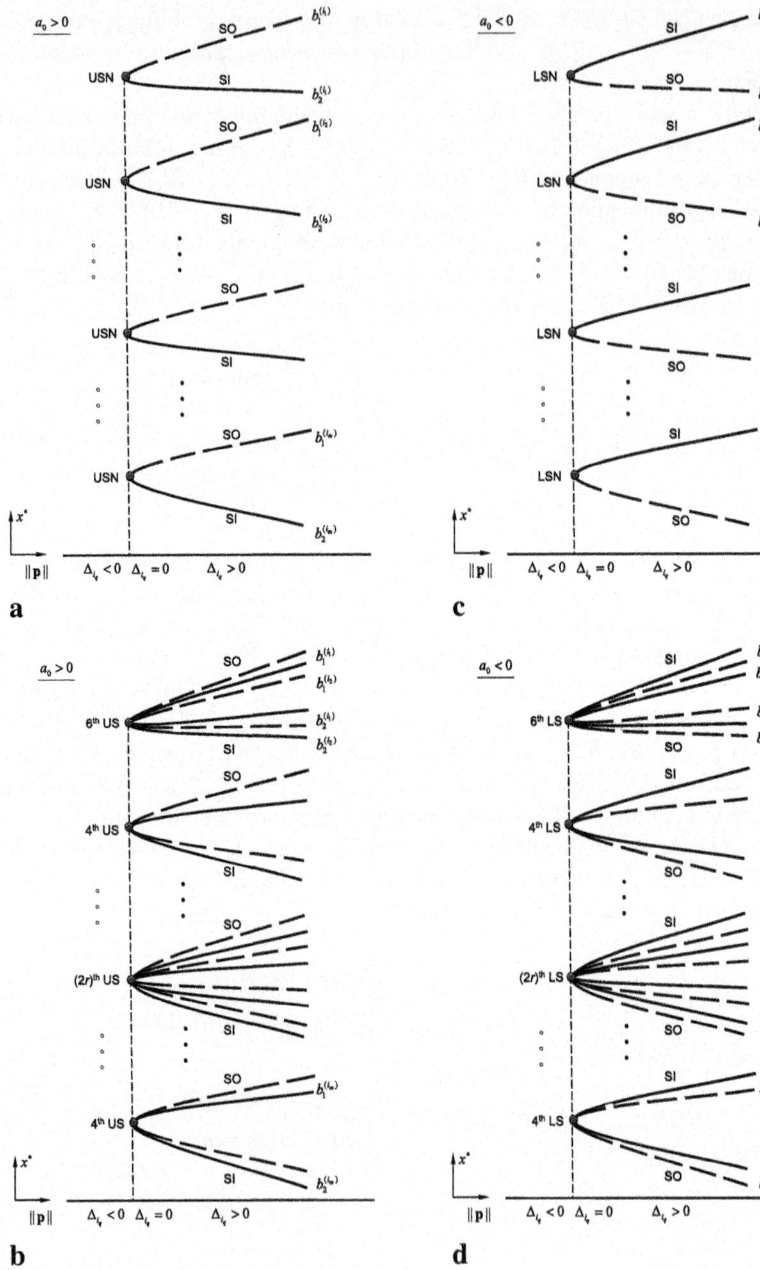

Fig. 6.1 (**a**) m-USN parallel bifurcation ($a_0 > 0$), (**b**) m-LSN parallel bifurcation ($a_0 < 0$), (**c**) $((2l_1)^{\text{th}}\text{US} : (2l_2)^{\text{th}}\text{US} : \cdots : (2l_s)^{\text{th}}\text{US})$ parallel bifurcation ($a_0 > 0$), (**d**) $((2l_1)^{\text{th}}\text{LS} : (2l_2)^{\text{th}}\text{LS} : \cdots : (2l_s)^{\text{th}}\text{LS})$ parallel bifurcation ($a_0 < 0$) in a $(2m)^{\text{th}}$-degree polynomial system. LS: lower saddle, US: upper saddle, SI: sink, SO: source. Stable and unstable equilibriums are represented by solid and dashed curves, respectively. The bifurcation points are marked by circular symbols

The $(2m)^{\text{th}}$-order lower-saddle-node appearing bifurcation for m-pairs of sink and source equilibriums is

$$(2m)^{\text{th}}\text{order LS} \xrightarrow[\text{appearing bifurcation}]{\text{cluster of } m\text{-quadratics}} \begin{cases} \text{SI, for } x^* = a_{2m}, \\ \text{SO, for } x^* = a_{2m-1}, \\ \vdots \\ \text{SI, for } x^* = a_2, \\ \text{SO, for } x^* = a_1. \end{cases} \tag{6.102}$$

The $(2m)^{\text{th}}$-order upper-saddle-node appearing bifurcation with m-pairs of simple sources and sinks is a *sprinkler-spraying cluster* of the m-pairs of simple sources and sinks. The $(2m)^{\text{th}}$-order lower-saddle-node *appearing* bifurcation with m-pairs of equilibriums is also a *sprinkler-spraying cluster* of the m-pairs of simple sources and sinks. Thus, the $(2m)^{\text{th}}$-order USN appearing bifurcation ($a_0 > 0$) and $(2m)^{\text{th}}$-order LSN bifurcation ($a_0 < 0$) are presented in Fig. 6.2a, b, respectively. The $(2m)^{\text{th}}$-order upper-saddle-node appearing bifurcation is named the $(2m)^{\text{th}}$-order USN *sprinkler-spaying* appearing bifurcation, and the $(2m)^{\text{th}}$-order lower-saddle-node appearing bifurcation is named the $(2m)^{\text{th}}$-order LSN *sprinkler-spraying*-appearing bifurcation.

A series of the saddle-node bifurcations are aligned up with varying parameters, which is formed a special pattern. For m-quadratics in the $(2m)^{\text{th}}$-order polynomial system, the following conditions should be satisfied:

$$\begin{aligned} &B_i \approx B_j \; i,j \in \{1, 2, \ldots, n\}; i \neq j, \\ &\Delta_i > \Delta_{i+1} \; (i = 1, 2, \ldots, n; n \leq m), \\ &\Delta_i = 0 \text{ with } \|\mathbf{p}_i\| < \|\mathbf{p}_{i+1}\|. \end{aligned} \tag{6.103}$$

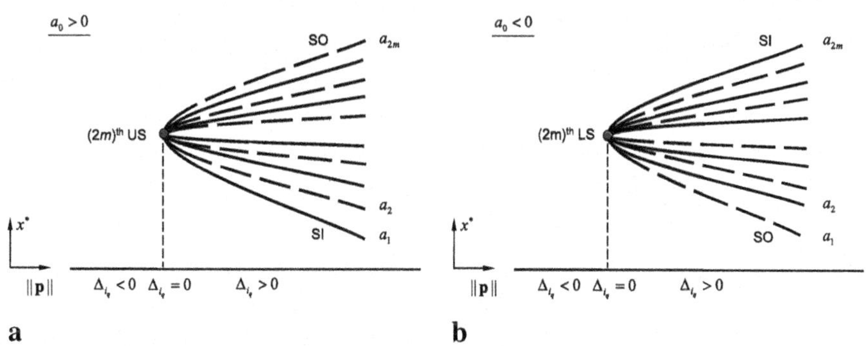

Fig. 6.2 (**a**) $(2m)^{\text{th}}$-order USN bifurcation ($a_0 > 0$), (**b**) $(2m)^{\text{th}}$-order LSN bifurcation ($a_0 < 0$) in the $(2m)^{\text{th}}$ polynomial system. LS: lower saddle, US: upper saddle, SI: sink, SO: source. Stable and unstable equilibriums are represented by solid and dashed curves, respectively. The bifurcation points are marked by circular symbols

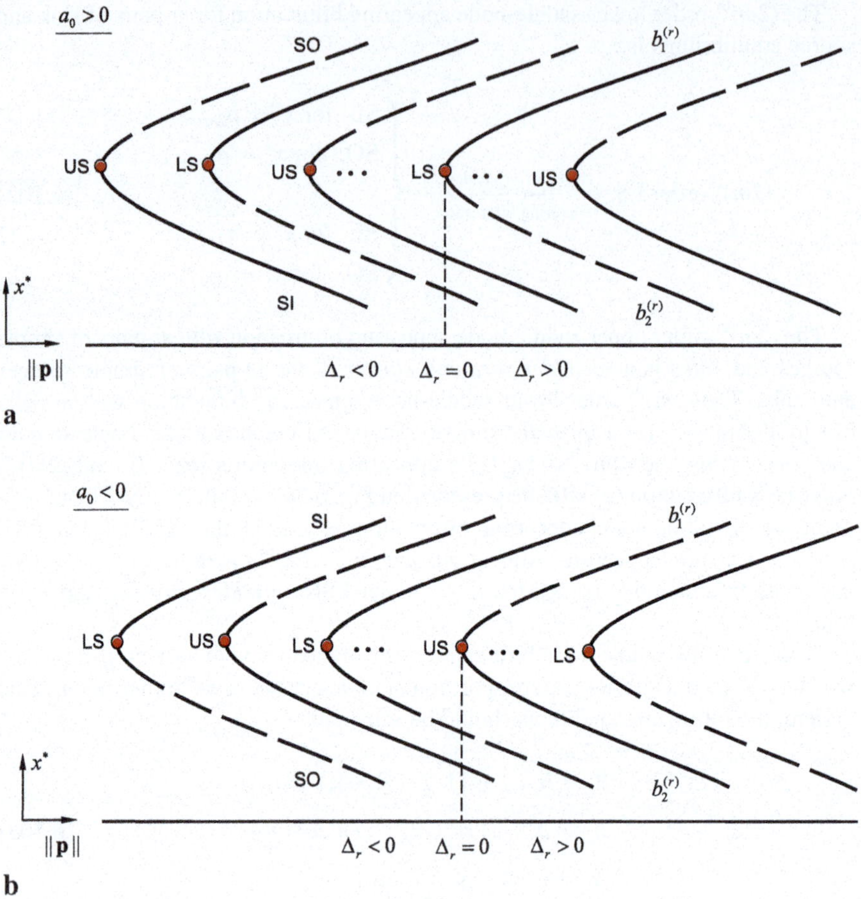

Fig. 6.3 (**a**) m - (US - LS - US - $\cdots$) series bifurcation ($a_0 > 0$), (**b**) m - (US - LS - US - $\cdots$) series bifurcation ($a_0 < 0$) in the $(2m)^{\text{th}}$-degree polynomial system. LS: lower saddle, US: upper saddle, SI: sink, SO: source. Stable and unstable equilibriums are represented by solid and dashed curves, respectively. The bifurcation points are marked by circular symbols

Thus, a series of m - (USN - LSN - USN - $\cdots$) *appearing* bifurcations ($a_0 > 0$) and a series of m - (LSN - USN - LSN - $\cdots$) *appearing* bifurcations ($a_0 < 0$) are presented in Fig. 6.3a, b. The bifurcation scenario is formed by the swapping pattern of USN and LSN appearing bifurcations. Such a bifurcation scenario is like the fish scale. Thus, such a bifurcation-swapping pattern of the USN and LSN is called the *fish-scale appearing* bifurcation in the $(2m)^{\text{th}}$-degree polynomial nonlinear system. There are two swapping bifurcations: (i) the USN-LSN *fish-scale appearing* bifurcation and (ii) the LSN-USN *fish-scale* appearing bifurcation.

6.2.2 Switching Bifurcations

Consider the roots of $x^2 + B_i x + C_i = 0$ as

$$\begin{aligned}&B_i = -(b_1^{(i)} + b_2^{(i)}), \Delta_i = (b_1^{(i)} - b_2^{(i)})^2 \geq 0,\\&x_{1,2}^{(i)} = b_{1,2}^{(i)}, \Delta_i > 0 \text{ if } b_1^{(i)} \neq b_2^{(i)} (i = 1,2,\ldots,n);\\&\left.\begin{aligned}&B_i \neq B_j \; (i,j = 1,2,\ldots,n; i \neq j)\\&\Delta_i = 0 \text{ at } b_1^{(i)} = b_2^{(i)} (i = 1,2,\ldots,n)\end{aligned}\right\} \text{at bifurcation.}\end{aligned} \tag{6.104}$$

The second-order singularity bifurcation is for the *switching* of a pair of simple sink and source. There are two *switching* bifurcations for $i \in \{1,2,\ldots,n\}$:

$$2^{\text{nd}}\text{order US} \xrightarrow[\text{switching bifurcation}]{i^{\text{th}}\text{quadratic factor}} \begin{cases}\text{SO}, & \text{for } a_{2i} = b_2^{(i)} \to b_1^{(i)},\\ \text{SI}, & \text{for } a_{2i-1} = b_1^{(i)} \to b_2^{(i)}.\end{cases} \tag{6.105}$$

$$2^{\text{nd}}\text{order LS} \xrightarrow[\text{switching bifurcation}]{i^{\text{th}}\text{quadratic factor}} \begin{cases}\text{SI}, & \text{for } a_{2i} = b_2^{(i)} \to b_1^{(i)},\\ \text{SO}, & \text{for } a_{2i-1} = b_1^{(i)} \to b_2^{(i)}.\end{cases} \tag{6.106}$$

A set of m-paralleled pairs of different simple upper-saddle-node *switching* bifurcations in the $(2m)^{\text{th}}$-degree polynomial nonlinear system is called the m-upper-saddle-node (m-USN) *parallel switching* bifurcation. Such a bifurcation is also called the m-upper-saddle-node (m-USN) *antenna-switching* bifurcation. For non-switching point, $\Delta_i > 0$ at $b_1^{(i)} \neq b_2^{(i)}$ ($i = 1, 2, \ldots, n$). At the bifurcation point, $\Delta_i = 0$ at $b_1^{(i)} = b_2^{(i)}$ ($i = 1, 2, \ldots, n$). The m-USN *parallel switching* bifurcation is

$$m\text{-USN}\begin{cases}\text{US} \xrightarrow[\textit{switching}]{m^{\text{th}}\textit{bifurcation}} \begin{cases}\text{SO} \downarrow \text{SI}, & \text{for } b_2^{(m)} = a_{2m} \downarrow a_{2m-1},\\ \text{SI} \uparrow \text{SO}, & \text{for } b_1^{(m)} = a_{2m-1} \uparrow a_{2m};\end{cases}\\ \vdots\\ \text{US} \xrightarrow[\textit{switching}]{i^{\text{th}}\textit{bifurcation}} \begin{cases}\text{SO} \downarrow \text{SI}, & \text{for } b_2^{(i)} = a_{2i} \downarrow a_{2i-1},\\ \text{SI} \uparrow \text{SO}, & \text{for } b_1^{(i)} = a_{2i-1} \uparrow a_{2i};\end{cases}\\ \vdots\\ \text{US} \xrightarrow[\textit{switching}]{1^{\text{st}}\textit{bifurcation}} \begin{cases}\text{SO} \downarrow \text{SI}, & \text{for } b_2^{(1)} = a_2 \downarrow a_1,\\ \text{SI} \uparrow \text{SO}, & \text{for } b_1^{(1)} = a_1 \uparrow a_2.\end{cases}\end{cases} \tag{6.107}$$

Similarly, a set of paralleled different simple lower-saddle bifurcations is called the m-lower-saddle-node (m-LSN) *parallel switching* bifurcation for the $(2m)^{\text{th}}$-degree polynomial nonlinear system. The lower-saddle-node *switching* bifurcation is also called the m-lower-saddle-node (m-LSN) *antenna switching* bifurcation. For non-switching point, $\Delta_i > 0$ at $b_1^{(i)} \neq b_2^{(i)}$ $(i = 1, 2, \ldots, n)$. At the bifurcation point, $\Delta_i = 0$ at $b_1^{(i)} = b_2^{(i)}$ $(i = 1, 2, \ldots, n)$. The m-LSN *antenna switching* bifurcation is

$$
m\text{-LSN}\begin{cases}
\text{LS} \xrightarrow[\textit{switching}]{m^{\text{th}}\textit{bifurcation}} \begin{cases} \text{SI} \downarrow \text{SO}, & \text{for } b_2^{(m)} = a_{2m} \downarrow a_{2m-1}, \\ \text{SO} \uparrow \text{SI}, & \text{for } b_1^{(m)} = a_{2m-1} \uparrow a_{2m}; \end{cases} \\
\vdots \\
\text{LS} \xrightarrow[\textit{switching}]{i^{\text{th}}\textit{bifurcation}} \begin{cases} \text{SI} \downarrow \text{SO}, & \text{for } b_2^{(i)} = a_{2i} \downarrow a_{2i-1}, \\ \text{SO} \uparrow \text{SI}, & \text{for } b_1^{(i)} = a_{2i-1} \uparrow a_{2i}; \end{cases} \\
\vdots \\
\text{LS} \xrightarrow[\textit{switching}]{1^{\text{st}}\textit{bifurcation}} \begin{cases} \text{SI} \downarrow \text{SO}, & \text{for } b_2^{(1)} = a_2 \downarrow a_1, \\ \text{SO} \uparrow \text{SI}, & \text{for } b_1^{(1)} = a_1 \uparrow a_2. \end{cases}
\end{cases}
\tag{6.108}
$$

Consider a switching bifurcation for a bundle of sink and source equilibriums with the following conditions:

$$
\begin{aligned}
&B_i = -(b_1^{(i)} + b_2^{(i)}), \Delta_i = (b_1^{(i)} - b_2^{(i)})^2 \geq 0, \\
&x_{1,2}^{(i)} = b_{1,2}^{(i)}, \Delta_i > 0 \text{ if } b_1^{(i)} \neq b_2^{(i)} (i = 1, 2, \ldots, n); \\
&\left.\begin{array}{l} B_i = B_j \; (i, j \in \{1, 2, \ldots, n\}; i \neq j) \\ \Delta_i = 0 \text{ at } b_1^{(i)} = b_2^{(i)} \; (i = 1, 2, \ldots, n) \end{array}\right\} \text{at bifurcation.}
\end{aligned}
\tag{6.109}
$$

Thus, the $(2l)^{\text{th}}$-order *switching* bifurcation can be for a bundle of simple sinks and sources. Two $(2l)^{\text{th}}$-order switching bifurcations for $l \in \{1, 2, \ldots, s\}$ are

$$
(2l)^{\text{th}}\text{order US} \xrightarrow[\text{switching bifurcation}]{\text{a bundle of } (2l)\text{-equilibriums}} \begin{cases} \text{SO}, & \text{for } a_{2s_l} \to b_{2s_l}, \\ \text{SI}, & \text{for } a_{2s_l - 1} \to b_{2s_l - 1}, \\ \vdots & \\ \text{SO}, & \text{for } a_{2s_1} \to b_{2s_1}, \\ \text{SI}, & \text{for } a_{2s_1 - 1} \to b_{2s_1 - 1}. \end{cases}
\tag{6.110}
$$

$$
(2l)^{\text{th}}\text{order LS} \xrightarrow[\text{switching bifurcation}]{\text{a bundle of } (2l)\text{-equilibriums}} \begin{cases} \text{SI}, & \text{for } a_{2s_l} \to b_{2s_l}, \\ \text{SO}, & \text{for } a_{2s_l - 1} \to b_{2s_l - 1}, \\ \vdots & \\ \text{SI}, & \text{for } a_{2s_1} \to b_{2s_1}, \\ \text{SO}, & \text{for } a_{2s_1 - 1} \to b_{2s_1 - 1}. \end{cases}
\tag{6.111}
$$

where $\Delta_{ij} = (a_i - a_j)^2 = (b_i - b_j)^2 = 0$ with $B_i = B_j$ $(i, j = 2s_1 - 1, 2s_1, \ldots, 2s_l - 1, 2s_l)$ and

$$\begin{aligned} \{a_{2s_1-1}, a_{2s_1}, \ldots, a_{2s_l-1}, a_{2s_l}\} &\underset{\text{before bifurcation}}{\subset} \text{sort}\{b_1^{(1)}, b_2^{(1)}, \ldots, b_1^{(n)}, b_2^{(n)}\}, \\ \{b_{2s_1-1}, b_{2s_1}, \ldots, b_{2s_l-1}, b_{2s_l}\} &\underset{\text{after bifurcation}}{\subset} \text{sort}\{b_1^{(1)}, b_2^{(1)}, \ldots, b_1^{(n)}, b_2^{(n)}\}. \end{aligned} \tag{6.112}$$

The $(2l-1)^{\text{th}}$-order *switching* bifurcation can be for a bundle of simple sinks and sources. Two $(2l-1)^{\text{th}}$-order switching bifurcations for $l \in \{1, 2, \cdots, s\}$ are

$$(2l-1)^{\text{th}}\text{order SO} \xrightarrow[\text{switching bifurcation}]{\text{a bundle of } (2l-1)\text{-equilibriums}} \begin{cases} \text{SO}, & \text{for } a_{2s_l-1} \to b_{2s_l-1}, \\ \vdots & \\ \text{SI}, & \text{for } a_{2s_1} \to b_{2s_1}, \\ \text{SO}, & \text{for } a_{2s_1-1} \to b_{2s_1-1}. \end{cases} \tag{6.113}$$

$$(2l-1)^{\text{th}}\text{order SI} \xrightarrow[\text{switching bifurcation}]{\text{a bundle of } (2l-1)\text{-equilibriums}} \begin{cases} \text{SI}, & \text{for } a_{2s_l-1} \to b_{2s_l-1}, \\ \vdots & \\ \text{SO}, & \text{for } a_{2s_1} \to b_{2s_1}, \\ \text{SI}, & \text{for } a_{2s_1-1} \to b_{2s_1-1}. \end{cases} \tag{6.114}$$

where $\Delta_{ij} = (a_i - a_j)^2 = (b_i - b_j)^2 = 0$ with $B_i = B_j$ $(i, j = 2s_1 - 1, 2s_1, \ldots, 2s_l - 1)$ and

$$\begin{aligned} \{a_{2s_1-1}, a_{2s_1}, \ldots, a_{2s_l-1}\} &\underset{\text{before bifurcation}}{\subset} \text{sort}\{b_1^{(1)}, b_2^{(1)}, \ldots, b_1^{(n)}, b_2^{(n)}\}, \\ \{b_{2s_1-1}, b_{2s_1}, \ldots, b_{2s_l-1}\} &\underset{\text{after bifurcation}}{\subset} \text{sort}\{b_1^{(1)}, b_2^{(1)}, \ldots, b_1^{(n)}, b_2^{(n)}\}. \end{aligned} \tag{6.115}$$

A set of paralleled, different, higher order upper-saddle-node switching bifurcations with multiplicity is the $((\alpha_1)^{\text{th}}\text{XX} : (\alpha_2)^{\text{th}}\text{XX} : \cdots : (\alpha_s)^{\text{th}}\text{XX})$ *parallel switching* bifurcation in the $(2m)^{\text{th}}$-degree polynomial system. At the *straw-bundle switching* bifurcation, $\Delta_i = 0$ $(i = 1, 2, \ldots, n)$ and $B_i = B_j$ $(i, j \in \{1, 2, \ldots, n\}; i \neq j)$. Thus, the *parallel straw-bundle switching* bifurcation is

$$\begin{aligned} &((\alpha_1)^{\text{th}}\text{XX} : (\alpha_2)^{\text{th}}\text{XX} : \cdots : (\alpha_s)^{\text{th}}\text{XX})\text{-switching} \\ &= \begin{cases} (\alpha_s)^{\text{th}} \text{ order XX switching}, \\ \vdots \\ (\alpha_2)^{\text{th}} \text{ order XX switching}, \\ (\alpha_1)^{\text{th}} \text{ order XX switching}; \end{cases} \end{aligned} \tag{6.116}$$

where

$$\begin{aligned} &\alpha_i \in \{2l_i, 2l_i - 1\} \text{ with } \Sigma_{i=1}^{s}\alpha_i = 2m, \\ &\text{and } \mathrm{XX} \in \{\mathrm{US}, \mathrm{LS}, \mathrm{SO}, \mathrm{SI}\}. \end{aligned} \tag{6.117}$$

The $(2l_i)^{\text{th}}$US for $(i = 1, 2, \cdots, s)$ with sinks and sources is the $(2l_i)^{\text{th}}$-order upper saddle for a switching of l_i-pairs of simple sinks and sources. With different orders of l_i-pairs of simple sinks and sources, the $(2l_i)^{\text{th}}$USN switching bifurcation possesses different *straw-bundle switching* for a bundle of stable and unstable equilibriums. The $(2l_1 : 2l_2 : \cdots : 2l_s)^{\text{th}}$ USN bifurcation is called the $(2l_1 : 2l_2 : \cdots : 2l_s)^{\text{th}}$ USN *straw-bundle switching* bifurcation:

$$(2l_1 : 2l_2 : \cdots : 2l_s)^{\text{th}}\text{USN switching} = \begin{cases} (2l_s)^{\text{th}} \text{ order USN switching,} \\ \vdots \\ (2l_2)^{\text{th}} \text{ order USN switching,} \\ (2l_1)^{\text{th}} \text{ order USN switching.} \end{cases} \tag{6.118}$$

If $l_i = 1$ for $i = 1, 2, \ldots, m$ with $n = m$, the simple upper-saddle-node parallel switching bifurcation or the upper-saddle-node *antenna switching* bifurcation is recovered.

Similarly, a set of paralleled different lower-saddle switching bifurcations with multiplicity is the $((2l_1)^{\text{th}}\text{LS} : (2l_2)^{\text{th}}\text{LS} : \cdots : (2l_s)^{\text{th}}\text{LS})$ parallel switching bifurcation in the $(2m)^{\text{th}}$-degree polynomial system. Thus, the $(2l_1 : 2l_2 : \cdots : 2l_s)^{\text{th}}$ LSN switching bifurcation is also called the $(2l_1 : 2l_2 : \cdots : 2l_s)^{\text{th}}$ LSN *straw-bundle* switching bifurcation. Again, at the LSN *straw-bundle switching* bifurcation, $\Delta_i = 0$ $(i = 1, 2, \ldots, n)$ and $B_i = B_j$ $(i, j \in \{1, 2, \ldots, n\}; i \neq j)$. Thus, the LSN *straw-bundle* switching bifurcation is

$$(2l_1 : 2l_2 : \cdots : 2l_s)^{\text{th}}\text{LSN switching} = \begin{cases} (2l_s)^{\text{th}} \text{ order LSN switching,} \\ \vdots \\ (2l_2)^{\text{th}} \text{ order LSN switching,} \\ (2l_1)^{\text{th}} \text{ order LSN switching.} \end{cases} \tag{6.119}$$

The set of m-upper-saddle-node (m-USN) *parallel switching* bifurcation is equivalent to the set of $(2 : 2 : \cdots : 2)^{\text{nd}}$-USN bifurcations. The set of m-lower-saddle-node (m-LSN) *parallel switching* bifurcation is equivalent to the set of $(2 : 2 : \cdots : 2)^{\text{nd}}$-LSN bifurcations. Such two sets of parallel switching bifurcations are presented in Fig. 6.4a, b for $a_0 > 0$ and $a_0 < 0$, respectively. A set of paralleled $(3^{\text{rd}}\text{SO} : 2^{\text{nd}}\text{LS} : \cdots : 4^{\text{th}}\text{LS} : \cdots : 3^{\text{rd}}\text{SI})$ *switching* bifurcations for SI and SO equilibriums is presented in Fig. 6.4c for $a_0 > 0$. However, for $a_0 < 0$, the set of $(3^{\text{rd}}\text{SI} : 2^{\text{nd}}\text{US} : \cdots : 4^{\text{th}}\text{US} : \cdots : 3^{\text{rd}}\text{SI})$ *switching* bifurcations for sources and sinks is presented in Fig. 6.4d.

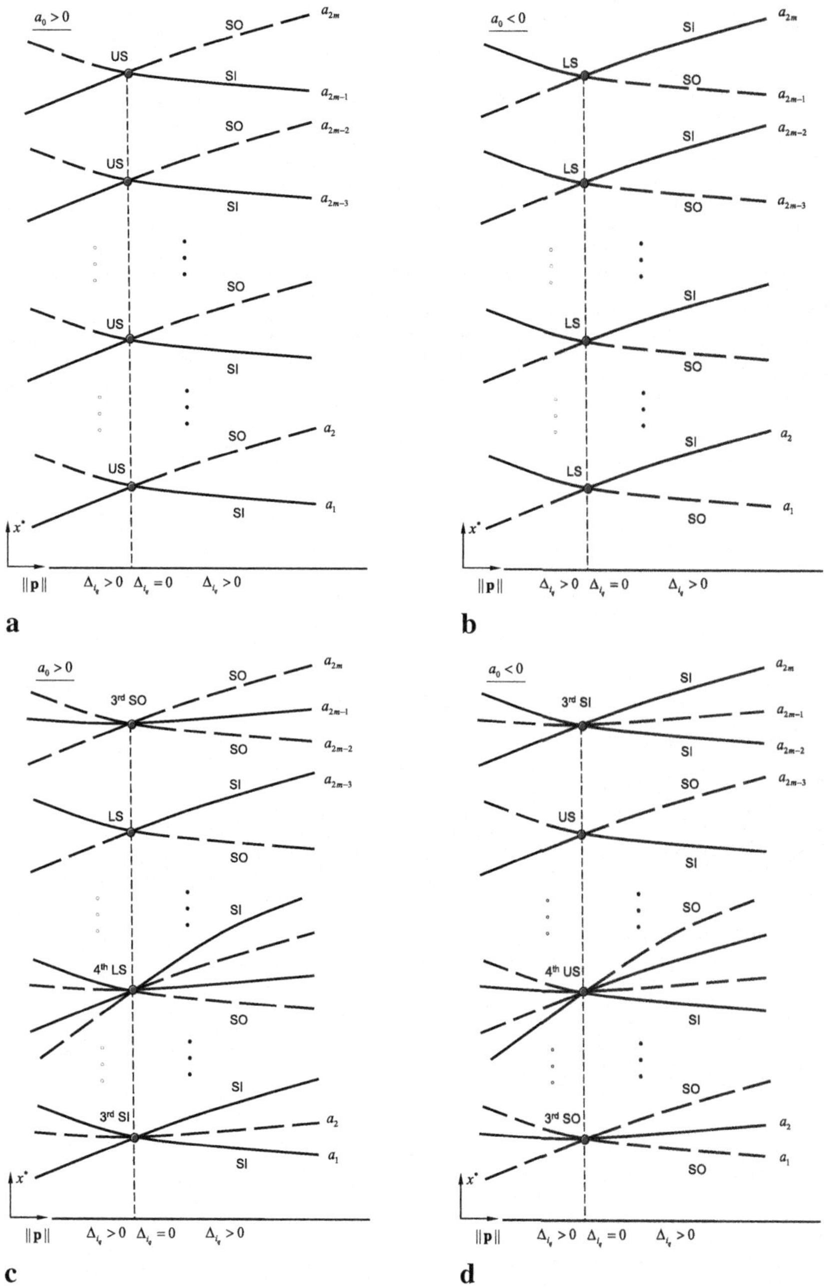

Fig. 6.4 Stability and bifurcations of equilibriums in a 1-dimensional ($a_0 < 0$), $(2m)^{\text{th}}$-degree polynomial system: (**a**) m-USN parallel switching bifurcation ($a_0 > 0$), (**b**) m-LSN parallel switching bifurcation ($a_0 < 0$), (**c**) (3^{rd}SO : 2^{nd}LS : $\cdots$: 3^{rd}SI) parallel switching bifurcation ($a_0 > 0$), (**d**) (3^{rd}SI : 2^{nd}US : $\cdots$: 3^{rd}SO) parallel switching bifurcation ($a_0 < 0$). LS: lower saddle, US: upper saddle, SI: sink, SO: source. Stable and unstable equilibriums are represented by solid and dashed curves, respectively. The bifurcation points are marked by circular symbols

6.2.3 Switching and Appearing Bifurcations

Consider a $(2m)^{\text{th}}$-degree 1-dimensional polynomial system in a form of

$$\dot{x} = a_0 Q(x) \prod_{i=1}^{2n_1} (x - c_i) \prod_{j=1}^{n_2} (x^2 + B_j x + C_j). \tag{6.120}$$

Without loss of generality, a function of $Q(x) > 0$ is either a polynomial function or a non-polynomial function. The roots of $x^2 + B_j x + C_j = 0$ are

$$b_{1,2}^{(j)} = -\frac{1}{2}B_j \pm \frac{1}{2}\sqrt{\Delta_j}, \Delta_j = B_j^2 - 4C_j \geq 0 \; (j = 1, 2, \ldots, n_2); \tag{6.121}$$

either

$$\begin{aligned}
&\{a_1^-, a_2^-, \ldots, a_{2n_1}^-\} = \text{sort}\{c_1, c_2, \ldots, c_{2n_1}\}, a_s^- \leq a_{s+1}^- \text{ before bifurcation} \\
&\{a_1^+, a_2^+, \ldots, a_{2n_3}^+\} = \text{sort}\{c_1, \ldots, c_{2n_1}; b_1^{(1)}, b_2^{(1)}, \ldots, b_1^{(n_2)}, b_2^{(n_2)}\}, \\
&a_s^+ \leq a_{s+1}^+, n_3 = n_1 + n_2 \text{ after bifurcation;}
\end{aligned} \tag{6.122}$$

or

$$\begin{aligned}
&\{a_1^-, a_2^-, \ldots, a_{2n_3}^-\} = \text{sort}\{c_1, c_2, \ldots, c_{2n_1}; b_1^{(1)}, b_2^{(1)}, \ldots, b_1^{(n_2)}, b_2^{(n_2)}\}, \\
&a_s^- \leq a_{s+1}^-, n_3 = n_1 + n_2 \text{ before bifurcation;} \\
&\{a_1^+, a_2^+, \ldots, a_{2n_1}^+\} = \text{sort}\{c_1, \ldots, c_{2n_1}\}, a_s^+ \leq a_{s+1}^+ \text{ after bifurcation;}
\end{aligned} \tag{6.123}$$

and

$$\left.\begin{aligned}
&B_{j_1} = B_{j_2} = \cdots = B_{j_s} \; (j_{k_1} \in \{1, 2, \ldots, n\}; j_{k_1} \neq j_{k_2}) \\
&(k_1, k_2 \in \{1, 2, \ldots, s\}; k_1 \neq k_2) \\
&\Delta_j = 0 \; (j \in U \subset \{1, 2, \ldots, n_2\}) \\
&c_i \neq -\frac{1}{2}B_j \; (i = 1, 2, \ldots, 2n_1, j = 1, 2, \ldots, n_2)
\end{aligned}\right\} \text{at bifurcation.} \tag{6.124}$$

Consider a just before bifurcation of $((\alpha_1^-)^{\text{th}}XX_1^- : (\alpha_2^-)^{\text{th}}XX_2^- : \ldots : (\alpha_{s_1}^-)^{\text{th}}XX_{s_1}^-)$ for simple sources and sinks. For $\alpha_i^- = 2l_i^- - 1$, $XX_i^- \in \{SO, SI\}$ and for

$\alpha_i^- = 2l_i^-$, $\mathrm{XX}_i^- \in \{\mathrm{US}, \mathrm{LS}\}$ $(i = 1, 2, \ldots, s_1)$. The detailed structures are as follows:

$$\left.\begin{array}{c}\mathrm{SI}\\ \mathrm{SO}\\ \vdots\\ \mathrm{SO}\\ \mathrm{SI}\end{array}\right\} \to (2l_i^- - 1)^{\mathrm{th}}\mathrm{SI}, \text{ and } \left.\begin{array}{c}\mathrm{SO}\\ \mathrm{SI}\\ \vdots\\ \mathrm{SI}\\ \mathrm{SO}\end{array}\right\} \to (2l_i^- - 1)^{\mathrm{th}}\mathrm{SO};$$
$$\left.\begin{array}{c}\mathrm{SO}\\ \mathrm{SI}\\ \vdots\\ \mathrm{SO}\\ \mathrm{SI}\end{array}\right\} \to (2l_i^-)^{\mathrm{th}}\mathrm{US}, \text{ and } \left.\begin{array}{c}\mathrm{SI}\\ \mathrm{SO}\\ \vdots\\ \mathrm{SI}\\ \mathrm{SO}\end{array}\right\} \to (2l_i^-)^{\mathrm{th}}\mathrm{LS}. \tag{6.125}$$

The bifurcation set of $((\alpha_1^-)^{\mathrm{th}}\mathrm{XX}_1^- : (\alpha_2^-)^{\mathrm{th}}\mathrm{XX}_2^- : \ldots : (\alpha_{s_1}^-)^{\mathrm{th}}\mathrm{XX}_{s_1}^-)$ at the same parameter point is called a *left-parallel-bundle* switching bifurcation.

Consider a just after bifurcation of $((\alpha_1^+)^{\mathrm{th}}\mathrm{XX}_1^+ : (\alpha_2^+)^{\mathrm{th}}\mathrm{XX}_2^+ : \ldots : (\alpha_{s_2}^+)^{\mathrm{th}}\mathrm{XX}_{s_2}^+)$ for simple sources and sinks. For $\alpha_i^+ = 2l_i^+ - 1$, $\mathrm{XX}_i^+ \in \{\mathrm{SO}, \mathrm{SI}\}$ and for $\alpha_i^+ = 2l_i^+$, $\mathrm{XX}_i^- \in \{\mathrm{US}, \mathrm{LS}\}$. The four detailed structures are as follows:

$$(2l_i^+ - 1)^{\mathrm{th}}\mathrm{SI} \to \left\{\begin{array}{c}\mathrm{SI}\\ \mathrm{SO}\\ \vdots\\ \mathrm{SO}\\ \mathrm{SI}\end{array}\right., \text{ and } (2l_i^+ - 1)^{\mathrm{th}}\mathrm{SO} \to \left\{\begin{array}{c}\mathrm{SO}\\ \mathrm{SI}\\ \vdots\\ \mathrm{SI}\\ \mathrm{SO}\end{array}\right.; \tag{6.126a}$$

$$(2l_i^+)^{\mathrm{th}}\mathrm{US} \to \left\{\begin{array}{c}\mathrm{SO}\\ \mathrm{SI}\\ \vdots\\ \mathrm{SO}\\ \mathrm{SI}\end{array}\right., \text{ and } (2l_i^+)^{\mathrm{th}}\mathrm{LS} \to \left\{\begin{array}{c}\mathrm{SI}\\ \mathrm{SO}\\ \vdots\\ \mathrm{SI}\\ \mathrm{SO}\end{array}\right.. \tag{6.126b}$$

The bifurcation set of $((\alpha_1^+)^{\mathrm{th}}\mathrm{XX}_1^+ : (\alpha_2^+)^{\mathrm{th}}\mathrm{XX}_2^+ : \ldots : (\alpha_{s_2}^+)^{\mathrm{th}}\mathrm{XX}_{s_2}^+)$ at the same parameter point is called a *right-parallel-bundle* switching bifurcation.

(i) For the just before and after bifurcation structure, if there exists a relation of

$$\begin{aligned}&(\alpha_i^-)^{\text{th}}\text{XX}_i^- = (\alpha_j^+)^{\text{th}}\text{XX}_j^+ = \alpha^{\text{th}}\text{XX}, \text{for } x^* = a_i^- = a_j^+ \\ &(i \in \{1,2,\ldots,s_1\}, j \in \{1,2,\ldots,s_2\}), \text{XX} \in \{\text{US}, \text{LS}, \text{SO}, \text{SI}\}\end{aligned} \tag{6.127}$$

then the bifurcation is a α^{th}XX *switching* bifurcation for simple equilibriums.

(ii) Just for the just before bifurcation structure, if there exists a relation of

$$\begin{aligned}&(2l_i^-)^{\text{th}}\text{XX}_i^- = (2l)^{\text{th}}\text{XX}, \text{for } x^* = a_i^- = a_i \\ &(i \in \{1,2,\ldots,s_1\}, \text{XX} \in \{\text{US}, \text{LS}\}\end{aligned} \tag{6.128}$$

then the bifurcation is a $(2l)^{\text{th}}$XX *left-appearing* (or *right-vanishing*) bifurcation for simple equilibriums.

(iii) Just for the just after bifurcation structure, if there exists a relation of

$$\begin{aligned}&(2l_i^+)^{\text{th}}\text{XX}_i^+ = (2l)^{\text{th}}\text{XX}, \text{for } x^* = a_i^+ = a_i \\ &(i \in \{1,2,\ldots,s_1\}), \text{XX} \in \{\text{US}, \text{LS}\}\end{aligned} \tag{6.129}$$

then the bifurcation is a $(2l)^{\text{th}}$XX *right-appearing* (or *left-vanishing*) bifurcation for simple equilibriums.

(iv) For the just before and after bifurcation structure, if there exists a relation of

$$\begin{aligned}&(\alpha_i^-)^{\text{th}}\text{XX}_i^- \neq (\alpha_j^+)^{\text{th}}\text{XX}_j^+ \text{ for } x^* = a_i^- = a_j^+ \\ &\text{XX}_i^-, \text{XX}_j^+ \in \{\text{US}, \text{LS}, \text{SO}, \text{SI}\} \\ &(i \in \{1,2,\ldots,s_1\}, j \in \{1,2,\ldots,s_2\}),\end{aligned} \tag{6.130}$$

then there are two *flower-bundle* switching bifurcations of simple equilibriums:

(iv_1) For $\alpha_j = \alpha_i + 2l$, the bifurcation is called a α_j^{th}XX right *flower-bundle* switching bifurcation for α_i to α_j-simple equilibriums with the appearance (birth) of $2l$-simple equilibriums.

(iv_2) For $\alpha_j = \alpha_i - 2l$, the bifurcation is called a α_i^{th}XX left *flower-bundle* switching bifurcation for α_i to α_j-simple equilibriums with the vanishing (death) of $2l$-simple equilibriums.

A general parallel switching bifurcation is

$$\begin{aligned}&((\alpha_1^-)^{\text{th}}\text{XX}_1^- : (\alpha_2^-)^{\text{th}}\text{XX}_2^- : \ldots : (\alpha_{s_1}^-)^{\text{th}}\text{XX}_{s_1}^-) \xrightarrow[\textit{bifucation}]{\textit{switching}} \\ &((\alpha_1^+)^{\text{th}}\text{XX}_1^+ : (\alpha_2^+)^{\text{th}}\text{XX}_2^+ : \ldots : (\alpha_{s_1}^+)^{\text{th}}\text{XX}_{s_2}^+).\end{aligned} \tag{6.131}$$

Such a general, parallel switching bifurcation consists of the *left* and *right* parallel-bundle switching bifurcations.

If the *left* and *right* parallel-bundle switching bifurcations are same in a parallel *flower-bundle* switching bifurcation, i.e.,

$$\begin{aligned}&(\alpha_i^-)^{\text{th}}\text{XX}_i^- = (\alpha_i^+)^{\text{th}}\text{XX}_i^+ = \alpha^{\text{th}}\text{XX}, \\ &\text{for } x^* = a_i^- = a_i^+ \ (i = 1, 2, \ldots, s\}\end{aligned} \tag{6.132}$$

then the parallel *flower-bundle* switching bifurcation becomes a parallel *straw-bundle switching* bifurcation of $((\alpha_1)^{\text{th}}\text{XX} : (\alpha_2)^{\text{th}}\text{XX} : \cdots : (\alpha_s)^{\text{th}}\text{XX})$.

If the *left* and *right* parallel-bundle switching bifurcations are different in a parallel *flower-bundle* switching bifurcation, i.e.,

$$\begin{aligned}&(\alpha_i^-)^{\text{th}}\text{XX}_i^- = (2l_i^-)^{\text{th}}\text{XX}, (\alpha_j^+)^{\text{th}}\text{XX}_j^+ = (2l_j^+)^{\text{th}}\text{YY}, \\ &\text{for } x^* = a_i^- \neq a_i^+ \ (i = 1, 2, \ldots, s\} \\ &\text{XX} \in \{\text{US}, \text{LS}\}, \text{YY} \in \{\text{US}, \text{LS}\}\end{aligned} \tag{6.133}$$

then the parallel *flower-bundle* switching bifurcation becomes a combination of two independent left and right parallel appearing bifurcations:

(i) A $((2l_1^-)^{\text{th}}\text{XX}_1^- : (2l_2^-)^{\text{th}}\text{XX}_2^- : \cdots : (2l_{s_1}^-)^{\text{th}}\text{XX}_{s_1}^-)$-*left parallel sprinkler-spraying-appearing (or right vanishing)* bifurcation
(ii) A $((2l_1^+)^{\text{th}}\text{XX}_1^+ : (2l_2^+)^{\text{th}}\text{XX}_2^+ : \cdots : (2l_{s_2}^+)^{\text{th}}\text{XX}_{s_2}^+)$-*right parallel sprinkler-spraying-appearing (or left vanishing)* bifurcation.

The $(6^{\text{th}}\text{US} : 4^{\text{th}}\text{LS} : \cdots : 4^{\text{th}}\text{US} : \text{SI})$ *appearing* bifurcation for $a_0 > 0$ is presented in Fig. 6.5a. Compared to the case of $a_0 > 0$, the bifurcation and stability conditions of equilibriums for $a_0 < 0$ will be swapped. The $(6^{\text{th}}\text{LS} : 4^{\text{th}}\text{US} : \cdots : 4^{\text{th}}\text{LS} : \text{SO})$ *parallel appearing* bifurcation is shown in Fig. 6.5b.

Such a kind of bifurcation is like a *waterfall appearing* bifurcation. The *switching* and *appearing* bifurcations of equilibriums exist at the same parameter. A set of paralleled, different *switching and appearing* bifurcations of higher order equilibriums is also named the $(l_1^{\text{th}}\text{XX} : l_2^{\text{th}}\text{XX} : \cdots : l_s^{\text{th}}\text{XX})$ parallel *switching and appearing* bifurcation in the $(2m)^{\text{th}}$-degree polynomial system. The l_ith XX *switching* and

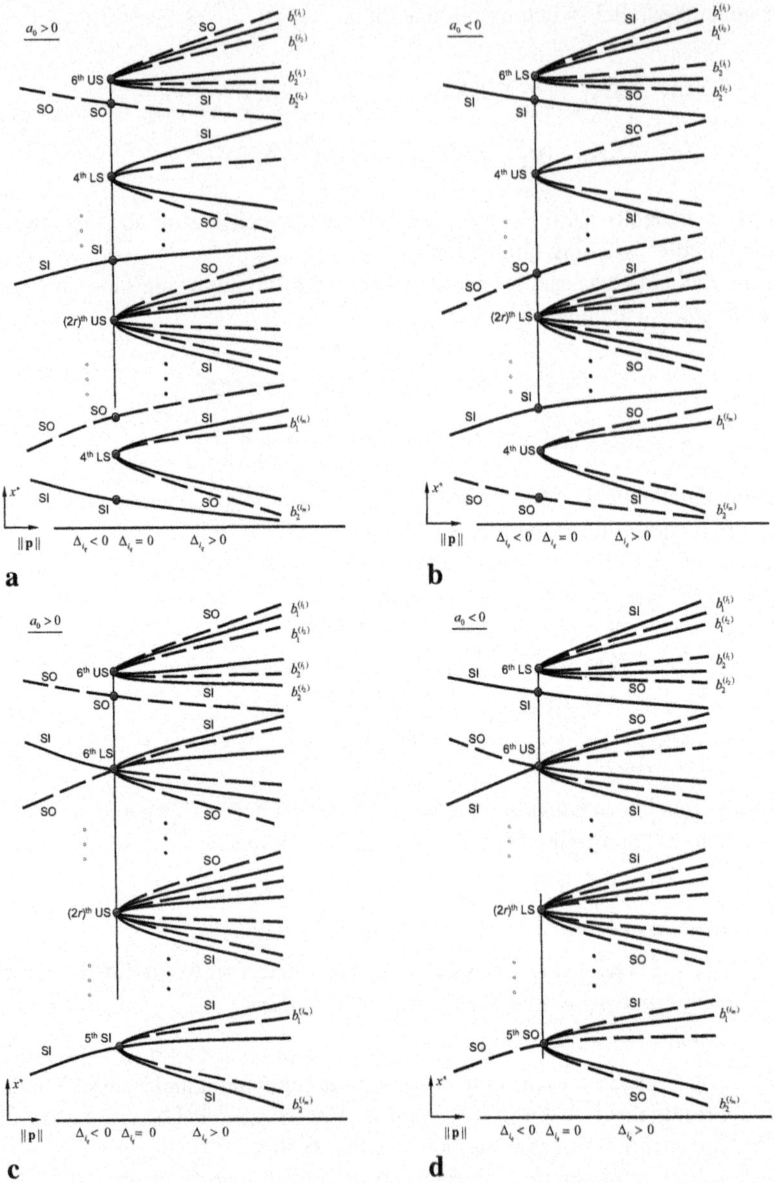

Fig. 6.5 Stability and bifurcation of equilibriums in a $(2m)^{\text{th}}$-degree polynomial system: (**a**) (6^{th}US : SO : 4^{th}LS : $\cdots$: SI) appearing bifurcation ($a_0 > 0$). (**b**) (6^{th}LS : SI : 4^{th}US : $\cdots$: SO) appearing bifurcation ($a_0 < 0$). (**c**) (6^{th}US : SO : 6^{th}LS : $\cdots$: 5^{th}SI) switching/appearing bifurcation ($a_0 > 0$). (**d**) (6^{th}LS : SI : 6^{th}US : $\cdots$: 5^{th}SO) switching/appearing bifurcation ($a_0 < 0$). LS: lower saddle, US: upper saddle, SI: sink, SO: source. Stable and unstable equilibriums are represented by solid and dashed curves, respectively. The bifurcation points are marked by circular symbols

appearing bifurcation possesses different *clusters* of stable and unstable equilibriums *before* and *after* the bifurcation.

The set of ($5^{\text{th}}\text{SI} : \cdots : \text{SO} : 6^{\text{th}}\text{US}$) *flower-bundle switching* bifurcation for SI and SO equilibriums is presented in Fig. 6.5c for $a_0 > 0$. Such a *flower-bundle* switching bifurcation is from (SI : SO : SI : SO) to ($5^{\text{th}}\text{SI} : \cdots : \text{SO} : 6^{\text{th}}\text{US}$) with a *waterfall appearing*. The set of ($5^{\text{th}}\text{SO} : \cdots : \text{SI} : 6^{\text{th}}\text{LS}$) *flower-bundle switching* bifurcation for SI and SO equilibriums is presented in Fig. 6.5d for $a_0 < 0$. Such a *flower-bundle* switching bifurcation is from (SO : SI : SO : SI) to ($5^{\text{th}}\text{SI} : \cdots : \text{SO} : 6^{\text{th}}\text{US}$) with a *waterfall appearing*. After the bifurcation, the *waterfall* equilibrium birth can be observed. The equilibriums before such a bifurcation are much less than after the bifurcation.

6.3 Higher Order Equilibrium Bifurcations

The afore-discussed appearing and switching bifurcations in the $(2m)^{\text{th}}$-degree polynomial system are relative to simple sources and sinks. As in Luo (92019), the higher order singularity bifurcations in the $(2m)^{\text{th}}$-degree polynomial system can be for higher order equilibriums (i.e., sink, source, upper saddle, lower saddle).

6.3.1 Appearing Bifurcations

Consider a $(2m)^{\text{th}}$-degree polynomial system as

$$\dot{x} = a_0 Q(x) \prod_{i=1}^{s} \left(x^2 + B_i x + C_i\right)^{\alpha_i}, \tag{6.134}$$

where $\alpha_i \in \{2l - 1, 2l\}$. Without loss of generality, a function of $Q(x) > 0$ is either a polynomial function or a non-polynomial function. The roots of $x^2 + B_i x + C_i = 0$ are

$$\begin{aligned} &b_{1,2}^{(i)} = -\frac{1}{2}B_i \pm \frac{1}{2}\sqrt{\Delta_i}, \Delta_i = B_i^2 - 4C_i \geq 0; \\ &\{a_1, a_2, \ldots, a_{2s-1}, a_{2s}\} = \text{sort}\{b_1^{(1)}, b_2^{(1)}, \ldots, b_1^{(s)}, b_2^{(s)}\}, a_j \leq a_{j+1}. \end{aligned} \tag{6.135}$$

There are four higher order bifurcations as follows:

$$\begin{aligned}&(2(2l_i-1))^{\text{th}}\text{order US}\xrightarrow[\text{appearing bifurcation}]{(2l_i-1)^{\text{th}}\text{order quadratics}}\\&\begin{cases}(2l_i-1)^{\text{th}}\text{ order SO}, & x^*=a_{2i},\\(2l_i-1)^{\text{th}}\text{ order SI}, & x^*=a_{2i-1};\end{cases}\end{aligned}\tag{6.136}$$

$$\begin{aligned}&(2(2l_i-1))^{\text{th}}\text{ order LS}\xrightarrow[\text{appearing bifurcation}]{(2l_i-1)^{\text{th}}\text{order quadratics}}\\&\begin{cases}(2l_i-1)^{\text{th}}\text{ order SI}, & x^*=a_{2i},\\(2l_i-1)^{\text{th}}\text{ order SO}, & x^*=a_{2i-1};\end{cases}\end{aligned}\tag{6.137}$$

$$\begin{aligned}&(2(2l_i))^{\text{th}}\text{ order US}\xrightarrow[\text{appearing bifurcation}]{(2l_i)^{\text{th}}\text{-order quadratics}}\\&\begin{cases}(2l_i)^{\text{th}}\text{ order US}, & x^*=a_{2i},\\(2l_i)^{\text{th}}\text{ order US}, & x^*=a_{2i-1};\end{cases}\end{aligned}\tag{6.138}$$

$$\begin{aligned}&(2(2l_i))^{\text{th}}\text{ order LS}\xrightarrow[\text{appearing bifurcation}]{(2l_i)^{\text{th}}\text{-order quadratics}}\\&\begin{cases}(2l_i)^{\text{th}}\text{ order LS}, & x^*=a_{2i},\\(2l_i)^{\text{th}}\text{ order LS}, & x^*=a_{2i-1}.\end{cases}\end{aligned}\tag{6.139}$$

(i) For $\alpha_i=2l_i-1$, the $(2(2l_i-1))^{\text{th}}$-order upper-saddle (US) *appearing* bifurcation is for the onset of the $(2l_i-1)^{\text{th}}$-order source (SO) ($x^*=a_{2i}$) and the $(2l-1)^{\text{th}}$-order sink (SI) ($x^*=a_{2i-1}$) with $a_{2i}>a_{2i-1}$ for $a_0>0$.
(ii) For $\alpha_i=2l_i-1$, the $(2(2l_i-1))^{\text{th}}$-order lower-saddle (LS) *appearing* bifurcation is for the onset of the $(2l_i-1)^{\text{th}}$-order sink (SI) ($x^*=a_{2i}$) and the $(2l_i-1)^{\text{th}}$-order source (SO) ($x^*=a_{2i-1}$) with $a_{2i}>a_{2i-1}$ for $a_0<0$.
(iii) For $\alpha_i=2l_i$, the $(2(2l_i))^{\text{th}}$-order upper-saddle (US) *appearing* bifurcation is for the onset of two $(2l_i)^{\text{th}}$-order upper saddles (US) ($x^*=a_{2i-1},a_{2i}$) with $a_{2i}\neq a_{2i-1}$ for $a_0>0$.
(iv) For $\alpha_i=2l_i$, the $(2(2l_i))^{\text{th}}$-order lower-saddle (LS) *appearing* bifurcation is for the onset of two $(2l_i)^{\text{th}}$-order lower saddles (LS) ($x^*=a_{2i-1},a_{2i}$) with $a_{2i}\neq a_{2i-1}$ for $a_0<0$.

From the higher order singular bifurcation conditions, in a $(2m)^{\text{th}}$-degree polynomial system, the higher order saddle-node bifurcations for appearing and switching of the higher order equilibriums are discussed herein. A set of paralleled different higher order upper-saddle *appearing* bifurcations in the $(2m)^{\text{th}}$-degree polynomial nonlinear system is called the $((2\alpha_1)^{\text{th}}\text{US}:(2\alpha_2)^{\text{th}}\text{US}:\cdots:(2\alpha_s)^{\text{th}}\text{US})$ *parallel appearing* bifurcation for $a_0>0$.

Define

$$\left((2\alpha_1)^{\text{th}}\text{US} : (2\alpha_2)^{\text{th}}\text{US} : \cdots : (2\alpha_s)^{\text{th}}\text{US}\right) = (2\alpha_1 : 2\alpha_2 : \cdots : 2\alpha_s)^{\text{th}}\text{US} \quad (6.140)$$

where $\alpha_i \in \{2l_i - 1, 2l_i\}$ for $i = 1, 2, \ldots, s$. Such an *appearing* bifurcation is called the $(2\alpha_1 : 2\alpha_2 : \cdots : 2\alpha_s)^{\text{th}}$US *teeth comb appearing* bifurcation.

Similarly, a set of paralleled different higher order lower-saddle *appearing* bifurcations in the $(2m)^{\text{th}}$-degree polynomial nonlinear system is called the $((2\alpha_1)^{\text{th}}\text{LS} : (2\alpha_2)^{\text{th}}\text{LS} : \cdots : (2\alpha_s)^{\text{th}}\text{LS})$ *parallel appearing* bifurcation for $a_0 < 0$.

Define

$$\left((2\alpha_1)^{\text{th}}\text{LS} : (2\alpha_2)^{\text{th}}\text{LS} : \cdots : (2\alpha_s)^{\text{th}}\text{LS}\right) = (2\alpha_1 : 2\alpha_2 : \cdots : 2\alpha_s)^{\text{th}}\text{LS} \quad (6.141)$$

where $\alpha_i \in \{2l_i - 1, 2l_i\}$ for $i = 1, 2, \cdots, s$. Such an appearing bifurcation is called the $(2\alpha_1 : 2\alpha_2 : \cdots : 2\alpha_s)^{\text{th}}$LS *teeth comb appearing* bifurcation.

Consider a 1-dimensional polynomial system as

$$\dot{x} = a_0 Q(x) \prod_{i=1}^{n} \left(x^2 + B_i x + C_i\right)^{\alpha_i}. \quad (6.142)$$

where $\alpha_i \in \{2r_i - 1, 2r_i\}$ $(i = 1, 2, \cdots, n)$. Without loss of generality, a function of $Q(x) > 0$ is either a polynomial function or a non-polynomial function. The roots of $x^2 + B_i x + C_i = 0$ are

$$\begin{aligned}
&x_{1,2}^{(i)} = -\frac{1}{2}B_i \pm \frac{1}{2}\sqrt{\Delta_i}, \Delta_i = B_i^2 - 4C_i \geq 0;\\
&B_i = B_j \; (i, j = 1, 2, \cdots, n; i \neq j)\\
&\{a_1, a_2, \cdots, a_{2l}\} = \text{sort}\{x_1^{(1)}, x_2^{(1)}, x_1^{(2)}, x_2^{(2)}, \cdots, x_1^{(r)}, x_2^{(r)}\}, a_i \leq a_{i+1}.
\end{aligned} \quad (6.143)$$

The higher order singularity bifurcation can be for a cluster of higher order sinks, sources, upper saddles, and lower saddles. There are four higher order bifurcations as follows:

For the higher order upper-saddle *appearing* bifurcation, the cluster of higher order sinks, sources, upper saddles, and lower saddles is given by the following two cases:

(i) The $(2(2l - 1))^{\text{th}}$order US *spraying-appearing* bifurcation for a cluster of higher order sinks, sources, upper saddles, and lower saddles is

$$(2(2l-1))^{\text{th}}\text{order US} \xrightarrow[\text{appearing bifurcation}]{\text{a cluster of } 2n\text{-XX}} \begin{cases} (\alpha_{2n})^{\text{th}} \text{ order XX for } x^* = a_{2n}, \\ (\alpha_{2n-1})^{\text{th}}\text{order XX for } x^* = a_{2n-1}, \\ \vdots \\ (\alpha_1)^{\text{th}}\text{order XX for } x^* = a_1; \end{cases} \tag{6.144}$$

where $2(2l-1) = \sum_{i=1}^{n} \alpha_i$ and the minimum and maximum equilibriums satisfy

$$\begin{aligned} (\alpha_{2n})^{\text{th}} \text{ order XX} &= \begin{cases} (2r_{2n})^{\text{th}} \text{ order US}, & \text{for } \alpha_{2n} = 2r_n, \\ (2r_{2n}-1)^{\text{th}} \text{ order SO}, & \text{for } \alpha_{2n} = 2r_n - 1; \end{cases} \\ (\alpha_1)^{\text{th}} \text{ order XX} &= \begin{cases} (2r_1)^{\text{th}} \text{ order US}, & \text{for } \alpha_1 = 2r_1, \\ (2r_1-1)^{\text{th}} \text{ order SO}, & \text{for } \alpha_1 = 2r_1 - 1. \end{cases} \end{aligned} \tag{6.145}$$

(ii) The $(2(2l))^{\text{th}}$order US *spraying-appearing* bifurcation for a cluster of higher order sinks, sources, upper saddles, and lower saddles is

$$(2(2l))^{\text{th}}\text{order US} \xrightarrow[\text{appearing bifurcation}]{\text{a cluster of } 2n\text{-XX}} \begin{cases} (\alpha_{2n})^{\text{th}} \text{ order XX for } x^* = a_{2n}, \\ (\alpha_{2n-1})^{\text{th}}\text{order XX for } x^* = a_{2n-1}, \\ \vdots \\ (\alpha_1)^{\text{th}}\text{order XX for } x^* = a_1; \end{cases} \tag{6.146}$$

where $2(2l) = \sum_{i=1}^{n} \alpha_i$ and the minimum and maximum equilibriums satisfy

$$\begin{aligned} (\alpha_{2n})^{\text{th}} \text{ order XX} &= \begin{cases} (2r_{2n})^{\text{th}} \text{ order US}, & \text{for } \alpha_{2n} = 2r_n, \\ (2r_{2n}-1)^{\text{th}} \text{ order SO}, & \text{for } \alpha_{2n} = 2r_n - 1; \end{cases} \\ (\alpha_1)^{\text{th}} \text{ order XX} &= \begin{cases} (2r_1)^{\text{th}} \text{ order US}, & \text{for } \alpha_1 = 2r_1, \\ (2r_1-1)^{\text{th}} \text{ order SI}, & \text{for } \alpha_1 = 2r_1 - 1. \end{cases} \end{aligned} \tag{6.147}$$

For the higher order lower-saddle bifurcation, the cluster of the higher order equilibriums is given by the following two cases.

(iii) The $(2(2l-1))^{\text{th}}$order LS *spraying-appearing* bifurcation for a cluster of higher order sinks, sources, upper saddles, and lower saddles is

$$(2(2l-1))^{\text{th}}\text{order LS} \xrightarrow[\text{appearing bifurcation}]{\text{a cluster of } 2n\text{-XX}} \begin{cases} (\alpha_{2n})^{\text{th}} \text{ order XX, for } x^* = a_{2n}, \\ (\alpha_{2n-1})^{\text{th}} \text{order XX, for } x^* = a_{2n-1}, \\ \vdots \\ (\alpha_1)^{\text{th}} \text{order XX, for } x^* = a_1; \end{cases} \tag{6.148}$$

where $2(2l-1) = \sum_{i=1}^{n} \alpha_i$ and the minimum and maximum equilibriums satisfy

$$\begin{aligned} (\alpha_{2n})^{\text{th}} \text{ order XX} &= \begin{cases} (2r_{2n})^{\text{th}} \text{ order LS}, & \text{for } \alpha_{2n} = 2r_n, \\ (2r_{2n}-1)^{\text{th}} \text{ order SI}, & \text{for } \alpha_{2n} = 2r_n - 1; \end{cases} \\ (\alpha_1)^{\text{th}} \text{ order XX} &= \begin{cases} (2r_1)^{\text{th}} \text{ order LS}, & \text{for } \alpha_1 = 2r_1, \\ (2r_1-1)^{\text{th}} \text{ order SI}, & \text{for } \alpha_1 = 2r_1 - 1. \end{cases} \end{aligned} \tag{6.149}$$

(iv) The $(2(2l))^{\text{th}}$order LS *spraying-appearing* bifurcation for a cluster of higher order sinks, sources, upper saddles, and lower saddles is

$$(2(2l))^{\text{th}}\text{order LS} \xrightarrow[\text{appearing bifurcation}]{\text{a cluster of } 2n\text{-XX}} \begin{cases} (\alpha_{2n})^{\text{th}} \text{ order XX, for } x^* = a_{2n}, \\ (\alpha_{2n-1})^{\text{th}} \text{order XX, for } x^* = a_{2n-1}, \\ \vdots \\ (\alpha_1)^{\text{th}} \text{order XX, for } x^* = a_1; \end{cases} \tag{6.150}$$

where $2(2l) = \sum_{i=1}^{n} \alpha_i$ and the minimum and maximum equilibriums satisfy

$$\begin{aligned} (\alpha_{2n})^{\text{th}} \text{ order XX} &= \begin{cases} (2r_{2n})^{\text{th}} \text{ order LS}, & \text{for } \alpha_{2n} = 2r_n, \\ (2r_{2n}-1)^{\text{th}} \text{ order SI}, & \text{for } \alpha_{2n} = 2r_n - 1; \end{cases} \\ (\alpha_1)^{\text{th}} \text{ order XX} &= \begin{cases} (2r_1)^{\text{th}} \text{ order LS}, & \text{for } \alpha_1 = 2r_1, \\ (2r_1-1)^{\text{th}} \text{ order SO}, & \text{for } \alpha_1 = 2r_1 - 1. \end{cases} \end{aligned} \tag{6.151}$$

A set of paralleled, different, higher order upper-saddle-node *appearing* bifurcations with multiplicity in the $(2m)^{\text{th}}$-degree polynomial system is the $((2\beta_1)^{\text{th}}\text{US} : (2\beta_2)^{\text{th}}\text{US} : \cdots : (2\beta_s)^{\text{th}}\text{US})$ parallel *appearing* bifurcation for clusters of higher order sinks, sources, upper saddles, and lower saddles. For the $(2\beta_i)^{\text{th}}\text{US}$

$$(2\beta_i)^{\text{th}}\text{US} = \begin{cases} (2(2l_i-1))^{\text{th}}\text{order US}, & \text{for } \beta_i = 2l_i - 1, \\ (2(2l_i))^{\text{th}} \text{ order US}, & \text{for } \beta_i = 2l_i; \end{cases} . \tag{6.152}$$

Similarly, the following notation is introduced as

$$((2\beta_1)^{\text{th}}\text{US} : (2\beta_2)^{\text{th}}\text{US} : \cdots : (2\beta_s)^{\text{th}}\text{US}) = (2\beta_1 : 2\beta_2 : \cdots : 2\beta_s)^{\text{th}}\text{US}. \quad (6.153)$$

Thus, the paralleled $(2\beta_1 : 2\beta_2 : \cdots : 2\beta_s)^{\text{th}}$ US spraying-*appearing* bifurcation is called the $(2\beta_1 : 2\beta_2 : \cdots : 2\beta_s)^{\text{th}}$ US *sprinkler-spraying-appearing* bifurcation for the higher order equilibriums. Similarly, a set of paralleled different lower-saddle appearing bifurcations for higher order singularity of equilibriums is called the $((2\beta_1)^{\text{th}}\text{LS} : (2\beta_2)^{\text{th}}\text{LS} : \cdots : (2\beta_s)^{\text{th}}\text{LS})$ parallel appearing bifurcation in the $(2m)^{\text{th}}$-degree polynomial system. Thus, the paralleled $(2\beta_1 : 2\beta_2 : \cdots : 2\beta_s)^{\text{th}}$ LS bifurcation is also called the $(2\beta_1 : 2\beta_2 : \cdots : 2\beta_s)^{\text{th}}$ LS *sprinkler-spraying*-appearing bifurcation for higher order equilibriums.

The $(2\alpha_1 : 2\alpha_2 : \cdots : 2\alpha_n)^{\text{th}}$US and $(2\alpha_1 : 2\alpha_2 : \cdots : 2\alpha_n)^{\text{th}}$LS *teeth comb appearing* bifurcations for the higher order singularity of equilibriums are presented in Figs. 6.6a, b for $a_0 > 0$ and $a_0 < 0$, respectively. The components of the *teeth comb appearing* bifurcation are

$$\begin{aligned}
&(2\alpha_j)^{\text{th}}\text{US} \xrightarrow[\text{appearing}]{\alpha_j = 2r_j} \begin{cases} (2r_j)^{\text{th}}\text{US} \\ (2r_j)^{\text{th}}\text{US} \end{cases} (j = i, n-1, \ldots), \\
&(2\alpha_k)^{\text{th}}\text{US} \xrightarrow[\text{appearing}]{\alpha_k = 2r_k - 1} \begin{cases} (2r_k - 1)^{\text{th}}\text{SO} \\ (2r_k - 1)^{\text{th}}\text{SI} \end{cases} (k = 1, n, \ldots);
\end{aligned} \quad (6.154)$$

and

$$\begin{aligned}
&(2\alpha_j)^{\text{th}}\text{LS} \xrightarrow[\text{appearing}]{\alpha_j = 2r_j} \begin{cases} (2r_j)^{\text{th}}\text{LS} \\ (2r_j)^{\text{th}}\text{LS} \end{cases} (j = i, n-1, \ldots), \\
&(2\alpha_k)^{\text{th}}\text{LS} \xrightarrow[\text{appearing}]{\alpha_k = 2r_k - 1} \begin{cases} (2r_k - 1)^{\text{th}}\text{SI} \\ (2r_k - 1)^{\text{th}}\text{SO} \end{cases} (k = 1, n, \ldots).
\end{aligned} \quad (6.155)$$

The $(2\beta_1 : 2\beta_2 : \cdots : 2\beta_n)^{\text{th}}$US and $(2\beta_1 : 2\beta_2 : \cdots : 2\beta_s)^{\text{th}}$LS *sprinkler-spraying-appearing* bifurcations for the higher order singularity of equilibriums are presented in Fig. 6.6c, d for $a_0 > 0$ and $a_0 < 0$, respectively. The components of the *sprinkler-spraying-appearing* bifurcation are

$$\begin{aligned}
&(2\beta_1 : 2\beta_2 : \cdots : 2\beta_n)^{\text{th}}\text{US} \\
&= ((2(2l_1 - 1) : \cdots : 2(2l_i) : \cdots : 2(2l_{n-1}) : 2(2l_n))^{\text{th}}\text{US}
\end{aligned} \quad (6.156)$$

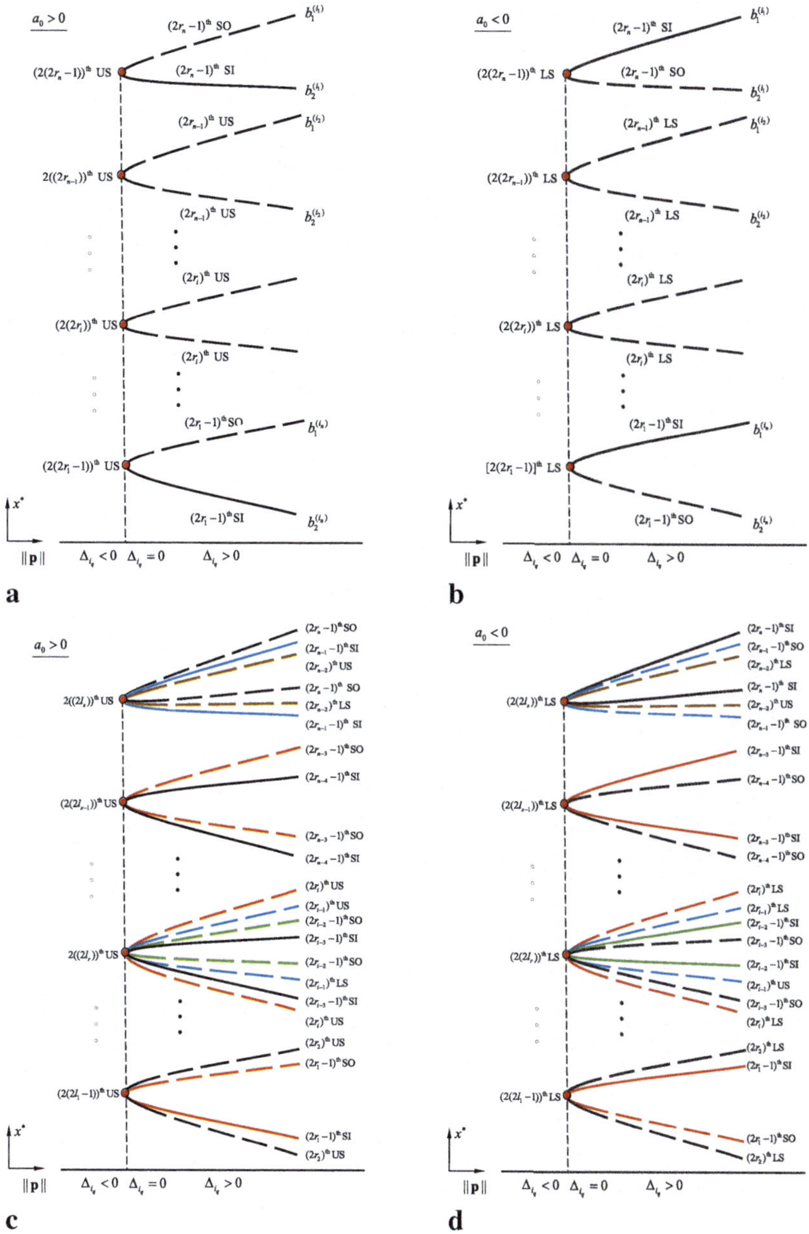

Fig. 6.6 The *teeth comb appearing* bifurcations of $(2(2r_1 - 1) : \cdots : 2(2r_{n-1}) : 2(2r_n - 1))^{\text{th}}$XX: (**a**) XX = US ($a_0 > 0$) and (**b**) XX = LS ($a_0 < 0$). The *sprinkler-spraying-appearing* bifurcations of $(2(2l_1 - 1) : \cdots : (2(2l_{n-1}) : 2(2l_n))^{\text{th}}$XX: (**c**) XX = US ($a_0 > 0$) and (**d**) XX = LS ($a_0 < 0$). LS: lower saddle, US: upper saddle, SI: sink, SO: source. Stable and unstable equilibriums are represented by solid and dashed curves, respectively. The bifurcation points are marked by circular symbols

and

$$\begin{aligned}&(2\beta_1 : 2\beta_2 : \cdots : 2\beta_n)^{\text{th}}\text{LS}\\&= ((2(2l_1 - 1) : \cdots : 2(2l_i) : \cdots : 2(2l_{n-1}) : 2(2l_n))^{\text{th}}\text{LS}\end{aligned} \tag{6.157}$$

For a cluster of m-quadratics, $B_i = B_j$ $(i, j \in \{1, 2, \cdots, n\}; i \neq j)$ and $\Delta_i = 0$ $(i = 1, 2, \cdots, n)$. The $(2m)^{\text{th}}$-order upper-saddle *appearing* bifurcation for n-pairs of the higher order singularity of equilibriums is

$$(2m)^{\text{th}}\text{order US} \xrightarrow[\text{appearing bifurcation}]{\text{a cluster of } 2n\text{-XX}} \begin{cases}(\alpha_{2n})^{\text{th}} \text{ order XX for } x^* = a_{2n},\\(\alpha_{2n-1})^{\text{th}}\text{order XX for } x^* = a_{2n-1},\\\vdots\\(\alpha_1)^{\text{th}}\text{order XX for } x^* = a_1;\end{cases} \tag{6.158}$$

where

$$2m = 2(2l) = \sum_{i=1}^{2n} \alpha_i, 2m = 2(2l-1) = \sum_{i=1}^{2n} \alpha_i. \tag{6.159}$$

The $(2m)^{\text{th}}$-order lower-saddle-node appearing bifurcation for higher order equilibriums is

$$(2m)^{\text{th}}\text{order LS} \xrightarrow[\text{appearing bifurcation}]{\text{a cluster of } 2n\text{-XX}} \begin{cases}(\alpha_{2n})^{\text{th}} \text{ order XX for } x^* = a_{2n},\\(\alpha_{2n-1})^{\text{th}}\text{order XX for } x^* = a_{2n-1},\\\vdots\\(\alpha_1)^{\text{th}}\text{order XX for } x^* = a_1.\end{cases} \tag{6.160}$$

The $(2m)^{\text{th}}$-order upper-saddle appearing bifurcation with n-pairs of higher order singularity of equilibriums is a *sprinkler-spraying cluster* of the n-pairs of higher order singularity of equilibriums. The $(2m)^{\text{th}}$-order lower-saddle *appearing* bifurcation with n-pairs of higher-order equilibriums is also a *sprinkler-spraying cluster* of the n-pairs of higher order singularity of equilibriums.

Thus, the $(2m)^{\text{th}}$-order US bifurcation $(a_0 > 0)$ and $(2m)^{\text{th}}$-order LS bifurcation $(a_0 < 0)$ are presented in Fig. 6.7a–d, respectively. The $(2m)^{\text{th}}$-order upper-saddle appearing bifurcation for higher order singularity of equilibriums is called the $(2m)^{\text{th}}$-order US *sprinkler-spraying-appearing* bifurcation, and the $(2m)^{\text{th}}$-order lower-saddle-node *appearing* bifurcation for higher order singularity of equilibriums is also called the $(2m)^{\text{th}}$-order LS *sprinkler-spraying*-appearing bifurcation.

A series of the saddle-node bifurcations for higher order singularity of equilibriums are aligned up with varying parameters, which is formed in a special pattern. For n-quadratics in the $(2m)^{\text{th}}$-order polynomial systems, the following conditions should be satisfied:

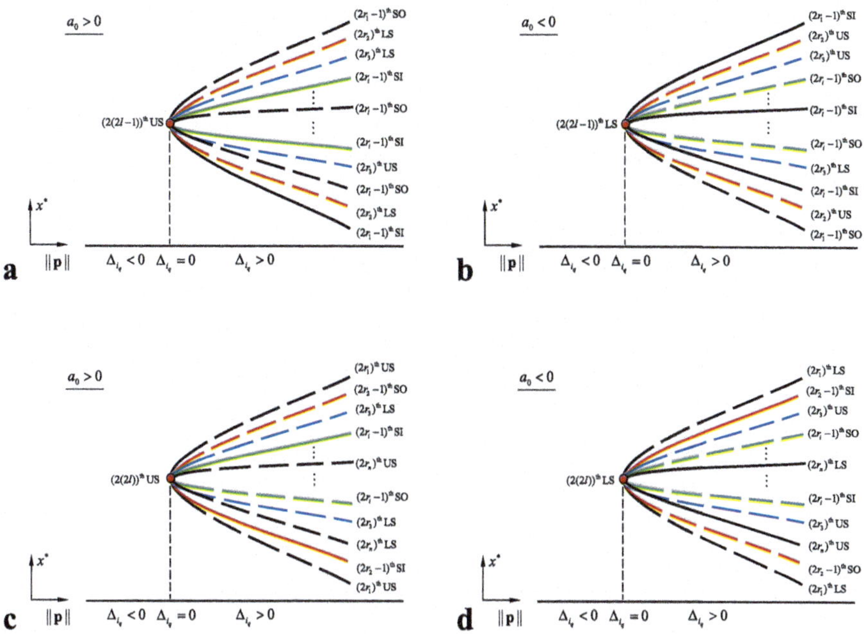

Fig. 6.7 Spraying appearing bifurcations for higher-order equilibriums in the $(2m)^{\text{th}}$ polynomial system: (**a**) $(2(2l-1))^{\text{th}}$ US spraying-appearing bifurcation ($a_0 > 0$), (**b**) $(2(2l-1))^{\text{th}}$ LS spraying appearing bifurcation ($a_0 < 0$), (**c**) $(2(2l))^{\text{th}}$ US spraying appearing bifurcation ($a_0 > 0$), (**d**) $(2(2l))^{\text{th}}$ US spraying appearing bifurcation ($a_0 < 0$). LS: lower-saddle, US: upper-saddle, SI: sink, SO: source. Stable and unstable equilibriums are represented by solid and dashed curves, respectively. The bifurcation points are marked by circular symbols

$$
\begin{aligned}
&B_i \approx B_j \; i,j \in \{1,2,\ldots,s\}; i \neq j,\\
&\Delta_i > \Delta_{i+1} \; (i = 1,2,\ldots,s; s \leq n < m),\\
&\Delta_i = 0 \text{ with } \|\mathbf{p}_i\| < \|\mathbf{p}_{i+1}\|.
\end{aligned}
\tag{6.161}
$$

The two series of the fish-scale switching bifurcations in Fig. 6.8a, c for $a_0 < 0$ have the following detailed structures:

$$
\left\{
\begin{array}{l}
(2(2r_1-1))^{\text{th}}\text{US} \rightarrow \left\{ \begin{array}{l} (2r_1-1)^{\text{th}}\text{SO},\\ (2r_1-1)^{\text{th}}\text{SI}; \end{array}\right.\\
(2(2r_2))^{\text{th}}\text{LS} \rightarrow \left\{ \begin{array}{l} (2r_2)^{\text{th}}\text{LS},\\ (2r_2)^{\text{th}}\text{LS}; \end{array}\right.\\
\vdots\\
(2(2r_n-1))^{\text{th}}\text{US} \rightarrow \left\{ \begin{array}{l} (2r_n-1)^{\text{th}}\text{SO},\\ (2r_n-1)^{\text{th}}\text{SI}; \end{array}\right.
\end{array}
\right.
\tag{6.162}
$$

and

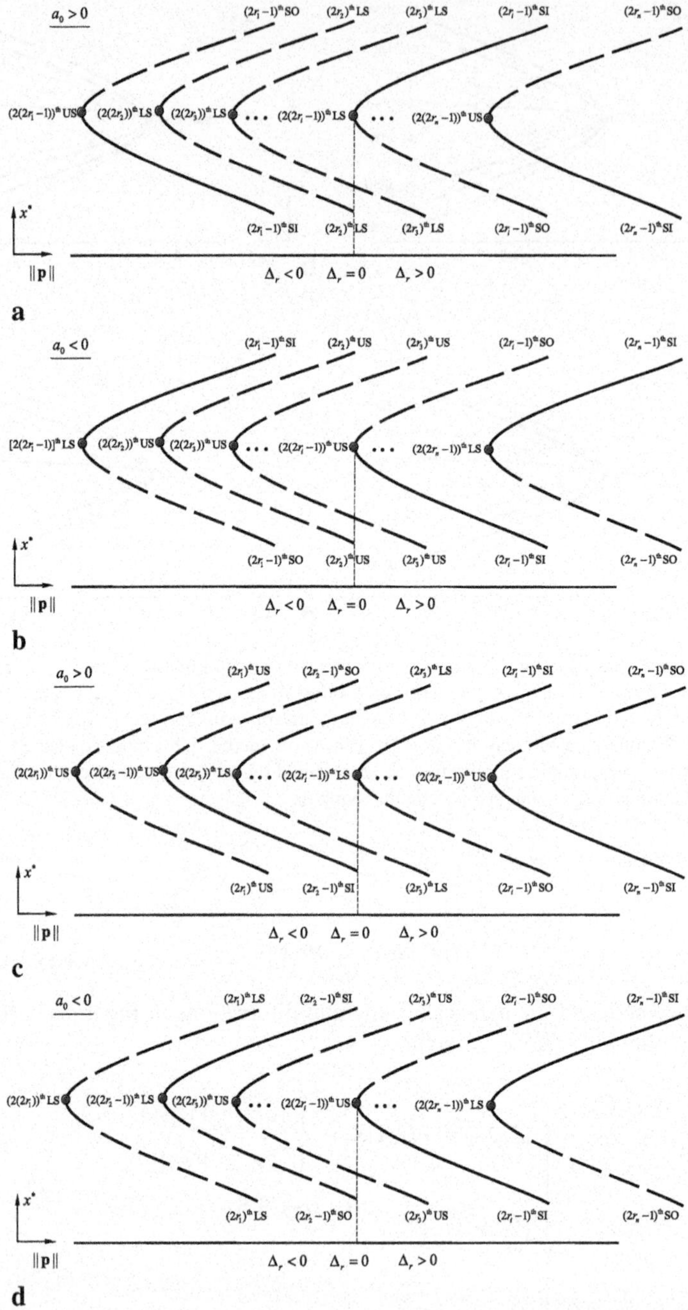

Fig. 6.8 The *fish-scale appearing* bifurcation patterns in a $(2m)^{\text{th}}$-degree polynomial system: (**a**) $(2(2r_1-1))^{\text{th}}$US - $(2(2r_2))^{\text{th}}$LS - $\cdots$ ($a_0 > 0$), (**b**) $(2(2r_1-1))^{\text{th}}$LS - $(2(2r_2))^{\text{th}}$US - $\cdots$ ($a_0 < 0$), (**c**) $(2(2r_1))^{\text{th}}$US - $(2(2r_2))^{\text{th}}$US - $\cdots$ ($a_0 > 0$), (**d**) $(2(2r_1))^{\text{th}}$LS - $(2(2r_2-1))^{\text{th}}$LS - $\cdots$ ($a_0 < 0$). LS: lower saddle, US: upper saddle, SI: sink, SO: source. Stable and unstable equilibriums are represented by solid and dashed curves, respectively. The bifurcation points are marked by circular symbols

$$\begin{cases} (2(2r_1))^{\text{th}}\text{US} \to \begin{cases} (2r_1)^{\text{th}}\text{US}, \\ (2r_1)^{\text{th}}\text{US}; \end{cases} \\ (2(2r_2-1))^{\text{th}}\text{US} \to \begin{cases} (2r_2-1)^{\text{th}}\text{SO}, \\ (2r_2-1)^{\text{th}}\text{SI}; \end{cases} \\ \vdots \\ (2(2r_n-1))^{\text{th}}\text{US} \to \begin{cases} (2r_n-1)^{\text{th}}\text{SO}, \\ (2r_n)^{\text{th}}\text{SI}. \end{cases} \end{cases} \tag{6.163}$$

Two series of fish-scale *appearing* bifurcations in Fig. 6.8b, d for $a_0 < 0$ have the following structures as

$$\begin{cases} (2(2r_1-1))^{\text{th}}\text{LS} \to \begin{cases} (2r_1-1)^{\text{th}}\text{SI}, \\ (2r_1-1)^{\text{th}}\text{SO}; \end{cases} \\ (2(2r_2))^{\text{th}}\text{US} \to \begin{cases} (2r_2)^{\text{th}}\text{US}, \\ (2r_2)^{\text{th}}\text{US}; \end{cases} \\ \vdots \\ (2(2r_n-1))^{\text{th}}\text{LS} \to \begin{cases} (2r_n-1)^{\text{th}}\text{SI}, \\ (2r_n-1)^{\text{th}}\text{SO}; \end{cases} \end{cases} \tag{6.164}$$

and

$$\begin{cases} (2(2r_1))^{\text{th}}\text{LS} \to \begin{cases} (2r_1)^{\text{th}}\text{LS}, \\ (2r_1)^{\text{th}}\text{LS}; \end{cases} \\ (2(2r_2-1))^{\text{th}}\text{LS} \to \begin{cases} (2r_2-1)^{\text{th}}\text{SI}, \\ (2r_2-1)^{\text{th}}\text{SO}; \end{cases} \\ \vdots \\ (2(2r_n-1))^{\text{th}}\text{LS} \to \begin{cases} (2r_n-1)^{\text{th}}\text{SI}, \\ (2r_n)^{\text{th}}\text{SO}. \end{cases} \end{cases} \tag{6.165}$$

The four *fish-scale appearing* bifurcation patterns for higher order equilibriums are different from the fish-scale appearing bifurcation patterns for simple equilibriums.

6.3.2 Switching Bifurcations

Consider the roots of $(x^2 + B_i x + C_i)^{\alpha_i} = 0$ as

$$\begin{aligned}
&B_i = -(b_1^{(i)} + b_2^{(i)}), \Delta_i = (b_1^{(i)} - b_2^{(i)})^2 \geq 0,\\
&x_{1,2}^{(i)} = b_{1,2}^{(i)}, \Delta_i > 0 \text{ if } b_1^{(i)} \neq b_2^{(i)} (i = 1, 2, \ldots, n);\\
&\left.\begin{aligned}&B_i \neq B_j \ (i, j = 1, 2, \ldots, n; i \neq j)\\ &\Delta_i = 0 \text{ at } b_1^{(i)} = b_2^{(i)} (i = 1, 2, \ldots, n)\end{aligned}\right\} \text{at bifurcation.}
\end{aligned} \tag{6.166}$$

The α_i^{th}-order singularity bifurcation is for the *switching* of a pair of higher order equilibriums (i.e., sinks, sources, upper saddles, and lower saddles). There are six *switching* bifurcations for $i \in \{1, 2, \ldots, n\}$:

$$(2l_i)^{\text{th}}\text{order US} \xrightarrow[\text{switching bifurcation}]{l_i = r_1^{(i)} + r_2^{(i)} - 1}
\begin{cases}
(2r_2^{(i)} - 1)^{\text{th}} \text{ order SO} \downarrow \text{SI}, \ \text{for } b_2^{(i)} = a_{2i} \downarrow a_{2i-1},\\
(2r_1^{(i)} - 1)^{\text{th}} \text{ order SI} \uparrow \text{SO}, \ \text{for } b_1^{(i)} = a_{2i-1} \uparrow a_{2i};
\end{cases} \tag{6.167}$$

$$(2l_i)^{\text{th}}\text{order LS} \xrightarrow[\text{switching bifurcation}]{l_i = r_1^{(i)} + r_2^{(i)} - 1}
\begin{cases}
(2r_2^{(i)} - 1)^{\text{th}} \text{ order SI} \downarrow \text{SO}, \ \text{for } b_2^{(i)} = a_{2i} \downarrow a_{2i-1},\\
(2r_1^{(i)} - 1)^{\text{th}} \text{ order SO} \uparrow \text{SI}, \ \text{for } b_1^{(i)} = a_{2i-1} \uparrow a_{2i};
\end{cases} \tag{6.168}$$

$$(2l_i)^{\text{th}}\text{order US} \xrightarrow[\text{switching bifurcation}]{l_i = r_1^{(i)} + r_2^{(i)}}
\begin{cases}
(2r_2^{(i)})^{\text{th}} \text{ order US} \downarrow \text{US}, \ \text{for } b_2^{(i)} = a_{2i} \downarrow a_{2i-1},\\
(2r_1^{(i)})^{\text{th}} \text{ order US} \uparrow \text{US} \ \text{for } b_1^{(i)} = a_{2i-1} \uparrow a_{2i};
\end{cases} \tag{6.169}$$

$$(2l_i)^{\text{th}}\text{order LS} \xrightarrow[\text{switching bifurcation}]{l_i = r_1^{(i)} + r_2^{(i)}}
\begin{cases}
(2r_2^{(i)})^{\text{th}} \text{ order LS} \downarrow \text{LS}, \ \text{for } b_2^{(i)} = a_{2i} \downarrow a_{2i-1},\\
(2r_1^{(i)})^{\text{th}} \text{ order LS} \uparrow \text{LS} \ \text{for } b_1^{(i)} = a_{2i-1} \uparrow a_{2i};
\end{cases} \tag{6.170}$$

$$(2l_i-1)^{\text{th}}\text{order SO}\xrightarrow[\text{switching bifurcation}]{l_i=r_1^{(i)}+r_2^{(i)}}\begin{cases}(2r_2^{(i)}-1)^{\text{th}}\text{ order SO}\downarrow\text{SO, for } b_2^{(i)}=a_{2i}\downarrow a_{2i-1},\\(2r_1^{(i)})^{\text{th}}\text{ order LS}\uparrow\text{US for } b_1^{(i)}=a_{2i-1}\uparrow a_{2i};\end{cases}\tag{6.171}$$

$$(2l_i-1)^{\text{th}}\text{order SI}\xrightarrow[\text{switching bifurcation}]{l_i=r_1^{(i)}+r_2^{(i)}}\begin{cases}(2r_2^{(i)}-1)^{\text{th}}\text{ order SI}\downarrow\text{SI, for } b_2^{(i)}=a_{2i}\downarrow a_{2i-1},\\(2r_1^{(i)})^{\text{th}}\text{ order US}\uparrow LS\text{ for } b_1^{(i)}=a_{2i-1}\uparrow a_{2i}.\end{cases}\tag{6.172}$$

A set of n-paralleled higher order XX *switching* bifurcations is called the $((a_1)^{\text{th}}\text{XX}:(a_2)^{\text{th}}\text{XX}:\ldots:(a_n)^{\text{th}}\text{XX})$ *parallel switching* bifurcation in the $(2m)^{\text{th}}$-degree polynomial nonlinear system. Such a bifurcation is also called the n-$((a_1)^{\text{th}}\text{XX}:(a_2)^{\text{th}}\text{XX}:\ldots:(a_n)^{\text{th}}\text{XX})$ *antenna switching* bifurcation. $\alpha_i\in\{2l_i,2l_i-1\}$ and XX $\in$ {SO, SI, US, LS}. For non-switching points, $\Delta_i>0$ at $b_1^{(i)}\neq b_2^{(i)}$ ($i=1,2,\ldots,n$). At the bifurcation point, $\Delta_i=0$ at $b_1^{(i)}=b_2^{(i)}$ ($i=1,2,\ldots,n$). The n-$((a_1)^{\text{th}}\text{XX}:(a_2)^{\text{th}}\text{XX}:\ldots:(a_n)^{\text{th}}\text{XX})$ *parallel antenna switching* bifurcation is

$$\begin{cases}\alpha_n^{\text{th}}\text{XX}_n\xrightarrow[\textit{switching}]{n^{\text{th}}\textit{bifurcation}}\begin{cases}(r_2^{(n)})^{\text{th}}\text{XX}_2^{(n)}\downarrow\text{YY}_1^{(n)}, & \text{for } b_2^{(n)}=a_{2n}\downarrow a_{2n-1},\\(r_1^{(n)})^{\text{th}}\text{XX}_1^{(n)}\uparrow\text{YY}_2^{(n)}, & \text{for } b_1^{(n)}=a_{2n-1}\uparrow a_{2n};\end{cases}\\ \vdots\\ \alpha_2^{\text{th}}\text{XX}_2\xrightarrow[\textit{switching}]{2^{\text{nd}}\textit{bifurcation}}\begin{cases}(r_2^{(2)})^{\text{th}}\text{XX}_2^{(2)}\downarrow\text{YY}_1^{(2)}, & \text{for } b_2^{(2)}=a_4\downarrow a_3,\\(r_1^{(2)})^{\text{th}}\text{XX}_1^{(2)}\uparrow\text{YY}_2^{(2)}, & \text{for } b_1^{(2)}=a_3\uparrow a_4;\end{cases}\\ \alpha_1^{\text{th}}\text{XX}_1\xrightarrow[\textit{switching}]{1^{st}\textit{bifurcation}}\begin{cases}(r_2^{(1)})^{\text{th}}\text{XX}_2^{(1)}\downarrow\text{YY}_1^{(1)}, & \text{for } b_2^{(1)}=a_2\downarrow a_1,\\(r_1^{(1)})^{\text{th}}\text{XX}_1^{(1)}\uparrow\text{YY}_2^{(1)}, & \text{for } b_1^{(1)}=a_1\uparrow a_2.\end{cases}\end{cases}\tag{6.173}$$

Such eight sets of parallel *switching* bifurcations of $((a_1)^{\text{th}}\text{XX}:(a_2)^{\text{th}}\text{XX}:\ldots:(a_n)^{\text{th}}\text{XX})$ are presented in Fig. 6.9(a, c, e, g) and (b, d, f, h) for $a_0>0$ and $a_0<0$, respectively. The eight switching bifurcation structures are as follows:

(a) $((2l_1)^{\text{th}}\text{US}:\cdots:(2l_{n-1}-1)^{\text{th}}\text{SO}:(2l_n)^{\text{th}}\text{US})$ for $a_0>0$,
(b) $((2l_1)^{\text{th}}\text{LS}:\cdots:(2l_{n-1}-1)^{\text{th}}\text{SI}:(2l_n)^{\text{th}}\text{LS})$ for $a_0<0$,
(c) $((2l_1)^{\text{th}}\text{LS}:\cdots:(2l_{n-1}-1)^{\text{th}}\text{SI}:(2l_n-1)^{\text{th}}\text{SO})$ for $a_0>0$,
(d) $((2l_1)^{\text{th}}\text{US}:\cdots:(2l_{n-1}-1)^{\text{th}}\text{SO}:(2l_n-1)^{\text{th}}\text{SI})$ for $a_0<0$,
(e) $((2l_1)^{\text{th}}\text{LS}:\cdots:(2l_{n-1}-1)^{\text{th}}\text{SI}:(2l_n-1)^{\text{th}}\text{SO})$ for $a_0>0$,
(f) $((2l_1)^{\text{th}}\text{US}:\cdots:(2l_{n-1}-1)^{\text{th}}\text{SI}:(2l_n-1)^{\text{th}}\text{SI})$ for $a_0<0$,
(g) $((2l_1)^{\text{th}}\text{US}:\cdots:(2l_{n-1}-1)^{\text{th}}\text{SO}:(2l_n)^{\text{th}}\text{US})$ for $a_0>0$,
(h) $((2l_1)^{\text{th}}\text{LS}:\cdots:(2l_{n-1}-1)^{\text{th}}\text{SI}:(2l_n)^{\text{th}}\text{LS})$ for $a_0<0$.

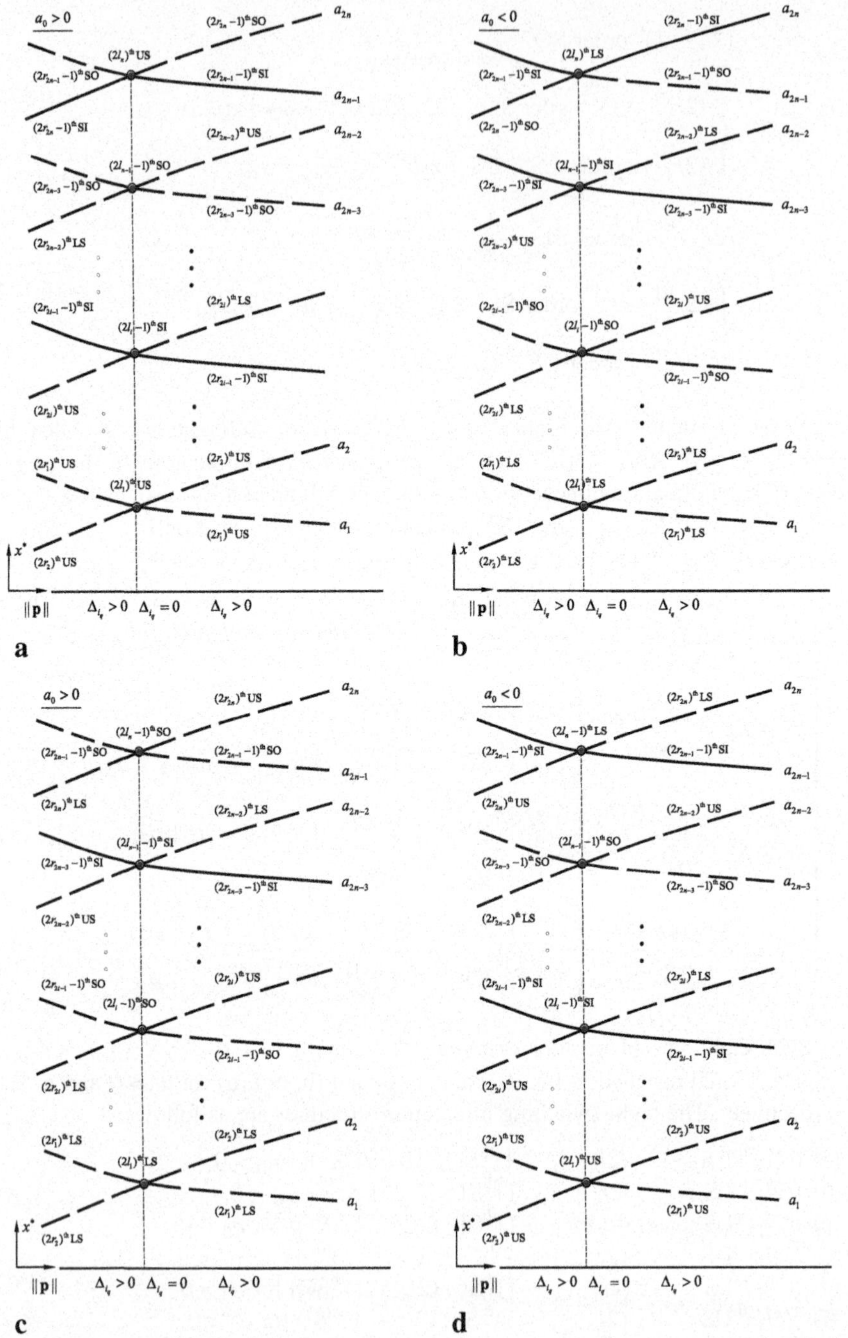

Fig. 6.9 Parallel *antenna* switching bifurcations for high-order equilibriums in a $(2m)^{\text{th}}$-degree polynomial system. $((a_1)^{\text{th}}\text{XX}:(a_2)^{\text{th}}\text{XX}:\ldots:(a_n)^{\text{th}}\text{XX})$: (**a**, **c**, **e**, **g**) for $a_0 > 0$. (**b**, **d**, **f**, **h**) for $a_0 < 0$. LS: lower saddle, US: upper saddle, SI: sink, SO: source. Stable and unstable equilibriums are represented by solid and dashed curves, respectively. The bifurcation points are marked by circular symbols

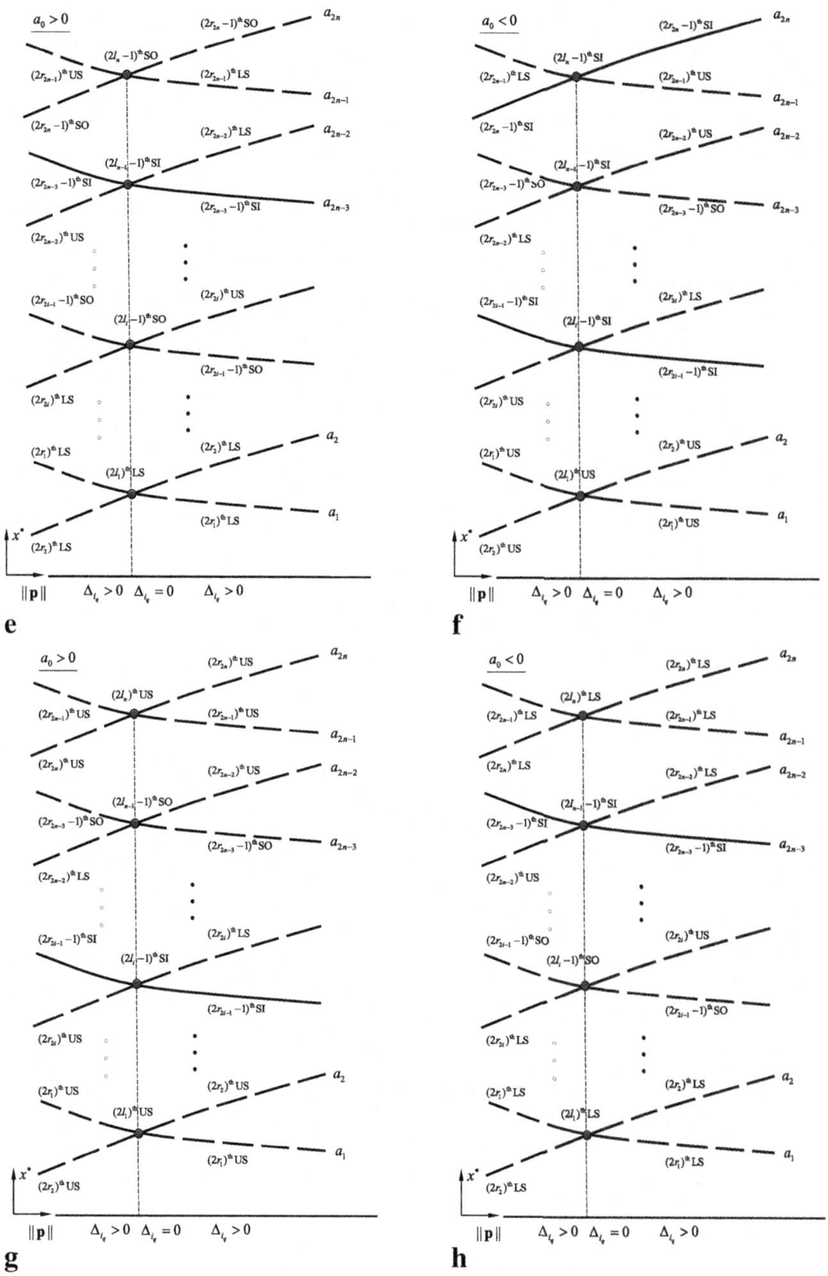

Fig. 6.9 (continued)

The same switching bifurcations with different higher order equilibriums are illustrated, which is different from the m-USN and m-LSN for simple sinks and sources.

Consider a switching bifurcation for a cluster of higher order equilibriums with the following conditions:

$$\left.\begin{array}{l} B_i = -(b_1^{(i)} + b_2^{(i)}),\, \Delta_i = (b_1^{(i)} - b_2^{(i)})^2 \geq 0, \\ x_{1,2}^{(i)} = b_{1,2}^{(i)},\, \Delta_i > 0 \text{ if } b_1^{(i)} \neq b_2^{(i)} (i = 1, 2, \ldots, n); \\ \left.\begin{array}{l} B_i = B_j \; (i, j \in \{1, 2, \ldots, n\}; i \neq j) \\ \Delta_i = 0 \text{ at } b_1^{(i)} = b_2^{(i)} \; (i = 1, 2, \ldots, n) \end{array}\right\} \text{at bifurcation.} \end{array}\right. \tag{6.174}$$

Thus, the $(\alpha_i)^{\text{th}}$-order *switching* bifurcation can be for a cluster of higher order equilibriums. The $(\alpha_i)^{\text{th}}$-order switching bifurcations for $i \in \{1, 2, \ldots, s\}$ are

$$\begin{array}{l} (\alpha_i)^{\text{th}} \text{order XX} \xrightarrow[\text{switching bifurcation}]{\alpha_i = \sum_{j=1}^{l_i} r_j^{(i)}} \\ \left\{\begin{array}{l} (r_s^{(i)})^{\text{th}} \text{ order } \text{XX}_{l_i}^{(i)} \downarrow \text{YY}_{l_i}^{(i)}, \quad \text{for } b_{l_i}^{(i)} \downarrow a_{l_i}^{(i)}, \\ \vdots \\ (r_j^{(i)})^{\text{th}} \text{ order } \text{XX}_j^{(i)} \downarrow \text{YY}_j^{(i)}, \quad \text{for } b_j^{(i)} \downarrow a_j^{(i)}, \\ \vdots \\ (r_1^{(i)})^{\text{th}} \text{ order } \text{XX}_1^{(i)} \uparrow \text{YY}_1^{(i)}, \quad \text{for } b_1^{(i)} \downarrow a_1^{(i)}, \end{array}\right. \end{array} \tag{6.175}$$

where

$$\left.\begin{array}{l} \{a_1^{(i)}, a_2^{(i)}, \ldots, a_{l_i-1}^{(i)}, a_{l_i}^{(i)}\} \underset{\text{before bifurcation}}{\subset} \text{sort}\{b_1^{(1)}, b_2^{(1)}, \ldots, b_1^{(n)}, b_2^{(n)}\}, \\ \{b_1^{(i)}, b_2^{(i)}, \ldots, b_{l_i-1}^{(i)}, b_{l_i}^{(i)}\} \underset{\text{After bifurcation}}{\subset} \text{sort}\{b_1^{(1)}, b_2^{(1)}, \ldots, b_1^{(n)}, b_2^{(n)}\}. \end{array}\right. \tag{6.176}$$

A set of paralleled, different, higher order upper-saddle-node switching bifurcations with multiplicity is the $((\alpha_1)^{\text{th}}\text{XX} : (\alpha_2)^{\text{th}}\text{XX} : \cdots : (\alpha_s)^{\text{th}}\text{XX})$ *parallel switching* bifurcation in the $(2m)^{\text{th}}$-degree polynomial system. At the *straw-bundle switching* bifurcation, $\Delta_i = 0$ $(i = 1, 2, \ldots, n)$ and $B_i = B_j$ $(i, j \in \{1, 2, \ldots, n\}; i \neq j)$. The *parallel straw-bundle switching* bifurcation for higher order equilibriums is

$$
\begin{aligned}
&((\alpha_1)^{\text{th}}\text{XX} : (\alpha_2)^{\text{th}}\text{XX} : \cdots : (\alpha_s)^{\text{th}}\text{XX})\text{-switching} \\
&= \begin{cases}
(\alpha_s)^{\text{th}} \text{ order XX switching,} \\
\vdots \\
(\alpha_2)^{\text{th}} \text{ order XX switching,} \\
(\alpha_1)^{\text{th}} \text{ order XX switching,}
\end{cases}
\end{aligned}
\tag{6.177}
$$

where

$$
\alpha_i \in \{2l_i, 2l_i - 1\} \text{ and XX} \in \{\text{US, LS, SO, SI}\}. \tag{6.178}
$$

Eight *parallel straw-bundle switching* bifurcations of $(\alpha_1^{\text{th}}\text{XX} : \alpha_2^{\text{th}}\text{XX} : \ldots : \alpha_n^{\text{th}}\text{XX})$ are presented in Figs. 6.10 and 6.11 for $a_0 > 0$ and $a_0 < 0$, respectively.

6.3.3 Appearing and Switching Bifurcations

Consider a $(2m)^{\text{th}}$-degree polynomial system in a form of

$$
\dot{x} = a_0 Q(x) \prod_{i=1}^{2n_1} (x - c_i)^{\alpha_i} \prod_{j=1}^{n_2} (x^2 + B_j x + C_j)^{\alpha_j}. \tag{6.179}
$$

Without loss of generality, a function of $Q(x) > 0$ is either a polynomial function or a non-polynomial function. The roots of $x^2 + B_j x + C_j = 0$ are

$$
b_{1,2}^{(j)} = -\frac{1}{2}B_j \pm \frac{1}{2}\sqrt{\Delta_j}, \Delta_j = B_j^2 - 4C_j \geq 0 \; (j = 1, 2, \ldots, n_2); \tag{6.180}
$$

either

$$
\begin{aligned}
&\{a_1^-, a_2^-, \ldots, a_{2n_1}^-\} = \text{sort}\{c_1, c_2, \ldots, c_{2n_1}\}, a_s^- \leq a_{s+1}^- \text{ before bifurcation} \\
&\{a_1^+, a_2^+, \ldots, a_{2n_3}^+\} = \text{sort}\{c_1, \ldots, c_{2n_1}; b_1^{(1)}, b_2^{(1)}, \ldots, b_1^{(n_2)}, b_2^{(n_2)}\}, \\
&a_s^+ \leq a_{s+1}^+, n_3 = n_1 + n_2 \text{ after bifurcation;}
\end{aligned}
\tag{6.181}
$$

or

$$
\begin{aligned}
&\{a_1^-, a_2^-, \ldots, a_{2n_3}^-\} = \text{sort}\{c_1, c_2, \ldots, c_{2n_1}; b_1^{(1)}, b_2^{(1)}, \ldots, b_1^{(n_2)}, b_2^{(n_2)}\}, \\
&a_s^- \leq a_{s+1}^-, n_3 = n_1 + n_2 \text{ before bifurcation;} \\
&\{a_1^+, a_2^+, \ldots, a_{2n_1}^+\} = \text{sort}\{c_1, \ldots, c_{2n_1}\}, a_s^+ \leq a_{s+1}^+ \text{ after bifurcation;}
\end{aligned}
\tag{6.182}
$$

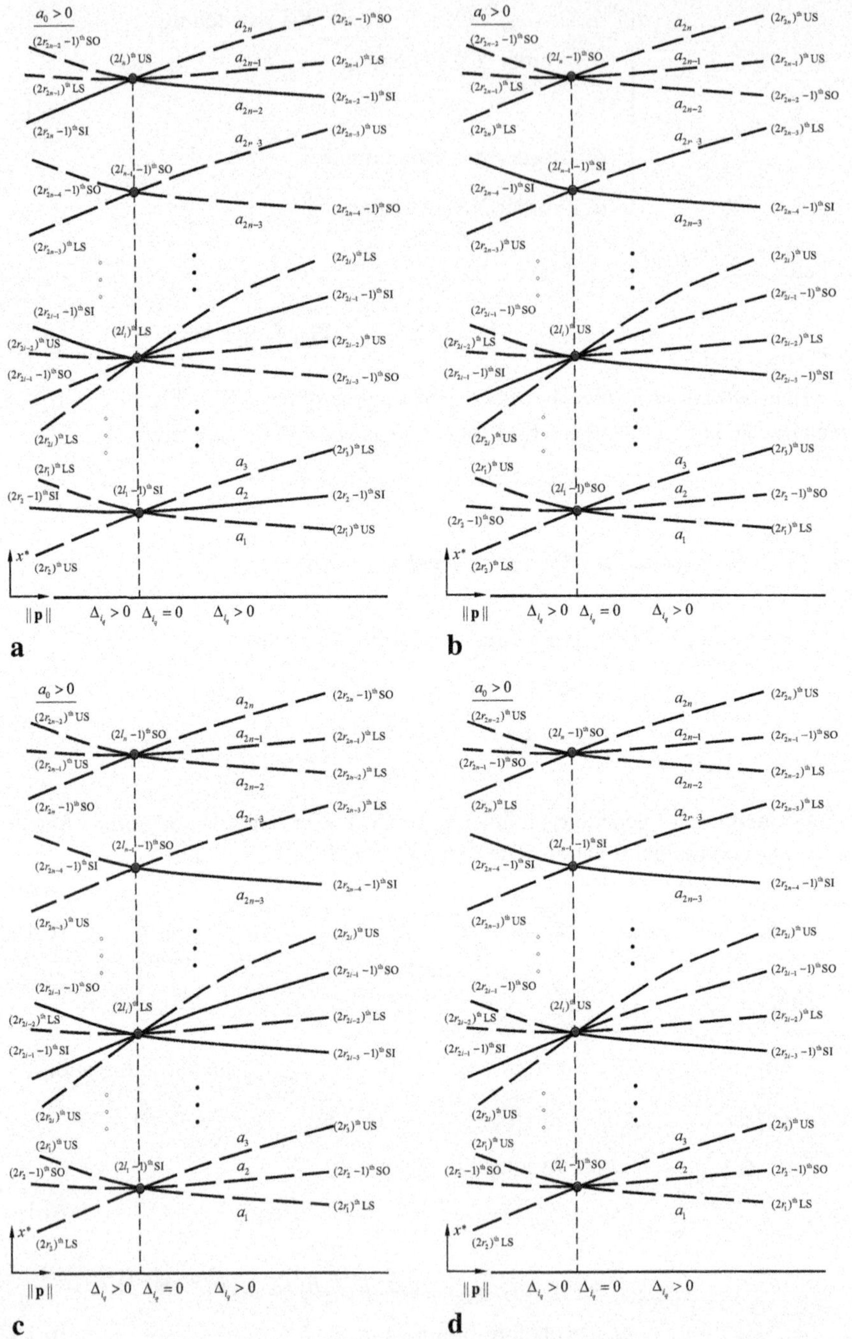

Fig. 6.10 **(a–d)** Four types of (r_1th XX : r_2th XX : $\cdots$: r_mth XX) parallel switching bifurcation for $a_0 > 0$ in the $(2m)^{\text{th}}$-degree polynomial system. LS: lower saddle, US: upper saddle, SI: sink, SO: source. Stable and unstable equilibriums are represented by solid and dashed curves, respectively. The bifurcation points are marked by circular symbols

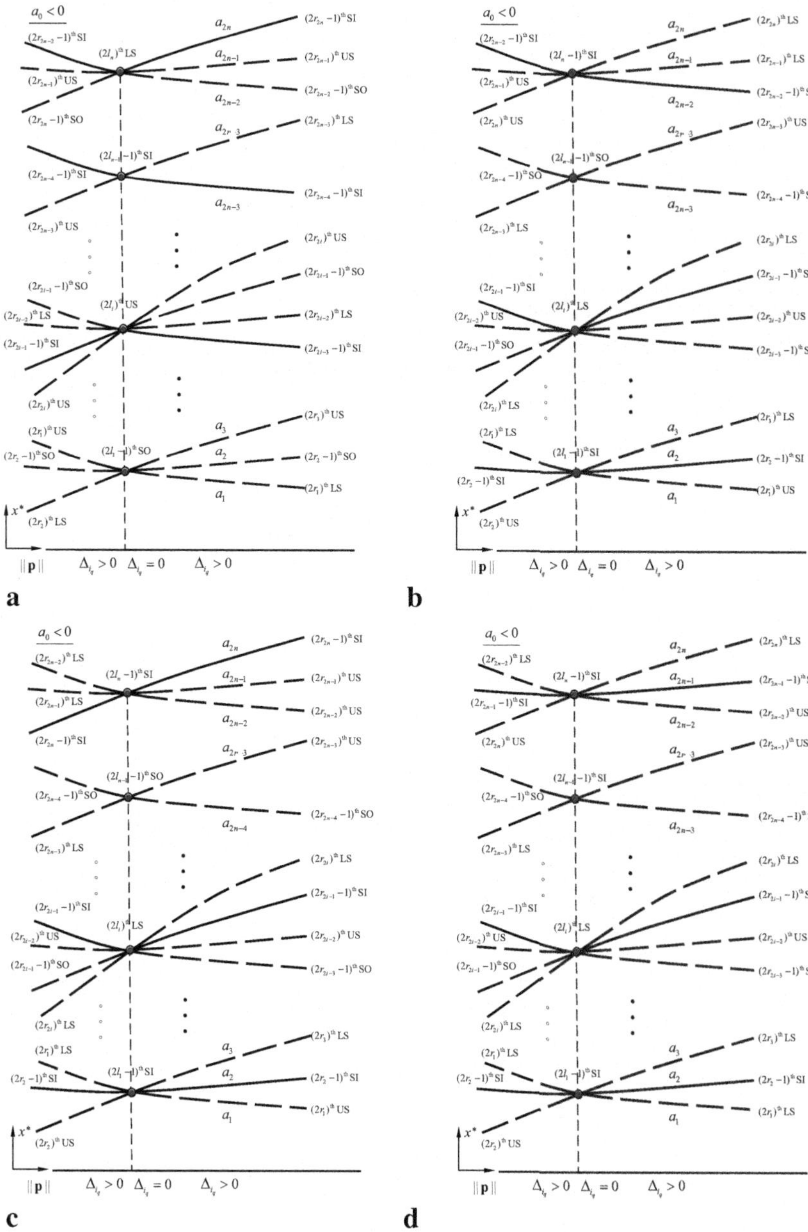

Fig. 6.11 **(a–d)** Four types of (r_1th XX : r_2th XX : $\cdots$: r_mth XX) parallel switching bifurcation for $a_0 < 0$ in the $(2m)^{\text{th}}$-degree polynomial system. LS: lower saddle, US: upper saddle, SI: sink, SO: source. Stable and unstable equilibriums are represented by solid and dashed curves, respectively. The bifurcation points are marked by circular symbols

and

$$\left.\begin{array}{l} B_{j_1} = B_{j_2} = \cdots = B_{j_s} \ (j_{k_1} \in \{1, 2, \ldots, n\}; j_{k_1} \neq j_{k_2}) \\ (k_1, k_2 \in \{1, 2, \ldots, s\}; k_1 \neq k_2) \\ \Delta_j = 0 \ (j \in U \subset \{1, 2, \ldots, n_2\} \\ c_i \neq -\frac{1}{2} B_j \ (i = 1, 2, \ldots, 2n_1, j = 1, 2, \ldots, n_2) \end{array}\right\} \text{at bifurcation.} \tag{6.183}$$

Consider a just before bifurcation of $((\beta_1^-)^{\text{th}}\text{XX}_1^- : (\beta_2^-)^{\text{th}}\text{XX}_2^- : \ldots : (\beta_{s_1}^-)^{\text{th}}\text{XX}_{s_1}^-)$ for higher order equilibriums. For $\beta_i^- = 2l_i^- - 1$, $\text{XX}_i^- \in \{\text{SO}, \text{SI}\}$ and for $\alpha_i^- = 2l_i^-$, $\text{XX}_i^- \in \{\text{US}, \text{LS}\}$ $(i = 1, 2, \ldots, s_1)$. The detailed structures are as follows:

$$\left.\begin{array}{l} (r_{s_i}^{(i)-})^{\text{th}} \text{ order } \text{XX}_{s_i}^{(i)-}, \ x^* = a_{k_{s_i}}^{(i)-}, \\ \vdots \\ (r_j^{(i)-})^{\text{th}} \text{ order } \text{XX}_j^{(i)-}, x^* = a_j^{(i)-} \\ \vdots \\ (r_1^{(i)-})^{\text{th}} \text{ order } \text{XX}_1^{(i)-}, x^* = a_j^{(i)-} \end{array}\right\} \xrightarrow[\text{switching bifurcation}]{\beta_i^- = \sum_{j=1}^{k_{s_i}} r_j^{(i)-}} (\beta_i^-)^{\text{th}} \text{order } \text{XX}^{(i)-}. \tag{6.184}$$

The bifurcation set of $((\beta_1^-)^{\text{th}}\text{XX}_1^- : (\beta_2^-)^{\text{th}}\text{XX}_2^- : \ldots : (\beta_{s_1}^-)^{\text{th}}\text{XX}_{s_1}^-)$ at the same parameter point is called a *left-parallel-straw-bundle* switching bifurcation.

Consider a just after bifurcation of $((\beta_1^+)^{\text{th}}\text{XX}_1^+ : (\beta_2^+)^{\text{th}}\text{XX}_2^+ : \ldots : (\beta_{s_2}^+)^{\text{th}}\text{XX}_{s_2}^+)$ for simple sources and sinks. For $\beta_i^+ = 2l_i^+ - 1$, $\text{XX}_i^+ \in \{\text{SO}, \text{SI}\}$ and for $\beta_i^+ = 2l_i^+$, $\text{XX}_i^+ \in \{\text{US}, \text{LS}\}$. The detailed structures are as follows:

$$(\beta_i^+)^{\text{th}} \text{order } \text{XX}^{(i)+} \xrightarrow[\text{switching bifurcation}]{\beta_i^- = \sum_{j=1}^{s_i} r_j^{(i)+}} \left\{\begin{array}{l} (r_{s_i}^{(i)+})^{\text{th}} \text{ order } \text{XX}_{s_i}^{(i)+}, \ x^* = a_{k_i}^{(i)+}, \\ \vdots \\ (r_j^{(i)+})^{\text{th}} \text{ order } \text{XX}_j^{(i)+}, x^* = a_j^{(i)+} \\ \vdots \\ (r_1^{(i)+})^{\text{th}} \text{ order } \text{XX}_1^{(i)+}, x^* = a_j^{(i)+}. \end{array}\right. \tag{6.185}$$

The bifurcation set of $((\beta_1^+)^{\text{th}}\text{XX}_1^+ : (\beta_2^+)^{\text{th}}\text{XX}_2^+ : \ldots : (\beta_{s_2}^+)^{\text{th}}\text{XX}_{s_2}^+)$ at the same parameter point is called a *right-parallel-straw-bundle* switching bifurcation.

(i) For the just before and after bifurcation structure, if there exists a relation of

$$\begin{aligned}&(\beta_i^-)^{\text{th}}\text{XX}_i^- = (\beta_j^+)^{\text{th}}\text{XX}_j^+ = (\beta_j)^{\text{th}}\text{XX}, \text{ for } x^* = a_i^- = a_j^+ \\ &(i,j \in \{1,2,\ldots,k\}), \text{XX} \in \{\text{US, LS, SO, SI}\}\end{aligned} \tag{6.186}$$

then the bifurcation is a $(\beta_j)^{\text{th}}\text{XX}_i$ *switching* bifurcation for higher order equilibriums.

(ii) Just for the just before bifurcation structure, if there exists a relation of

$$\begin{aligned}&(2l_i^-)^{\text{th}}\text{XX}_i^- = (2l_i)^{\text{th}}\text{XX}, \text{ for } x^* = a_i^- = a_i \\ &(i \in \{1,2,\ldots,s_1\}, \text{XX} \in \{\text{US, LS}\}\end{aligned} \tag{6.187}$$

then the bifurcation is a $(2l)^{\text{th}}$XX *left-appearing* (or *right-vanishing*) bifurcation for higher order equilibriums.

(iii) Just for the just after bifurcation structure, if there exists a relation of

$$\begin{aligned}&(2l_i^+)^{\text{th}}\text{XX}_i^+ = (2l_i)^{\text{th}}\text{XX}, \text{ for } x^* = a_i^+ = a_i \\ &(i \in \{1,2,\ldots,s_1\}), \text{XX} \in \{\text{US, LS}\}\end{aligned} \tag{6.188}$$

then the bifurcation is a $(2l)^{\text{th}}$XX *right-appearing* (or *left-vanishing*) bifurcation for higher order equilibriums.

(iv) For the just before and after bifurcation structure, if there exists a relation of

$$\begin{aligned}&(\beta_i^-)^{\text{th}}\text{XX}_i^- \neq (\beta_j^+)^{\text{th}}\text{XX}_j^+ \text{ for } x^* = a_i^- = a_j^+ \\ &\text{XX}_i^-, \text{XX}_j^+ \in \{\text{US, LS, SO, SI}\} \\ &(i \in \{1,2,\ldots,s_1\}, j \in \{1,2,\ldots,s_2\}),\end{aligned} \tag{6.189}$$

then two *flower-bundle* switching bifurcations of higher order equilibriums are as follows.

(iv_1) For $\beta_j = \beta_i + 2l$, the bifurcation is called a $(\beta_j^+)^{\text{th}}\text{XX}_j^+$ right *flower-bundle* switching bifurcation for the $(\beta_i^-)^{\text{th}}\text{XX}_i^-$ to $(\beta_j^+)^{\text{th}}\text{XX}_j^+$ switching of *higher order* equilibriums with the appearance (or birth) of $(2l)^{\text{th}}$XX right-appearing (or left-vanishing) bifurcation.

(iv_2) For $\beta_j = \beta_i - 2l$, the bifurcation is called a $(\beta_i^-)^{\text{th}}\text{XX}_i^-$ left *flower-bundle* switching bifurcation for the $(\beta_i^-)^{\text{th}}\text{XX}_i^-$ to $(\beta_j^+)^{\text{th}}\text{XX}_j^+$ switching of higher order equilibriums with the vanishing (or death) of $(2l)^{\text{th}}$XX *left-appearing* (or right-vanishing) bifurcation.

A general parallel switching bifurcation is

$$
\begin{aligned}
&((\beta_1^-)^{\text{th}}\text{XX}_1^- : (\beta_2^-)^{\text{th}}\text{XX}_2^- : \ldots : (\beta_{s_1}^-)^{\text{th}}\text{XX}_{s_1}^-) \xrightarrow[\textit{bifucation}]{\textit{switching}} \\
&((\beta_1^+)^{\text{th}}\text{XX}_1^+ : (\beta_2^+)^{\text{th}}\text{XX}_2^+ : \ldots : (\beta_{s_2}^+)^{\text{th}}\text{XX}_{s_2}^+).
\end{aligned} \tag{6.190}
$$

Such a general, parallel switching bifurcation consists of the *left* and *right* parallel-bundle switching bifurcations for higher order equilibriums.

If the *left* and *right* parallel-bundle switching bifurcations are same in a parallel *flower-bundle* switching bifurcation, i.e.,

$$
\begin{aligned}
&(\beta_i^-)^{\text{th}}\text{XX}_i^- = (\beta_i^+)^{\text{th}}\text{XX}_i^+ = \beta^{\text{th}}\text{XX}, \\
&\text{for } x^* = a_i^- = a_i^+ \ (i = 1, 2, \ldots, s\}
\end{aligned} \tag{6.191}
$$

then the parallel *flower-bundle* switching bifurcation becomes a parallel *straw-bundle switching* bifurcation of $((\alpha_1)^{\text{th}}\text{XX} : (\beta_2)^{\text{th}}\text{XX} : \cdots : (\beta_s)^{\text{th}}\text{XX})$.

If the *left* and *right* parallel-bundle switching bifurcations are different in a parallel *flower-bundle* switching bifurcation, i.e.,

$$
\begin{aligned}
&(\alpha_i^-)^{\text{th}}\text{XX}_i^- = (2l_j^-)^{\text{th}}\text{XX}, (\alpha_j^+)^{\text{th}}\text{XX}_j^+ = (2l_j^+)^{\text{th}}\text{YY}, \\
&\text{for } x^* = a_i^- \neq a_j^+ \ (i = 1, 2, \ldots, s_1; j = 1, 2, \ldots, s_2), \\
&\text{XX} \in \{\text{US}, \text{LS}\}, \text{YY} \in \{\text{US}, \text{LS}\},
\end{aligned} \tag{6.192}
$$

then the parallel *flower-bundle* switching bifurcation for higher order equilibriums becomes a combination of two independent left and right parallel appearing bifurcations:

(i) a $((2l_1^-)^{\text{th}}\text{XX}_1^- : (2l_2^-)^{\text{th}}\text{XX}_2^- : \cdots : (2l_{s_1}^-)^{\text{th}}\text{XX}_{s_1}^-)$ *left parallel sprinkler-spraying-appearing (or right-vanishing)* bifurcation and
(ii) a $((2l_1^+)^{\text{th}}\text{XX}_1^+ : (2l_2^+)^{\text{th}}\text{XX}_2^+ : \cdots : (2l_{s_2}^+)^{\text{th}}\text{XX}_{s_2}^+)$ *right parallel sprinkler-spraying-appearing (or left-vanishing)* bifurcation.

The parallel switching and appearing bifurcations for higher order equilibriums are presented in Fig. 6.12a–d. The *waterfall* appearing bifurcations and the *flower-bundle* switching bifurcations for higher order equilibriums are presented.

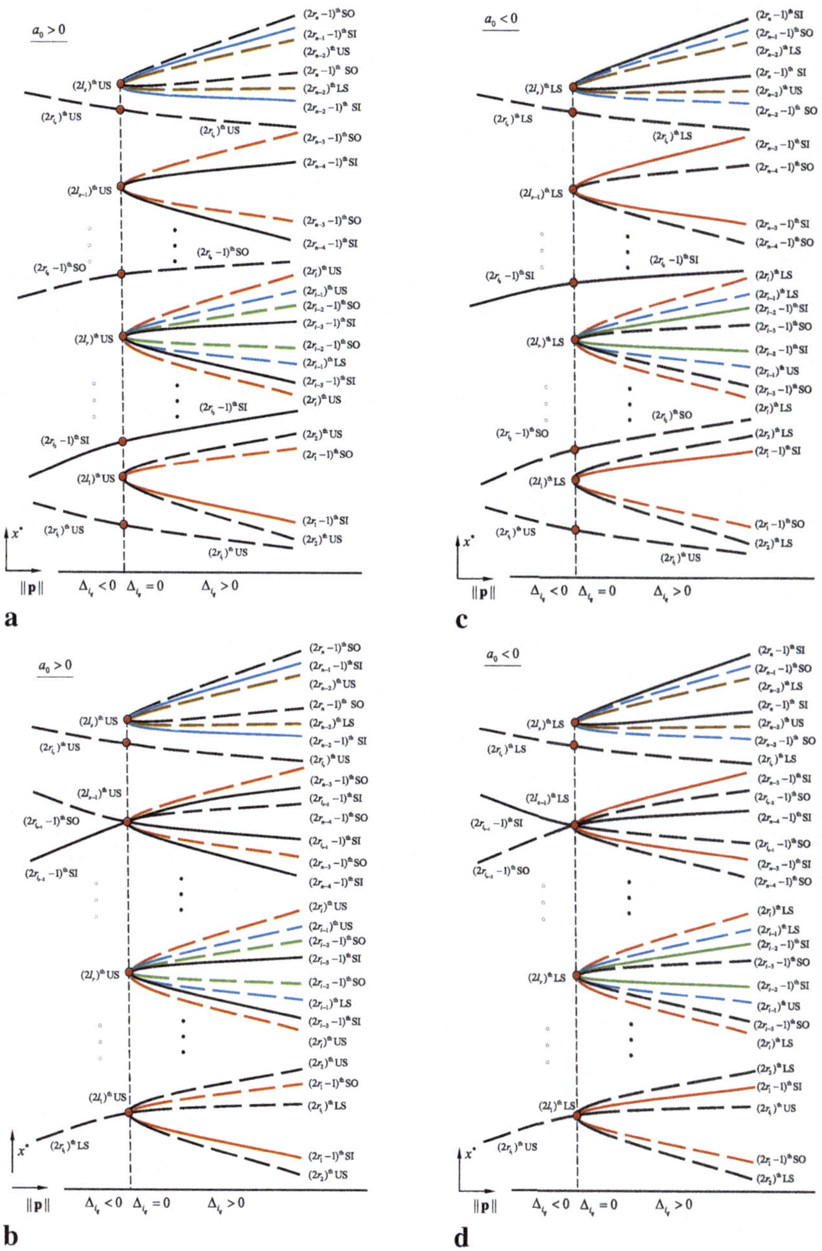

Fig. 6.12 (r_1th XX : r_2th XX : $\cdots$: r_nth XX) parallel switching/appearing bifurcations ($a_0 > 0$): (**a**) without switching, and (**b**) with switching. The (r_1th XX : r_2th XX : $\cdots$: r_nth XX) parallel switching/appearing bifurcation ($a_0 < 0$): (**c**) without switching, and (**d**) with switching. LS: lower saddle, US: upper saddle, SI: sink, SO: source. Stable and unstable equilibriums are represented by solid and dashed curves, respectively. The bifurcation points are marked by circular symbols

Reference

Luo, A.C.J., 2020, The stability and bifurcation of the $(2m)^{\text{th}}$ degree polynomial systems, *Journal of Vibration Testing and System Dynamics*, 4(1), pp. 1–42.

Chapter 7
$(2m+1)^{\text{th}}$-Degree Polynomial Systems

In this chapter, the global stability and bifurcations of equilibriums in the $(2m+1)^{\text{th}}$-degree polynomial systems are discussed for a better understanding of the complexity of bifurcations and stability of equilibriums in such a $(2m+1)^{\text{th}}$-degree polynomial system. The appearing and switching bifurcations are presented for simple equilibriums and higher-order equilibriums. The *broom-appearing* bifurcations, *broom-spraying-appearing* and *broom-sprinkler-spraying-appearing* bifurcations for simple and higher-order equilibriums are presented. The *antenna-switching* bifurcations for simple and higher-order equilibriums are discussed and the parallel *straw-bundle-switching* and *flower-bundle-switching* bifurcations for simple and higher order equilibriums are also presented.

7.1 Global Stability and Bifurcations

In a similar fashion of low-degree polynomial systems, the global stability and bifurcation of equilibriums in the $(2m+1)^{\text{th}}$-degree polynomial nonlinear systems are discussed as in Luo (2019). The stability and bifurcation of each individual equilibrium are analyzed from the local analysis in Chapters 1 and 2.

Definition 7.1 Consider a $(2m+1)^{\text{th}}$-degree polynomial nonlinear system

$$\begin{aligned} \dot{x} &= A_0(\mathbf{p})x^{2m+1} + A_1(\mathbf{p})x^{2m} + \ldots + A_{2m-1}(\mathbf{p})x^2 + A_{2m}x + A_{2m+1}(\mathbf{p}) \\ &= a_0(\mathbf{p})(x - a(\mathbf{p}))\left[x^2 + B_1(\mathbf{p})x + C_1(\mathbf{p})\right] \ldots \left[x^2 + B_m(\mathbf{p})x + C_m(\mathbf{p})\right] \end{aligned} \tag{7.1}$$

where $A_0(\mathbf{p}) \neq 0$, and

$$\mathbf{p} = (p_1, p_2, \ldots, p_m)^{\mathrm{T}}. \tag{7.2}$$

A. C. J. Luo, *Bifurcation and Stability in Nonlinear Dynamical Systems*, Nonlinear Systems and Complexity 28, https://doi.org/10.1007/978-3-030-22910-8_7

(i) If

$$\Delta_i = B_i^2 - 4C_i < 0 \text{ for } i = 1, 2, \ldots, m, \tag{7.3}$$

the $(2m+1)^{\text{th}}$-degree polynomial system has one equilibrium of $x^* = a$, and the corresponding standard form is

$$\dot{x} = a_0(x-a)[(x+\frac{1}{2}B_1)^2 + \frac{1}{4}(-\Delta_1)] \ldots [(x+\frac{1}{2}B_m)^2 + \frac{1}{4}(-\Delta_m)]. \tag{7.4}$$

The flow of such a system with one equilibrium is called a single-equilibrium flow.

(a) If $a_0 > 0$, the equilibrium flow with $x^* = a$ is called a source flow.
(b) If $a_0 < 0$, the equilibrium flow with $x^* = a$ is called a sink flow.

(ii) If

$$\begin{aligned} &\Delta_i = B_i^2 - 4C_i > 0, i = i_1, i_2, \ldots, i_l \in \{1, 2, \ldots, m\}, \\ &\Delta_j = B_j^2 - 4C_j < 0, j = i_{l+1}, i_{l+2}, \ldots, i_m \in \{1, 2, \ldots, m\} \\ &\text{with } l \in \{0, 1, \ldots, m\}, \end{aligned} \tag{7.5}$$

the $(2m+1)^{\text{th}}$-degree polynomial nonlinear system has $(2l+1)$-equilibriums as

$$\begin{aligned} &x^* = b_1^{(i)} = -\frac{1}{2}(B_i + \sqrt{\Delta_i}), x^* = b_2^{(i)} = -\frac{1}{2}(B_i - \sqrt{\Delta_i}) \\ &i \in \{i_1, i_2, \ldots, i_l\} \subseteq \{1, 2, \ldots, m\}. \end{aligned} \tag{7.6}$$

(ii_1) If

$$\begin{aligned} &b_r^{(i)} \neq b_s^{(j)} \text{ for } r,s \in \{1,2\}; i,j = 1,2, \ldots, l \\ &\{a_1, a_2, \ldots, a_{2l}\} = \text{sort}\{a, b_1^{(1)}, b_2^{(1)}, \cdots, b_1^{(l)}, b_2^{(l)}\}, a_s < a_{s+1}, \end{aligned} \tag{7.7}$$

then the corresponding standard form is

$$\dot{x} = a_0 \prod_{i_1=1}^{2l+1} (x - a_{i_1}) \prod_{k=l+1}^{m} [(x+\frac{1}{2}B_{i_k})^2 + \frac{1}{4}(-\Delta_{i_k})]. \tag{7.8}$$

(a) If $a_0 > 0$, the simple equilibrium separatrix flow is called a (SO : SI : . . .: SO : SI : . . .SI : SO)-flow.

(b) If $a_0 < 0$, the simple equilibrium separatrix flow is called a (SI : SO : . . .: SI : SO : . . .SO : SI)-flow.

(ii$_2$) If

$$\begin{aligned}
&\{a_1, a_2, \ldots, a_{2l+1}\} = \mathrm{sort}\{a, b_1^{(1)}, b_2^{(1)}, \ldots, b_1^{(l)}, b_2^{(l)}\},\\
&a_{i_1} \equiv a_1 = \ldots = a_{l_1},\\
&a_{i_2} \equiv a_{l_1+1} = \ldots = a_{l_1+l_2},\\
&\vdots\\
&a_{i_r} \equiv a_{\Sigma_{i=1}^{r-1} l_i+1} = \ldots = a_{\Sigma_{i=1}^{r-1} l_i+l_r} = a_{2l+1}\\
&\text{with } \Sigma_{s=1}^{r} l_s = 2l+1,
\end{aligned} \tag{7.9}$$

then the corresponding standard form is

$$\dot{x} = a_0 \prod_{s=1}^{r} (x - a_{i_s})^{l_s} \prod_{k=l+1}^{m} [(x + \frac{1}{2}B_{i_k})^2 + \frac{1}{4}(-\Delta_{i_k})]. \tag{7.10}$$

The equilibrium separatrix flow is called a (l_1th XX : l_2th XX : . . . : l_rth XX)-flow.

(a) For $a_0 > 0$ and $p = 1, 2, \ldots, r$,

$$l_p\text{th XX} = \begin{cases} (2r_p - 1)^{\text{th}}\text{order source}, & \text{for } \alpha_p = 2M_p - 1, l_p = 2r_p - 1;\\ (2r_p - 1)^{\text{th}}\text{order sink}, & \text{for } \alpha_p = 2M_p, l_p = 2r_p - 1;\\ (2r_p)^{\text{th}}\text{order lower-saddle}, & \text{for } \alpha_p = 2M_p - 1, l_p = 2r_p;\\ (2r_p)^{\text{th}}\text{order upper-saddle}, & \text{for } \alpha_p = 2M_p, l_p = 2r_p, \end{cases} \tag{7.11}$$

where

$$\alpha_p = \sum\nolimits_{s=p}^{r} l_s. \tag{7.12}$$

(b) For $a_0 < 0$ and $p = 1, 2, \cdots, r$,

$$l_p\text{th XX} = \begin{cases} (2r_p - 1)^{\text{th}}\text{order sink}, & \text{for } \alpha_p = 2M_p - 1, l_p = 2r_p - 1;\\ (2r_p - 1)^{\text{th}}\text{order source}, & \text{for } \alpha_p = 2M_p, l_p = 2r_p - 1;\\ (2r_p)^{\text{th}}\text{order upper-saddle}, & \text{for } \alpha_p = 2M_p - 1, l_p = 2r_p;\\ (2r_p)^{\text{th}}\text{order lower-saddle}, & \text{for } \alpha_p = 2M_p, l_p = 2r_p. \end{cases} \tag{7.13}$$

(c) The equilibrium of $x^* = a_{i_p}$ for $(l_p > 1)$-repeated equilibrium switching is called an l_pth XX bifurcation of $(l_{p_1}\text{th XX} : l_{p_2}\text{th XX} : \ldots : l_{p_\beta}\text{th XX})$ equilibrium switching at a point $\mathbf{p} = \mathbf{p}_1 \in \partial\Omega_{12}$, and the bifurcation condition is

$$\begin{aligned} &a_{i_p} \equiv a_{\Sigma_{i=1}^{p-1} l_i+1} = \ldots = a_{\Sigma_{i=1}^{p-1} l_i+l_p}, \\ &a^{\pm}_{\Sigma_{i=1}^{p-1} l_i+1} \neq \ldots \neq a^{\pm}_{\Sigma_{i=1}^{p-1} l_i+l_p};\ l_p = \sum\nolimits_{i=1}^{\beta} l_{p_i}. \end{aligned} \tag{7.14}$$

(iii) If

$$\begin{aligned} &\Delta_i = B_i^2 - 4C_i = 0, i \in \{i_{11}, i_{12}, \ldots, i_{1s}\} \subseteq \{i_1, i_2, \ldots, i_l\} \subseteq \{1, 2, \ldots, m\}, \\ &\Delta_k = B_k^2 - 4C_k > 0, k \in \{i_{21}, i_{22}, \ldots, i_{2r}\} \subseteq \{i_1, i_2, \ldots, i_l\} \subseteq \{1, 2, \ldots, m\}, \\ &\Delta_j = B_j^2 - 4C_j < 0,\ \ j \in \{i_{l+1}, i_{l+2}, \ldots, i_m\} \subseteq \{1, 2, \ldots, m\}, \end{aligned} \tag{7.15}$$

the $(2m + 1)^{\text{th}}$-degree polynomial nonlinear system has $(2l + 1)$-equilibriums as

$$\begin{aligned} &x^* = b_1^{(i)} = -\frac{1}{2}B_i, x^* = b_2^{(i)} = -\frac{1}{2}B_i \text{ for } i \in \{i_{11}, i_{12}, \ldots, i_{1s}\}, \\ &x^* = b_1^{(k)} = -\frac{1}{2}(B_k + \sqrt{\Delta_k}), x^* = b_2^{(k)} = -\frac{1}{2}(B_k - \sqrt{\Delta_k}) \\ &\text{for } k \in \{i_{21}, i_{22}, \ldots, i_{2r}\}. \end{aligned} \tag{7.16}$$

If

$$\begin{aligned} &\{a_1, a_2, \ldots, a_{2l+1}\} = \text{sort}\left\{a, b_1^{(1)}, b_2^{(1)}, \ldots, b_1^{(l)}, b_2^{(l)}\right\}, \\ &a_{i_1} \equiv a_1 = \ldots = a_{l_1}, \\ &a_{i_2} \equiv a_{l_1+1} = \ldots = a_{l_1+l_2}, \\ &\vdots \\ &a_{i_r} \equiv a_{\Sigma_{i=1}^{r-1} l_i+1} = \ldots = a_{\Sigma_{i=1}^{r-1} l_i+l_r} = a_{2l+1} \\ &\text{with } \Sigma_{s=1}^{r} l_s = 2l + 1, \end{aligned} \tag{7.17}$$

then the corresponding standard form is

$$\dot{x} = a_0 \prod_{s=1}^{r} (x - a_{i_s})^{l_s} \prod_{k=l+1}^{m} [(x + \frac{1}{2}B_{i_k})^2 + \frac{1}{4}(-\Delta_{i_k})]. \tag{7.18}$$

The equilibrium separatrix flow is called a (l_1th XX : l_2th XX : … : l_rth XX)-flow.

(a) The equilibrium of $x^* = a_{i_p}$ for ($l_p > 1$)-repeated equilibriums appearing or vanishing is called an l_pth XX bifurcation of equilibrium at a point $\mathbf{p} = \mathbf{p}_1 \in \partial\Omega_{12}$, and the bifurcation condition is

$$
\begin{aligned}
& a_{i_p} \equiv a_{\Sigma_{i=1}^{p-1} l_i+1} = \ldots = a_{\Sigma_{i=1}^{p-1} l_i+l_p} = -\frac{1}{2}B_{i_q}, \\
& \text{with } \Delta_{i_q} = B_{i_q}^2 - 4C_{i_q} = 0 \ (i_q \in \{i_1, i_2, \ldots, i_l\}), \\
& a^+_{\Sigma_{i=1}^{p-1} l_i+1} \neq \ldots \neq a^+_{\Sigma_{i=1}^{p-1} l_i+l_p} \text{ or } a^-_{\Sigma_{i=1}^{p-1} l_i+1} \neq \ldots \neq a^-_{\Sigma_{i=1}^{p-1} l_i+l_p}.
\end{aligned}
\tag{7.19}
$$

(b) The equilibrium of $x^* = a_{i_q}$ for ($l_q > 1$)-repeated equilibriums switching is called an l_qth XX bifurcation of (l_{q_1}thXX : l_{q_2}thXX : … : l_{q_β}thXX) equilibrium switching at a point $\mathbf{p} = \mathbf{p}_1 \in \partial\Omega_{12}$, and the switching bifurcation condition is

$$
\begin{aligned}
& a_{i_q} \equiv a_{\Sigma_{i=1}^{q-1} l_i+1} = \ldots = a_{\Sigma_{i=1}^{q-1} l_i+l_p}, \\
& a^\pm_{\Sigma_{i=1}^{q-1} l_i+1} \neq \ldots \neq a^\pm_{\Sigma_{i=1}^{q-1} l_i+l_q};\ l_q = \Sigma_{i=1}^{\beta} l_{q_i}.
\end{aligned}
\tag{7.20}
$$

(c) The equilibrium of $x^* = a_{i_p}$ for $(l_{p_1} \geq 1)$-repeated equilibriums appearance/vanishing and $(l_{p_2} \geq 2)$ repeated equilibriums switching of $(l_{p_{21}}\text{thXX} : l_{p_{22}}\text{thXX} : \ldots : l_{p_{2\beta}}\text{thXX})$ is called an $l_p$thXX bifurcation of equilibrium at a point $\mathbf{p} = \mathbf{p}_1 \in \partial\Omega_{12}$, and the bifurcation condition is

$$
\begin{aligned}
& a_{i_p} \equiv a_{\Sigma_{i=1}^{p-1} l_i+1} = \cdots = a_{\Sigma_{i=1}^{p-1} l_i+l_p} \\
& \text{with } \Delta_{i_q} = B_{i_q}^2 - 4C_{i_q} = 0 \quad (i_q \in \{i_1, i_2, \ldots, i_l\}) \\
& a^+_{\Sigma_{i=1}^{p-1} l_i+j_1} \neq \cdots \neq a^+_{\Sigma_{i=1}^{p-1} l_i+j_{p_1}} \text{ or } a^-_{\Sigma_{i=1}^{p_1-1} l_i+j_1} \neq \cdots \neq a^-_{\Sigma_{i=1}^{p_1-1} l_i+j_{p_1}}, \\
& \text{for } \{j_1, j_2, \ldots, j_{p_1}\} \subseteq \{1, 2, \ldots, l_p\}, \\
& a^\pm_{\Sigma_{i=1}^{p-1} l_i+k_1} \neq \cdots \neq a^\pm_{\Sigma_{i=1}^{p-1} l_i+k_{p2}} \\
& \text{for } \{k_1, k_2, \ldots, k_{p_2}\} \subseteq \{1, 2, \ldots, l_p\}, \\
& \text{with } l_{p_1} + l_{p_2} = l_p;\ l_{p_2} = \Sigma_{i=1}^{\beta} l_{p_{2i}}
\end{aligned}
\tag{7.21}
$$

(iv) If

$$
\Delta_i = B_i^2 - 4C_i > 0 \ \text{ for } i = 1,2, \ldots, m \tag{7.22}
$$

the $(2m + 1)^{\text{th}}$-degree polynomial nonlinear system has $(2m+1)$-equilibriums as

$$x^* = b_1^{(i)} = -\frac{1}{2}(B_i + \sqrt{\Delta_i}), x^* = b_2^{(i)} = -\frac{1}{2}(B_i - \sqrt{\Delta_i}) \tag{7.23}$$

for $i = 1, 2, \ldots, m$.

(iv_1) If

$$\begin{aligned} &b_r^{(i)} \neq b_s^{(j)} \text{ for } r, s \in \{1,2\}; i, j = 1, 2, \ldots, m \\ &\{a_1, a_2, \ldots, a_{2m}\} = \text{sort}\left\{a, b_1^{(1)}, b_2^{(1)}, \ldots, b_1^{(m)}, b_2^{(m)}\right\} \ (a_k < a_{k+1}), \end{aligned} \tag{7.24}$$

then the corresponding standard form is

$$\dot{x} = a_0(x - a_1)(x - a_2)(x - a_3)(x - a_4)\ldots(x - a_{2m})(x - a_{2m+1}). \tag{7.25}$$

This flow is formed with all the simple equilibriums.

(a) If $a_0 > 0$, the separatrix flow with $(2m + 1)$ equilibriums is called a (SO : SI : …: SO : SI : …SI : SO)-flow.
(b) If $a_0 < 0$, the separatrix flow with $(2m + 1)$ equilibriums is called a (SI : SO : …: SI : SO : …SO : SI)-flow.

(iv_2) If

$$\begin{aligned} &\{a_1, a_2 \ldots, a_{2m+1}\} = \text{sort}\{a, b_1^{(1)}, b_2^{(1)}, \ldots, b_1^{(m)}, b_2^{(m)}\}, \\ &a_{i_1} \equiv a_1 = \ldots = a_{l_1}, \\ &a_{i_2} \equiv a_{l_1+1} = \ldots = a_{l_1+l_2}, \\ &\vdots \\ &a_{i_r} \equiv a_{\Sigma_{i=1}^{r-1} l_i+1} = \ldots = a_{\Sigma_{i=1}^{r-1} l_i + l_r} = a_{2m+1} \\ &\text{with } \Sigma_{s=1}^{r} l_s = 2m + 1, \end{aligned} \tag{7.26}$$

then the corresponding standard form is

$$\dot{x} = a_0 \prod_{s=1}^{r} (x - a_{i_s})^{l_s}. \tag{7.27}$$

The equilibrium separatrix flow is called a (l_1th XX : l_2th XX : … : l_rth XX)-flow. The equilibrium of $x^* = a_{i_p}$ for l_p-repeated equilibriums switching is called an l_pth XX bifurcation of (l_{p_1}th XX : l_{p_2}th XX : … : l_{p_β}th XX) equilibrium switching at a point $\mathbf{p} = \mathbf{p}_1 \in \partial\Omega_{12}$, and the switching bifurcation condition is

$$\begin{aligned} &a_{i_p} \equiv a_{\Sigma_{i=1}^{p-1} l_i+1} = \ldots = a_{\Sigma_{i=1}^{p-1} l_i+l_p}, \\ &a^{\pm}_{\Sigma_{i=1}^{p-1} l_i+1} \neq \ldots \neq a^{\pm}_{\Sigma_{i=1}^{p-1} l_i+l_p}; \; l_p = \sum\nolimits_{i=1}^{\beta} l_{p_i}. \end{aligned} \tag{7.28}$$

Definition 7.2 Consider a $(2m+1)^{\text{th}}$-degree polynomial nonlinear system as

$$\begin{aligned} \dot{x} &= A_0(\mathbf{p})x^{2m+1} + A_1(\mathbf{p})x^{2m} + \ldots + A_{2m-1}(\mathbf{p})x^2 + A_{2m}x + A_{2m+1}(\mathbf{p}) \\ &= a_0(\mathbf{p})(x - a(\mathbf{p}))\prod_{i=1}^{n}[x^2 + B_i(\mathbf{p})x + C_i(\mathbf{p})]^{q_i} \end{aligned} \tag{7.29}$$

where $A_0(\mathbf{p}) \neq 0$, and

$$\mathbf{p} = (p_1, p_2, \ldots, p_m)^{\mathrm{T}}, m = \sum\nolimits_{i=1}^{n} q_i. \tag{7.30}$$

(i) If

$$\Delta_i = B_i^2 - 4C_i < 0 \;\text{ for } i = 1,2, \ldots, n \tag{7.31}$$

the $(2m+1)^{\text{th}}$-degree polynomial nonlinear system has one equilibrium of $x^* = a$, and the corresponding standard form is

$$\dot{x} = a_0(x-a)\prod_{i=1}^{n}[(x + \frac{1}{2}B_i)^2 + \frac{1}{4}(-\Delta_i)]^{q_i}. \tag{7.32}$$

The flow of such a system with one equilibrium is called a single equilibrium flow.

(a) If $a_0 > 0$, the equilibrium flow of $x^* = a$ is called a source flow.
(b) If $a_0 < 0$, the equilibrium flow of $x^* = a$ is called a sink flow.

(ii) If

$$\begin{aligned} &\Delta_i = B_i^2 - 4C_i > 0, i \in \{i_1, i_2, \ldots, i_l\} \subseteq \{1, 2, \ldots, n\}, \\ &\Delta_j = B_j^2 - 4C_j < 0, j \in \{i_{l+1}, i_{l+2}, \ldots, i_n\} \subseteq \{1, 2, \ldots, n\} \end{aligned} \tag{7.33}$$

the $(2m + 1)^{\text{th}}$-degree polynomial nonlinear system has $(2l + 1)$-equilibriums as

$$\begin{aligned} &x^* = b_1^{(i)} = -\frac{1}{2}(B_i + \sqrt{\Delta_i}), x^* = b_2^{(i)} = -\frac{1}{2}(B_i - \sqrt{\Delta_i}) \\ &i \in \{i_1, i_2, \ldots, i_l\} \subseteq \{1, 2, \ldots, n\}. \end{aligned} \tag{7.34}$$

(ii_1) If

$$\begin{aligned}&b_r^{(i)} \neq b_s^{(j)} \text{ for } r, s \in \{1,2\}; i,j = 1,2,\ldots,l\\&\{a_1, a_2, \ldots, a_{2l+1}\} = \text{sort}\{a, \underbrace{b_1^{(1)}, b_2^{(1)}}_{q_1 \text{ sets}}, \ldots, \underbrace{b_1^{(r)}, b_2^{(r)}}_{q_r \text{ sets}}\}, a_s \leq a_{s+1},\end{aligned} \tag{7.35}$$

then the corresponding standard form is

$$\dot{x} = a_0 \prod_{s=1}^{2l+1} (x - a_s)^{l_s} \prod_{k=l+1}^{n} [(x + \tfrac{1}{2}B_{i_k})^2 + \tfrac{1}{4}(-\Delta_{i_k})]^{q_{i_k}} \tag{7.36}$$
with $l_s \in \{q_{i_1}, q_{i_2}, \cdots, q_{i_l}, 1\}$.

The equilibrium separatrix flow is called a ($l_1$thXX : $l_2$thXX :...: l_{2l+1}th XX)-flow.

(a) For $a_0 > 0$ and $p = 1, 2, \ldots, 2l + 1$,

$$l_p\text{th XX} = \begin{cases} (2r_p - 1)^{\text{th}}\text{order source}, & \text{for } \alpha_p = 2M_p - 1, l_p = 2r_p - 1;\\ (2r_p - 1)^{\text{th}}\text{order sink}, & \text{for } \alpha_p = 2M_p, l_p = 2r_p - 1;\\ (2r_p)^{\text{th}}\text{order lower-saddle}, & \text{for } \alpha_p = 2M_p - 1, l_p = 2r_p;\\ (2r_p)^{\text{th}}\text{order upper-saddle}, & \text{for } \alpha_p = 2M_p, l_p = 2r_p, \end{cases} \tag{7.37}$$

where

$$\alpha_p = \sum\nolimits_{s=p}^{2l+1} l_s. \tag{7.38}$$

(b) For $a_0 < 0$ and $p = 1, 2, \ldots, 2l + 1$,

$$l_p\text{th XX} = \begin{cases} (2r_p - 1)^{\text{th}}\text{order sink}, & \text{for } \alpha_p = 2M_p - 1, l_p = 2r_p - 1;\\ (2r_p - 1)^{\text{th}}\text{order source}, & \text{for } \alpha_p = 2M_p, l_p = 2r_p - 1;\\ (2r_p)^{\text{th}}\text{order upper-saddle}, & \text{for } \alpha_p = 2M_p - 1, l_p = 2r_p;\\ (2r_p)^{\text{th}}\text{order lower-saddle}, & \text{for } \alpha_p = 2M_p, l_p = 2r_p. \end{cases} \tag{7.39}$$

(ii$_2$) If

$$\{a_1, a_2, \ldots, a_{2l+1}\} = \text{sort}\{a, \underbrace{b_1^{(1)}, b_2^{(1)}}_{q_1 \; sets}, \ldots, \underbrace{b_1^{(r)}, b_2^{(r)}}_{q_r \; sets}\},$$
$$a_{i_1} \equiv a_1 = \ldots = a_{l_1},$$
$$a_{i_2} \equiv a_{l_1+1} = \ldots = a_{l_1+l_2}, \tag{7.40}$$
$$\vdots$$
$$a_{i_r} \equiv a_{\Sigma_{i=1}^{r-1} l_i+1} = \ldots = a_{\Sigma_{i=1}^{r-1} l_i + l_r} = a_{2l+1}$$
$$\text{with } \Sigma_{s=1}^{r} l_s = 2l+1,$$

then the corresponding standard form is

$$\dot{x} = a_0 \prod_{s=1}^{r} (x - a_{i_s})^{l_s} \prod_{k=l+1}^{n} [(x + \frac{1}{2}B_{i_k})^2 + \frac{1}{4}(-\Delta_{i_k})]^{q_{i_k}}. \tag{7.41}$$

The equilibrium separatrix flow is called an (l_1th XX : l_2th XX : ... : l_rth XX)-flow.

(a) For $a_0 > 0$ and $s = 1, 2, \ldots, r$,

$$l_p\text{th XX} = \begin{cases} (2r_p - 1)^{\text{th}} \text{order source,} \;\; \text{for } \alpha_p = 2M_p - 1, l_p = 2r_p - 1; \\ (2r_p - 1)^{\text{th}} \text{order sink,} \;\; \text{for } \alpha_p = 2M_p, l_p = 2r_p - 1; \\ (2r_p)^{\text{th}} \text{order lower-saddle,} \;\; \text{for } \alpha_p = 2M_p - 1, l_p = 2r_p; \\ (2r_p)^{\text{th}} \text{order upper-saddle,} \;\; \text{for } \alpha_p = 2M_p, l_p = 2r_p, \end{cases} \tag{7.42}$$

where

$$\alpha_p = \sum\nolimits_{s=p}^{r} l_s. \tag{7.43}$$

(b) For $a_0 < 0$ and $p = 1, 2, \ldots, r$,

$$l_p\text{th XX} = \begin{cases} (2r_p - 1)^{\text{th}} \text{order sink,} \;\; \text{for } \alpha_p = 2M_p - 1, l_p = 2r_p - 1; \\ (2r_p - 1)^{\text{th}} \text{order source,} \;\; \text{for } \alpha_p = 2M_p, l_p = 2r_p - 1; \\ (2r_p)^{\text{th}} \text{order upper-saddle,} \;\; \text{for } \alpha_p = 2M_p - 1, l_p = 2r_p; \\ (2r_p)^{\text{th}} \text{order lower-saddle,} \;\; \text{for } \alpha_p = 2M_p, l_p = 2r_p. \end{cases} \tag{7.44}$$

(c) The equilibrium of $x^* = a_{i_p}$ for ($l_p > 1$)-repeated equilibriums switching is called an l_pth XX bifurcation of (l_{p_1}th XX : l_{p_2}th XX : ... : l_{p_β}th XX) equilibrium switching at a point $\mathbf{p} = \mathbf{p}_1 \in \partial\Omega_{12}$, and the switching bifurcation condition is

$$\begin{aligned} &a_{i_p} \equiv a_{\Sigma_{i=1}^{p-1} l_i+1} = \ldots = a_{\Sigma_{i=1}^{p-1} l_i+l_p}, \\ &a^{\pm}_{\Sigma_{i=1}^{p-1} l_i+1} \neq \ldots \neq a^{\pm}_{\Sigma_{i=1}^{p-1} l_i+l_p};\ l_p = \sum\nolimits_{i=1}^{\beta} l_{p_i}. \end{aligned} \tag{7.45}$$

(iii) If

$$\begin{aligned} &\Delta_i = B_i^2 - 4C_i = 0, i \in \{i_{11}, i_{12}, \ldots, i_{1s}\} \subseteq \{i_1, i_2, \ldots, i_l\} \subseteq \{1,2,\ldots,n\}, \\ &\Delta_k = B_k^2 - 4C_k > 0, k \in \{i_{21}, i_{22}, \ldots, i_{2r}\} \subseteq \{i_1, i_2, \ldots, i_l\} \subseteq \{1,2,\ldots,n\}, \\ &\Delta_j = B_j^2 - 4C_j < 0,\ j \in \{i_{l+1}, i_{l+2}, \ldots, i_n\} \subseteq \{1,2,\ldots,n\} \text{ with } i \neq j \neq k, \end{aligned} \tag{7.46}$$

the $(2m+1)^{\text{th}}$-degree polynomial nonlinear system has $(2l + 1)$-equilibriums as

$$\begin{aligned} &x^* = b_1^{(i)} = -\frac{1}{2}B_i, x^* = b_2^{(i)} = -\frac{1}{2}B_i \text{ for } i \in \{i_{11}, i_{12}, \ldots, i_{1s}\}, \\ &x^* = b_1^{(k)} = -\frac{1}{2}(B_k + \sqrt{\Delta_k}), x^* = b_2^{(k)} = -\frac{1}{2}(B_k - \sqrt{\Delta_k}) \\ &\text{for } k \in \{i_{21}, i_{22}, \ldots, i_{2r}\}. \end{aligned} \tag{7.47}$$

If

$$\begin{aligned} &\{a_1, a_2, \ldots, a_{2l+1}\} = \text{sort}\{a, b_1^{(1)}, b_2^{(1)}, \ldots, b_1^{(l)}, b_2^{(l)}\}, \\ &a_{i_1} \equiv a_1 = \ldots = a_{l_1}, \\ &a_{i_2} \equiv a_{l_1+1} = \ldots = a_{l_1+l_2}, \\ &\vdots \\ &a_{i_r} \equiv a_{\Sigma_{i=1}^{r-1} l_i+1} = \ldots = a_{\Sigma_{i=1}^{r-1} l_i+l_r} = a_{2l+1} \\ &\text{with } \Sigma_{s=1}^{r} l_s = 2l + 1, \end{aligned} \tag{7.48}$$

then the corresponding standard form is

$$\dot{x} = a_0 \prod_{s=1}^{r} (x - a_{i_s})^{l_s} \prod_{k=l+1}^{n} [(x + \frac{1}{2}B_{i_k})^2 + \frac{1}{4}(-\Delta_{i_k})]^{q_{i_k}}. \tag{7.49}$$

The equilibrium separatrix flow is called an (l_1th XX : l_2th XX : . . . : l_rth XX)-flow.

(a) The equilibrium of $x^* = a_{i_p}$ for ($l_p > 1$)-repeated equilibriums appearing or vanishing is called an l_pth XX bifurcation of equilibrium at a point $\mathbf{p} = \mathbf{p}_1 \in \partial\Omega_{12}$, and the bifurcation condition is

$$
\begin{aligned}
& a_{i_p} \equiv a_{\Sigma_{i=1}^{p-1} l_i+1} = \ldots = a_{\Sigma_{i=1}^{p-1} l_i+l_p} = -\frac{1}{2} B_{i_q} \\
& \text{with } \Delta_{i_q} = B_{i_q}^2 - 4C_{i_q} = 0 \ \left(i_q \in \{i_1, i_2, \ldots, i_l\}\right), \\
& a^{+}_{\Sigma_{i=1}^{p-1} l_i+1} \neq \ldots \neq a^{+}_{\Sigma_{i=1}^{p-1} l_i+l_p} \text{ or } a^{-}_{\Sigma_{i=1}^{p-1} l_i+1} \neq \ldots \neq a^{-}_{\Sigma_{i=1}^{p-1} l_i+l_p}.
\end{aligned} \tag{7.50}
$$

(b) The equilibrium of $x^* = a_{i_q}$ for ($l_q > 1$)-repeated equilibriums switching is called an l_qth XX bifurcation of (l_{q_1}thXX : l_{q_2}thXX : … : l_{q_β}thXX) equilibrium switching at a point $\mathbf{p} = \mathbf{p}_1 \in \partial\Omega_{12}$, and the bifurcation condition is

$$
\begin{aligned}
& a_{i_q} \equiv a_{\Sigma_{i=1}^{q-1} l_i+1} = \ldots = a_{\Sigma_{i=1}^{q-1} l_i+l_p}, \\
& a^{\pm}_{\Sigma_{i=1}^{q-1} l_i+1} \neq \ldots \neq a^{\pm}_{\Sigma_{i=1}^{q-1} l_i+l_q};\ l_p = \sum\nolimits_{i=1}^{\beta} l_{p_i}.
\end{aligned} \tag{7.51}
$$

(iv) If

$$
\Delta_i = B_i^2 - 4C_i > 0 \ \text{ for } i = 1,2, \ldots, n \tag{7.52}
$$

the $(2m + 1)^{\text{th}}$-degree polynomial nonlinear system has $(2n + 1)$-equilibriums as

$$
\begin{aligned}
& x^* = b_1^{(i)} = -\frac{1}{2}(B_i + \sqrt{\Delta_i}), x^* = b_2^{(i)} = -\frac{1}{2}(B_i - \sqrt{\Delta_i}) \\
& \text{for } i = 1, 2, \ldots, n.
\end{aligned} \tag{7.53}
$$

(iv$_1$) If

$$
\begin{aligned}
& b_r^{(i)} \neq b_s^{(j)} \text{ for } r, s \in \{1, 2\}, (i,j = 1, 2, \ldots, n); \\
& \{a_1, a_2, \ldots, a_{2n+1}\} = \text{sort}\{a, \underbrace{b_1^{(1)}, b_2^{(1)}}_{q_1 \text{ sets}}, \ldots, \underbrace{b_1^{(n)}, b_2^{(n)}}_{q_n \text{ sets}}\} \ (a_s \leq a_{s+1}),
\end{aligned} \tag{7.54}
$$

then the corresponding standard form is

$$
\dot{x} = a_0 \prod_{s=1}^{2n+1} (x - a_s)^{l_s} \text{ with } l_s \in \{q_{i_1}, q_{i_2}, \ldots, q_{i_n}, 1\}. \tag{7.55}
$$

The equilibrium separatrix flow is called an ($l_1$thXX : $l_2$thXX :…: l_{2n+1}th XX)-flow.

(a) For $a_0 > 0$ and $p = 1, 2, \ldots, 2n + 1$,

$$l_p\text{th XX} = \begin{cases} (2r_p - 1)^{\text{th}}\text{order source}, & \text{for } \alpha_p = 2M_p - 1, l_p = 2r_p - 1; \\ (2r_p - 1)^{\text{th}}\text{order sink}, & \text{for } \alpha_p = 2M_p, l_p = 2r_p - 1; \\ (2r_p)^{\text{th}}\text{order lower-saddle}, & \text{for } \alpha_p = 2M_p - 1, l_p = 2r_p; \\ (2r_p)^{\text{th}}\text{order upper-saddle}, & \text{for } \alpha_p = 2M_p, l_p = 2r_p, \end{cases} \tag{7.56}$$

where

$$\alpha_p = \sum\nolimits_{s=p}^{2n+1} l_s. \tag{7.57}$$

(b) For $a_0 < 0$ and $p = 1, 2, \ldots, 2n + 1$,

$$l_p\text{th XX} = \begin{cases} (2r_p - 1)^{\text{th}}\text{order sink}, & \text{for } \alpha_p = 2M_p - 1, l_p = 2r_p - 1; \\ (2r_p - 1)^{\text{th}}\text{order source}, & \text{for } \alpha_p = 2M_p, l_p = 2r_p - 1; \\ (2r_p)^{\text{th}}\text{order upper-saddle}, & \text{for } \alpha_p = 2M_p - 1, l_p = 2r_p; \\ (2r_p)^{\text{th}}\text{order lower-saddle}, & \text{for } \alpha_p = 2M_p, l_p = 2r_p. \end{cases} \tag{7.58}$$

(iv_2) If

$$\begin{aligned} &\{a_1, a_2, \ldots, a_{2n+1}\} = \text{sort}\{a, \underbrace{b_1^{(1)}, b_2^{(1)}}_{q_1 \text{ sets}}, \ldots, \underbrace{b_1^{(n)}, b_2^{(n)}}_{q_n \text{ sets}}\}, \\ &a_{i_1} \equiv a_1 = \ldots = a_{l_1}, \\ &a_{i_2} \equiv a_{l_1+1} = \ldots = a_{l_1+l_2}, \\ &\vdots \\ &a_{i_r} \equiv a_{\Sigma_{i=1}^{r-1} l_i+1} = \ldots = a_{\Sigma_{i=1}^{r-1} l_i+l_r} = a_{2n+1}, \\ &\text{with } \Sigma_{s=1}^{r} l_s = 2n + 1, \end{aligned} \tag{7.59}$$

then the corresponding standard form is

$$\dot{x} = a_0 \prod_{s=1}^{r} (x - a_{i_s})^{l_s}. \tag{7.60}$$

The equilibrium separatrix flow is called an (l_1th XX : l_2th XX : … : l_rth XX)-flow. The equilibrium of $x^* = a_{i_p}$ for l_p-repeated equilibriums switching is called an l_pth XX switching bifurcation of $(l_{p_1}\text{th XX} : l_{p_2}\text{th XX} : \ldots : l_{p_\beta}\text{th XX})$ equilibrium at a point $\mathbf{p} = \mathbf{p}_1 \in \partial\Omega_{12}$, and the switching bifurcation condition is

$$\begin{aligned}&a_{i_p} \equiv a_{\Sigma_{i=1}^{p-1} l_i+1} = \ldots = a_{\Sigma_{i=1}^{p-1} l_i+l_p},\\&a^{\pm}_{\Sigma_{i=1}^{p-1} l_i+1} \neq \ldots \neq a^{\pm}_{\Sigma_{i=1}^{p-1} l_i+l_p};\ l_p = \sum\nolimits_{i=1}^{\beta} l_{p_i}.\end{aligned} \tag{7.61}$$

Definition 7.3 Consider a 1-dimensional, $(2m + 1)^{\text{th}}$-degree polynomial nonlinear dynamical system

$$\begin{aligned}\dot{x} &= A_0(\mathbf{p})x^{2m+1} + A_1(\mathbf{p})x^{2m} + \ldots + A_{2m-1}(\mathbf{p})x^2 + A_{2m}x + A_{2m+1}(\mathbf{p})\\&= a_0(\mathbf{p})\prod_{s=1}^{r}(x - c_{i_s}(\mathbf{p}))^{l_s}\prod_{i=r+1}^{n}[x^2 + B_i(\mathbf{p})x + C_i(\mathbf{p})]^{q_i}\end{aligned} \tag{7.62}$$

where $A_0(\mathbf{p}) \neq 0$, and

$$\sum\nolimits_{s=1}^{r} l_s = 2l + 1,\ \sum\nolimits_{i=r+1}^{n} q_i = (m - l), \mathbf{p} = (p_1, p_2, \ldots, p_m)^{\mathrm{T}}. \tag{7.63}$$

(i) If

$$\begin{aligned}&\Delta_i = B_i^2 - 4C_i < 0 \ \text{ for } i = r + 1, r + 2, \ldots, n,\\&\{a_1, a_2, \ldots, a_r\} = \text{sort}\{c_1, c_2, \ldots, c_r\} \text{ with } a_i < a_{i+1}\end{aligned} \tag{7.64}$$

the $(2m+1)^{\text{th}}$-degree polynomial system has equilibriums of $x^* = a_{i_s}(\mathbf{p})$ ($s = 1, 2, \ldots, r$), and the corresponding standard form is

$$\dot{x} = a_0(\mathbf{p})\prod_{j=1}^{r}(x - a_{i_j})^{l_j}\prod_{i=r+1}^{n}[(x + \frac{1}{2}B_i)^2 + \frac{1}{4}(-\Delta_i)]^{l_i}. \tag{7.65}$$

The equilibrium separatrix flow is called an (l_1th XX : l_2th XX : … : l_rth XX)-flow.

(a) For $a_0 > 0$ and $s = 1, 2, \ldots, r$,

$$l_p\text{th XX} = \begin{cases}(2r_p - 1)^{\text{th}}\text{order source}, & \text{for } \alpha_p = 2M_p - 1, l_p = 2r_p - 1;\\(2r_p - 1)^{\text{th}}\text{order sink}, & \text{for } \alpha_p = 2M_p, l_p = 2r_p - 1;\\(2r_p)^{\text{th}}\text{order lower-saddle}, & \text{for } \alpha_p = 2M_p - 1, l_p = 2r_p;\\(2r_p)^{\text{th}}\text{order upper-saddle}, & \text{for } \alpha_p = 2M_p, l_p = 2r_p,\end{cases} \tag{7.66}$$

where

$$\alpha_p = \sum\nolimits_{s=p}^{r} l_s. \tag{7.67}$$

(b) For $a_0 < 0$ and $p = 1, 2, \ldots, r$,

$$l_p\text{th XX} = \begin{cases} (2r_p - 1)^{\text{th}}\text{order sink}, & \text{for } \alpha_p = 2M_p - 1, l_p = 2r_p - 1; \\ (2r_p - 1)^{\text{th}}\text{order source}, & \text{for } \alpha_p = 2M_p, l_p = 2r_p - 1; \\ (2r_p)^{\text{th}}\text{order upper-saddle}, & \text{for } \alpha_p = 2M_p - 1, l_p = 2r_p; \\ (2r_p)^{\text{th}}\text{order lower-saddle}, & \text{for } \alpha_p = 2M_p, l_p = 2r_p. \end{cases} \tag{7.68}$$

(ii) If

$$\begin{aligned} &\Delta_i = B_i^2 - 4C_i > 0, i = j_1, j_2, \ldots, j_s \in \{l+1, l+2, \ldots, n\}, \\ &\Delta_j = B_j^2 - 4C_j < 0, j = j_{s+1}, j_{s+2}, \ldots, j_n \in \{l+1, l+2, \ldots, n\} \\ &\text{with s} \in \{1, \ldots, n-l\}, \end{aligned} \tag{7.69}$$

the $(2m+1)^{\text{th}}$-degree polynomial nonlinear system has $2n_2$-equilibriums as

$$\begin{aligned} &x^* = b_1^{(i)} = -\frac{1}{2}(B_i + \sqrt{\Delta_i}), x^* = b_2^{(i)} = -\frac{1}{2}(B_i - \sqrt{\Delta_i}) \\ &i \in \{j_1, j_2, \ldots, j_{n_1}\} \subseteq \{l+1, l+2, \ldots, n\}. \end{aligned} \tag{7.70}$$

If

$$\begin{aligned} &\{a_1, a_2, \ldots, a_{2n_2+1}\} = \text{sort}\{c_1, c_2, \ldots, c_{2l+1}, \underbrace{b_1^{(r+1)}, b_2^{(r+1)}}_{q_{r+1}\ sets}, \ldots, \underbrace{b_1^{(n_1)}, b_2^{(n_1)}}_{q_{n_1}\ sets}\}, \\ &a_{i_1} \equiv a_1 = \ldots = a_{l_1}, \\ &a_{i_2} \equiv a_{l_1+1} = \ldots = a_{l_1+l_2}, \\ &\vdots \\ &a_{i_{n_1}} \equiv a_{\Sigma_{i=1}^{n_1-1} l_i+1} = \ldots = a_{\Sigma_{i=1}^{n_1-1} l_i + l_{n_1}} = a_{2n_2+1} \\ &\text{with } \Sigma_{s=1}^{n_1} l_s = 2n_2 + 1, \end{aligned} \tag{7.71}$$

then the corresponding standard form is

$$\dot{x} = a_0 \prod_{s=1}^{n_1} (x - a_{i_s})^{l_s} \prod_{i=n_2+1}^{n} [(x + \frac{1}{2} B_i)^2 + \frac{1}{4}(-\Delta_i)]^{q_i}. \tag{7.72}$$

The equilibrium separatrix flow is called an (l_1th XX : l_2th XX : . . . : l_{n_1}th XX)-flow.

(a) For $a_0 > 0$ and $p = 1, 2, \ldots, r, r + 1, \ldots, n_1$,

$$l_p\text{th XX} = \begin{cases} (2r_p - 1)^{\text{th}}\text{order source}, & \text{for } \alpha_p = 2M_p - 1, l_p = 2r_p - 1; \\ (2r_p - 1)^{\text{th}}\text{order sink}, & \text{for } \alpha_p = 2M_p, l_p = 2r_p - 1; \\ (2r_p)^{\text{th}}\text{order lower-saddle}, & \text{for } \alpha_p = 2M_p - 1, l_p = 2r_p; \\ (2r_p)^{\text{th}}\text{order upper-saddle}, & \text{for } \alpha_p = 2M_p, l_p = 2r_p, \end{cases} \tag{7.73}$$

where

$$\alpha_p = \sum\nolimits_{s=p}^{n_1} l_s. \tag{7.74}$$

(b) For $a_0 < 0$ and $p = 1, 2, \ldots, r, r + 1, \ldots, n_1$,

$$l_p\text{th XX} = \begin{cases} (2r_p - 1)^{\text{th}}\text{order sink}, & \text{for } \alpha_p = 2M_p - 1, l_p = 2r_p - 1; \\ (2r_p - 1)^{\text{th}}\text{order source}, & \text{for } \alpha_p = 2M_p, l_p = 2r_p - 1; \\ (2r_p)^{\text{th}}\text{order upper-saddle}, & \text{for } \alpha_p = 2M_p - 1, l_p = 2r_p; \\ (2r_p)^{\text{th}}\text{order lower-saddle}, & \text{for } \alpha_p = 2M_p, l_p = 2r_p. \end{cases} \tag{7.75}$$

(c) The equilibrium of $x^* = a_{i_p}$ for ($l_p > 1$)-repeated equilibriums switching is called an l_pth XX switching bifurcation of $(l_{p_1}\text{th XX} : l_{p_2}\text{th XX} : \ldots : l_{p_\beta}\text{th XX})$ equilibrium at a point $\mathbf{p} = \mathbf{p}_1 \in \partial\Omega_{12}$, and the bifurcation condition is

$$\begin{aligned} & a_{i_p} \equiv a_{\Sigma_{i=1}^{p-1} l_i + 1} = \ldots = a_{\Sigma_{i=1}^{p-1} l_i + l_p}, \\ & a^{\pm}_{\Sigma_{i=1}^{p-1} l_i + 1} \neq \ldots \neq a^{\pm}_{\Sigma_{i=1}^{p-1} l_i + l_p}; \; l_p = \sum\nolimits_{i=1}^{\beta} l_{p_i}. \end{aligned} \tag{7.76}$$

(iii) If

$$\begin{aligned}
&\Delta_i = B_i^2 - 4C_i = 0,\\
&\text{for } i \in \{i_{11}, i_{12}, \ldots, i_{1s}\} \subseteq \{i_{l+1}, i_{l+2}, \ldots, i_{n_2}\} \subseteq \{l+1, l+2, \ldots, n\},\\
&\Delta_k = B_k^2 - 4C_k > 0,\\
&\text{for } k \in \{i_{21}, i_{22}, \ldots, i_{2r}\} \subseteq \{i_{l+1}, i_{l+2}, \ldots, i_{n_2}\} \subseteq \{l+1, l+2, \ldots, n\},\\
&\Delta_j = B_j^2 - 4C_j < 0,\\
&\text{for } j \in \{i_{n_2+1}, i_{n_2+2}, \ldots, i_n\} \subseteq \{l+1, l+2, \ldots, n\},
\end{aligned} \tag{7.77}$$

the $(2m+1)^{\text{th}}$-degree polynomial nonlinear system has $(2n_2+1)$-equilibriums as

$$\begin{aligned}
&x^* = b_1^{(i)} = -\frac{1}{2}B_i, x^* = b_2^{(i)} = -\frac{1}{2}B_i \text{ for } i \in \{i_{11}, i_{12}, \ldots, i_{1s}\},\\
&x^* = b_1^{(k)} = -\frac{1}{2}(B_k + \sqrt{\Delta_k}), x^* = b_2^{(k)} = -\frac{1}{2}(B_k - \sqrt{\Delta_k})\\
&\text{for } i \in \{i_{21}, i_{22}, \ldots, i_{2r}\}.
\end{aligned} \tag{7.78}$$

If

$$\begin{aligned}
&\{a_1, a_2, \ldots, a_{2n_2+1}\} = \text{sort}\{a, c_1, c_2, \ldots, c_{2l}, \underbrace{b_1^{(r)}, b_2^{(r)}}_{q_r \text{ sets}}, \ldots, \underbrace{b_1^{(n_1)}, b_2^{(n_1)}}_{q_{n_1} \text{ sets}}\},\\
&a_{i_1} \equiv a_1 = \ldots = a_{l_1},\\
&a_{i_2} \equiv a_{l_1+1} = \ldots = a_{l_1+l_2},\\
&\vdots\\
&a_{i_{n_1}} \equiv a_{\Sigma_{i=1}^{n_1-1} l_i+1} = \ldots = a_{\Sigma_{i=1}^{n_1-1} l_i + l_{n_1}} = a_{2n_2+1}\\
&\text{with } \Sigma_{s=1}^{n_1} l_s = 2n_2 + 1,
\end{aligned} \tag{7.79}$$

then the corresponding standard form is

$$\dot{x} = a_0 \prod_{s=1}^{n_1} (x - a_{i_s})^{l_s} \prod_{i=n_2+1}^{n} [(x + \frac{1}{2}B_i)^2 + \frac{1}{4}(-\Delta_i)]^{q_i}. \tag{7.80}$$

The equilibrium separatrix flow is called an l_1th XX : l_2th XX : ... : l_{n_1}th XX)-flow.

(a) The equilibrium of $x^* = a_{i_p}$ for $(q_p > 1)$-repeated equilibriums appearing or vanishing is called an l_pth XX bifurcation of equilibrium at a point $\mathbf{p} = \mathbf{p}_1 \in \partial\Omega_{12}$, and the bifurcation condition is

$$
\begin{aligned}
&a_{i_p} \equiv a_{\Sigma_{i=1}^{p-1} l_i+1} = \ldots = a_{\Sigma_{i=1}^{p-1} l_i+l_p} = -\frac{1}{2}B_{i_q} \\
&\text{with } \Delta_{i_q} = B_{i_q}^2 - 4C_{i_q} = 0 \left(i_q \in \{i_1, i_2, \ldots, i_l\}\right) \\
&a^{+}_{\Sigma_{i=1}^{p-1} q_i+1} \neq \ldots \neq a^{+}_{\Sigma_{i=1}^{p-1} q_i+q_p} \text{ or } a^{-}_{\Sigma_{i=1}^{p-1} q_i+1} \neq \ldots \neq a^{-}_{\Sigma_{i=1}^{p-1} q_i+q_p}.
\end{aligned}
\tag{7.81}
$$

(b) The equilibrium of $x^* = a_{i_q}$ for $(l_p > 1)$-repeated equilibriums switching is called an l_pth XX bifurcation of $(l_{p_1}$th XX : l_{p_2}th XX : … : l_{p_β}th XX) equilibrium switching at a point $\mathbf{p} = \mathbf{p}_1 \in \partial\Omega_{12}$, and the bifurcation condition is

$$
\begin{aligned}
&a_{i_q} \equiv a_{\Sigma_{i=1}^{q-1} l_i+1} = \ldots = a_{\Sigma_{i=1}^{q-1} l_i+l_p}, \\
&a^{\pm}_{\Sigma_{i=1}^{q-1} l_i+1} \neq \ldots \neq a^{\pm}_{\Sigma_{i=1}^{q-1} l_i+l_q}, \; l_p = \sum\nolimits_{i=1}^{\beta} l_{p_i}.
\end{aligned}
\tag{7.82}
$$

(c) The equilibrium of $x^* = a_{i_p}$ for $\left(l_{p_1} \geq 1\right)$-repeated equilibriums appearing/vanishing and $\left(l_{p_2} \geq 2\right)$-repeated equilibriums switching of $\left(l_{p_{21}}\text{th XX} :\right.$ $l_{p_{22}}$th XX : … : $l_{p_{2\beta}}$th XX) is called an q_pth XX bifurcation of equilibrium at a point $\mathbf{p} = \mathbf{p}_1 \in \partial\Omega_{12}$, and the bifurcation condition is

$$
\begin{aligned}
&a_{i_p} \equiv a_{\Sigma_{i=1}^{p-1} q_i+1} = \ldots = a_{\Sigma_{i=1}^{p-1} q_i+q_p} \\
&\text{with } \Delta_{i_q} = B_{i_q}^2 - 4C_{i_q} = 0 \left(i_q \in \{i_1, i_2, \ldots, i_l\}\right) \\
&a^{+}_{\Sigma_{i=1}^{p-1} q_i+j_1} \neq \ldots \neq a^{+}_{\Sigma_{i=1}^{p-1} q_i+j_{p_1}} \text{ or } a^{-}_{\Sigma_{i=1}^{p_1-1} q_i+j_1} \neq \ldots \neq a^{-}_{\Sigma_{i=1}^{p_1-1} q_i+j_{p_1}}, \\
&\text{for } \{j_1 j_2 \cdots j_{p_1}\} \subseteq \left\{1, 2, \ldots, q_p\right\}, \\
&a^{\pm}_{\Sigma_{i=1}^{p-1} q_i+k_1} \neq \ldots \neq a^{\pm}_{\Sigma_{i=1}^{p-1} q_i+k_{p_2}} \\
&\text{for } \left\{k_1, k_2, \ldots, k_{p_2}\right\} \subseteq \left\{1, 2, \ldots, q_p\right\}, \\
&\text{with } l_{p_1} + l_{p_2} = l_p.
\end{aligned}
\tag{7.83}
$$

(iv) If

$$
\Delta_i = B_i^2 - 4C_i > 0 \;\text{ for } i = l+1, l+2, \ldots, n \tag{7.84}
$$

the $(2m+1)^{\text{th}}$-degree polynomial nonlinear system has $(2m + 1)$-equilibriums as

$$
\begin{aligned}
&x^* = b_1^{(i)} = -\frac{1}{2}(B_i + \sqrt{\Delta_i}), x^* = b_2^{(i)} = -\frac{1}{2}(B_i - \sqrt{\Delta_i}) \\
&\text{for } i = l+1, l+2, \ldots, n.
\end{aligned}
\tag{7.85}
$$

If

$$\{a_1, a_2, \ldots, a_{2m+1}\} = \mathrm{sort}\{c_1, c_2, \ldots, c_{2l+1}, \underbrace{b_1^{(r+1)}, b_2^{(r+1)}}_{q_{r+1}\text{ sets}}, \ldots, \underbrace{b_1^{(n)}, b_2^{(n)}}_{q_n\text{sets}}\},$$

$$\begin{aligned} &a_{i_1} \equiv a_1 = \ldots = a_{l_1}, \\ &a_{i_2} \equiv a_{l_1+1} = \ldots = a_{l_1+l_2}, \\ &\vdots \\ &a_{i_n} \equiv a_{\Sigma_{i=1}^{n-1} l_i+1} = \ldots = a_{\Sigma_{i=1}^{n-1} l_i+l_r} = a_{2m+1} \end{aligned} \tag{7.86}$$

with $\Sigma_{s=1}^{n} l_s = 2m+1$,

then the corresponding standard form is

$$\dot{x} = a_0 \prod_{s=1}^{r} (x - a_{i_s})^{l_s}. \tag{7.87}$$

The equilibrium separatrix flow is called an (l_1th XX : l_2th XX : ... : l_rth XX)-flow. The equilibrium of $x^* = a_{i_p}$ for l_p-repeated equilibriums switching is called an l_pth XX switching bifurcation of $(l_{p_1}$th XX : l_{p_2}th XX : ... : l_{p_β}th XX) equilibrium at a point $\mathbf{p} = \mathbf{p}_1 \in \partial\Omega_{12}$, and the bifurcation condition is

$$\begin{aligned} &a_{i_p} \equiv a_{\Sigma_{i=1}^{p-1} l_i+1} = \ldots = a_{\Sigma_{i=1}^{p-1} l_i+l_p}, \\ &a^{\pm}_{\Sigma_{i=1}^{p-1} l_i+1} \neq \ldots \neq a^{\pm}_{\Sigma_{i=1}^{p-1} l_i+l_p}; \; l_p = \sum\nolimits_{i=1}^{\beta} l_{p_i}. \end{aligned} \tag{7.88}$$

7.2 Simple Equilibrium Bifurcations

To illustrate the bifurcations in the (2m+1)th-degree polynomial system as in Luo (2019), the detailed discussion with graphical illustrations will be presented as follows.

7.2.1 *Appearing Bifurcations*

Consider a (2m+1)th-degree polynomial system as

$$\dot{x} = a_0 Q(x)(x - a) \prod_{i=1}^{n} (x^2 + B_i x + C_i) \tag{7.89}$$

where without loss of generality, a function of $Q(x) > 0$ is either a polynomial function or a non-polynomial function. The roots of $x^2 + B_i x + C_i = 0$ are

$$\begin{aligned}
&b_{1,2}^{(i)} = -\frac{1}{2}B_i \pm \frac{1}{2}\sqrt{\Delta_i},\ \Delta_i = B_i^2 - 4C_i \geq 0\ (i = 1,2,\ldots,n);\\
&\{a_1, a_2, \ldots, a_{2l}\} \subset \text{sort}\{b_1^{(1)}, b_2^{(1)}, b_1^{(2)}, b_2^{(2)}, \ldots, b_1^{(n)}, b_2^{(n)}\}, a_s \leq a_{s+1};\\
&\left.\begin{array}{l} B_i \neq B_j\ (i,j = 1,2,\ldots,n; i \neq j) \\ \Delta_i = 0\ (i = 1,2,\ldots,n) \end{array}\right\} \text{at bifurcation.}
\end{aligned} \tag{7.90}$$

The second-order singularity bifurcation is for the birth of a pair of simple sink and source. There are two *appearing* bifurcations for $i \in \{1,2,\ldots,n\}$:

$$2^{\text{nd}}\text{order US} \xrightarrow[\text{appearing bifurcation}]{i^{\text{th}}\text{quadratic factor}} \begin{cases} \text{SO}, & \text{for } x^* = a_{2i}, \\ \text{SI}, & \text{for } x^* = a_{2i-1}. \end{cases} \tag{7.91}$$

$$2^{\text{nd}}\text{order LS} \xrightarrow[\text{appearing bifurcation}]{i^{\text{th}}\text{quadratic factor}} \begin{cases} \text{SI}, & \text{for } x^* = a_{2i}, \\ \text{SO}, & \text{for } x^* = a_{2i-1}. \end{cases} \tag{7.92}$$

If $x^* = a \neq -\frac{1}{2}B_i\ (i \in \{1,2,\ldots,m\})$, the equilibrium of $x* = a$ breaks a cluster of *teeth comb* appearing bifurcations to two parts. The *teeth comb* appearing bifurcation generated by the m-pairs of quadratics becomes a *broom appearing* bifurcation. The two *broom appearing* bifurcations are

$$\begin{aligned}
&\text{SO}\ (x^* = a) \xrightarrow[\text{appearing bifurcation}]{l_1 + l_2 = m} (l_1\text{-LSN} : \text{SO} : l_2\text{-USN})\\
&= \begin{cases} l_2\text{-USN, for } x^* \in \{a_{2j}, a_{2j+1}, i = l_1 + 1 \ldots l_1 + l_2\}, \\ \text{SO, for } x^* = a = a_{2(l_1+1)-1} \\ l_1\text{-LSN, for } x^* \in \{a_{2i-1}, a_{2i}, i = 1,2,\ \ldots, l_1\} \end{cases}
\end{aligned} \tag{7.93}$$

and

$$\begin{aligned}
&\text{SI}\ (x^* = a) \xrightarrow[m\text{-appearing bifurcation}]{l_1 + l_2 = m} (l_1\text{-USN} : \text{SO} : l_2\text{-LSN})\\
&= \begin{cases} l_2\text{-LSN, for } x^* \in \{a_{2j}, a_{2j+1} : j = l_1 + 1 \ldots l_1 + l_2\}, \\ \text{SI, for } x^* = a = a_{2(l_1+1)-1} \\ l_1\text{-USN, for } x^* \in \{a_{2i-1}, a_{2i} : i = 1,2,\ \ldots, l_1\} \end{cases}
\end{aligned} \tag{7.94}$$

where the l_j-USN and l_j-LSN ($j = 1,2$) are

$$
l_j\text{-USN}\begin{cases}
\text{US}\xrightarrow[\text{appearing}]{(l_j+s_j)^{\text{th}}\text{bifurcation}}\begin{cases}\text{SO}, & \text{for } x^*=a_{2(s_j+l_j)+\delta_j^2},\\ \text{SI}, & \text{for } x^*=a_{2(s_j+l_j)-1+\delta_j^2};\end{cases}\\
\vdots\\
\text{US}\xrightarrow[\text{appearing}]{(s_j)^{\text{th}}\text{bifurcation}}\begin{cases}\text{SO}, & \text{for } x^*=a_{2s_j+\delta_j^2},\\ \text{SI}, & \text{for } x^*=a_{2s_j-1+\delta_j^2}\end{cases}
\end{cases}
\tag{7.95}
$$

$$
l_j\text{-LSN}\begin{cases}
\text{LS}\xrightarrow[\text{appearing}]{(l_j+s_j)^{\text{th}}\text{bifurcation}}\begin{cases}\text{SI}, & \text{for } x^*=a_{2(s_j+l_j)+\delta_j^2},\\ \text{SO}, & \text{for } x^*=a_{2(s_j+l_j)-1+\delta_j^2};\end{cases}\\
\vdots\\
\text{LS}\xrightarrow[\text{appearing}]{(s_j)^{\text{th}}\text{bifurcation}}\begin{cases}\text{SI}, & \text{for } x^*=a_{2s_j+\delta_j^2},\\ \text{SO}, & \text{for } x^*=a_{2s_j-1+\delta_j^2}\end{cases}
\end{cases}
\tag{7.96}
$$

for $s_j \in \{0, 1, 2, \ldots, m\}$ and $0 \le l_j \le m$ with $0 \le l_j \le m$.

Four special broom appearing bifurcations are

$$
\text{SO }(x^*=a)\rightarrow\begin{cases}
\text{SO}\rightarrow\text{SO}, \text{for } x^*=a=a_{2m+1},\\
m\text{-LSN}\begin{cases}
\text{LS}\xrightarrow[\text{appearing}]{m^{\text{th}}\text{bifurcation}}\begin{cases}\text{SI}, & \text{for } x^*=a_{2m},\\ \text{SO}, & \text{for } x^*=a_{2m-1};\end{cases}\\
\vdots\\
\text{LS}\xrightarrow[\text{appearing}]{1^{\text{st}}\text{bifurcation}}\begin{cases}\text{SI}, & \text{for } x^*=a_{2},\\ \text{SO}, & \text{for } x^*=a_{1}.\end{cases}
\end{cases}
\end{cases}
\tag{7.97}
$$

$$
\text{SI }(x^*=a)\rightarrow\begin{cases}
\text{SI}\rightarrow\text{SI}, \text{for } x^*=a=a_{2m+1},\\
m\text{-USN}\begin{cases}
\text{US}\xrightarrow[\text{appearing}]{m^{\text{th}}\text{bifurcation}}\begin{cases}\text{SO}, & \text{for } x^*=a_{2m},\\ \text{SI}, & \text{for } x^*=a_{2m-1};\end{cases}\\
\vdots\\
\text{US}\xrightarrow[\text{appearing}]{1^{\text{st}}\text{bifurcation}}\begin{cases}\text{SO}, & \text{for } x^*=a_{2},\\ \text{SI}, & \text{for } x^*=a_{1}.\end{cases}
\end{cases}
\end{cases}
\tag{7.98}
$$

and

$$
\text{SO}\ (x^* = a) \to \begin{cases} m\text{-USN} \begin{cases} \text{US} \xrightarrow[\text{appearing}]{m^{\text{th}}\text{bifurcation}} \begin{cases} \text{SO}, & \text{for } x^* = a_{2m+1}, \\ \text{SI}, & \text{for } x^* = a_{2m}; \end{cases} \\ \vdots \\ \text{US} \xrightarrow[\text{appearing}]{1^{\text{st}}\text{bifurcation}} \begin{cases} \text{SO}, & \text{for } x^* = a_3, \\ \text{SI}, & \text{for } x^* = a_2; \end{cases} \end{cases} \\ \text{SO} \to \text{SO}, \text{for } x^* = a = a_1. \end{cases} \tag{7.99}
$$

$$
\text{SI}\ (x^* = a) \to \begin{cases} m\text{-LSN} \begin{cases} \text{LS} \xrightarrow[\text{appearing}]{m^{\text{th}}\text{bifurcation}} \begin{cases} \text{SI}, & \text{for } x^* = a_{2m+1}, \\ \text{SO}, & \text{for } x^* = a_{2m}; \end{cases} \\ \vdots \\ \text{LS} \xrightarrow[\text{appearing}]{1^{\text{st}}\text{bifurcation}} \begin{cases} \text{SI}, & \text{for } x^* = a_3, \\ \text{SO}, & \text{for } x^* = a_2; \end{cases} \end{cases} \\ \text{SI} \to \text{SI}, \text{for } x^* = a = a_1. \end{cases} \tag{7.100}
$$

If $x^* = a = -\frac{1}{2}B_i$ ($i \in \{1, 2, \ldots, m\}$), the equilibrium of $x^* = a$ possesses a third-order SI or SO switching bifurcation (or pitchfork bifurcation). The teeth comb appearing bifurcation generated by the m-pairs of quadratics becomes a *broom appearing* bifurcation. The two *broom appearing* bifurcations are

$$
\begin{aligned}
&\text{SO}\ (x^* = a) \xrightarrow[\text{appearing bifurcation}]{m = l_1 + l_2 + 1} (l_1\text{-LSN} : 3^{\text{rd}}\text{SO} : l_2\text{-USN}) \\
&= \begin{cases} l_2\text{-USN, for } x^* \in \{a_{2j}, a_{2j+1}, j = l_1 + 2 \ldots l_1 + l_2\}, \\ 3^{\text{rd}}\text{SO} \to \begin{cases} \text{SO, for } x^* = a_{2(l_1+2)-1} \\ \text{SI, for } x^* = a = a_{2(l_1+1)} \\ \text{SO, for } x^* = a_{2(l_1+1)-1} \end{cases} \\ l_1\text{-LSN, for } x^* \in \{a_{2i-1}, a_{2i}, i = 1, 2, \ldots, l_1\} \end{cases}
\end{aligned} \tag{7.101}
$$

and

$$
\begin{aligned}
&\text{SI}\ (x^* = a) \xrightarrow[\text{appearing bifurcation}]{m = l_1 + l_2 + 1} (l_1\text{-USN} : 3^{\text{rd}}\text{SI} : l_2\text{-LSN}) \\
&= \begin{cases} l_2\text{-LSN, for } x^* \in \{a_{2j}, a_{2j+1}, j = l_1 + 2 \ldots l_1 + l_2\}, \\ 3^{\text{rd}}\text{SI} \to \begin{cases} \text{SI, for } x^* = a_{2(l_1+2)-1} \\ \text{SO, for } x^* = a = a_{2(l_1+1)} \\ \text{SI, for } x^* = a_{2(l_1+1)-1} \end{cases} \\ l_1\text{-USN, for } x^* \in \{a_{2i-1}, a_{2i}, i = 1, 2, \ldots, l_1\}. \end{cases}
\end{aligned} \tag{7.102}
$$

Consider an appearing bifurcation for a cluster of sink and source equilibriums with the following conditions:

$$\left.\begin{array}{l} B_i = B_j\ (i,j \in \{1,2,\ldots,n\}, i \neq j) \\ \Delta_j = 0\ (i = 1,2,\ldots,n) \end{array}\right\} \text{ at bifurcation.} \tag{7.103}$$

Thus, the $(2l)^{\text{th}}$-order appearing bifurcation is for a cluster of simple sinks and sources. Two $(2l)^{\text{th}}$-order *appearing* bifurcations for $l \in \{1,2,\ldots,s\}$ are

$$(2l)^{\text{th}}\text{order USN} \xrightarrow[\text{appearing bifurcation}]{\text{cluster of } l\text{-quadratics}} \begin{cases} \text{SO}, & \text{for } x^* = a_{2s_l}, \\ \text{SI}, & \text{for } x^* = a_{2s_l-1}, \\ \vdots & \\ \text{SO}, & \text{for } x^* = a_{2s_1}, \\ \text{SI}, & \text{for } x^* = a_{2s_1-1} \end{cases} \tag{7.104}$$

$$(2l)^{\text{th}}\text{order LSN} \xrightarrow[\text{appearing bifurcation}]{\text{cluster of } l\text{-quadratics}} \begin{cases} \text{SI}, & \text{for } x^* = a_{2s_l}, \\ \text{SO}, & \text{for } x^* = a_{2s_l-1}, \\ \vdots & \\ \text{SI}, & \text{for } x^* = a_{2s_1}, \\ \text{SO}, & \text{for } x^* = a_{2s_1-1}. \end{cases} \tag{7.105}$$

If $x^* = a \neq -\frac{1}{2}B_i\ (i \in \{1,2,\ldots,n\})$, the equilibrium of $x^* = a$ breaks a cluster of *sprinkler-spraying* appearing bifurcations to two parts. The sprinkler-spraying appearing bifurcation generated by the m-pairs of quadratics becomes a *broom-sprinkler-spraying appearing* bifurcation. The two broom-sprinkler-spraying *appearing* bifurcations are

$$\begin{aligned} &\text{SO}\,(x^* = a) \xrightarrow[\text{appearing bifurcation}]{m = m_1 + m_2} (r_1\text{-LSG} : \text{SO} : r_2\text{-USG}) \\ &= \begin{cases} r_2\text{-USG} \to \begin{cases} (2l_{r_2}^{(2)})^{\text{th}}\text{USN}\,(x^* = a_{r_1+r_2+1}), \\ \vdots \\ (2l_1^{(2)})^{\text{th}}\text{USN}\,(x^* = a_{r_1+2}), \end{cases} \\ \text{SO}\,(a = a_{r_1+1}) \to \text{SO}\,(a = a_{2(m_1+1)-1}), \\ r_1\text{-LSG} \to \begin{cases} (2l_{r_1}^{(1)})^{\text{th}}\text{LSN}\,(x^* = a_{r_1}), \\ \vdots \\ (2l_1^{(1)})^{\text{th}}\text{LSN}\,(x^* = a_1) \end{cases} \end{cases} \end{aligned} \tag{7.106}$$

and

$$\mathrm{SI}\ (x^* = a) \xrightarrow[\text{appearing bifurcation}]{m=m_1+m_2} (r_1\text{-USG} : \mathrm{SI} : r_2\text{-LSG})$$
$$= \begin{cases} r_2\text{-LSG} \to \begin{cases} \left(2l_{r_2}^{(2)}\right)^{\text{th}} \mathrm{LSN}\ (x^* = a_{r_1+r_2+1}), \\ \vdots \\ \left(2l_1^{(2)}\right)^{\text{th}} \mathrm{LSN}\ (x^* = a_{r_1+2}), \end{cases} \\ \mathrm{SI}\ (a = a_{r_1+1}) \to \mathrm{SI}\ \left(a = a_{2(m_1+1)-1}\right), \\ r_1\text{-USG} \to \begin{cases} \left(2l_{r_1}^{(1)}\right)^{\text{th}} \mathrm{USN}\ (x^* = a_{r_1}), \\ \vdots \\ \left(2l_1^{(1)}\right)^{\text{th}} \mathrm{USN}\ (x^* = a_1) \end{cases} \end{cases} \tag{7.107}$$

for $m_1 = \sum_{i=1}^{r_1} l_i^{(1)}, m_2 = \sum_{j=1}^{r_2} l_j^{(2)}$; and the acronyms USG and LSG are the upper-saddle-node and lower-saddle-node bifurcation groups, respectively.

Four special broom-sprinkler-spraying appearing bifurcations are

$$\mathrm{SO}\ (x^* = a) \xrightarrow[\text{appearing bifurcation}]{m=\sum_{i=1}^{r} l_i} \begin{cases} \mathrm{SO}\ (a = a_{2r+1}) \to \mathrm{SO}\ (a = a_{2m+1}), \\ r\text{-LSG} \to \begin{cases} (2l_r)^{\text{th}} \mathrm{LSN}\ (x^* = a_r), \\ \vdots \\ (2l_1)^{\text{th}} \mathrm{LSN}\ (x^* = a_1), \end{cases} \end{cases} \tag{7.108}$$

$$\mathrm{SI}\ (x^* = a) \xrightarrow[\text{appearing bifurcation}]{m=\sum_{i=1}^{r} l_i} \begin{cases} \mathrm{SI}\ (a = a_{2r+1}) \to \mathrm{SI}\ (a = a_{2m+1}), \\ r\text{-USG} \to \begin{cases} (2l_r)^{\text{th}} \mathrm{USN}\ (x^* = a_r), \\ \vdots \\ (2l_1)^{\text{th}} \mathrm{USN}\ (x^* = a_1), \end{cases} \end{cases} \tag{7.109}$$

and

$$\mathrm{SO}\ (x^* = a) \xrightarrow[\text{appearing bifurcation}]{m=\sum_{i=1}^{r} l_i} \begin{cases} r\text{-USG} \to \begin{cases} (2l_r)^{\text{th}} \mathrm{USN}\ (x^* = a_{r+1}), \\ \vdots \\ (2l_1)^{\text{th}} \mathrm{USN}\ (x^* = a_2), \end{cases} \\ \mathrm{SO}\ (a = a_1) \to \mathrm{SO}\ (a = a_1), \end{cases} \tag{7.110}$$

$$\text{SI}\,(x^* = a) \xrightarrow[\text{appearing bifurcation}]{m=\sum_{i=1}^{r} l_i} \begin{cases} r\text{-LSG} \to \begin{cases} (2l_r)^{\text{th}}\text{LSN}\,(x^* = a_{r+1}), \\ \vdots \\ (2l_1)^{\text{th}}\text{LSN}\,(x^* = a_2), \end{cases} \\ \text{SI}\,(a = a_1) \to \text{SO}\,(a = a_1) \end{cases} \tag{7.111}$$

If $x^* = a = -\frac{1}{2}B_i$ $(i \in \{1, 2, \ldots, l\})$, the equilibrium of $x^* = a$ possesses a $(2l+1)^{\text{th}}$-order SI or SO switching bifurcation (or broom-switching bifurcation). The sprinkler-spraying appearing bifurcation generated by the m-pairs of quadratics becomes a *broom-sprinkler-spraying switching* bifurcation. The two *broom-switching* bifurcations are

$$\begin{aligned} &\text{SO}\,(x^* = a) \xrightarrow[\text{switching bifurcation}]{m=m_1+m_2+l} (r_1\text{-LSG} : (2l+1)^{\text{rd}}\text{SO} : r_2\text{-USG}) \\ &= \begin{cases} r_2\text{-USG} \to \begin{cases} (2l_{r_2}^{(2)})^{\text{th}}\text{USN}\,(x^* = a_{r_2+r_1+1}), \\ \vdots \\ (2l_1^{(2)})^{\text{th}}\text{USN}\,(x^* = a_{r_1+2}), \end{cases} \\ (2l+1)^{\text{rd}}\text{SO}\,(a = a_{r_1+1}); \\ r_1\text{-LSG} \to \begin{cases} (2l_{r_1}^{(1)})^{\text{th}}\text{LSN}\,(x^* = a_{r_1}), \\ \vdots \\ (2l_1^{(1)})^{\text{th}}\text{LSN}\,(x^* = a_1) \end{cases} \end{cases} \end{aligned} \tag{7.112}$$

and

$$\begin{aligned} &\text{SI}\,(x^* = a) \xrightarrow[\text{switching bifurcation}]{m=m_1+m_2+l} (r_1\text{-USG} : (2l+1)^{\text{rd}}\text{SI} : r_2\text{-LSG}) \\ &= \begin{cases} r_2\text{-LSG} \to \begin{cases} (2l_{r_2}^{(2)})^{\text{th}}\text{LSN}\,(x^* = a_{r_2+r_1+1}), \\ \vdots \\ (2l_1^{(2)})^{\text{th}}\text{LSN}\,(x^* = a_{r_1+2}), \end{cases} \\ (2l+1)^{\text{rd}}\text{SI}\,(a = a_{r_1+1}); \\ r_1\text{-USG} \to \begin{cases} (2l_{r_1}^{(1)})^{\text{th}}\text{USN}\,(x^* = a_{r_1}), \\ \vdots \\ (2l_1^{(1)})^{\text{th}}\text{USN}\,(x^* = a_1), \end{cases} \end{cases} \end{aligned} \tag{7.113}$$

where

$$(2l+1)^{\text{th}}\text{order SO } (x^*=a) \xrightarrow[\text{appearing bifurcation}]{\text{cluster of } l\text{-quadratics}} \begin{cases} \text{SO}, & \text{for } x^* = a_{2s_l+1}, \\ \text{SI}, & \text{for } x^* = a_{2s_l}, \\ \vdots & \\ \text{SI}, & \text{for } x^* = a_{2s_1}, \\ \text{SO}, & \text{for } x^* = a_{2s_1-1} \end{cases} \tag{7.114}$$

$$(2l+1)^{\text{th}}\text{order SI } (x^*=a) \xrightarrow[\text{appearing bifurcation}]{\text{cluster of } l\text{-quadratics}} \begin{cases} \text{SI}, & \text{for } x^* = a_{2s_l+1}, \\ \text{SO}, & \text{for } x^* = a_{2s_l}, \\ \vdots & \\ \text{SO}, & \text{for } x^* = a_{2s_1}, \\ \text{SI}, & \text{for } x^* = a_{2s_1-1} \end{cases} \tag{7.115}$$

where $x^* = a \in \{a_{2s_1-1}, \ldots, a_{2s_l}, a_{2s_l+1}\}$.

In Fig. 7.1a, b, the simple switching with two *teeth comb appearing* bifurcations is presented for $a_0 > 0$ and $a_0 < 0$, respectively. The two bifurcation structures are

(a) SO $\to$ (l_1 - LSN : SO : l_2 - USN),
(b) SI $\to$ (l_1 - USN : SI : l_2 - LSN)

with $l_1 + l_2 = m$. In Fig. 7.1c, d, the third-order pitchfork switching bifurcation with two *teeth comb appearing* bifurcations is presented for $a_0 > 0$ and $a_0 < 0$, respectively. The two bifurcation structures are

(c) SO $\to$ (l_1 - LSN : 3rdSO : l_2 - USN),
(d) SI $\to$ (l_1 - USN : 3rdSI : l_2 - LSN)

with $l_1 + l_2 = m - 1$.

In Fig. 7.2a, b, the simple switching with two *sprinkler-spraying-appearing* bifurcations is presented for $a_0 > 0$ and $a_0 < 0$, respectively. The two bifurcation structures are

(a) SO $\to$ (r_1 - LSG : SO : r_2 - USG),
(b) SI $\to$ (r_1 - USG : SI : r_2 - LSG)

with $r_1 + r_2 + 1 = n$ and $m_1 + m_2 = m$ where $m_1 = \sum_{i=1}^{r_1} l_i^{(1)}, m_2 = \sum_{j=1}^{r_2} l_j^{(2)}$. In Fig. 7.2c, d, the $(2l + 1)^{\text{th}}$-order broom-switching with two *sprinkler-spraying-appearing* bifurcations is presented for $a_0 > 0$ and $a_0 < 0$, respectively. The two bifurcation structures are

(c) SO $\to$ (r_1 - LSG : $(2l + 1)^{\text{th}}$SO : r_2 - USG),
(d) SI $\to$ (r_1 - USG : $(2l + 1)^{\text{th}}$SI : r_2 - LSG)

with $r_1 + r_2 + 1 = n$ and $m_1 + m_2 + l = m$ where $m_1 = \sum_{i=1}^{r_1} l_i^{(1)}, m_2 = \sum_{j=1}^{r_2} l_j^{(2)}$.

For a cluster of m-quadratics, $B_i = B_j$ ($i,j \in \{1, 2, \ldots, m\}; i \neq j$) and $\Delta_i = 0$ ($i \in \{1, 2, \ldots, m\}$). The $(2m)^{\text{th}}$-order upper-saddle-node appearing bifurcation for m-pairs of sink and source equilibriums is

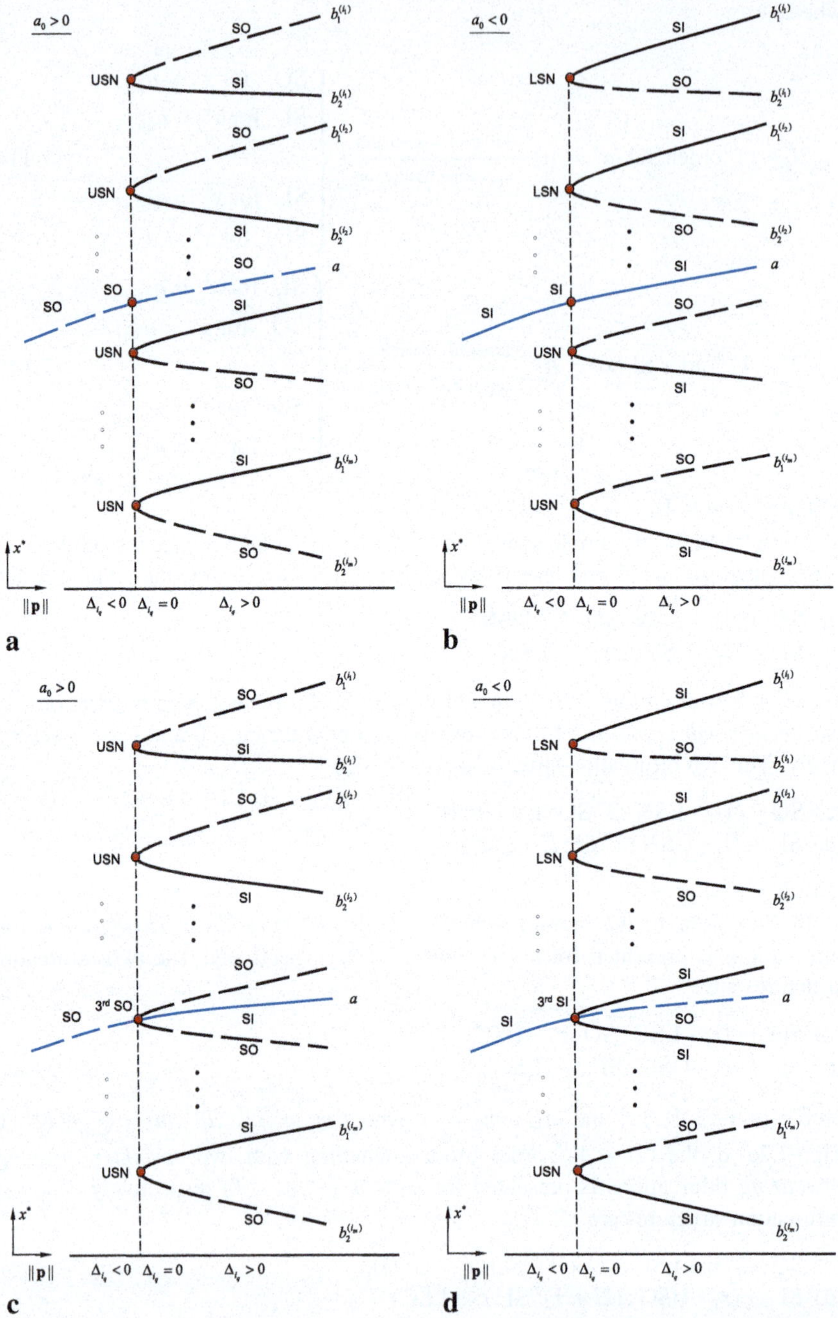

Fig. 7.1 Simple broom-switching bifurcations: (**a**) (US : . . . : US : SO : LS : . . . : LS) ($a_0 > 0$), (**b**) (LS : . . . : LS : SI : US : . . . : US) ($a_0 < 0$), (**c**) (US : . . . : US : 3rdSO : LS : . . . : LS) ($a_0 > 0$), (**d**) (LS : . . . : LS : 3rdSI : US : . . . : US) ($a_0 < 0$) in a $(2m + 1)^{\text{th}}$-degree polynomial system. LS: lower saddle, US: upper saddle, SI: sink, SO: source. Stable and unstable equilibriums are represented by solid and dashed curves, respectively. The bifurcation points are marked by circular symbols

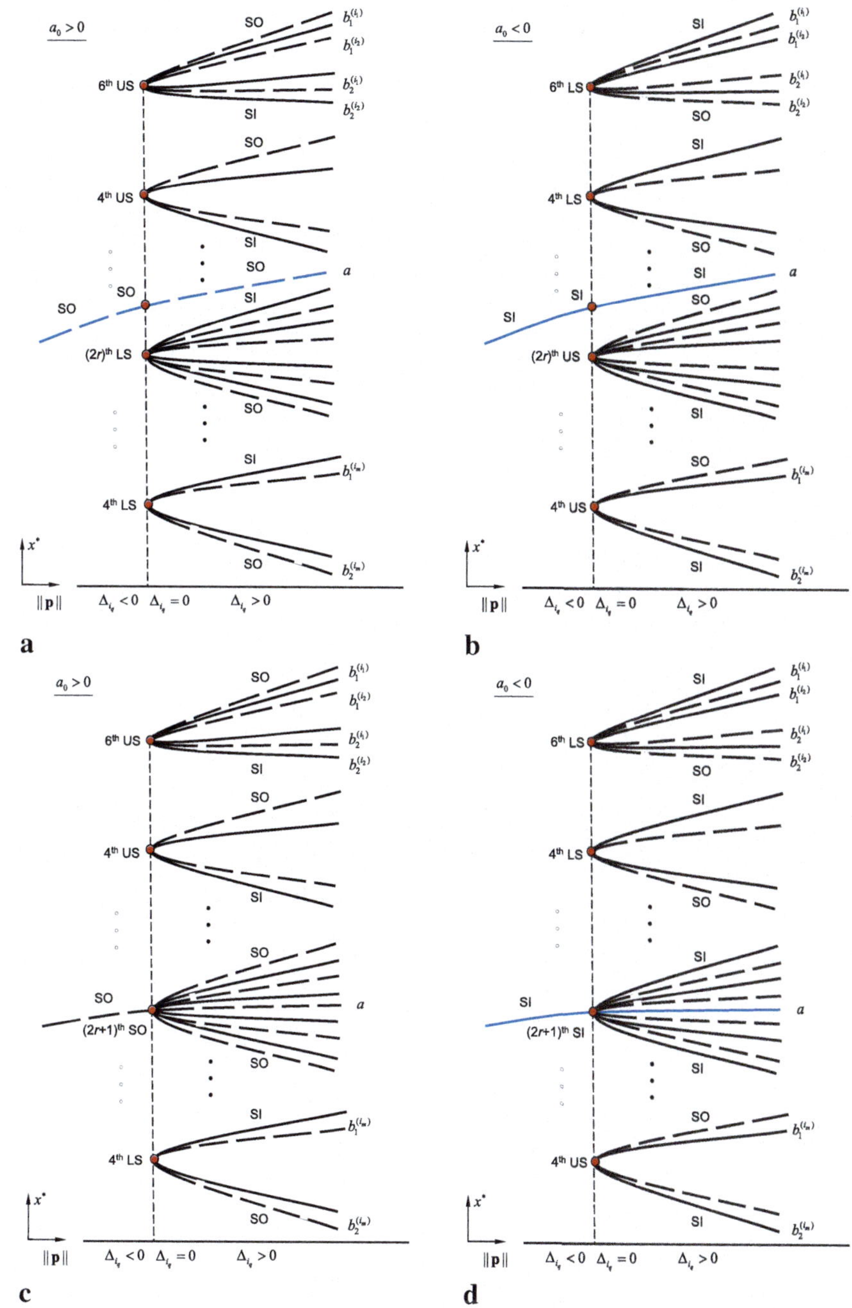

Fig. 7.2 Broom appearing bifurcation: (**a**) (r_1 - LSN : SO : r_2 - USN) ($a_0 > 0$). (**b**) (r_1 - USN : SI : r_2 - LSN) ($a_0 < 0$); broom-sprinkler-spraying-switching bifurcation. (**c**) (r_1 - LSG : $(2l_k + 1)^{\text{th}}$SO : r_2 - USG) ($a_0 > 0$). (**d**) (r_1 - LSG : $(2l_k + 1)^{\text{th}}$SO : r_2 - USG) ($a_0 < 0$) in a $(2m + 1)^{\text{th}}$-degree polynomial system. LS: lower saddle, US: upper saddle, SI: sink, SO: source. Stable and unstable equilibriums are represented by solid and dashed curves, respectively. The bifurcation points are marked by circular symbols

$$(2m)^{\text{th}} \text{ order US} \xrightarrow[\text{appearing bifurcation}]{\text{cluster of } m\text{-quadratics}} \begin{cases} \text{SO}, & \text{for } x^* = a_{2m}, \\ \text{SI}, & \text{for } x^* = a_{2m-1}; \\ \vdots & \\ \text{SO}, & \text{for } x^* = a_2, \\ \text{SI}, & \text{for } x^* = a_1. \end{cases} \tag{7.116}$$

The $(2m)^{\text{th}}$-order lower-saddle-node appearing bifurcation for m-pairs of sink and source equilibriums is

$$(2m)^{\text{th}} \text{ order LS} \xrightarrow[\text{appearing bifurcation}]{\text{cluster of } m\text{-quadratics}} \begin{cases} \text{SI}, & \text{for } x^* = a_{2m}, \\ \text{SO}, & \text{for } x^* = a_{2m-1}; \\ \vdots & \\ \text{SI}, & \text{for } x^* = a_2, \\ \text{SO}, & \text{for } x^* = a_1. \end{cases} \tag{7.117}$$

There are four simple switching and $(2m)^{\text{th}}$-order saddle-node appearing bifurcations: The two switching bifurcations of SO $\to$ ($(2m)^{\text{th}}$US : SO) and SI $\to$ ($(2m)^{\text{th}}$LS : SI) with two $(2m)^{\text{th}}$-order USN and LSN spraying-appearing bifurcations are

$$\text{SO } (x^* = a) \to \begin{cases} \text{SO} \to \text{SO, for } x^* = a = a_{2m+1} \\ (2m)^{\text{th}} \text{ order LSN} \to \begin{cases} \text{SI}, & \text{for } x^* = a_{2m}, \\ \text{SO}, & \text{for } x^* = a_{2m-1}; \\ \vdots & \\ \text{SI}, & \text{for } x^* = a_2, \\ \text{SO}, & \text{for } x^* = a_1 \end{cases} \end{cases} \tag{7.118}$$

$$\text{SI } (x^* = a) \to \begin{cases} \text{SI} \to \text{SI, for } x^* = a = a_{2m+1} \\ (2m)^{\text{th}} \text{ order USN} \to \begin{cases} \text{SO}, & \text{for } x^* = a_{2m}, \\ \text{SI}, & \text{for } x^* = a_{2m-1}; \\ \vdots & \\ \text{SO}, & \text{for } x^* = a_2, \\ \text{SI}, & \text{for } x^* = a_1 \end{cases} \end{cases} \tag{7.119}$$

and the two switching bifurcations of SO $\to$ (SO : $(2m)^{\text{th}}$US) and SI $\to$ (SI: $(2m)^{\text{th}}$LS) with two $(2m)^{\text{th}}$-order USN and LSN spraying-appearing bifurcations are

$$
\text{SO}\ (x^* = a) \to \begin{cases} (2m)^{\text{th}} \text{ order USN} \to \begin{cases} \text{SO}, & \text{for } x^* = a_{2m+1}, \\ \text{SI}, & \text{for } x^* = a_{2m}; \\ \vdots & \\ \text{SO}, & \text{for } x^* = a_3, \\ \text{SI}, & \text{for } x^* = a_2; \end{cases} \\ \text{SO} \to \text{SO}, \text{for } x^* = a = a_1 \end{cases} \tag{7.120}
$$

$$
\text{SI}\ (x^* = a) \to \begin{cases} (2m)^{\text{th}} \text{ order LSN} \to \begin{cases} \text{SI}, & \text{for } x^* = a_{2m+1}, \\ \text{SO}, & \text{for } x^* = a_{2m}; \\ \vdots & \\ \text{SI}, & \text{for } x^* = a_3, \\ \text{SO}, & \text{for } x^* = a_2; \end{cases} \\ \text{SI} \to \text{SI}, \text{for } x^* = a = a_1. \end{cases} \tag{7.121}
$$

The $(2m+1)^{\text{th}}$-order source broom-switching bifurcation is

$$
\text{SO}(x^* = a) \xrightarrow{\text{switching}} (2m+1)^{\text{th}} \text{ order SO} \begin{cases} \text{SO}, & \text{for } x^* = a_{2m+1}, \\ \text{SI}, & \text{for } x^* = a_{2m}, \\ \vdots & \\ \text{SI}, & \text{for } x^* = a_2, \\ \text{SO}, & \text{for } x^* = a_1. \end{cases} \tag{7.122}
$$

The $(2m+1)^{\text{th}}$-order sink *broom-switching* bifurcation is

$$
\text{SI}\ (x^* = a_1) \xrightarrow{\text{switching}} (2m+1)^{\text{th}} \text{ order SI} \begin{cases} \text{SI}, & \text{for } x^* = a_{2m+1}, \\ \text{SO}, & \text{for } x^* = a_{2m}, \\ \vdots & \\ \text{SO}, & \text{for } x^* = a_2, \\ \text{SI}, & \text{for } x^* = a_1. \end{cases} \tag{7.123}
$$

The switching bifurcation consists of a simple switching and the $(2m)^{\text{th}}$-order saddle-node appearing bifurcation with m-pairs of source and sinks. The $(2m)^{\text{th}}$-order saddle-node appearing bifurcation is a *sprinkler-spraying cluster* of the m-pairs of sources and sinks. Thus, the four switching bifurcations of

$\text{SO} \to (\text{SO} : (2m)^{\text{th}}\text{US})$ for $a_0 > 0$,
$\text{SI} \to (\text{SI} : (2m)^{\text{th}}\text{LS})$ for $a_0 < 0$,
$\text{SO} \to ((2m)^{\text{th}}\text{US} : \text{SO})$ for $a_0 > 0$,
$\text{SI} \to ((2m)^{\text{th}}\text{LS} : \text{SI})$ for $a_0 < 0$

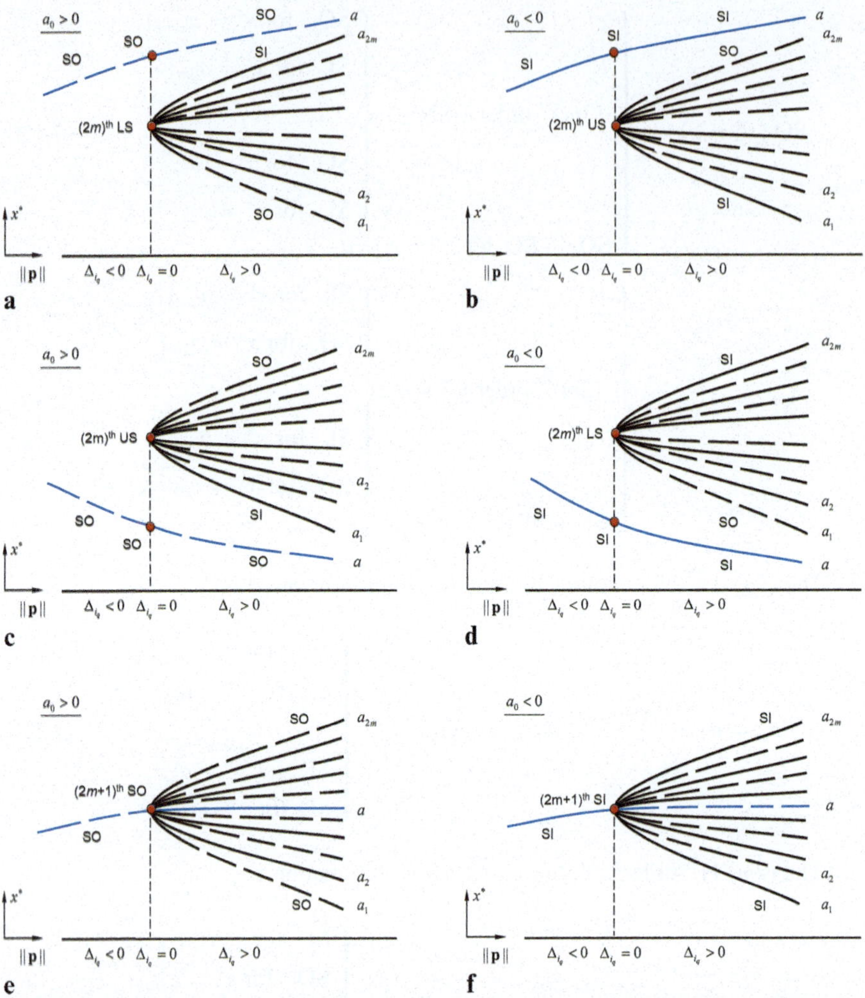

Fig. 7.3 (**a**) (SO : $(2m)^{\text{th}}$LS)-switching bifurcation ($a_0 > 0$), (**b**) (SI : $(2m)^{\text{th}}$SO)-switching bifurcation ($a_0 < 0$), (**c**) ($(2m)^{\text{th}}$US : SO)-switching bifurcation ($a_0 > 0$), (**d**) ($(2m)^{\text{th}}$LS : SI)-switching bifurcation ($a_0 < 0$), (**e**) $(2m + 1)^{\text{th}}$ SO broom appearing bifurcation ($a_0 > 0$), (**f**) $(2m + 1)^{\text{th}}$ SI broom appearing bifurcation ($a_0 < 0$) in the $(2m + 1)^{\text{th}}$-degree polynomial system. LS: lower saddle, US: upper saddle, SI: sink, SO: source. Stable and unstable equilibriums are represented by solid and dashed curves, respectively. The bifurcation points are marked by circular symbols

are presented in Fig. 7.3a–d, respectively. The $(2m + 1)^{\text{th}}$-order source switching bifurcation is named the $(2m{+}1)^{\text{th}}$SO broom-switching bifurcation and the $(2m + 1)^{\text{th}}$-order sink switching bifurcation is named the $(2m + 1)^{\text{th}}$SI broom-switching bifurcation. Such a $(2m + 1)^{\text{th}}$XX *broom-switching* bifurcation is from simple equilibrium to a $(2m + 1)^{\text{th}}$XX *broom*-switching bifurcation. The two broom-switching bifurcations of

SO $\to (2m + 1)^{\text{th}}$SO for $a_0 > 0$,
SI $\to (2m + 1)^{\text{th}}$SI for $a_0 < 0$

are presented in Fig. 7.3e, f, respectively.

A series of the third-order source and sink bifurcations is aligned up with varying parameters. Such a special pattern is from m-quadratics in the $(2m+1)^{\text{th}}$-order polynomial systems, and the following conditions should be satisfied:

$$\left.\begin{array}{l} a(\mathbf{p}_i) = -\frac{1}{2}B_i \text{ and } a(\mathbf{p}_j) = -\frac{1}{2}B_j \\ B_i \approx B_j \; i,j \in \{1,2,\ldots,n\}; i \neq j, \\ \Delta_i > \Delta_{i+1} \; (i = 1,2,\ldots,n; n \leq m), \\ \Delta_i = 0 \text{ with } \|\mathbf{p}_i\| < \|\mathbf{p}_{i+1}\|. \end{array}\right. \tag{7.124}$$

Thus, a series of m - (3^{rd}SO - 3^{rd}SI - 3^{rd}SO - ...) *switching* bifurcations ($a_0 > 0$) and a series of m - (3^{rd}SI - 3^{rd}SO - 3^{rd}SI - ...) *switching* bifurcations ($a_0 < 0$) are presented in Fig. 7.4a, b. The bifurcation scenario is formed by the swapping pattern of third SI and third SO switching bifurcations. Such a bifurcation scenario is like the fish bone. Thus, such a bifurcation swapping pattern of third SI and third SO switching bifurcations is called the *fish-bone switching* bifurcation in the $(2m+1)^{\text{th}}$-degree polynomial nonlinear system. There are two swaps of the third SI and third SO bifurcations: (a) the third SO-third SI *fish-bone switching* bifurcation and (b) the third SI-third SO *fish-bone switching* bifurcation.

7.2.2 Switching Bifurcations

In the $(2m + 1)^{\text{th}}$-order polynomial system, among the possible $(2m + 1)$ roots, there are two roots to satisfy $x^2 + B_i x + C_i = 0$ with

$$\begin{array}{l} B_i = -(b_1^{(i)} + b_2^{(i)}), \Delta_i = (b_1^{(i)} - b_2^{(i)})^2 \geq 0, \\ x_{1,2}^{(i)} = b_{1,2}^{(i)}, \Delta_i > 0 \text{ if } b_1^{(i)} \neq b_2^{(i)} \, (i = 1,2,\ldots,n); \\ \left.\begin{array}{l} B_i \neq B_j \; (i,j = 1,2,\ldots,n; i \neq j) \\ \Delta_i = 0 \text{ at } b_1^{(i)} = b_2^{(i)} \, (i = 1,2,\ldots,n) \end{array}\right\} \text{at bifurcation.} \end{array} \tag{7.125}$$

The second-order singularity bifurcation is for the *switching* of a pair of simple sink and source. There are two *switching* bifurcations for $i \in \{1,2,\ldots,n\}$:

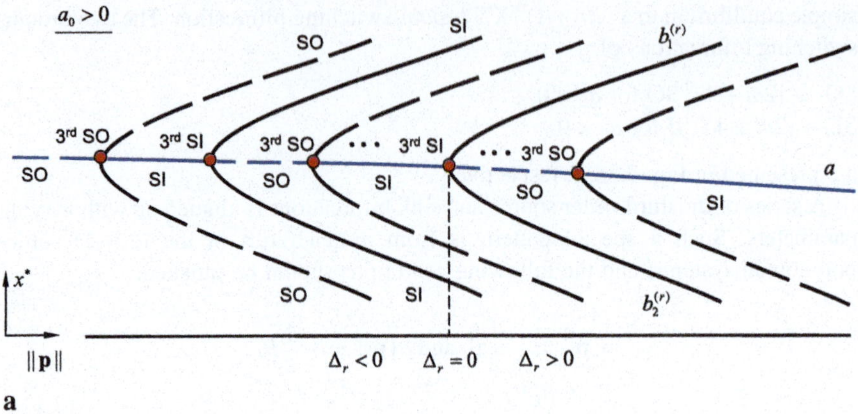

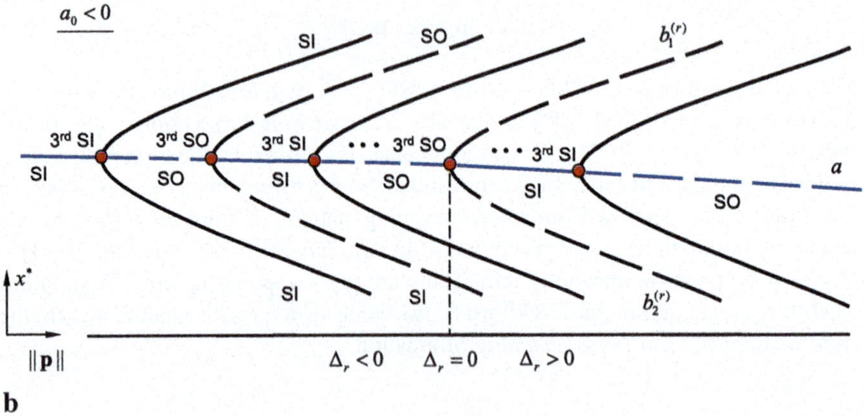

Fig. 7.4 (**a**) m - (3^{rd}SO - 3^{rd}SI - 3^{rd}SO...) series bifurcation ($a_0 > 0$), (**b**) m - (3^{rd}SI - 3^{rd}SO - 3^{rd}SI...) series switching bifurcation ($a_0 < 0$) in the $(2m+1)^{\text{th}}$-degree polynomial system. SI: sink, SO: source. Stable and unstable equilibriums are represented by solid and dashed curves, respectively. The bifurcation points are marked by circular symbols

$$2^{\text{nd}}\text{order US} \xrightarrow[\text{switching bifurcation}]{i^{\text{th}}\text{quadratic factor}} \begin{cases} \text{SO}, & \text{for } a_{2i} = b_2^{(i)} \to b_1^{(i)}, \\ \text{SI}, & \text{for } a_{2i-1} = b_1^{(i)} \to b_2^{(i)} \end{cases} \tag{7.126}$$

$$2^{\text{nd}}\text{order LS} \xrightarrow[\text{switching bifurcation}]{i^{\text{th}}\text{quadratic factor}} \begin{cases} \text{SI}, & \text{for } a_{2i} = b_2^{(i)} \to b_1^{(i)}, \\ \text{SO}, & \text{for } a_{2i-1} = b_1^{(i)} \to b_2^{(i)}. \end{cases} \tag{7.127}$$

For non-switching point, $\Delta_i > 0$ at $b_1^{(i)} \neq b_2^{(i)}$ $(i = 1, 2, \ldots, n)$. At the bifurcation point, $\Delta_i = 0$ at $b_1^{(i)} = b_2^{(i)}$ $(i = 1, 2, \ldots, n)$. The l-USN *parallel switching* bifurcation for $s_i \in \{0, 1, \ldots, m\}$ $(i = 1, 2, \ldots, l)$ is

$$
l\text{-USN}\begin{cases}
\text{US}\xrightarrow[\textit{switching}]{s_l^{\text{th}}\textit{bifurcation}}\begin{cases}\text{SO}\downarrow\text{SI}, & \text{for } b_2^{(s_l)}=a_{2s_l}\downarrow a_{2s_l-1},\\ \text{SI}\uparrow\text{SO}, & \text{for } b_1^{(s_l)}=a_{2s_l-1}\uparrow a_{2s_l};\end{cases}\\
\vdots\\
\text{US}\xrightarrow[\textit{switching}]{s_1^{\text{th}}\textit{bifurcation}}\begin{cases}\text{SO}\downarrow\text{SI}, & \text{for } b_2^{(s_1)}=a_{2s_1}\downarrow a_{2s_1-1},\\ \text{SI}\uparrow\text{SO}, & \text{for } b_1^{(s_1)}=a_{2s_1-1}\uparrow a_{2s_1}.\end{cases}
\end{cases}
\tag{7.128}
$$

The l-LSN *antenna-switching* bifurcation for $s_i \in \{0,1,\ \ldots, m\}$ $(i=1,2,\ldots,l)$ is

$$
l\text{-LSN}\begin{cases}
\text{LS}\xrightarrow[\textit{switching}]{s_l^{\text{th}}\textit{bifurcation}}\begin{cases}\text{SI}\downarrow\text{SO}, & \text{for } b_2^{(s_l)}=a_{2s_l}\downarrow a_{2s_l-1},\\ \text{SO}\uparrow\text{SI}, & \text{for } b_1^{(s_l)}=a_{2s_l-1}\uparrow a_{2s_l};\end{cases}\\
\vdots\\
\text{LS}\xrightarrow[\textit{switching}]{s_1^{\text{th}}\textit{bifurcation}}\begin{cases}\text{SI}\downarrow\text{SO}, & \text{for } b_2^{(s_1)}=a_{2s_1}\downarrow a_{2s_1-1},\\ \text{SO}\uparrow\text{SI}, & \text{for } b_1^{(s_1)}=a_{2s_1-1}\uparrow a_{2s_1}.\end{cases}
\end{cases}
\tag{7.129}
$$

Two *antenna*-switching bifurcation structures exist for the $(2m+1)^{\text{th}}$-order polynomial system. The $(l_1$ - LSN : SO : l_2 - USN)-switching bifurcation for $a_0 > 0$ is

$$
(l_1\text{-LSN} : \text{SO} : l_2\text{-USN}) \xrightarrow{l_1+l_2=m} \begin{cases} l_2\text{-USN} \\ \text{SO}\rightarrow\text{SO}, \\ l_1\text{-LSN} \end{cases}
\tag{7.130}
$$

and the $(l_1$ - USN : SI : l_2 - LSN)-switching bifurcation for $a_0 < 0$ is

$$
(l_1\text{-USN} : \text{SI} : l_2\text{-LSN}) \xrightarrow{l_1+l_2=m} \begin{cases} l_2\text{-LSN}, \\ \text{SI}\rightarrow\text{SI}, \\ l_1\text{-USN}. \end{cases}
\tag{7.131}
$$

As in the $(2m+1)^{\text{th}}$-order polynomial system, consider a switching bifurcation for a bundle of sink and source equilibriums with the following conditions:

$$
\left.\begin{array}{l} B_i = B_j\ (i,j\in\{1,2,\ \ldots,n\}; i\neq j) \\ \Delta_i = 0 \text{ at } b_1^{(i)} = \mathrm{b}_2^{(i)}\ (i=1,2,\ \ldots,n) \end{array}\right\} \text{at bifurcation.}
\tag{7.132}
$$

Two $(2l)^{\text{th}}$-order switching bifurcations for $l \in \{1,2,\ldots,s\}$ are

$$(2l)^{\text{th}}\text{order US} \xrightarrow[\text{switching bifurcation}]{\text{a bundle of }(2l)\text{-equilibriums}} \begin{cases} \text{SO}, & \text{for } a_{2s_l} \to b_{2s_l}, \\ \text{SI}, & \text{for } a_{2s_l-1} \to b_{2s_l-1}, \\ \vdots & \\ \text{SO}, & \text{for } a_{2s_1} \to b_{2s_1}, \\ \text{SI}, & \text{for } a_{2s_1-1} \to b_{2s_1-1} \end{cases} \tag{7.133}$$

$$(2l)^{\text{th}}\text{order LS} \xrightarrow[\text{switching bifurcation}]{\text{a bundle of }(2l)\text{-equilibriums}} \begin{cases} \text{SI}, & \text{for } a_{2s_l} \to b_{2s_l}, \\ \text{SO}, & \text{for } a_{2s_l-1} \to b_{2s_l-1}, \\ \vdots & \\ \text{SI}, & \text{for } a_{2s_1} \to b_{2s_1}, \\ \text{SO}, & \text{for } a_{2s_1-1} \to b_{2s_1-1} \end{cases} \tag{7.134}$$

where $\Delta_{ij} = (a_i - a_j)^2 = (b_i - b_j)^2 = 0$ with $B_i = B_j\,(i,j = 2s_1 - 1, 2s_1, \ldots, 2s_l - 1, 2s_l)$ and

$$\begin{aligned} &\{a_{2s_1-1}, a_{2s_1}, \ldots, a_{2s_l-1}, a_{2s_l}\} \underset{\text{before bifurcation}}{\subset} \text{sort}\{b_1^{(1)}, b_2^{(1)}, \ldots, b_1^{(n)}, b_2^{(n)}, a\}, \\ &\{b_{2s_1-1}, b_{2s_1}, \ldots, b_{2s_l-1}, b_{2s_l}\} \underset{\text{after bifurcation}}{\subset} \text{sort}\{b_1^{(1)}, b_2^{(1)}, \ldots, b_1^{(n)}, b_2^{(n)}, a\}. \end{aligned} \tag{7.135}$$

Two $(2l + 1)^{\text{th}}$-order switching bifurcations for $l \in \{1, 2, \ldots, s\}$ are

$$(2l+1)^{\text{th}}\text{order SO} \xrightarrow[\text{switching bifurcation}]{\text{a bundle of }(2l+1)\text{-equilibriums}} \begin{cases} \text{SO}, & \text{for } a_{2s_l+1} \to b_{2s_l+1}, \\ \vdots & \\ \text{SI}, & \text{for } a_{2s_1} \to b_{2s_1}, \\ \text{SO}, & \text{for } a_{2s_1-1} \to b_{2s_1-1} \end{cases} \tag{7.136}$$

$$(2l+1)^{\text{th}}\text{order SI} \xrightarrow[\text{switching bifurcation}]{\text{a bundle of }(2l+1)\text{-equilibriums}} \begin{cases} \text{SI}, & \text{for } a_{2s_l+1} \to b_{2s_l+1}, \\ \vdots & \\ \text{SO}, & \text{for } a_{2s_1} \to b_{2s_1}, \\ \text{SI}, & \text{for } a_{2s_1-1} \to b_{2s_1-1} \end{cases} \tag{7.137}$$

where $\Delta_{ij} = (a_i - a_j)^2 = (b_i - b_j)^2 = 0$ with $B_i = B_j\,(i,j = 2s_1 - 1, 2s_1, \ldots, 2s_l + 1)$ and

$$\begin{aligned} &\{a_{2s_1-1}, a_{2s_1}, \ldots, a_{2s_l+1}\} \underset{\text{before bifurcation}}{\subset} \text{sort}\{b_1^{(1)}, b_2^{(1)}, \ldots, b_1^{(n)}, b_2^{(n)}, a\}, \\ &\{b_{2s_1-1}, b_{2s_1}, \ldots, b_{2s_l+1}\} \underset{\text{After bifurcation}}{\subset} \text{sort}\{b_1^{(1)}, b_2^{(1)}, \ldots, b_1^{(n)}, b_2^{(n)}, a\}. \end{aligned} \tag{7.138}$$

A set of paralleled, different, higher order upper-saddle-node switching bifurcations with multiplicity is the $((\alpha_1)^{\text{th}}\text{XX} : (\alpha_2)^{\text{th}}\text{XX} : \ldots : (\alpha_s)^{\text{th}}\text{XX})$ *parallel switching* bifurcation in the $(2m+1)^{\text{th}}$-degree polynomial system. At the *straw-bundle switching* bifurcation, $\Delta_i = 0$ $(i = 1, 2, \ldots, n)$ and $B_i = B_j$ $(i, j \in \{1, 2, \ldots, n\}; i \neq j)$. Thus, the *parallel straw-bundle switching* bifurcation is

$$\begin{aligned} &((\alpha_1)^{\text{th}}\text{XX} : (\alpha_2)^{\text{th}}\text{XX} : \ldots : (\alpha_s)^{\text{th}}\text{XX})\text{-switching} \\ &= \begin{cases} (\alpha_s)^{\text{th}} \text{ order XX switching}, \\ \vdots \\ (\alpha_2)^{\text{th}} \text{ order XX switching}, \\ (\alpha_1)^{\text{th}} \text{ order XX switching} \end{cases} \end{aligned} \tag{7.139}$$

where

$$\begin{aligned} &\alpha_i \in \{2l_i, 2l_i - 1\} \text{ with } \textstyle\sum_{i=1}^{s} \alpha_i = 2m + 1; \\ &\text{and XX} \in \{\text{US}, \text{LS}, \text{SO}, \text{SI}\}. \end{aligned} \tag{7.140}$$

The $(2l_1^{(j)} : 2l_2^{(j)} : \ldots : 2l_s^{(j)})^{\text{th}}$ USN parallel switching bifurcation is called the $(2l_1^{(j)} : 2l_2^{(j)} : \ldots : 2l_s^{(j)})^{\text{th}}$ USN parallel *straw-bundle switching* bifurcation:

$$\begin{aligned} s_j\text{-USG} &= (2l_1^{(j)} : 2l_2^{(j)} : \ldots : 2l_{s_j}^{(j)})^{\text{th}}\text{USN switching} \\ &= \begin{cases} (2l_s^{(j)})^{\text{th}} \text{ order USN switching}, \\ \vdots \\ (2l_2^{(j)})^{\text{th}} \text{ order USN switching}, \\ (2l_1^{(j)})^{\text{th}} \text{ order USN switching}. \end{cases} \end{aligned} \tag{7.141}$$

The $(2l_1^{(j)} : 2l_2^{(j)} : \ldots : 2l_s^{(j)})^{\text{th}}$ LSN parallel switching bifurcation is called the $(2l_1^{(j)} : 2l_2^{(j)} : \ldots : 2l_s^{(j)})^{\text{th}}$ LSN parallel *straw-bundle switching* bifurcation:

$$\begin{aligned} s_j\text{-LSG} &= (2l_1^{(j)} : 2l_2^{(j)} : \ldots : 2l_{s_j}^{(j)})^{\text{th}}\text{LSN switching} \\ &= \begin{cases} (2l_s^{(j)})^{\text{th}} \text{ order LSN switching}, \\ \vdots \\ (2l_2^{(j)})^{\text{th}} \text{ order LSN switching}, \\ (2l_1^{(j)})^{\text{th}} \text{ order LSN switching}. \end{cases} \end{aligned} \tag{7.142}$$

The $(s_1\text{-LSG} : (2l_1^{(2)}+1)^{\text{th}}\text{SO} : s_3\text{-USG})$-switching bifurcation for $a_0 > 0$ is

$$(s_1\text{-LSG} : (2l_1^{(2)}+1)^{\text{th}}\text{SO} : s_3\text{-USG}) = \begin{cases} (2l_1^{(3)} : \ldots : 2l_{s_2}^{(3)})^{\text{th}}\text{-USN}, \\ (2l_1^{(2)}+1)^{\text{th}}\text{SO}, \\ (2l_1^{(1)} : \ldots : 2l_{s_1}^{(1)})^{\text{th}}\text{-LSN} \end{cases} \quad (7.143)$$

and the $(s_1\text{-USG} : (2l_1^{(2)}+1)^{\text{th}}\text{SI} : s_3\text{-LSG})$-switching bifurcation for $a_0 < 0$ is

$$(s_1\text{-USG} : (2l_1^{(2)}+1)^{\text{th}}\text{SI} : s_3\text{-LSG}) = \begin{cases} (2l_1^{(3)} : \ldots : 2l_{s_3}^{(3)})^{\text{th}}\text{-LSN}, \\ (2l_1^{(2)}+1)^{\text{th}}\text{SI}, \\ (2l_1^{(1)} : \ldots : 2l_{s_1}^{(1)})^{\text{th}}\text{-USN}. \end{cases} \quad (7.144)$$

The two (l_1 - USN : SO : l_2 - LSN) and (l_1 - LSN : SI : l_2 - USN) parallel switching bifurcations ($l_1 + l_2 = m$) are presented in Fig. 7.5a, b for $a_0 > 0$ and $a < 0$, respectively. A set of (3^{rd}SO : … : SI : 3^{rd}SO) parallel, *switching* bifurcations for SI and SO equilibriums is presented in Fig. 7.5c for $a_0 > 0$. However, for $a_0 < 0$, the set of (3^{rd}SI : ⋯ : SO : 3^{rd}SI) *switching* bifurcations for sources and sinks is presented in Fig. 7.5d.

7.2.3 Switching and Appearing Bifurcations

Consider a $(2m+1)^{\text{th}}$-degree polynomial system in a form of

$$\dot{x} = a_0 Q(x) \prod_{i=1}^{2n_1+1} (x - c_i) \prod_{j=1}^{n_2} (x^2 + B_j x + C_j). \quad (7.145)$$

Without loss of generality, a function of $Q(x) > 0$ is either a polynomial function or a non-polynomial function. The roots of $x^2 + B_j x + C_j = 0$ are

$$b_{1,2}^{(j)} = -\frac{1}{2}B_j \pm \frac{1}{2}\sqrt{\Delta_j}, \Delta_j = B_j^2 - 4C_j \geq 0 \; (j = 1, 2, \ldots, n_2); \quad (7.146)$$

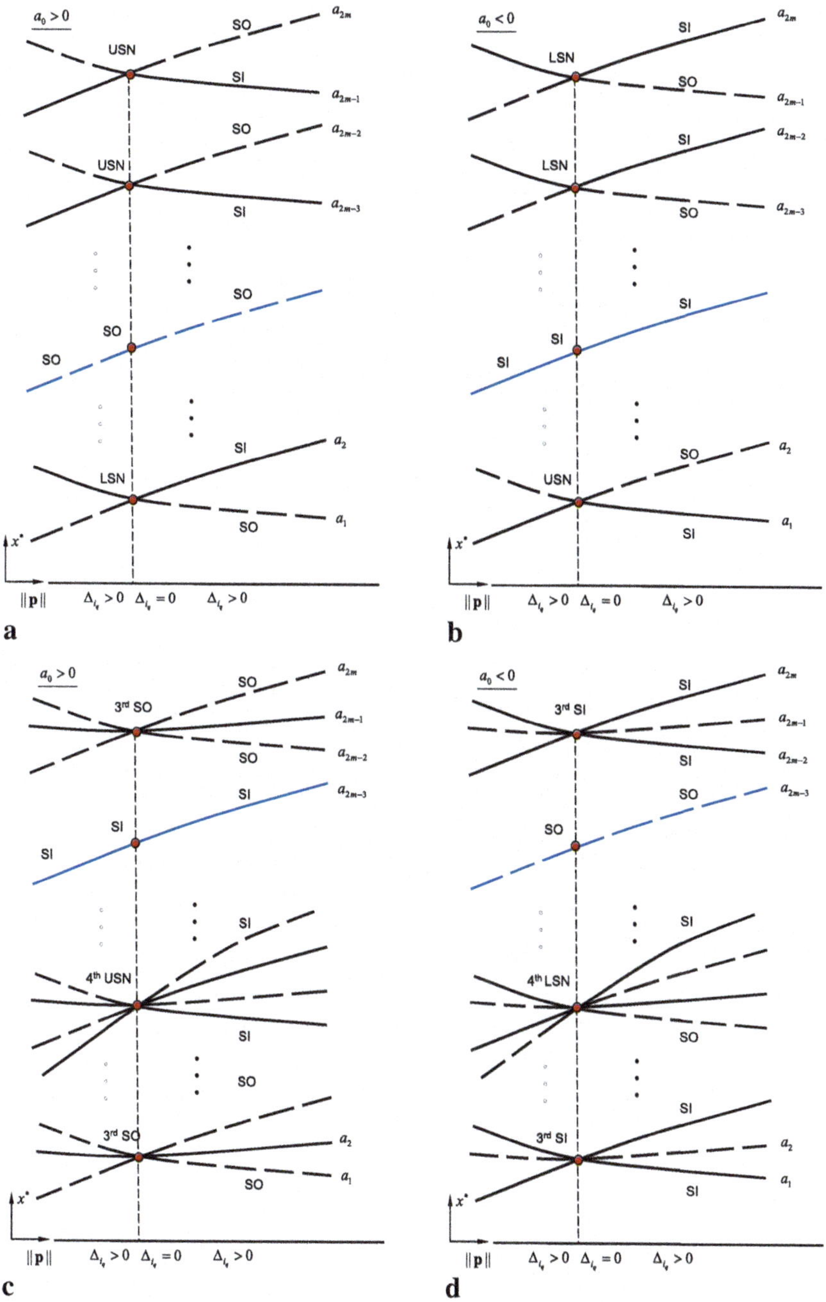

Fig. 7.5 Parallel switching bifurcations: (**a**) (l_1 - USG : SO : l_2 - USG) ($a_0 > 0$), (**b**) (l_1 - USG : SI : l_2 - LSG) ($a_0 < 0$), (**c**) (3^{rd}SI : . . . : USN : 3^{rd}SO) ($a_0 > 0$), (**d**) (3^{rd}SO : . . . : LSN : 3^{rd}SI) ($a_0 < 0$) in the $(2m + 1)^{th}$-degree polynomial nonlinear system. LSN: lower saddle node, USN: upper saddle node, SI: sink, SO: source. Stable and unstable equilibriums are represented by solid and dashed curves, respectively. The bifurcation points are marked by circular symbols

either

$$
\begin{aligned}
&\{a_1^-, a_2^-, \ldots, a_{2n_1+1}^-\} = \mathrm{sort}\{c_1, c_2, \ldots, c_{2n_1+1}\}, \\
&a_s^- \le a_{s+1}^- \text{ before bifurcation;} \\
&\{a_1^+, a_2^+, \ldots, a_{2n_3+1}^+\} = \mathrm{sort}\{c_1, \ldots, c_{2n_1+1}; b_1^{(1)}, b_2^{(1)}, \ldots, b_1^{(n_2)}, b_2^{(n_2)}\}, \\
&a_s^+ \le a_{s+1}^+, n_3 = n_1 + n_2 \text{ after bifurcation}
\end{aligned}
\tag{7.147}
$$

or

$$
\begin{aligned}
&\{a_1^-, a_2^-, \ldots, a_{2n_3+1}^-\} = \mathrm{sort}\{c_1, c_2, \ldots, c_{2n_1}; b_1^{(1)}, b_2^{(1)}, \ldots, b_1^{(n_2)}, b_2^{(n_2)}, a\}, \\
&a_s^- \le a_{s+1}^-, n_3 = n_1 + n_2 \text{ before bifurcation;} \\
&\{a_1^+, a_2^+, \ldots, a_{2n_1+1}^+\} = \mathrm{sort}\{c_1, \ldots, c_{2n_1}, a\}, \\
&a_s^+ \le a_{s+1}^+ \text{ after bifurcation}
\end{aligned}
\tag{7.148}
$$

and

$$
\left.
\begin{aligned}
&B_{j_1} = B_{j_2} = \ldots = B_{j_s}\ (j_{k_1} \in \{1, 2, \ldots, n\}; j_{k_1} \ne j_{k_2}) \\
&(k_1, k_2 \in \{1, 2, \ldots, s\}; k_1 \ne k_2) \\
&\Delta_j = 0\ (j \in U \subset \{1, 2, \ldots, n_2\} \\
&c_i \ne -\frac{1}{2}B_j\ (i = 1, 2, \ldots, 2n_1, j = 1, 2, \ldots, n_2)
\end{aligned}
\right\} \text{at bifurcation.} \tag{7.149}
$$

Consider a just before bifurcation of $((\alpha_1^-)^{\mathrm{th}}\mathrm{XX}_1^- : (\alpha_2^-)^{\mathrm{th}}\mathrm{XX}_2^- : \ldots :(\alpha_{s_1}^-)^{\mathrm{th}}\mathrm{XX}_{s_1}^-)$ with $\sum_{i=1}^{s_1}\alpha_i = 2m_1 + 1$ for simple sources and sinks in the $(2m+1)^{\mathrm{th}}$-degree polynomial system. For $\alpha_i^- = 2l_i^- - 1$, $\mathrm{XX}_i^- \in \{\mathrm{SO}, \mathrm{SI}\}$ and for $\alpha_i^- = 2l_i^-$, $\mathrm{XX}_i^- \in \{\mathrm{US}, \mathrm{LS}\}$ $(i = 1, 2, \ldots, s_1)$. The detailed structures are as follows:

$$
\begin{array}{l}
\left.\begin{array}{c}\text{SI}\\ \text{SO}\\ \vdots\\ \text{SO}\\ \text{SI}\end{array}\right\} \to (2l_i^- - 1)^{\text{th}}\text{SI}, \quad \text{and} \quad \left.\begin{array}{c}\text{SO}\\ \text{SI}\\ \vdots\\ \text{SI}\\ \text{SO}\end{array}\right\} \to (2l_i^- - 1)^{\text{th}}\text{SO};\\
\left.\begin{array}{c}\text{SO}\\ \text{SI}\\ \vdots\\ \text{SO}\\ \text{SI}\end{array}\right\} \to (2l_i^-)^{\text{th}}\text{US}, \quad \text{and} \quad \left.\begin{array}{c}\text{SI}\\ \text{SO}\\ \vdots\\ \text{SI}\\ \text{SO}\end{array}\right\} \to (2l_i^-)^{\text{th}}\text{LS}.
\end{array}
\tag{7.150}
$$

The bifurcation set of $((\alpha_1^-)^{\text{th}}\text{XX}_1^- : (\alpha_2^-)^{\text{th}}\text{XX}_2^- : \ldots : (\alpha_{s_1}^-)^{\text{th}}\text{XX}_{s_1}^-)$ at the same parameter point is called a *left-parallel-bundle* switching bifurcation.

Consider a just after bifurcation of $((\alpha_1^+)^{\text{th}}\text{XX}_1^+ : (\alpha_2^+)^{\text{th}}\text{XX}_2^+ : \ldots : (\alpha_{s_2}^+)^{\text{th}}\text{XX}_{s_2}^+)$ with $\sum_{i=1}^{s_2}\alpha_i^+ = 2m_2 + 1$ for simple sources and sinks in the $(2m+1)^{\text{th}}$-degree polynomial system. $\text{XX}_i^+ \in \{\text{SO}, \text{SI}\}$ for $\alpha_i^+ = 2l_i^+ - 1$, and $\text{XX}_i^- \in \{\text{US}, \text{LS}\}$ for $\alpha_i^+ = 2l_i^+$. The four detailed structures are as follows:

$$
\begin{array}{l}
(2l_i^+ - 1)^{\text{th}}\text{SI} \to \left\{\begin{array}{c}\text{SI}\\ \text{SO}\\ \vdots\\ \text{SO}\\ \text{SI}\end{array}\right., \quad \text{and} \quad (2l_i^+ - 1)^{\text{th}}\text{SO} \to \left\{\begin{array}{c}\text{SO}\\ \text{SI}\\ \vdots\\ \text{SI}\\ \text{SO}\end{array}\right.;\\
(2l_i^+)^{\text{th}}\text{US} \to \left\{\begin{array}{c}\text{SO}\\ \text{SI}\\ \vdots\\ \text{SO}\\ \text{SI}\end{array}\right., \quad \text{and} \quad (2l_i^+)^{\text{th}}\text{LS} \to \left\{\begin{array}{c}\text{SI}\\ \text{SO}\\ \vdots\\ \text{SI}\\ \text{SO}\end{array}\right..
\end{array}
\tag{7.151}
$$

The bifurcation set of $((\alpha_1^+)^{\text{th}}\text{XX}_1^+ : (\alpha_2^+)^{\text{th}}\text{XX}_2^+ : \ldots : (\alpha_{s_2}^+)^{\text{th}}\text{XX}_{s_2}^+)$ at the same parameter point is called a *right-parallel:bundle* switching bifurcation.

(i) For the just before and after bifurcation structure, if there exists a relation of

$$\begin{aligned}&(\alpha_i^-)^{\text{th}}\text{XX}_i^- = (\alpha_j^+)^{\text{th}}\text{XX}_j^+ = \alpha^{\text{th}}\text{XX}, \text{ for } x^* = a_i^- = a_j^+\\&(i \in \{1,2,\ldots,s_1\}, j \in \{1,2,\ldots,s_2\}), \text{XX} \in \{\text{US}, \text{LS}, \text{SO}, \text{SI}\}\end{aligned} \tag{7.152}$$

then the bifurcation is a α^{th}XX *switching* bifurcation for simple equilibriums.

(ii) Just for the just before bifurcation structure, if there exists a relation of

$$\begin{aligned}&(2l_i^-)^{\text{th}}\text{XX}_i^- = (2l)^{\text{th}}\text{XX}, \text{ for } x^* = a_i^- = a_i\\&(i \in \{1,2,\ldots,s_1\}, \text{XX} \in \{\text{US}, \text{LS}\}\end{aligned} \tag{7.153}$$

then the bifurcation is a$(2l)^{\text{th}}$XX *left-appearing* (or *right-vanishing*) bifurcation for simple equilibriums.

(iii) Just for the just after bifurcation structure, if there exists a relation of

$$\begin{aligned}&(2l_i^+)^{\text{th}}\text{XX}_i^+ = (2l)^{\text{th}}\text{XX}, \text{ for } x^* = a_i^+ = a_i\\&(i \in \{1,2,\ldots,s_1\}), \text{XX} \in \{\text{US}, \text{LS}\}\end{aligned} \tag{7.154}$$

then the bifurcation is a $(2l)^{\text{th}}$XX *right-appearing* (or *left-vanishing*) bifurcation for simple equilibriums.

(iv) For the just before and after bifurcation structure, if there exists a relation of

$$\begin{aligned}&(\alpha_i^-)^{\text{th}}\text{XX}_i^- \neq (\alpha_j^+)^{\text{th}}\text{XX}_j^+ \text{ for } x^* = a_i^- = a_j^+\\&\text{XX}_i^-, \text{XX}_j^+ \in \{\text{US}, \text{LS}, \text{SO}, \text{SI}\}\\&(i \in \{1,2,\ldots,s_1\}, j \in \{1,2,\ldots,s_2\}),\end{aligned} \tag{7.155}$$

then there are two *flower-bundle* switching bifurcations of simple equilibriums:

(iv_1) For $\alpha_j = \alpha_i + 2l$, the bifurcation is called a $(\alpha_i^+)^{\text{th}}\text{XX}_i^+$ right *flower-bundle* switching bifurcation for α_i to α_j-simple equilibriums with the appearance (birth) of $2l$-simple equilibriums.

(iv_2) For $\alpha_j = \alpha_i - 2l$, the bifurcation is called a $(\alpha_i^-)^{\text{th}}\text{XX}_i^-$ left *flower-bundle* switching bifurcation for α_i to α_j-simple equilibriums with the vanishing (death) of $2l$-simple equilibriums.

A general parallel switching bifurcation is

$$\begin{aligned}&((\alpha_1^-)^{\text{th}}\text{XX}_1^- : (\alpha_2^-)^{\text{th}}\text{XX}_2^- : \ldots : (\alpha_{s_1}^-)^{\text{th}}\text{XX}_{s_1}^-) \xrightarrow[\text{bifucation}]{\text{switching}}\\&((\alpha_1^+)^{\text{th}}\text{XX}_1^+ : (\alpha_2^+)^{\text{th}}\text{XX}_2^+ : \ldots : (\alpha_{s_2}^+)^{\text{th}}\text{XX}_{s_2}^+).\end{aligned} \tag{7.156}$$

Such a general, parallel switching bifurcation consists of the *left* and *right* parallel-bundle switching bifurcations.

If the *left* and *right* parallel-bundle switching bifurcations are same in a parallel *flower-bundle* switching bifurcation, i.e.,

$$\begin{aligned}&(\alpha_i^-)^{\text{th}}\text{XX}_i^- = (\alpha_i^+)^{\text{th}}\text{XX}_i^+ = (\alpha_i)^{\text{th}}\text{XX}_i\\&\text{for } x^* = a_i^- = a_i^+ \ (i = 1, 2, \ldots, s\}\end{aligned} \tag{7.157}$$

then the parallel *flower-bundle* switching bifurcation becomes a parallel *straw-bundle switching* bifurcation of $((\alpha_1)^{\text{th}}\text{XX} : (\alpha_2)^{\text{th}}\text{XX} : \ldots : (\alpha_s)^{\text{th}}\text{XX})$.

If the *left* and *right* parallel-bundle switching bifurcations are different in a parallel *flower-bundle* switching bifurcation, i.e.,

$$\begin{aligned}&(\alpha_i^-)^{\text{th}}\text{XX}_i^- = (2l_i^-)^{\text{th}}\text{XX}, (\alpha_j^+)^{\text{th}}\text{XX}_j^+ = (2l_j^+)^{\text{th}}\text{YY},\\&\text{for } x^* = a_i^- \neq a_i^+ \ (i = 1, 2, \ldots, s\}\\&\text{XX} \in \{\text{US}, \text{LS}\}, \text{YY} \in \{\text{US}, \text{LS}\}\end{aligned} \tag{7.158}$$

then the parallel *flower-bundle* switching bifurcation becomes a combination of two independent left and right parallel appearing bifurcations:

(i) A $((2l_1^-)^{\text{th}}\text{XX}_1^- : (2l_2^-)^{\text{th}}\text{XX}_2^- : \ldots : (2l_{s_1}^-)^{\text{th}}\text{XX}_{s_1}^-)$-*left parallel sprinkler-spraying-appearing (or right vanishing)* bifurcation

(ii) A $((2l_1^+)^{\text{th}}\text{XX}_1^+ : (2l_2^+)^{\text{th}}\text{XX}_2^+ : \ldots : (2l_{s_1}^+)^{\text{th}}\text{XX}_{s_2}^+)$-*right parallel sprinkler-spraying-appearing (or left vanishing)* bifurcation

The (4^{th}LS : ... : SO : 6^{th}US) *parallel appearing* bifurcation for $a_0 > 0$ is presented in Fig. 7.6a. The (4^{th}US : ... : SI : 6^{th}LS) *parallel appearing* bifurcation for $a_0 < 0$ is shown in Fig. 7.6b. Such a kind of bifurcation is also like a *waterfall appearing* bifurcation. The (5^{th}SO : ... : 6^{th}US : 6^{th}US) parallel, *flower-bundle switching* bifurcation for SI and SO equilibriums is presented in Fig. 7.6c for $a_0 > 0$. Such a parallel *flower-bundle* switching bifurcation is from (SO : SI : SO) to (5^{th}SO : ... : 6^{th}US : 6^{th}US) with a *waterfall appearance.* The set of (5^{th}SI : ... : 6^{th}LS : 6^{th}LS) *flower-bundle switching* bifurcation for SI and SO equilibriums is presented in Fig. 7.6d for $a_0 < 0$. Such a parallel *flower-bundle* switching bifurcation is from (SI : SO : SI) to (5^{th}SI : ... : 6^{th}LS : 6^{th}LS) with a *waterfall appearance.* After the bifurcation, the *waterfall* equilibrium birth can be observed. The equilibriums before such a bifurcation are much less than after the bifurcation.

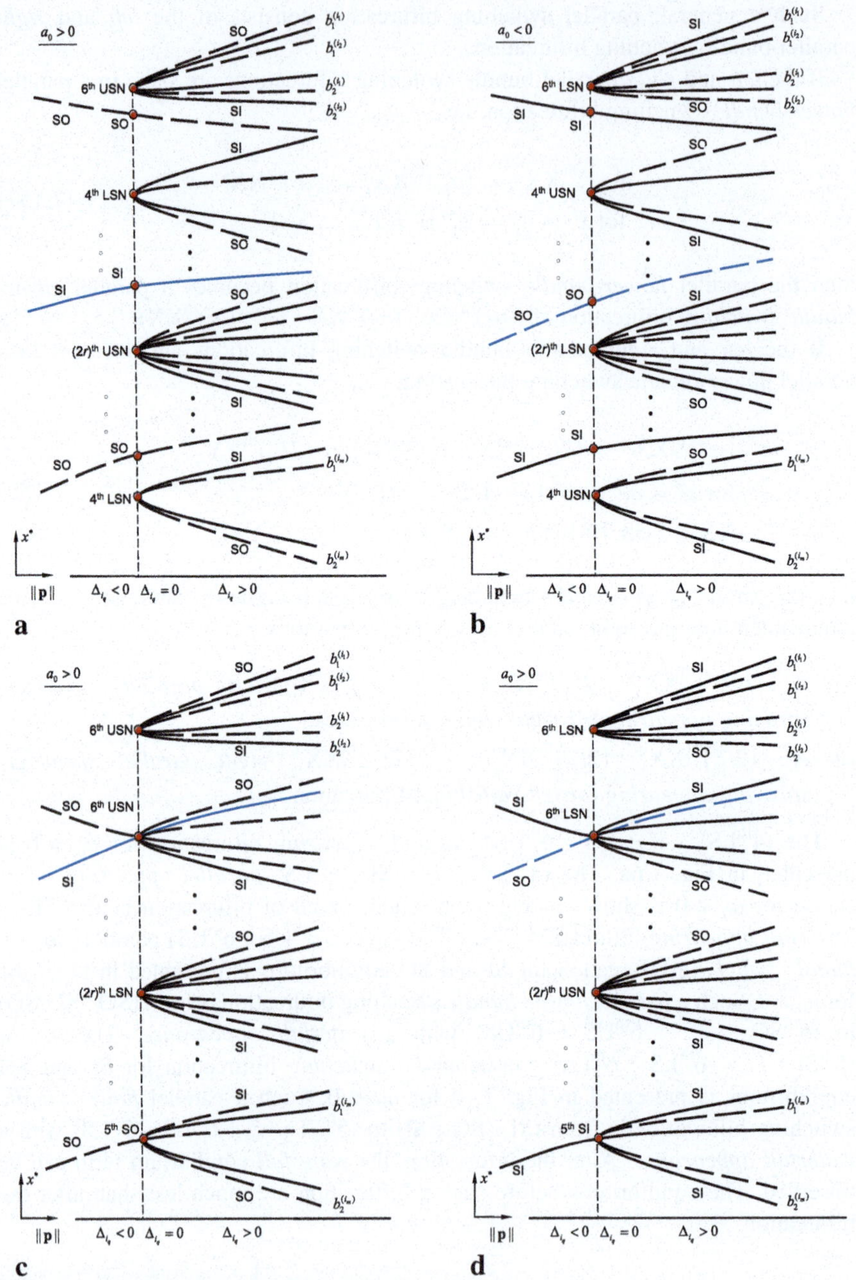

Fig. 7.6 Switching and appearing bifurcations. Simple switching: (**a**) (4^{th}LSN : . . . : SO : 6^{th}USN) ($a_0 > 0$), (**b**) (4^{th}USN : . . . : SI : 6^{th}LSN) ($a_0 < 0$). Higher order switching: (**c**) (5^{th}SI : . . . : 6^{th}USN : 6^{th}USN) ($a_0 > 0$), (**d**) (5^{th}SO : . . . : 6^{th}LSN : 6^{th}LSN) ($a_0 < 0$) in the $(2m+1)^{\text{th}}$-degree polynomial nonlinear system. LSN: lower saddle node, USN: upper saddle node, SI: sink, SO: source. Stable and unstable equilibriums are represented by solid and dashed curves, respectively. The bifurcation points are marked by circular symbols

7.3 Higher Order Equilibrium Bifurcations

The afore-discussed appearing and switching bifurcations in the $(2m+1)^{\text{th}}$-degree polynomial system are relative to the simple sources and sinks. As similar to the $(2m)^{\text{th}}$-degree polynomial system, the higher order singularity bifurcations in the $(2m+1)^{\text{th}}$-degree polynomial system can be for higher order sinks, sources, upper saddles, and lower saddles as in Luo (2019).

7.3.1 *Higher Order Equilibrium Bifurcations*

Consider a $(2m+1)^{\text{th}}$-degree polynomial system as

$$\dot{x} = a_0 Q(x)(x-a)\prod_{i=1}^{s}(x^2 + B_i x + C_i)^{\alpha_i}, \tag{7.159}$$

where $\alpha_i \in \{2l_i - 1, 2l_i\}$. Without loss of generality, a function of $Q(x) > 0$ is either a polynomial function or a non-polynomial function. The roots of $x^2 + B_i x + C_i = 0$ are

$$\begin{aligned} &b_{1,2}^{(i)} = -\frac{1}{2}B_i \pm \frac{1}{2}\sqrt{\Delta_i}, \Delta_i = B_i^2 - 4C_i \geq 0;\\ &\{a_1, a_2, \ldots, a_{2s-1}, a_{2s}, a_{2s+1}\} = \text{sort}\{b_1^{(1)}, b_2^{(1)}, \ldots, b_1^{(s)}, b_2^{(s)}, a\}, a_j \leq a_{j+1}. \end{aligned} \tag{7.160}$$

For $a \neq -\frac{1}{2}B_i$ $(i = 1, 2, \ldots, s)$, there are four higher order bifurcations as follows:

$$\begin{aligned} &(2(2l_i-1))^{\text{th}}\text{order US} \xrightarrow[\text{appearing bifurcation}]{(2l_i-1)^{\text{th}}\text{order quadratics}} \\ &\begin{cases} (2l_i-1)^{\text{th}} \text{ order SO}, & x^* = b_2^{(i)}, \\ (2l_i-1)^{\text{th}} \text{ order SI}, & x^* = b_1^{(i)}; \end{cases} \end{aligned} \tag{7.161}$$

$$\begin{aligned} &(2(2l_i-1))^{\text{th}} \text{ order LS} \xrightarrow[\text{appearing bifurcation}]{(2l_i-1)^{\text{th}}\text{order quadratics}} \\ &\begin{cases} (2l_i-1)^{\text{th}} \text{ order SI}, & x^* = b_2^{(i)}, \\ (2l_i-1)^{\text{th}} \text{ order SO}, & x^* = b_1^{(i)}; \end{cases} \end{aligned} \tag{7.162}$$

$$
\begin{aligned}
&(2(2l_i))^{\text{th}} \text{ order US} \xrightarrow[\text{appearing bifurcation}]{(2l_i)^{\text{th}}\text{-order quadratics}} \\
&\begin{cases} (2l_i)^{\text{th}} \text{ order US}, \ x^* = b_2^{(i)}, \\ (2l_i)^{\text{th}} \text{ order US}, \ x^* = b_1^{(i)}; \end{cases}
\end{aligned}
\tag{7.163}
$$

$$
\begin{aligned}
&(2(2l_i))^{\text{th}} \text{ order LS} \xrightarrow[\text{appearing bifurcation}]{(2l_i)^{\text{th}}\text{-order quadratics}} \\
&\begin{cases} (2l_i)^{\text{th}} \text{ order LS}, x^* = b_2^{(i)}, \\ (2l_i)^{\text{th}} \text{ order LS}, x^* = b_1^{(i)}. \end{cases}
\end{aligned}
\tag{7.164}
$$

(i) For $\alpha_i = 2l_i - 1$, the $(2(2l_i - 1))^{\text{th}}$-order upper-saddle (US) *appearing* bifurcation is for the onset of the $(2l_i - 1)^{\text{th}}$-order source (SO) ($x^* = b_2^{(i)}$) and the $(2l_i - 1)^{\text{th}}$-order sink (SI) ($x^* = b_1^{(i)}$) with $b_2^{(i)} > b_1^{(i)}$.

(ii) For $\alpha_i = 2l_i - 1$, the $(2(2l_i - 1))^{\text{th}}$-order lower-saddle (LS) *appearing* bifurcation is for the onset of the $(2l_i - 1)^{\text{th}}$-order sink (SI) ($x^* = b_2^{(i)}$) and the $(2l_i - 1)^{\text{th}}$-order source (SI) ($x^* = b_1^{(i)}$) with $b_2^{(i)} > b_1^{(i)}$.

(iii) For $\alpha_i = 2l_i$, the $(2(2l_i))^{\text{th}}$-order upper-saddle (US) *appearing* bifurcation is for the onset of two $(2l_i)^{\text{th}}$-order upper saddles (US) ($x^* = b_1^{(i)}, b_2^{(i)}$) with $b_2^{(i)} > b_1^{(i)}$.

(iv) For $\alpha_i = 2l_i$, the $(2(2l_i))^{\text{th}}$ order lower-saddle (LS) *appearing* bifurcation is for the onset of two $(2l_i)^{\text{th}}$-order lower saddles (LS) ($x^* = b_1^{(i)}, b_2^{(i)}$) with $b_2^{(i)} > b_1^{(i)}$.

The equilibrium of $x^* = a \neq -\frac{1}{2}B_i$ $(i = 1, 2, \ldots, s)$ breaks a cluster of *teeth comb* appearing bifurcations of higher order equilibrium to two parts. The *teeth comb* appearing bifurcation generated by the s-pairs of quadratics becomes a *broom appearing* bifurcation for higher order equilibriums. The two *broom appearing* bifurcations for higher order equilibriums are

$$
\text{SO}\ (x^* = a) \xrightarrow[\text{appearing bifurcation}]{\sum_{j=1}^{2}\sum_{i=1}^{s_j}\alpha_i^{(j)}=m}
\begin{cases} ((2\alpha_1^{(2)})^{\text{th}}\text{US}:\ldots:(2\alpha_{s_2}^{(2)})^{\text{th}}\text{US}), \\ \text{SO, for } x^* = a = a_{2(s_1+1)-1}, \\ ((2\alpha_1^{(1)})^{\text{th}}\text{LS}:\ldots:(2\alpha_{s_1}^{(1)})^{\text{th}}\text{LS}) \end{cases}
\tag{7.165}
$$

and

$$
\text{SI}\ (x^* = a) \xrightarrow[\text{appearing bifurcation}]{\sum_{j=1}^{2}\sum_{i=1}^{s_j}\alpha_i^{(j)}=m}
\begin{cases} ((2\alpha_1^{(2)})^{\text{th}}\text{LS}:\ldots:(2\alpha_{s_2}^{(2)})^{\text{th}}\text{LS}), \\ \text{SI, for } x^* = a = a_{2(s_1+1)-1}, \\ ((2\alpha_1^{(1)})^{\text{th}}\text{US}:\ldots:(2\alpha_{s_1}^{(1)})^{\text{th}}\text{US}) \end{cases}
\tag{7.166}
$$

where

$$((2\alpha_1^{(j)})^{\text{th}}\text{US}:\ldots:(2\alpha_{s_j}^{(j)})^{\text{th}}\text{US}) = \begin{cases} (2\alpha_{s_j}^{(j)})^{\text{th}}\text{US} \to \begin{cases} (\alpha_{s_j}^{(j)})^{\text{th}}\text{XX}, \\ (\alpha_{s_j}^{(j)})^{\text{th}}\text{XX}; \end{cases} \\ \vdots \\ (2\alpha_1^{(j)})^{\text{th}}\text{US} \to \begin{cases} (\alpha_1^{(j)})^{\text{th}}\text{XX}, \\ (\alpha_1^{(j)})^{\text{th}}\text{XX}; \end{cases} \end{cases} \tag{7.167}$$

$$((2\alpha_1^{(j)})^{\text{th}}\text{LS}:\ldots:(2\alpha_{s_j}^{(j)})^{\text{th}}\text{LS}) = \begin{cases} (2\alpha_{s_j}^{(j)})^{\text{th}}\text{LS} \to \begin{cases} (\alpha_{s_j}^{(j)})^{\text{th}}\text{XX}, \\ (\alpha_{s_j}^{(j)})^{\text{th}}\text{XX}; \end{cases} \\ \vdots \\ (2\alpha_1^{(j)})^{\text{th}}\text{LS} \to \begin{cases} (\alpha_1^{(j)})^{\text{th}}\text{XX}, \\ (\alpha_1^{(j)})^{\text{th}}\text{XX} \end{cases} \end{cases} \tag{7.168}$$

for $j = 1,2$.

Four special broom appearing bifurcations for higher order equilibriums are

$$\text{SO } (x^* = a) \xrightarrow[\text{appearing bifurcation}]{\sum_{i=1}^{s}\alpha_i = m} \begin{cases} \text{SO, for } x^* = a = a_{2s+1}, \\ ((2\alpha_1)^{\text{th}}\text{LS}:\ldots:(2\alpha_s)^{\text{th}}\text{LS}) \end{cases} \tag{7.169}$$

$$\text{SI } (x^* = a) \xrightarrow[\text{appearing bifurcation}]{\sum_{i=1}^{s}\alpha_i = m} \begin{cases} \text{SO, for } x^* = a = a_{2s+1}, \\ ((2\alpha_1)^{\text{th}}\text{US}:\ldots:(2\alpha_s)^{\text{th}}\text{US}) \end{cases} \tag{7.170}$$

and

$$\text{SO } (x^* = a) \xrightarrow[\text{appearing bifurcation}]{\sum_{i=1}^{s}\alpha_i = m} \begin{cases} ((2\alpha_1)^{\text{th}}\text{US}:\ldots:(2\alpha_s)^{\text{th}}\text{US}) \\ \text{SO, for } x^* = a = a_1 \end{cases} \tag{7.171}$$

$$\text{SI } (x^* = a) \xrightarrow[\text{appearing bifurcation}]{\sum_{i=1}^{s}\alpha_i = m} \begin{cases} ((2\alpha_1)^{\text{th}}\text{LS}:\ldots:(2\alpha_s)^{\text{th}}\text{LS}) \\ \text{SI, for } x^* = a = a_1. \end{cases} \tag{7.172}$$

For $a = -\frac{1}{2}B_i$ $(i \in \{1, 2, \ldots, s\})$, there are four higher order bifurcations as follows:

$$\mathrm{SO}(x^* = a) \to (2(2l_i - 1) + 1)^{\text{th}}\mathrm{SO} = \begin{cases} (2l_i - 1)^{\text{th}} \text{ order SO}, x^* = b_2^{(i)}, \\ \mathrm{SI}, x^* = a, \\ (2l_i - 1)^{\text{th}} \text{ order SO}, x^* = b_1^{(i)}; \end{cases} \tag{7.173}$$

$$\mathrm{SI}\ (x^* = a) \to (2(2l_i - 1) + 1)^{\text{th}}\mathrm{SI} = \begin{cases} (2l_i - 1)^{\text{th}} \text{ order SI}, \ x^* = b_2^{(i)}, \\ \mathrm{SO}, \ x^* = a, \\ (2l_i - 1)^{\text{th}} \text{ order SI}, \ x^* = b_1^{(i)}; \end{cases} \tag{7.174}$$

$$\mathrm{SI}\ (x^* = a) \to (2(2l_i) + 1)^{\text{th}}\mathrm{SO} = \begin{cases} (2l_i)^{\text{th}} \text{ order US}, \ x^* = b_2^{(i)}, \\ \mathrm{SO}, \ x^* = a, \\ (2l_i)^{\text{th}} \text{ order LS}, \ x^* = b_1^{(i)}; \end{cases} \tag{7.175}$$

$$\mathrm{SI}\ (x^* = a) \to (2(2l_i) + 1)^{\text{th}}\mathrm{SI} = \begin{cases} (2l_i)^{\text{th}} \text{ order LS}, \ x^* = b_2^{(i)}, \\ \mathrm{SI}, \ x^* = a, \\ (2l_i)^{\text{th}} \text{ order US}, \ x^* = b_1^{(i)}. \end{cases} \tag{7.176}$$

(i) For $\alpha_i = 2l_i - 1$, the $(2(2l_i - 1) + 1)^{\text{th}}$-order source (SO) *switching* bifurcation is with the $(2l_i - 1)^{\text{th}}$-order source (SO) $(x^* = b_2^{(i)})$ and the $(2l - 1)^{\text{th}}$-order sink (SI) $(x^* = b_1^{(i)})$ with $b_2^{(i)} > a > b_1^{(i)}$.
(ii) For $\alpha_i = 2l_i - 1$, the $(2(2l_i - 1) + 1)^{\text{th}}$-order sink (SI) *switching* bifurcation is with the $(2l_i - 1)^{\text{th}}$-order sink (SI) $(x^* = b_2^{(i)})$ and the $(2l_i - 1)^{\text{th}}$-order source (SO) $(x^* = b_1^{(i)})$ with $b_2^{(i)} > a > b_1^{(i)}$.
(iii) For $\alpha_i = 2l_i$, the $(2(2l_i) + 1)^{\text{th}}$-order source (SO) *switching* bifurcation is with the $(2l_i)^{\text{th}}$-order upper saddle (US) $(x^* = b_2^{(i)})$ and the $(2l_i)^{\text{th}}$-order upper saddles (LS) $(x^* = b_1^{(i)})$ with $b_2^{(i)} > a > b_1^{(i)}$.
(iv) For $\alpha_i = 2l_i$, the $(2(2l_i) + 1)^{\text{th}}$-order sink (SI) *switching* bifurcation is with the $(2l_i)^{\text{th}}$-order upper saddle (LS) $(x^* = b_2^{(i)})$ and the $(2l_i)^{\text{th}}$-order upper saddles (US) $(x^* = b_1^{(i)})$ with $b_2^{(i)} > a > b_1^{(i)}$.

If $x^* = a = -\frac{1}{2}B_i$ $(i \in \{1, 2, \ldots, m\})$, the equilibrium of $x^* = a$ possesses a $(2(2l_i - 1) + 1)^{\text{th}}$ and $(2(2l_i) + 1)^{\text{th}}$-order SI or SO switching bifurcations (or pitchfork bifurcations) for higher order equilibriums. The teeth comb appearing bifurcation generated by the m-pairs of quadratics becomes a *broom-switching* bifurcation. Such a *broom-switching* bifurcation consists of a *pitchfork-switching* bifurcation and two teeth comb appearing bifurcations in the $(2m + 1)$th-degree polynomial system. Four *broom-switching* bifurcations for higher order equilibriums are

$$\mathrm{SO}\,(x^*=a)\xrightarrow[\text{appearing bifurcation}]{\sum_{j=1}^{2}\sum_{i=1}^{s_j}\alpha_i^{(j)}+2l_{s_1+1}=m}\begin{cases}((2\alpha_1^{(2)})^{\text{th}}\mathrm{US}:\ldots:(2\alpha_{s_2}^{(2)})^{\text{th}}\mathrm{US}),\\ (2(2l_{s_1+1})+1)^{\text{th}}\mathrm{SO}\begin{cases}(2l_{s_1+1})^{\text{th}}\mathrm{US},\\ \mathrm{SO},x^*=a,\\ (2l_{s_1+1})^{\text{th}}\mathrm{LS},\end{cases}\\ ((2\alpha_1^{(1)})^{\text{th}}\mathrm{LS}:\ldots:(2\alpha_{s_1}^{(1)})^{\text{th}}\mathrm{LS});\end{cases} \tag{7.177}$$

$$\mathrm{SO}\,(x^*=a)\xrightarrow[\text{appearing bifurcation}]{\sum_{j=1}^{2}\sum_{i=1}^{s_j}\alpha_i^{(j)}+2l_{s_1+1}-1=m}\begin{cases}((2\alpha_1^{(2)})^{\text{th}}\mathrm{US}:\ldots:(2\alpha_{s_2}^{(2)})^{\text{th}}\mathrm{US}),\\ (2(2l_{s_1+1}-1)+1)^{\text{th}}\mathrm{SO}\begin{cases}(2l_{s_1+1}-1)^{\text{th}}\mathrm{SO},\\ \mathrm{SI},x^*=a,\\ (2l_{s_1+1}-1)^{\text{th}}\mathrm{SO},\end{cases}\\ ((2\alpha_1^{(1)})^{\text{th}}\mathrm{LS}:\ldots:(2\alpha_{s_1}^{(1)})^{\text{th}}\mathrm{LS});\end{cases} \tag{7.178}$$

and

$$\mathrm{SI}\,(x^*=a)\xrightarrow[\text{appearing bifurcation}]{\sum_{j=1}^{2}\sum_{i=1}^{s_j}\alpha_i^{(j)}+2l_{s_1+1}=m}\begin{cases}((2\alpha_1^{(2)})^{\text{th}}\mathrm{LS}:\ldots:(2\alpha_{s_2}^{(2)})^{\text{th}}\mathrm{LS}),\\ (2(2l_{s_1+1})+1)^{\text{th}}\mathrm{SI}\begin{cases}(2l_{s_1+1})^{\text{th}}\mathrm{LS},\\ \mathrm{SI},x^*=a,\\ (2l_{s_1+1})^{\text{th}}\mathrm{US},\end{cases}\\ ((2\alpha_1^{(1)})^{\text{th}}\mathrm{US}:\ldots:(2\alpha_{s_1}^{(1)})^{\text{th}}\mathrm{US});\end{cases} \tag{7.179}$$

$$\mathrm{SI}\,(x^*=a)\xrightarrow[\text{appearing bifurcation}]{\sum_{j=1}^{2}\sum_{i=1}^{s_j}\alpha_i^{(j)}+2l_{s_1+1}-1=m}\begin{cases}((2\alpha_1^{(2)})^{\text{th}}\mathrm{LS}:\ldots:(2\alpha_{s_2}^{(2)})^{\text{th}}\mathrm{LS}),\\ (2(2l_{s_1+1}-1)+1)^{\text{th}}\mathrm{SI}\begin{cases}(2l_{s_1+1}-1)^{\text{th}}\mathrm{SI},\\ \mathrm{SO},x^*=a,\\ (2l_{s_1+1}-1)^{\text{th}}\mathrm{SI};\end{cases}\\ ((2\alpha_1^{(1)})^{\text{th}}\mathrm{US}:\ldots:(2\alpha_{s_1}^{(1)})^{\text{th}}\mathrm{US}).\end{cases} \tag{7.180}$$

Consider a $(2m+1)^{\text{th}}$-degree polynomial system as

$$\dot{x} = a_0 Q(x)(x-a)\prod_{i=1}^{n}(x^2 + B_i x + C_i)^{\alpha_i} \tag{7.181}$$

where $\alpha_i \in \{2r_i - 1, 2r_i\}$ $(i = 1, 2, \ldots, n)$. Without loss of generality, a function of $Q(x) > 0$ is either a polynomial function or a non-polynomial function. The roots of $x^2 + B_i x + C_i = 0$ are

$$\begin{aligned}
&x_{1,2}^{(i)} = -\frac{1}{2}B_i \pm \frac{1}{2}\sqrt{\Delta_i}, \Delta_i = B_i^2 - 4C_i \geq 0;\\
&B_i = B_j \; (i, j \in \{1, 2, \ldots, n\}; i \neq j)\\
&\{a_1, a_2, \ldots, a_{2n+1}\} \subset \text{sort}\{x_1^{(1)}, x_2^{(1)}, x_1^{(2)}, x_2^{(2)}, \ldots, x_1^{(n)}, x_2^{(n)}, a\}, a_i \leq a_{i+1}.
\end{aligned} \tag{7.182}$$

The higher order singularity bifurcation can be for a cluster of higher order equilibriums. There are four higher order bifurcations as follows:

(i) The $(2(2l-1))^{\text{th}}$-order upper-saddle (US) *spraying-appearing* bifurcation for a cluster of higher order sinks, sources, upper saddles, and lower saddles is

$$\begin{aligned}
&(2\beta)^{\text{th}}\text{US} = (2(2l-1))^{\text{th}}\text{order US}\\
&\xrightarrow[\text{appearing bifurcation}]{\text{a cluster of } 2n\text{-XX}}
\begin{cases}
(\alpha_{2n})^{\text{th}} \text{ order XX for } x^* = a_{2n},\\
(\alpha_{2n-1})^{\text{th}}\text{order XX for } x^* = a_{2n-1},\\
\vdots\\
(\alpha_1)^{\text{th}}\text{order XX for } x^* = a_1;
\end{cases}
\end{aligned} \tag{7.183}$$

where $2(2l-1) = \sum_{i=1}^{n}\alpha_i$ and

$$\begin{aligned}
&(\alpha_{2n})^{\text{th}} \text{ order XX} = \begin{cases}
(2r_{2n})^{\text{th}} \text{ order US}, & \text{for } \alpha_{2n} = 2r_n,\\
(2r_{2n}-1)^{\text{th}} \text{ order SO}, & \text{for } \alpha_{2n} = 2r_n - 1;
\end{cases}\\
&(\alpha_1)^{\text{th}} \text{ order XX} = \begin{cases}
(2r_1)^{\text{th}} \text{ order US}, & \text{for } \alpha_1 = 2r_1,\\
(2r_1-1)^{\text{th}} \text{ order SO}, & \text{for } \alpha_1 = 2r_1 - 1.
\end{cases}
\end{aligned} \tag{7.184}$$

(ii) The $(2(2l))^{\text{th}}$-order upper-saddle (US) *spraying-appearing* bifurcation for a cluster of higher order sinks, sources, upper saddles, and lower saddles is

$$(2\beta)^{\text{th}}\text{US} = (2(2l))^{\text{th}}\text{order US} \xrightarrow[\text{appearing bifurcation}]{\text{a cluster of } 2n\text{-XX}} \begin{cases} (\alpha_{2n})^{\text{th}} \text{ order XX for } x^* = a_{2n}, \\ (\alpha_{2n-1})^{\text{th}}\text{order XX for } x^* = a_{2n-1}, \\ \vdots \\ (\alpha_1)^{\text{th}}\text{order XX for } x^* = a_1 \end{cases} \tag{7.185}$$

where $2(2l) = \sum_{i=1}^{n} \alpha_i$ and

$$\begin{aligned} (\alpha_{2n})^{\text{th}} \text{ order XX} &= \begin{cases} (2r_{2n})^{\text{th}} \text{ order US}, & \text{for } \alpha_{2n} = 2r_n, \\ (2r_{2n} - 1)^{\text{th}} \text{ order SO}, & \text{for } \alpha_{2n} = 2r_n - 1; \end{cases} \\ (\alpha_1)^{\text{th}} \text{ order XX} &= \begin{cases} (2r_1)^{\text{th}} \text{ order US}, & \text{for } \alpha_1 = 2r_1, \\ (2r_1 - 1)^{\text{th}} \text{ order SI}, & \text{for } \alpha_1 = 2r_1 - 1. \end{cases} \end{aligned} \tag{7.186}$$

For the higher order lower-saddle bifurcation, the cluster of the higher order equilibriums is given by the following two cases.

(iii) The $(2(2l-1))^{\text{th}}$-order lower-saddle (LS) *spraying-appearing* bifurcation for a cluster of higher order sinks, sources, upper saddles, and lower saddles is

$$(2\beta)^{\text{th}}\text{LS} = (2(2l-1))^{\text{th}}\text{order LS} \xrightarrow[\text{appearing bifurcation}]{\text{a cluster of } 2n\text{-XX}} \begin{cases} (\alpha_{2n})^{\text{th}} \text{ order XX, for } x^* = a_{2n}, \\ (\alpha_{2n-1})^{\text{th}}\text{order XX, for } x^* = a_{2n-1}, \\ \vdots \\ (\alpha_1)^{\text{th}}\text{order XX, for } x^* = a_1 \end{cases} \tag{7.187}$$

where $2(2l-1) = \sum_{i=1}^{n} \alpha_i$ and

$$\begin{aligned} (\alpha_{2n})^{\text{th}} \text{ order XX} &= \begin{cases} (2r_{2n})^{\text{th}} \text{ order LS}, & \text{for } \alpha_{2n} = 2r_n, \\ (2r_{2n} - 1)^{\text{th}} \text{ order SI}, & \text{for } \alpha_{2n} = 2r_n - 1; \end{cases} \\ (\alpha_1)^{\text{th}} \text{ order XX} &= \begin{cases} (2r_1)^{\text{th}} \text{ order LS}, & \text{for } \alpha_1 = 2r_1, \\ (2r_1 - 1)^{\text{th}} \text{ order SI}, & \text{for } \alpha_1 = 2r_1 - 1. \end{cases} \end{aligned} \tag{7.188}$$

(iv) The $(2(2l))^{\text{th}}$-order lower-order *spraying-appearing* bifurcation for a cluster of higher order sinks, sources, upper saddles, and lower saddles is

$$
\begin{aligned}
&(2\beta)^{\text{th}}\text{LS} = (2(2l))^{\text{th}}\text{order LS}\\
&\xrightarrow[\text{appearing bifurcation}]{\text{a cluster of } 2n\text{-XX}}
\begin{cases}
(\alpha_{2n})^{\text{th}} \text{ order XX, for } x^* = a_{2n},\\
(\alpha_{2n-1})^{\text{th}}\text{order XX, for } x^* = a_{2n-1},\\
\vdots\\
(\alpha_1)^{\text{th}}\text{order XX, for } x^* = a_1
\end{cases}
\end{aligned}
\tag{7.189}
$$

where $2(2l) = \sum_{i=1}^{n}\alpha_i$ and

$$
\begin{aligned}
&(\alpha_{2n})^{\text{th}} \text{ order XX} = \begin{cases}
(2r_{2n})^{\text{th}} \text{ order LS}, & \text{for } \alpha_{2n} = 2r_n,\\
(2r_{2n}-1)^{\text{th}} \text{ order SI}, & \text{for } \alpha_{2n} = 2r_n - 1;
\end{cases}\\
&(\alpha_1)^{\text{th}} \text{ order XX} = \begin{cases}
(2r_1)^{\text{th}} \text{ order LS}, & \text{for } \alpha_1 = 2r_1,\\
(2r_1-1)^{\text{th}} \text{ order SO}, & \text{for } \alpha_1 = 2r_1 - 1.
\end{cases}
\end{aligned}
\tag{7.190}
$$

If $x^* = a \neq -\frac{1}{2}B_i$ $(i \in \{1, 2, \ldots, n\})$, the equilibrium of $x^* = a$ breaks a cluster of *sprinkler-spraying* appearing bifurcations for higher order equilibriums to two parts. The sprinkler-spraying appearing bifurcation generated by the m-pairs of quadratics becomes a *broom-sprinkler-spraying-appearing* bifurcation. The two *broom-sprinkler*-spraying-*appearing* bifurcations in the $(2m+1)^{\text{th}}$-degree polynomial system are

$$
\text{SO}\ (x^* = a) \xrightarrow[\text{appearing bifurcation}]{m = m_1 + m_2}
\begin{cases}
((2\beta_1^{(2)})^{\text{th}}\text{US} : \ldots : (2\beta_{r_2}^{(2)})^{\text{th}}\text{US}),\\
\text{SO}\ (a = a_{r_1+1}) \to \text{SO}\ (a = a_{2(m_1+1)-1}),\\
((2\beta_1^{(1)})^{\text{th}}\text{LS} : \ldots : (2\beta_{r_1}^{(1)})^{\text{th}}\text{LS})
\end{cases}
\tag{7.191}
$$

and

$$
\text{SI}\ (x^* = a) \xrightarrow[\text{appearing bifurcation}]{m = m_1 + m_2}
\begin{cases}
((2\beta_1^{(2)})^{\text{th}}\text{LS} : \ldots : (2\beta_{r_2}^{(2)})^{\text{th}}\text{LS}),\\
\text{SI}\ (a = a_{r_1+1}) \to \text{SO}\ (a = a_{2(m_1+1)-1}),\\
((2\beta_1^{(1)})^{\text{th}}\text{US} : \ldots : (2\beta_{r_1}^{(1)})^{\text{th}}\text{US})
\end{cases}
\tag{7.192}
$$

for $m_1 = \sum_{i=1}^{r_1}\beta_i^{(1)}$, $m_2 = \sum_{j=1}^{r_2}\beta_j^{(2)}$; and the acronyms USG and LSG are the upper-saddle-node bifurcation group and lower-saddle-node bifurcation group, respectively.

Four special broom-sprinkler-spraying-appearing bifurcations of the $(2m+1)^{\text{th}}$-degree polynomial system are

$$\mathrm{SO}\,(x^*=a)\xrightarrow[\text{appearing bifurcation}]{m=\sum_{i=1}^{r}\beta_i}\begin{cases}\mathrm{SO}\,(a=a_{2m+1})\to\mathrm{SO}\,(a=a_{2m+1}),\\((2\beta_1)^{\text{th}}\mathrm{LS}:\ldots:(2\beta_r)^{\text{th}}\mathrm{LS});\end{cases}\tag{7.193}$$

$$\mathrm{SI}\,(x^*=a)\xrightarrow[\text{appearing bifurcation}]{m=\sum_{i=1}^{r}\beta_i}\begin{cases}\mathrm{SI}\,(a=a_{2m+1})\to\mathrm{SI}\,(a=a_{2m+1}),\\((2\beta_1)^{\text{th}}\mathrm{US}:\ldots:(2\beta_r)^{\text{th}}\mathrm{US});\end{cases}\tag{7.194}$$

and

$$\mathrm{SO}\,(x^*=a)\xrightarrow[\text{appearing bifurcation}]{m=\sum_{i=1}^{r}\beta_i}\begin{cases}((2\beta_1)^{\text{th}}\mathrm{US}:\ldots:(2\beta_r)^{\text{th}}\mathrm{US}),\\\mathrm{SO}\,(a=a_1)\to\mathrm{SO}\,(a=a_1);\end{cases}\tag{7.195}$$

$$\mathrm{SI}\,(x^*=a)\xrightarrow[\text{appearing bifurcation}]{m=\sum_{i=1}^{r}\beta_i}\begin{cases}((2\beta_1)^{\text{th}}\mathrm{LS}:\ldots:(2\beta_r)^{\text{th}}\mathrm{LS}),\\\mathrm{SI}\,(a=a_1)\to\mathrm{SI}\,(a=a_1).\end{cases}\tag{7.196}$$

If $x^*=a=-\frac{1}{2}B_i$ $(i\in\{1,2,\ldots,l\})$, the equilibrium of $x^*=a$ possesses a $(2l+1)^{\text{th}}$-order SI or SO switching bifurcation (or broom-switching bifurcation) for higher order equilibriums. The sprinkler-spraying-appearing bifurcation generated by the m-pairs of quadratics becomes a *broom-sprinkler-spraying-switching* bifurcation. The two *broom-switching* bifurcations in the $(2m+1)^{\text{th}}$-degree polynomial system are

$$\mathrm{SO}\,(x^*=a)\xrightarrow[\text{switching bifurcation}]{m=m_1+m_2+\beta}$$

$$\begin{cases}(2\beta_1^{(2)}:\ldots:2\beta_{r_2}^{(2)})^{\text{th}}\mathrm{US}\xrightarrow[\textit{appearing}]{m_2=\sum_{j=1}^{r_2}\beta_j^{(2)}}\begin{cases}(2\beta_{r_2}^{(2)})^{\text{th}}\mathrm{US}\,(x^*=a_{r_1+r_2+1}),\\\vdots\\(2\beta_1^{(2)})^{\text{th}}\mathrm{US}\,(x^*=a_{r_1+2});\end{cases}\\(2\beta+1)^{\text{th}}\mathrm{SO}\,(a=a_{r_1+1});\\(2\beta_1^{(1)}:\ldots:2\beta_{r_1}^{(1)})^{\text{th}}\mathrm{LS}\xrightarrow[\textit{appearing}]{m_1=\sum_{i=1}^{r_1}\beta_i^{(1)}}\begin{cases}(2\beta_{r_1}^{(1)})^{\text{th}}\mathrm{LS}\,(x^*=a_{r_1}),\\\vdots\\(2\beta_1^{(1)})^{\text{th}}\mathrm{LS}\,(x^*=a_1);\end{cases}\end{cases}\tag{7.197}$$

and

$$
\mathrm{SI}\,(x^*=a)\xrightarrow[\text{switching bifurcation}]{m=m_1+m_2+\beta}
\begin{cases}
(2\beta_1^{(2)}:\ldots:2\beta_{r_2}^{(2)})^{\text{th}}\mathrm{LS}\xrightarrow[\textit{appearing}]{m_2=\sum_{j=1}^{r_2}\beta_j^{(2)}}
\begin{cases}
(2\beta_{r_2}^{(2)})^{\text{th}}\mathrm{LS}\ (x^*=a_{r_1+r_2+1}),\\
\vdots\\
(2\beta_1^{(2)})^{\text{th}}\mathrm{LS}\ (x^*=a_{r_1+2});
\end{cases}\\
(2\beta+1)^{\text{th}}\mathrm{SI}\ (a=a_{r_1+1});\\
(2\beta_1^{(1)}:\ldots:2\beta_{r_2}^{(1)})^{\text{th}}\mathrm{US}\xrightarrow[\textit{appearing}]{m_1=\sum_{i=1}^{r_1}\beta_i^{(1)}}
\begin{cases}
(2\beta_{r_1}^{(1)})^{\text{th}}\mathrm{US}\ (x^*=a_{r_1}),\\
\vdots\\
(2\beta_1^{(1)})^{\text{th}}\mathrm{US}\ (x^*=a_1)
\end{cases}
\end{cases}
\tag{7.198}
$$

where

$$
(2\beta+1)^{\text{th}}\text{order SO}(x^*=a)\xrightarrow[\text{appearing bifurcation}]{\text{cluster of } l\text{-quadratics}}
\begin{cases}
(\alpha_{2s_l+1})^{\text{th}}\mathrm{XX},\ \ \text{for } x^*=a_{2s_l+1},\\
(\alpha_{2s_l})^{\text{th}}\mathrm{XX},\ \ \text{for } x^*=a_{2s_l},\\
\vdots\\
(\alpha_{2s_1})^{\text{th}}\mathrm{XX},\ \ \text{for } x^*=a_{2s_1},\\
(\alpha_{2s_1-1})^{\text{th}}\mathrm{XX},\ \ \text{for } x^*=a_{2s_1-1}.
\end{cases}
\tag{7.199}
$$

$$
(2\beta+1)^{\text{th}}\text{order SI }(x^*=a)\xrightarrow[\text{appearing bifurcation}]{\text{cluster of } l\text{-quadratics}}
\begin{cases}
(\alpha_{2s_l+1})^{\text{th}}\mathrm{XX},\ \ \text{for } x^*=a_{2s_l+1},\\
(\alpha_{2s_l})^{\text{th}}\mathrm{XX},\ \ \text{for } x^*=a_{2s_l},\\
\vdots\\
(\alpha_{2s_1})^{\text{th}}\mathrm{XX},\ \ \text{for } x^*=a_{2s_1},\\
(\alpha_{2s_1-1})^{\text{th}}\mathrm{XX},\ \ \text{for } x^*=a_{2s_1-1}
\end{cases}
\tag{7.200}
$$

where $x^*=a\in\{a_{2s_1-1},\ldots,a_{2s_l},a_{2s_l+1}\}$ and $2\beta+1=\sum_{i=1}^{l}\alpha_{2s_i-1}+\alpha_{2s_i}+\alpha_{2s_l+1}$.

The two appearing bifurcations for the higher order singularity of equilibriums are

(i) SO $\rightarrow$ $((2\alpha_1)^{\text{th}}\text{LS} : \ldots : (2\alpha_i)^{\text{th}}\text{LS} : \text{SO} : \ldots : (2\alpha_{n-1})^{\text{th}}\text{US} : (2\alpha_n)^{\text{th}}\text{US})$,
(ii) SO $\rightarrow$ $((2\alpha_1)^{\text{th}}\text{US} : \ldots : (2\alpha_i)^{\text{th}}\text{US} : \text{SO} : \ldots : (2\alpha_{n-1})^{\text{th}}\text{LS} : (2\alpha_n)^{\text{th}}\text{LS})$,

as presented in Figs. 7.7a, b for $a_0 > 0$ and $a_0 < 0$, respectively. The *broom appearing* bifurcation for the higher order equilibriums are illustrated. The components of the *broom appearing* bifurcation are

$$\begin{aligned}
&(2\alpha_j)^{\text{th}}\text{US} \xrightarrow[\textit{appearing}]{\alpha_j=2r_j} \begin{cases} (2r_j)^{\text{th}}\text{US} \\ (2r_j)^{\text{th}}\text{US} \end{cases} (j = i, n-1, \ldots),\\
&(2\alpha_k)^{\text{th}}\text{US} \xrightarrow[\textit{appearing}]{\alpha_k=2r_k-1} \begin{cases} (2r_k-1)^{\text{th}}\text{SO} \\ (2r_k-1)^{\text{th}}\text{SI} \end{cases} (k = 1, n, \ldots);
\end{aligned} \tag{7.201}$$

and

$$\begin{aligned}
&(2\alpha_j)^{\text{th}}\text{LS} \xrightarrow[\textit{appearing}]{\alpha_j=2r_j} \begin{cases} (2r_j)^{\text{th}}\text{LS} \\ (2r_j)^{\text{th}}\text{LS} \end{cases} (j = i, n-1, \ldots),\\
&(2\alpha_k)^{\text{th}}\text{LS} \xrightarrow[\textit{appearing}]{\alpha_k=2r_k-1} \begin{cases} (2r_k-1)^{\text{th}}\text{SI} \\ (2r_k-1)^{\text{th}}\text{SO} \end{cases} (k = 1, n, \ldots).
\end{aligned} \tag{7.202}$$

The simple equilibrium does not interact with the bifurcation points. There are four switching and appearing bifurcations of the higher-order equilibriums.

The two *broom-sprinkler-spraying-switching* bifurcations for the higher order singularity of equilibriums are

(iii) $((2(2r_1-1)^{\text{th}}\text{LS} : \ldots : (2(2r_i)+1)\text{SO} : \ldots : (2(2r_{n-1}))^{\text{th}}\text{US} : (2(2r_n-1))^{\text{th}}\text{US})$,
(iv) $((2(2r_1-1)^{\text{th}}\text{US} : \ldots : (2(2r_i)+1)\text{SI} : \ldots : (2(2r_{n-1}))^{\text{th}}\text{LS} : (2(2r_n-1))^{\text{th}}\text{LS})$,
(v) $((2(2r_1-1)^{\text{th}}\text{LS} : \ldots : (2(2r_i))\text{LS} : \ldots : (2(2r_{n-1}))^{\text{th}}\text{LS} : (2(2r_n-1)+1)^{\text{th}}\text{SO})$,
(vi) $((2(2r_1-1)^{\text{th}}\text{US} : \ldots : (2(2r_i))\text{US} : \ldots : (2(2r_{n-1}))^{\text{th}}\text{US} : (2(2r_n-1)+1)^{\text{th}}\text{SI})$,

as presented in Figs. 7.7(c, e) and (d, g) for $a_0 > 0$ and $a_0 < 0$, respectively. The $(2(2r_i)+1)^{\text{th}}$SO and $(2(2r_i)+1)^{\text{th}}$SI switching bifurcations are

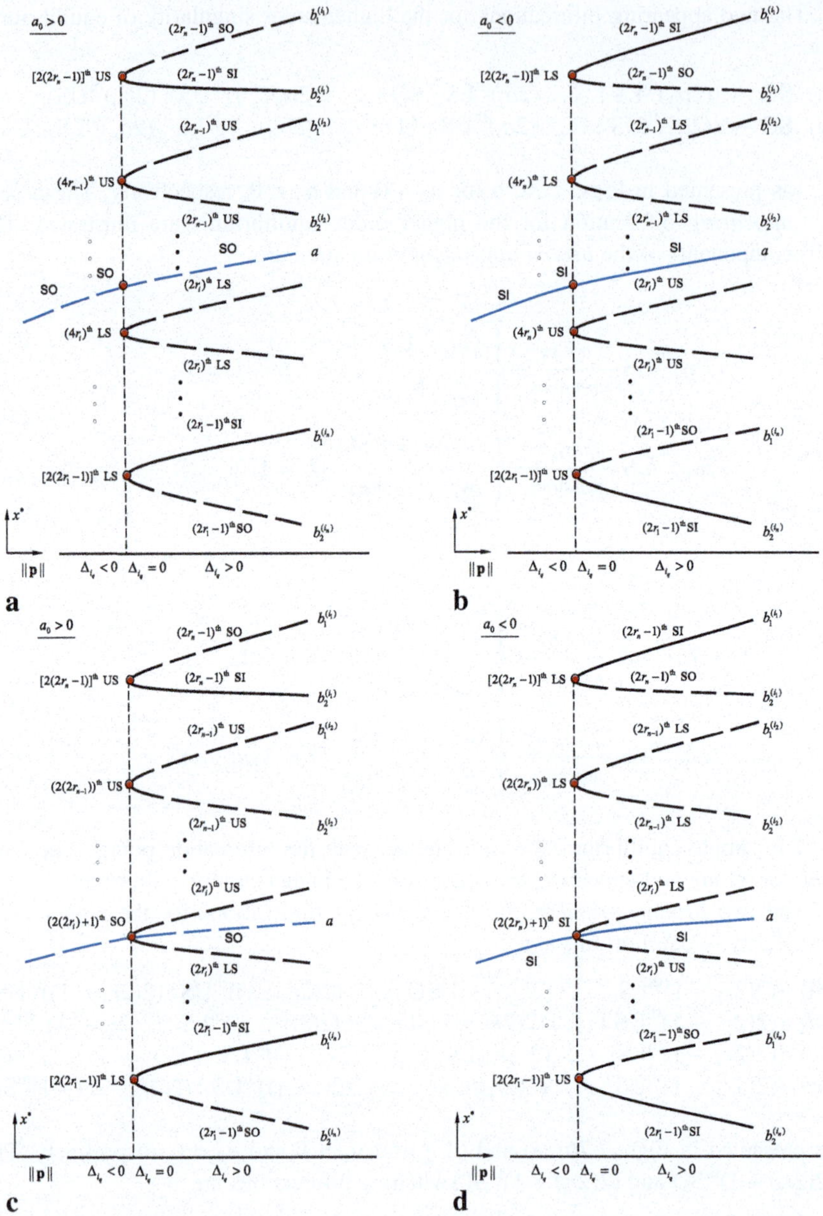

Fig. 7.7 Six bifurcations in a $(2m + 1)^{\text{th}}$-degree polynomial system. (**a**, **b**) Two broom appearing bifurcations. (**c**–**f**) Broom-switching bifurcations. LS: lower saddle, US: upper saddle, SI: sink, SO: source. Stable and unstable equilibriums are represented by solid and dashed curves, respectively. The bifurcation points are marked by circular symbols

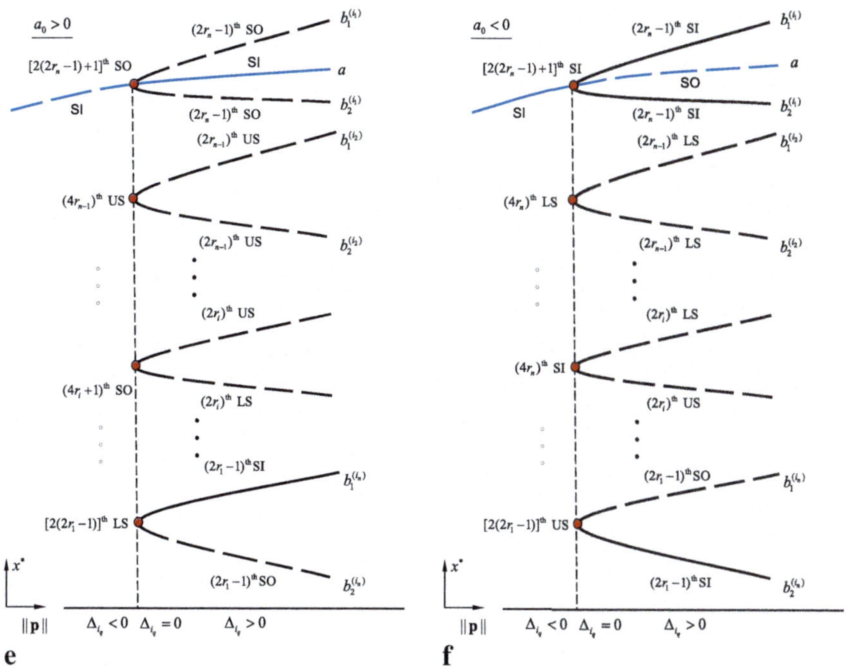

Fig. 7.7 (continued)

$$
\begin{aligned}
(2(2r_i)+1)^{\text{th}}\text{SO} &\xrightarrow[\text{appearing}]{\alpha_j=2r_j} \begin{cases} (2r_i)^{\text{th}}\text{US}, \\ \text{SO}, \\ (2r_i)^{\text{th}}\text{LS}, \end{cases} \\
(2(2r_i)+1)^{\text{th}}\text{SI} &\xrightarrow[\text{appearing}]{\alpha_j=2r_j} \begin{cases} (2r_i)^{\text{th}}\text{LS}, \\ \text{SI} \\ (2r_i)^{\text{th}}\text{US}, \end{cases}
\end{aligned}
\tag{7.203}
$$

and the $(2(2r_n-1)+1)^{\text{th}}$SO and $(2(2r_n-1)+1)^{\text{th}}$SI switching bifurcations are

$$
\begin{aligned}
(2(2r_n-1)+1)^{\text{th}}\text{SO} &\xrightarrow[\text{appearing}]{\alpha_j=2r_j} \begin{cases} (2r_n-1)^{\text{th}}\text{SO}, \\ \text{SI}, \\ (2r_n-1)^{\text{th}}\text{SO}, \end{cases} \\
(2(2r_n-1)+1)^{\text{th}}\text{SI} &\xrightarrow[\text{appearing}]{\alpha_j=2r_j} \begin{cases} (2r_n-1)^{\text{th}}\text{SI}, \\ \text{SO} \\ (2r_n-1)^{\text{th}}\text{SI}. \end{cases}
\end{aligned}
\tag{7.204}
$$

In Fig. 7.8a, b, the simple switching with two *sprinkler-spraying-appearing* bifurcations are presented for $a_0 > 0$ and $a_0 < 0$, respectively. The two bifurcation structures are

(i) $\text{SO} \to ((2\beta_1)^{\text{th}}\text{LS} : \ldots : (2\beta_i)^{\text{th}}\text{LS} : \text{SO} : \ldots : (2\beta_{n-1})^{\text{th}}\text{US} : (2\beta_n)^{\text{th}}\text{US})$,
(ii) $\text{SI} \to ((2\beta_1)^{\text{th}}\text{US} : \ldots : (2\beta_i)^{\text{th}}\text{US} : \text{SI} : \ldots : (2\beta_{n-1})^{\text{th}}\text{LS} : (2\beta_n)^{\text{th}}\text{LS})$,

where $m = \sum_{i=1}^{n}\beta_i$, $\beta_1 = (2l_1 - 1), \ldots, \beta_i = 2l_i, \ldots, \beta_{n-1} = 2l_{n-1}, \beta_n = 2l_n$. In Fig. 7.8 (c, e) and (d, f), the $(2l + 1)^{\text{th}}$-order *broom-switching* with two *sprinkler-spraying-appearing* bifurcations are presented for $a_0 > 0$ and $a_0 < 0$, respectively. The two bifurcation structures are:

(iii) $\text{SO} \to ((2\beta_1)^{\text{th}}\text{US} : \ldots : (2\beta_i + 1)^{\text{th}}\text{SO} : \ldots : (2\beta_{n-1})^{\text{th}}\text{US} : (2\beta_n)^{\text{th}}\text{US})$,
(iv) $\text{SI} \to ((2\beta_1)^{\text{th}}\text{LS} : \ldots : (2\beta_i + 1)^{\text{th}}\text{SI} : \ldots : (2\beta_{n-1})^{\text{th}}\text{LS} : (2\beta_n)^{\text{th}}\text{LS})$,
(v) $\text{SO} \to ((2\beta_1 + 1)^{\text{th}}\text{SO} : \ldots : (2\beta_i)^{\text{th}}\text{US} : \ldots : (2\beta_{n-1})^{\text{th}}\text{US} : (2\beta_n)^{\text{th}}\text{US})$,
(vi) $\text{SI} \to ((2\beta_1 + 1)^{\text{th}}\text{SI} : \ldots : (2\beta_i)^{\text{th}}\text{LS} : \ldots : (2\beta_{n-1})^{\text{th}}\text{LS} : (2\beta_n)^{\text{th}}\text{LS})$.

For a cluster of m-quadratics, $B_i = B_j$ $(i,j \in \{1,2,\ldots,m\}; i \neq j)$ and $\Delta_i = 0$ $(i \in \{1,2,\ldots,m\})$. The $(2m)^{\text{th}}$-order upper-saddle-node appearing bifurcation for higher-order equilibriums is

$$(2m)^{\text{th}} \text{ order US} \xrightarrow[\text{appearing bifurcation}]{\text{cluster of s-quadratics}} \begin{cases} (\alpha_{2s})^{\text{th}}\text{XX}, & \text{for } x^* = a_{2s}, \\ (\alpha_{2s-1})^{\text{th}}\text{XX}, & \text{for } x^* = a_{2s-1}, \\ \vdots \\ (\alpha_1)^{\text{th}}\text{XX}, & \text{for } x^* = a_1, \end{cases} \tag{7.205}$$

where $2m = \sum_{j=1}^{2s}\alpha_j$ and $m = (2l - 1), (2l)$:

$$\begin{aligned} (\alpha_1)^{\text{th}}\text{XX} &= \begin{cases} (2l_1)^{\text{th}}\text{LS}, & \text{for } \alpha_1 = 2l_1, \\ (2l_1 - 1)^{\text{th}}\text{SI}, & \text{for } \alpha_1 = 2l_1 - 1; \end{cases} \\ (\alpha_{2s})^{\text{th}}\text{XX} &= \begin{cases} (2l_{2s})^{\text{th}}\text{US}, & \text{for } \alpha_{2s} = 2l_{2s}, \\ (2l_{2s} - 1)^{\text{th}}\text{SI}, & \text{for } \alpha_{2s} = 2l_{2s} - 1. \end{cases} \end{aligned} \tag{7.206}$$

The $(2m)^{\text{th}}$-order lower-saddle-node appearing bifurcation for higher-order equilibriums is

$$(2m)^{\text{th}} \text{ order LS} \xrightarrow[\text{appearing bifurcation}]{\text{cluster of s-quadratics}} \begin{cases} (\alpha_{2s})^{\text{th}}\text{XX}, & \text{for } x^* = a_{2s}, \\ (\alpha_{2s-1})^{\text{th}}\text{XX}, & \text{for } x^* = a_{2s-1}; \\ \vdots \\ (\alpha_1)^{\text{th}}\text{XX}, & \text{for } x^* = a_1 \end{cases} \tag{7.207}$$

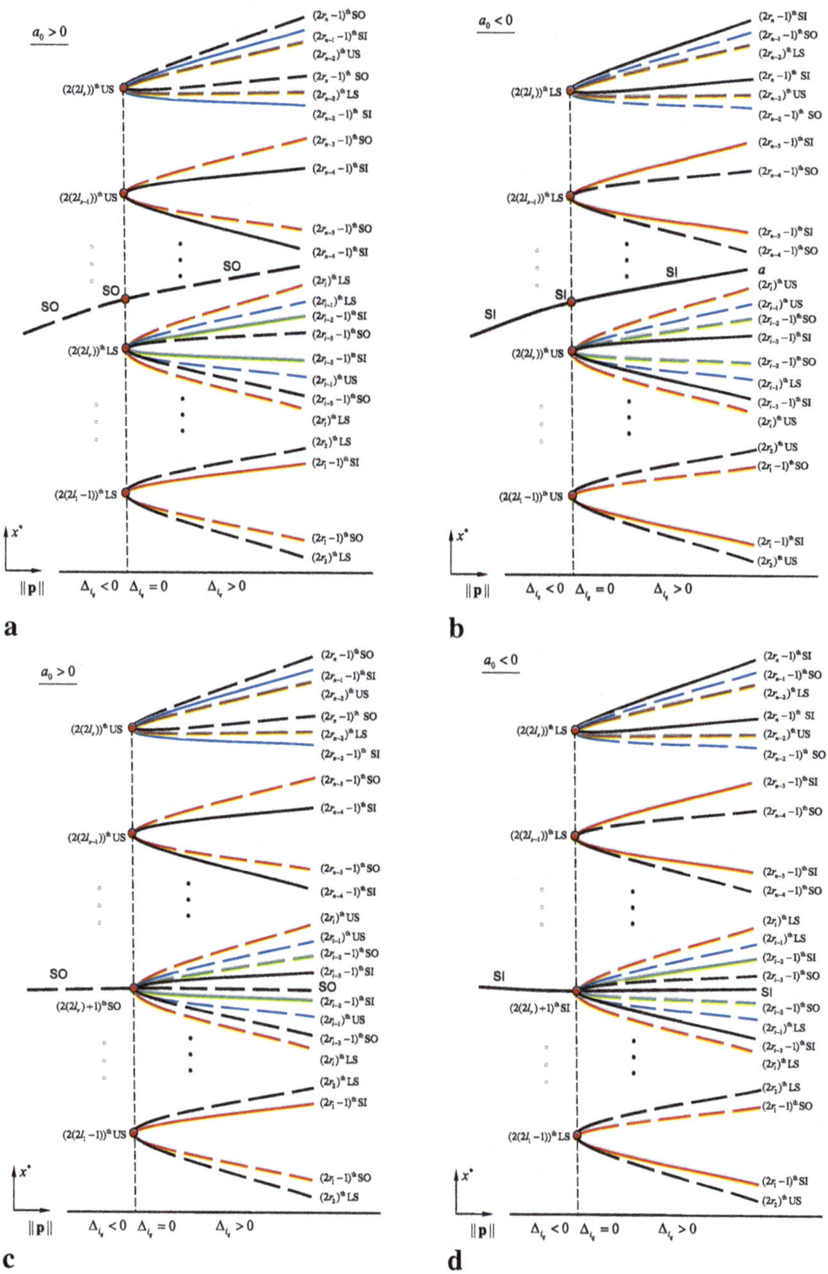

Fig. 7.8 Six types of bifurcations in a $(2m + 1)^{\text{th}}$-degree polynomial system. (**a**, **b**) Broom-sprinkler-spraying-appearing bifurcations, (**c–f**) broom-spraying-switching bifurcations with equilibrium clusters. LS: lower saddle, US: upper saddle, SI: sink, SO: source. Stable and unstable equilibriums are represented by solid and dashed curves, respectively. The bifurcation points are marked by circular symbols

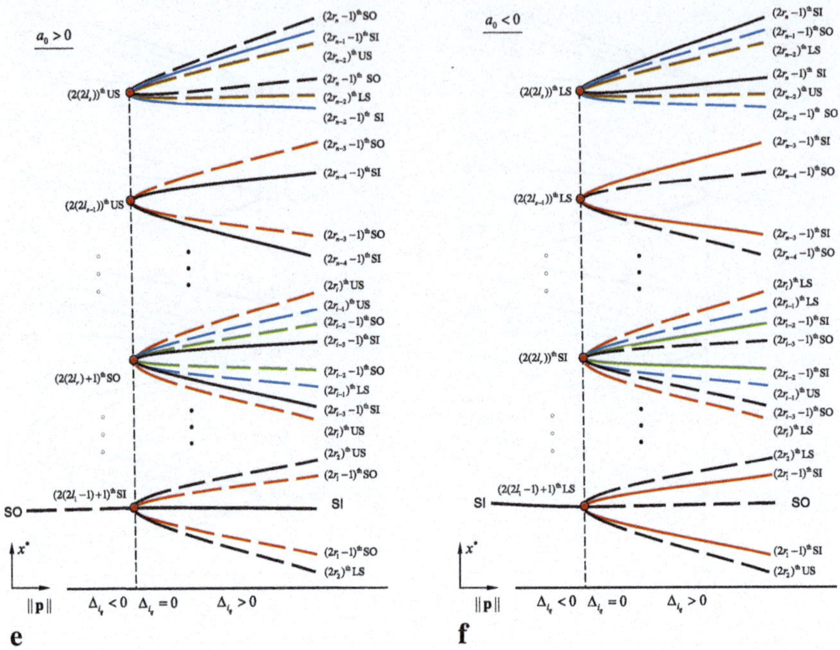

Fig. 7.8 (continued)

where

$$(\alpha_1)^{\text{th}}\text{XX} = \begin{cases} (2l_1)^{\text{th}}\text{US}, & \text{for } \alpha_1 = 2l_1 \\ (2l_1 - 1)^{\text{th}}\text{SO}, & \text{for } \alpha_1 = 2l_1 - 1 \end{cases}$$
$$(\alpha_{2s})^{\text{th}}\text{XX} = \begin{cases} (2l_{2s})^{\text{th}}\text{LS}, & \text{for } \alpha_{2s} = 2l_{2s} \\ (2l_{2s} - 1)^{\text{th}}\text{SI}, & \text{for } \alpha_{2s} = 2l_{2s} - 1 \end{cases} \tag{7.208}$$

There are four simple switching and $(2m)^{\text{th}}$-order saddle-node appearing bifurcations for higher order equilibriums: The two switching bifurcations of SO → $((2m)^{\text{th}}$US : SO) and SI → $((2m)^{\text{th}}$LS : SI) with two $(2m)^{\text{th}}$-order USN and LSN spraying-appearing bifurcations in the $(2m+1)^{\text{th}}$-degree polynomial system are

$$\text{SO}\ (x^* = a) \to \begin{cases} \text{SO} \to \text{SO, for } x^* = a = a_{2m+1}, \\ (2m)^{\text{th}} \text{ order LS} \end{cases} \tag{7.209}$$

$$\text{SI}\ (x^* = a) \to \begin{cases} \text{SI} \to \text{SI, for } x^* = a = a_{2m+1}, \\ (2m)^{\text{th}} \text{ order US} \end{cases} \tag{7.210}$$

and the two switching bifurcations of SO → (SO : $(2m)^{\text{th}}$US) and SI → (SI : $(2m)^{\text{th}}$LS) with two $(2m)^{\text{th}}$-order USN and LSN spraying-appearing bifurcations in the $(2m+1)^{\text{th}}$-degree polynomial system are

$$\text{SO}\,(x^* = a) \to \begin{cases} (2m)^{\text{th}} \text{ order US}, \\ \text{SO} \to \text{SO}, \text{for } x^* = a = a_1; \end{cases} \tag{7.211}$$

$$\text{SI}\,(x^* = a) \to \begin{cases} (2m)^{\text{th}} \text{ order LS}, \\ \text{SI} \to \text{SI}, \text{for } x^* = a = a_1. \end{cases} \tag{7.212}$$

The $(2m+1)^{\text{th}}$-order source broom-switching bifurcation for higher order equilibrium is

$$\text{SO}(x^* = a) \xrightarrow{\text{switching}} (2m+1)^{\text{th}} \text{ order SO} \begin{cases} (\alpha_{2s+1})^{\text{th}}\text{XX}, & \text{for } x^* = a_{2s+1}, \\ (\alpha_{2s})^{\text{th}}\text{XX}, & \text{for } x^* = a_{2s}, \\ \vdots \\ (\alpha_1)^{\text{th}}\text{XX}, \text{for } x^* = a_1 \end{cases} \tag{7.213}$$

where $2m+1 = \sum_{j=1}^{2s+1} \alpha_j$, $m = (2l-1), (2l)$ and

$$\begin{aligned} (\alpha_1)^{\text{th}}\text{XX} &= \begin{cases} (2l_1)^{\text{th}}\text{LS}, & \text{for } \alpha_1 = 2l_1, \\ (2l_1 - 1)^{\text{th}}\text{SI}, & \text{for } \alpha_1 = 2l_1 - 1; \end{cases} \\ (\alpha_{2s+1})^{\text{th}}\text{XX} &= \begin{cases} (2l_{2s+1})^{\text{th}}\text{US}, & \text{for } \alpha_{2s+1} = 2l_{2s+1}, \\ (2l_{2s+1} - 1)^{\text{th}}\text{SO}, & \text{for } \alpha_{2s+1} = 2l_{2s+1} - 1. \end{cases} \end{aligned} \tag{7.214}$$

The $(2m+1)^{\text{th}}$-order sink *broom-switching* bifurcation is

$$\text{SI}(x^* = a) \xrightarrow{\text{switching}} (2m+1)^{\text{th}} \text{ order SI} \begin{cases} (\alpha_{2s+1})^{\text{th}}\text{XX}, & \text{for } x^* = a_{2s+1}, \\ (\alpha_{2s})^{\text{th}}\text{XX}, & \text{for } x^* = a_{2s}, \\ \vdots \\ (\alpha_1)^{\text{th}}\text{XX}, \text{for } x^* = a_1 \end{cases} \tag{7.215}$$

where

$$
\begin{aligned}
&(\alpha_1)^{\text{th}}\text{XX} = \begin{cases} (2l_1)^{\text{th}}\text{US}, & \text{for } \alpha_1 = 2l_1, \\ (2l_1 - 1)^{\text{th}}\text{SO}, & \text{for } \alpha_1 = 2l_1 - 1; \end{cases} \\
&(\alpha_{2s+1})^{\text{th}}\text{XX} = \begin{cases} (2l_{2s+1})^{\text{th}}\text{LS}, & \text{for } \alpha_{2s+1} = 2l_{2s+1}, \\ (2l_{2s+1} - 1)^{\text{th}}\text{SI}, & \text{for } \alpha_{2s+1} = 2l_{2s+1} - 1. \end{cases}
\end{aligned}
\tag{7.216}
$$

The switching bifurcation consists of a simple switching and the $(2m)^{\text{th}}$-order saddle-node appearing bifurcation with m-pairs of source and sinks. The $(2m)^{\text{th}}$-order saddle-node appearing bifurcation is a *sprinkler-spraying cluster* of the m-pairs of sources and sinks. Thus, the four switching bifurcations of

SO → (SO : $(2m)^{\text{th}}$US) for higher order equilibriums for $a_0 > 0$,
SI → (SI : $(2m)^{\text{th}}$LS) for higher order equilibriums for $a_0 < 0$,
SO → ($(2m)^{\text{th}}$US : SO) for higher order equilibriums for $a_0 > 0$,
SI → ($(2m)^{\text{th}}$LS : SI) for higher order equilibrium for $a_0 < 0$

are presented in Fig. 7.9a–d, respectively. The $(2m + 1)^{\text{th}}$-order source switching bifurcation is named the $(2m + 1)^{\text{th}}$SO broom-sprinkle-spraying-switching bifurcation, and the $(2m + 1)^{\text{th}}$-order sink switching bifurcation is named the $(2m + 1)^{\text{th}}$SI broom-switching bifurcation. Such a $(2m + 1)^{\text{th}}$XX *broom-switching* bifurcation is from simple equilibrium to a $(2m + 1)^{\text{th}}$XX *broom*-switching bifurcation. The two broom-switching bifurcations for higher order equilibriums of

SO → $(2m + 1)^{\text{th}}$SO for higher order equilibrium for $a_0 > 0$,
SI → $(2m + 1))^{\text{th}}$SI for higher order equilibrium for $a_0 < 0$

are presented in Fig. 7.9e, f, respectively.

A series of the $(2\alpha_i + 1)^{\text{th}}$-order source and sink bifurcations are aligned up with varying parameters, which is formed in a special pattern. Such a special pattern is from m-quadratics in the $(2m + 1)^{\text{th}}$-degree polynomial system; the following conditions should be satisfied:

$$
\begin{aligned}
&a(\mathbf{p}_i) = -\frac{1}{2}B_i \text{ and } a(\mathbf{p}_j) = -\frac{1}{2}B_j \\
&B_i \approx B_j \ i,j \in \{1,2,\ldots,n\}; i \neq j, \\
&\Delta_i > \Delta_{i+1} \ (i = 1,2,\ldots,n; n \leq m), \\
&\Delta_i = 0 \text{ with } \|\mathbf{p}_i\| < \|\mathbf{p}_{i+1}\|.
\end{aligned}
\tag{7.217}
$$

Four series of *switching* bifurcations in the $(2m + 1)^{\text{th}}$-degree polynomial nonlinear system are

(i) $(2(2r_1 - 1) + 1)^{\text{th}}$SO - $(2(2r_2) + 1)^{\text{th}}$SI - … - $(2(2r_n - 1) + 1)^{\text{th}}$SO),
(ii) $(2(2r_1 - 1) + 1)^{\text{th}}$SI - $(2(2r_2) + 1)^{\text{th}}$SO - … - $(2(2r_n - 1) + 1)^{\text{th}}$SI),
(iii) $(2(2r_1) + 1)^{\text{th}}$SO - $(2(2r_2 - 1) + 1)^{\text{th}}$SO - … - $(2(2r_n - 1))^{\text{th}}$SO),
(iv) $(2(2r_1) + 1)^{\text{th}}$SI - $(2(2r_2 - 1) + 1)^{\text{th}}$SI - … - $(2(2r_n - 1) + 1)^{\text{th}}$SI),

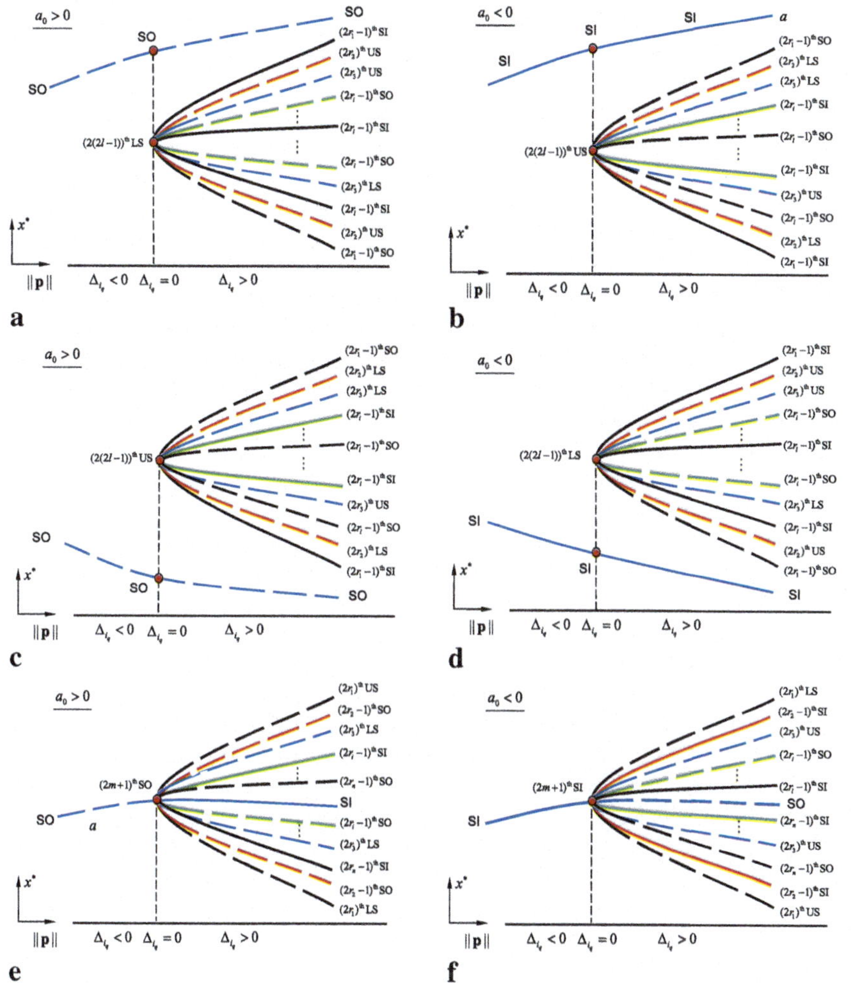

Fig. 7.9 Broom-switching bifurcations of equilibriums in $(2m + 1)^{\text{th}}$ polynomial system: (**a**) (SO : $(2m)^{\text{th}}$US)-appearing bifurcation ($a_0 > 0$), (**b**) (SI : $(2m)^{\text{th}}$US)-appearing bifurcation ($a_0 < 0$), (**c**) ($(2m)^{\text{th}}$US, SO)-appearing bifurcation ($a_0 > 0$), (**d**) ($(2m)^{\text{th}}$LS : SI)-appearing bifurcation ($a_0 < 0$). (**e**) $(2m + 1)^{\text{th}}$SO-switching-appearing bifurcation ($a_0 > 0$), (**f**) $(2m + 1)^{\text{th}}$SO-switching-appearing bifurcation ($a_0 < 0$). LS: lower saddle, US: upper saddle, SI: sink, SO: source. Stable and unstable equilibriums are represented by solid and dashed curves, respectively. The bifurcation points are marked by circular symbols

as presented in Fig. 7.10 (a, c)–(b, f) for ($a_0 > 0$) and ($a_0 < 0$), respectively. The swapping pattern of higher order sinks and sources switching bifurcations cannot be observed. Such a bifurcation scenario is like the fish bone for the higher order switching bifurcations for higher order equilibriums.

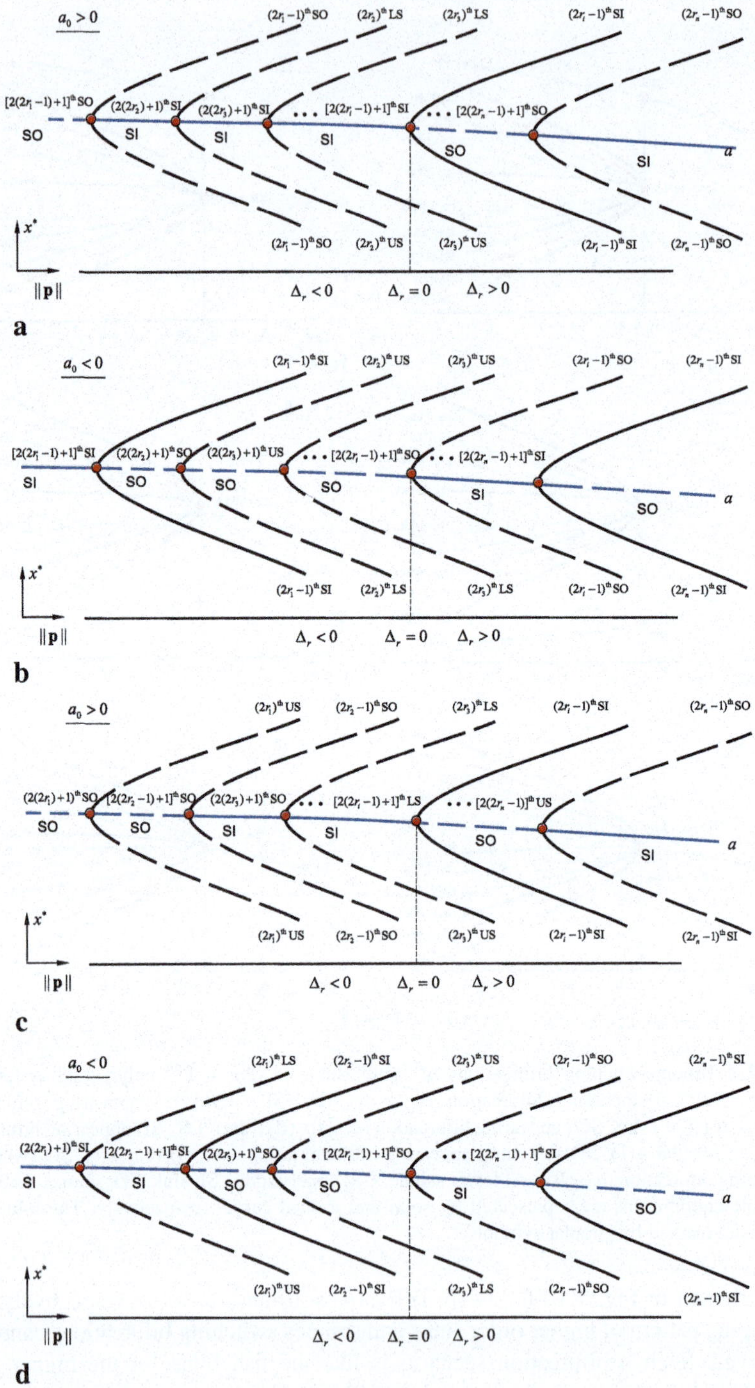

Fig. 7.10 Four series of switching bifurcations of equilibriums in a $(2m + 1)^{\text{th}}$ polynomial system: (**a**, **c**) for $a_0 > 0$, (**b**, **d**) for $a_0 < 0$. LS: lower saddle, US: upper saddle, SI: sink, SO: source. Stable and unstable equilibriums are represented by solid and dashed curves, respectively. The bifurcation points are marked by circular symbols

7.3.2 Switching Bifurcations

Consider the roots of $(x^2 + B_i x + C_i)^{\alpha_i} = 0$ as

$$\left.\begin{aligned} &B_i = -(b_1^{(i)} + b_2^{(i)}), \Delta_i = (b_1^{(i)} - b_2^{(i)})^2 \geq 0, \\ &x_{1,2}^{(i)} = b_{1,2}^{(i)}, \Delta_i > 0 \text{ if } \mathrm{b}_1^{(i)} \neq b_2^{(i)}\ (i = 1, 2, \ldots, n); \\ &\left.\begin{aligned} &B_i \neq B_j\ (i, j = 1, 2, \ldots, n; i \neq j) \\ &\Delta_i = 0 \text{ at } b_1^{(i)} = b_2^{(i)}\ (i = 1, 2, \ldots, n) \end{aligned}\right\} \text{at bifurcation.} \end{aligned}\right. \tag{7.218}$$

The $(\alpha_i)^{\text{th}}$-order singularity bifurcation is for the *switching* of a pair of higher order equilibriums (i.e., sinks, sources, upper saddles, and lower saddles). There are six *switching* bifurcations for $i \in \{1, 2, \ldots, n\}$:

$$(2l_i)^{\text{th}}\,\text{order US} \xrightarrow[\text{switching bifurcation}]{l_i = r_1^{(i)} + r_2^{(i)} - 1} \begin{cases} (2r_2^{(i)} - 1)^{\text{th}} \text{ order SO} \downarrow \text{SI, for } b_2^{(i)} = a_{2i} \downarrow a_{2i-1}, \\ (2r_1^{(i)} - 1)^{\text{th}} \text{ order SI} \uparrow \text{SO, for } b_1^{(i)} = a_{2i-1} \uparrow a_{2i}; \end{cases} \tag{7.219}$$

$$(2l_i)^{\text{th}}\,\text{order LS} \xrightarrow[\text{switching bifurcation}]{l_i = r_1^{(i)} + r_2^{(i)} - 1} \begin{cases} (2r_2^{(i)} - 1)^{\text{th}} \text{ order SI} \downarrow \text{SO, for } b_2^{(i)} = a_{2i} \downarrow a_{2i-1}, \\ (2r_1^{(i)} - 1)^{\text{th}} \text{ order SO} \uparrow \text{SI, for } b_1^{(i)} = a_{2i-1} \uparrow a_{2i}; \end{cases} \tag{7.220}$$

$$(2l_i)^{\text{th}}\,\text{order US} \xrightarrow[\text{switching bifurcation}]{l_i = r_1^{(i)} + r_2^{(i)}} \begin{cases} (2r_2^{(i)})^{\text{th}} \text{ order US} \downarrow \text{US, for } b_2^{(i)} = a_{2i} \downarrow a_{2i-1}, \\ (2r_1^{(i)})^{\text{th}} \text{ order US} \uparrow \text{US, for } \mathrm{b}_1^{(i)} = a_{2i-1} \uparrow a_{2i}; \end{cases} \tag{7.221}$$

$$(2l_i)^{\text{th}}\,\text{order LS} \xrightarrow[\text{switching bifurcation}]{l_i = r_1^{(i)} + r_2^{(i)}} \begin{cases} (2r_2^{(i)})^{\text{th}} \text{ order LS} \downarrow \text{LS, for } b_2^{(i)} = a_{2i} \downarrow a_{2i-1}, \\ (2r_1^{(i)})^{\text{th}} \text{ order LS} \uparrow \text{LS, for } \mathrm{b}_1^{(i)} = a_{2i-1} \uparrow a_{2i}; \end{cases} \tag{7.222}$$

$$
\begin{gathered}
(2l_i-1)^{\text{th}}\,\text{order SO}\xrightarrow[\text{switching bifurcation}]{l_i=r_1^{(i)}+r_2^{(i)}}\\
\begin{cases}
(2r_2^{(i)}-1)^{\text{th}}\,\text{order SO}\downarrow\text{SO, for } b_2^{(i)}=a_{2i}\downarrow a_{2i-1},\\
(2r_1^{(i)})^{\text{th}}\,\text{order LS}\uparrow\text{US for } b_1^{(i)}=a_{2i-1}\uparrow a_{2i};
\end{cases}
\end{gathered}
\tag{7.223}
$$

$$
\begin{gathered}
(2l_i-1)^{\text{th}}\,\text{order SI}\xrightarrow[\text{switching bifurcation}]{l_i=r_1^{(i)}+r_2^{(i)}}\\
\begin{cases}
(2r_2^{(i)}-1)^{\text{th}}\,\text{order SI}\downarrow\text{SI, for } b_2^{(i)}=a_{2i}\downarrow a_{2i-1},\\
(2r_1^{(i)})^{\text{th}}\,\text{order US}\uparrow\text{LS for } b_1^{(i)}=a_{2i-1}\uparrow a_{2i}.
\end{cases}
\end{gathered}
\tag{7.224}
$$

A set of n-paralleled higher order XX *switching* bifurcations is called the $((\alpha_1)^{\text{th}}\text{XX} : (\alpha_2)^{\text{th}}\text{XX} : \ldots : (\alpha_n)^{\text{th}}\text{XX})$ *parallel switching* bifurcation in the $(2m + 1)$th-degree polynomial nonlinear system. Such a bifurcation is also called the $((\alpha_1)^{\text{th}}\text{XX} : (\alpha_2)^{\text{th}}\text{XX} : \ldots : (\alpha_n)^{\text{th}}\text{XX})$ *antenna-switching* bifurcation. $\alpha_i \in \{2l_i, 2l_i - 1\}$ and $\text{XX} \in \{\text{SO, SI, US, LS}\}$. For non-switching points, $\Delta_i > 0$ at $b_1^{(i)} \neq b_2^{(i)}$ $(i = 1, 2, \ldots, n)$. At the bifurcation point, $\Delta_i = 0$ at $b_1^{(i)} = b_2^{(i)}$ $(i = 1, 2, \ldots, n)$. The *parallel antenna-switching* bifurcation for higher order equilibriums in the $(2m + 1)$th-degree polynomial system is

$$
\begin{cases}
((\alpha_1^{(2)})^{\text{th}}\text{XX}_1 : \ldots : (\alpha_{l_2}^{(2)})^{\text{th}}\text{XX}_{l_2})\\
\text{SI (or SO), for } x^* = a\\
((\alpha_1^{(1)})^{\text{th}}\text{XX}_1 : \ldots : (\alpha_{l_1}^{(1)})^{\text{th}}\text{XX}_{l_1})
\end{cases}
\tag{7.225}
$$

where

$$
\begin{gathered}
(\alpha_{s_i}^{(i)})^{\text{th}}\text{XX}_{s_i}\xrightarrow[\text{switching}]{s_i{}^{\text{th}}\text{bifurcation}}\\
\begin{cases}
(r_2^{(s_i)})^{\text{th}}\text{XX}_2^{(s_i)}\downarrow\text{YY}_1^{(s_i)},\ \text{for } b_2^{(s_i)}=a_{2s_i}^{(i)}\downarrow a_{2s_i-1}^{(i)},\\
(r_1^{(s_i)})^{\text{th}}\text{XX}_1^{(s_i)}\uparrow\text{YY}_2^{(s_i)},\ \text{for } b_1^{(s_i)}=a_{2s_i-1}^{(i)}\uparrow a_{2s_i}^{(i)};
\end{cases}\\
(s_i=1,2,\ldots,l_i,\ i=1,2).
\end{gathered}
\tag{7.226}
$$

Such eight sets of parallel *switching* bifurcations for higher order equilibrium are presented in Fig. 7.11(a, c, e, g) and (b, d, f, h) for $a_0 > 0$ and $a_0 < 0$, respectively. The eight switching bifurcation structures are as follows:

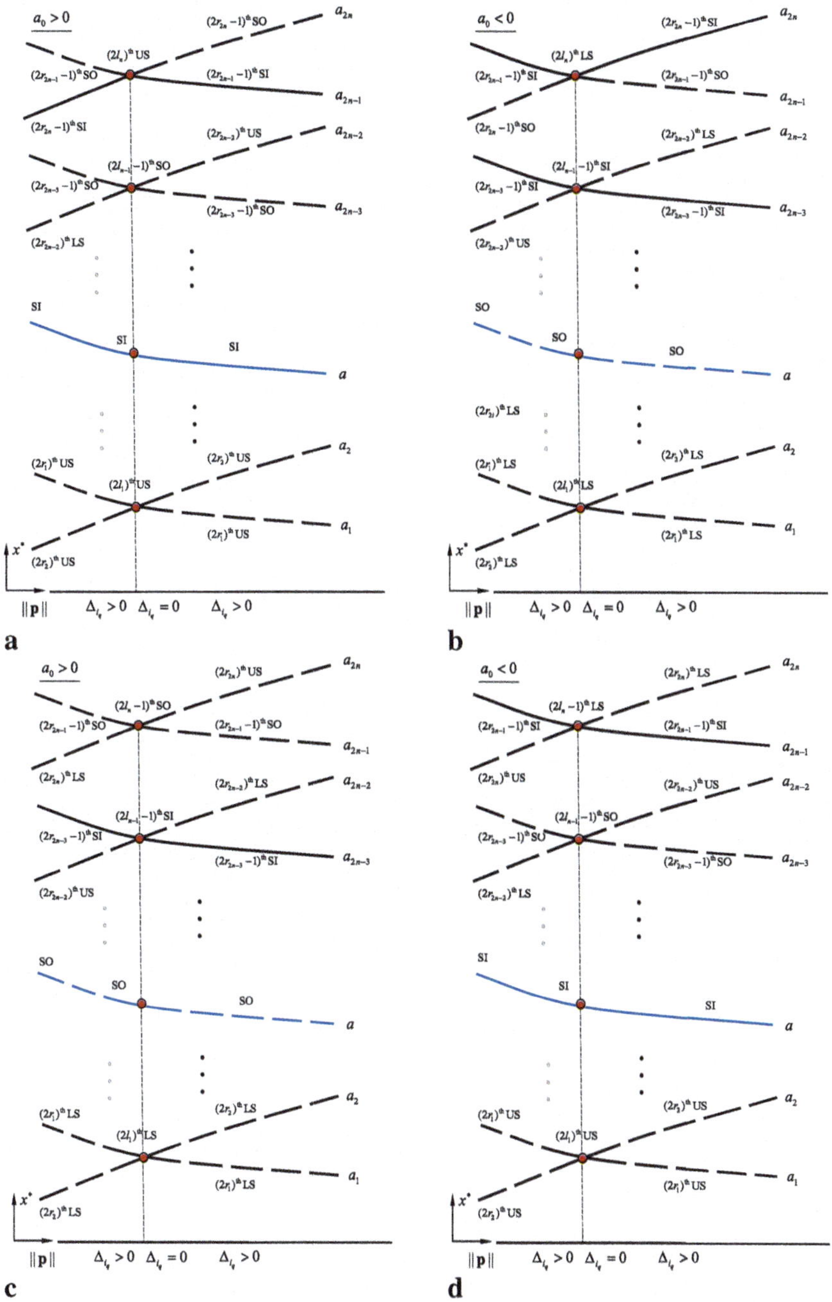

Fig. 7.11 Antenna parallel switching bifurcation of equilibriums for a $(2m + 1)^{\text{th}}$-degree polynomial nonlinear dynamical system. (**a**, **c**, **e**, **g**) Four parallel bifurcations for $a_0 > 0$. (**b**, **d**, **f**, **h**) Four parallel bifurcations for $a_0 < 0$. LS: lower saddle, US: upper saddle, SI: sink, SO: source. Stable and unstable equilibriums are represented by solid and dashed curves, respectively. The bifurcation points are marked by circular symbols

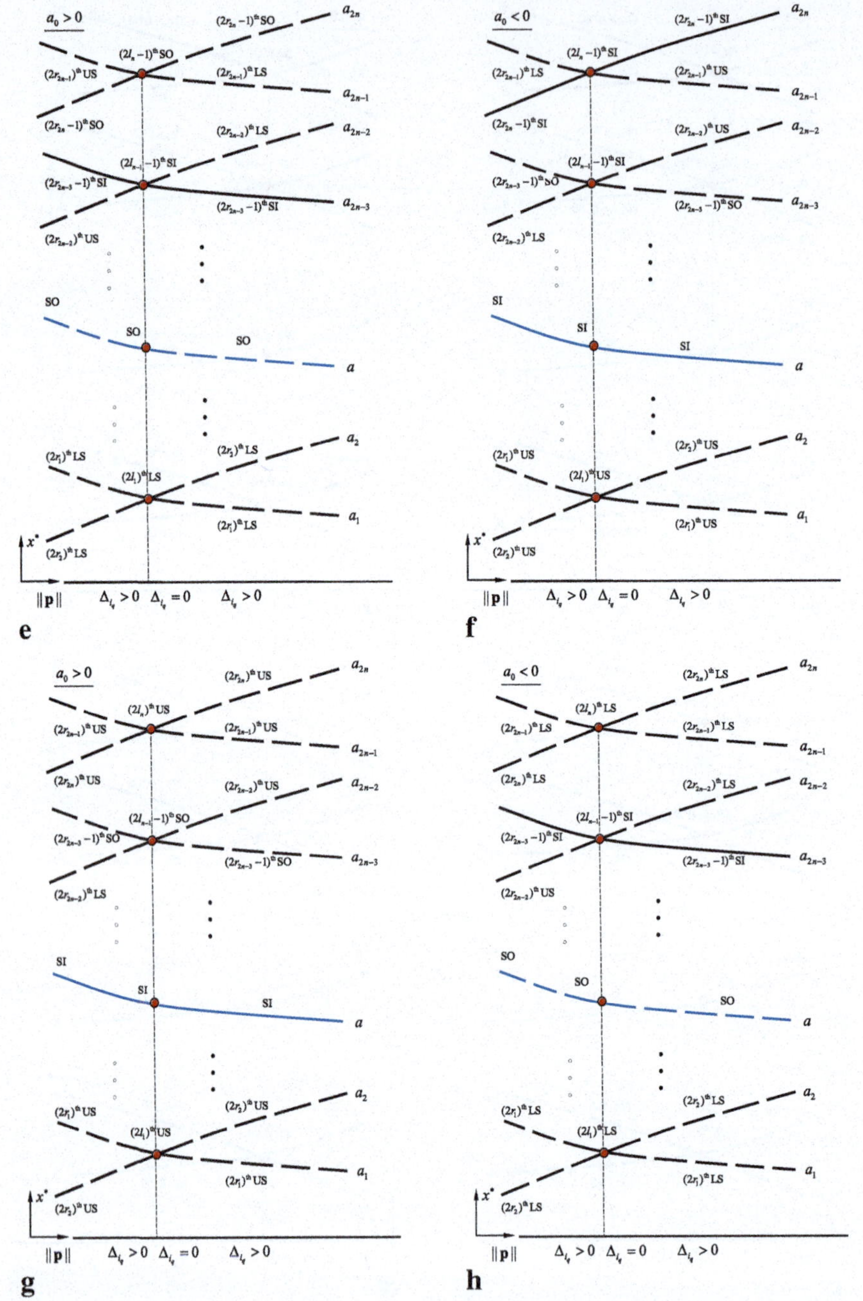

Fig. 7.11 (continued)

(i) $((2l_1)^{\text{th}}\text{US} : \ldots : \text{SI} : \ldots : (2l_{n-1}-1)^{\text{th}}\text{SO} : (2l_n)^{\text{th}}\text{US})$ for $a_0 > 0$,
(ii) $((2l_1)^{\text{th}}\text{LS} : \ldots : \text{SO} : \ldots : (2l_{n-1}-1)^{\text{th}}\text{SI} : (2l_n)^{\text{th}}\text{LS})$ for $a_0 < 0$,
(iii) $((2l_1)^{\text{th}}\text{LS} : \ldots : \text{SO} : \ldots : (2l_{n-1}-1)^{\text{th}}\text{SI} : (2l_n-1)^{\text{th}}\text{SO})$ for $a_0 > 0$,
(iv) $((2l_1)^{\text{th}}\text{US} : \ldots : \text{SI} : \ldots : (2l_{n-1}-1)^{\text{th}}\text{SO} : (2l_n-1)^{\text{th}}\text{SI})$ for $a_0 < 0$,
(v) $((2l_1)^{\text{th}}\text{LS} : \ldots : \text{SO} : \ldots : (2l_{n-1}-1)^{\text{th}}\text{SI} : (2l_n-1)^{\text{th}}\text{SO})$ for $a_0 > 0$,
(vi) $((2l_1)^{\text{th}}\text{US} : \ldots : \text{SI} : \ldots : (2l_{n-1}-1)^{\text{th}}\text{SI} : (2l_n-1)^{\text{th}}\text{SI})$ for $a_0 < 0$,
(vii) $((2l_1)^{\text{th}}\text{US} : \ldots : \text{SI} : \ldots : (2l_{n-1}-1)^{\text{th}}\text{SO} : (2l_n)^{\text{th}}\text{US})$ for $a_0 > 0$,
(viii) $((2l_1)^{\text{th}}\text{LS} : \ldots : \text{SO} : \ldots : (2l_{n-1}-1)^{\text{th}}\text{SI} : (2l_n)^{\text{th}}\text{LS})$ for $a_0 < 0$.

The switching bifurcations with different higher order equilibriums are similar to the (l_1 - LSN : SO : l_2 - USN) and (l_1 - USN : SI : l_2 - LSN) switching bifurcations for simple sinks and sources.

Consider a switching bifurcation for a cluster of higher order equilibriums with the following conditions:

$$\left.\begin{array}{l} B_i = B_j\,(i,j \in \{1,2,\ldots,n\}; i \neq j) \\ \Delta_i = 0\,\text{at}\,b_1^{(i)} = b_2^{(i)}\ (i = 1,2,\ldots,n) \end{array}\right\}\text{at bifurcation.} \tag{7.227}$$

Thus, the $(2l)^{\text{th}}$-order *switching* bifurcation can be for a cluster of higher order equilibriums. Two $(2l)^{\text{th}}$-order switching bifurcations for $l \in \{1,2,\ldots,s\}$ are

$$\begin{array}{l} (\alpha_i)^{\text{th}}\,\text{order}\,\text{XX} \xrightarrow[\text{switching bifurcation}]{\alpha_i=\sum_{j=1}^{s} r_j^{(i)}} \\ \left\{\begin{array}{l} (r_s^{(i)})^{\text{th}}\text{order}\,\text{XX}_s^{(i)} \downarrow \text{YY}_s^{(i)},\,\text{for}\,\text{b}_s^{(i)} \downarrow a_s^{(i)} \\ \vdots \\ (r_j^{(i)})^{\text{th}}\,\text{order}\,\text{XX}_j^{(i)} \downarrow \text{YY}_j^{(i)},\,\text{for}\,b_j^{(i)} \downarrow a_s^{(i)} \\ \vdots \\ (r_1^{(i)})^{\text{th}}\,\text{order}\,\text{XX}_1^{(i)} \uparrow \text{YY}_1^{(i)},\,\text{for}\,b_1^{(i)} \downarrow a_s^{(i)} \end{array}\right. \end{array} \tag{7.228}$$

where

$$\begin{array}{l} \{a_{2s_1-1}, a_{2s_1}, \ldots, a_{2s_l-1}, a_{2s_l}\} \underset{\text{before bifurcation}}{=} \text{sort}\{b_1^{(s_1)}, b_2^{(s_1)}, \ldots, b_1^{(s_l)}, b_2^{(s_l)}\}, \\ \{b_{2s_1-1}, b_{2s_1}, \ldots, b_{2s_l-1}, b_{2s_l}\} \underset{\text{after bifurcation}}{=} \text{sort}\{b_1^{(s_1)}, b_2^{(s_1)}, \ldots, b_1^{(s_l)}, b_2^{(s_l)}\}. \end{array} \tag{7.229}$$

A set of paralleled, different, higher order upper-saddle-node switching bifurcations with multiplicity is the $((\alpha_1)^{\text{th}}\text{XX} : (\alpha_2)^{\text{th}}\text{XX} : \ldots : (\alpha_s)^{\text{th}}\text{XX})$ *parallel switching* bifurcation in the $(2m+1)^{\text{th}}$-degree polynomial system. At the *straw-bundle switching* bifurcation, $\Delta_i = 0$ $(i = 1,2,\ldots,n)$ and $B_i = B_j$ $(i,j \in \{1,2,\ldots,n\}; i \neq j)$. The *parallel straw-bundle switching* bifurcation for higher order equilibriums is

$$
\begin{aligned}
&((\alpha_1)^{\text{th}}\text{XX} : (\alpha_2)^{\text{th}}\text{XX} : \ldots : (\alpha_s)^{\text{th}}\text{XX})\ \text{switching} \\
&= \begin{cases} (\alpha_s)^{\text{th}}\ \text{order XX switching}, \\ \vdots \\ (\alpha_2)^{\text{th}}\ \text{order XX switching}, \\ (\alpha_1)^{\text{th}}\ \text{order XX switching} \end{cases}
\end{aligned} \tag{7.230}
$$

where

$$
\alpha_i \in \{2l_i, 2l_i - 1\}\ \text{and}\ \text{XX} \in \{\text{US}, \text{LS}, \text{SO}, \text{SI}\}. \tag{7.231}
$$

Thus,

$$
\begin{cases} ((\alpha_1^{(2)})^{\text{th}}\text{XX} : (\alpha_2^{(2)})^{\text{th}}\text{XX} : \ldots : (\alpha_{s_2}^{(2)})^{\text{th}}\text{XX}) \\ \text{SI (or SO)} \\ ((\alpha_1^{(1)})^{\text{th}}\text{XX} : (\alpha_2^{(1)})^{\text{th}}\text{XX} : \ldots : (\alpha_{s_1}^{(1)})^{\text{th}}\text{XX}). \end{cases} \tag{7.232}
$$

Eight *parallel straw-bundle switching* bifurcations of $((\alpha_1)^{\text{th}}\text{XX} : (\alpha_2)^{\text{th}}\text{XX} : \ldots : (\alpha_n)^{\text{th}}\text{XX})$ are presented in Figs. 7.12 and 7.13 for $a_0 > 0$ and $a_0 < 0$, respectively.

7.3.3 Switching and Appearing Bifurcations

Consider a $(2m + 1)^{\text{th}}$-degree 1-dimensional polynomial system in a form of

$$
\dot{x} = a_0 Q(x) \prod_{i=1}^{n_1} (x - c_i)^{\alpha_i} \prod_{j=1}^{n_2} (x^2 + B_j x + C_j)^{\alpha_j}. \tag{7.233}
$$

where $\sum_{i=1}^{n_1} \alpha_i = 2s_1 + 1$. Without loss of generality, a function of $Q(x) > 0$ is either a polynomial function or a non-polynomial function. The roots of $x^2 + B_j x + C_j = 0$ are

$$
b_{1,2}^{(j)} = -\frac{1}{2}B_j \pm \frac{1}{2}\sqrt{\Delta_j}, \Delta_j = B_j^2 - 4C_j \geq 0\ (j = 1, 2, \ldots, n_2); \tag{7.234}
$$

either

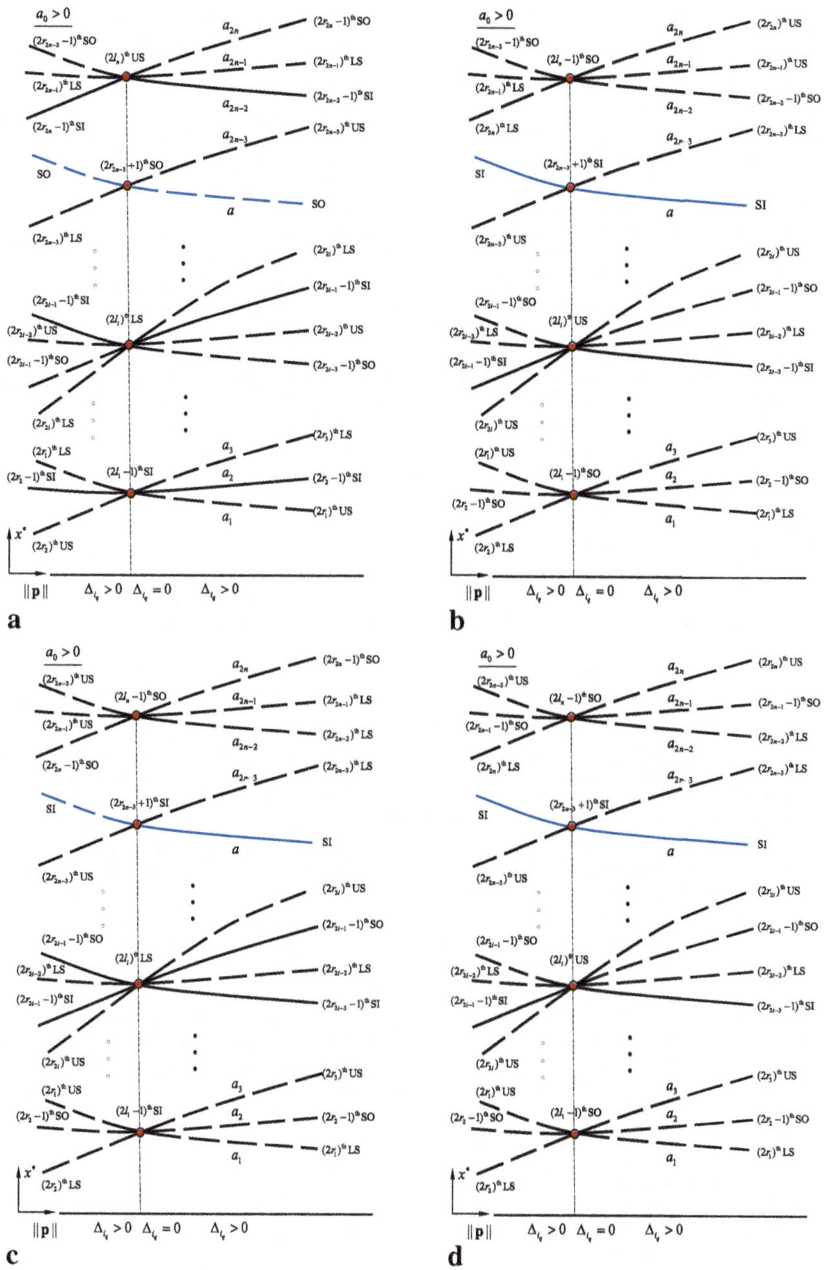

Fig. 7.12 **(a–d)** Four types of (r_1th XX : r_2th XX : . . . : r_nth XX) parallel switching bifurcation for $a_0 > 0$ in the $(2m + 1)^{\text{th}}$-degree polynomial system. LS: lower saddle, US: upper saddle, SI: sink, SO: source. Stable and unstable equilibriums are represented by solid and dashed curves, respectively. The bifurcation points are marked by circular symbols

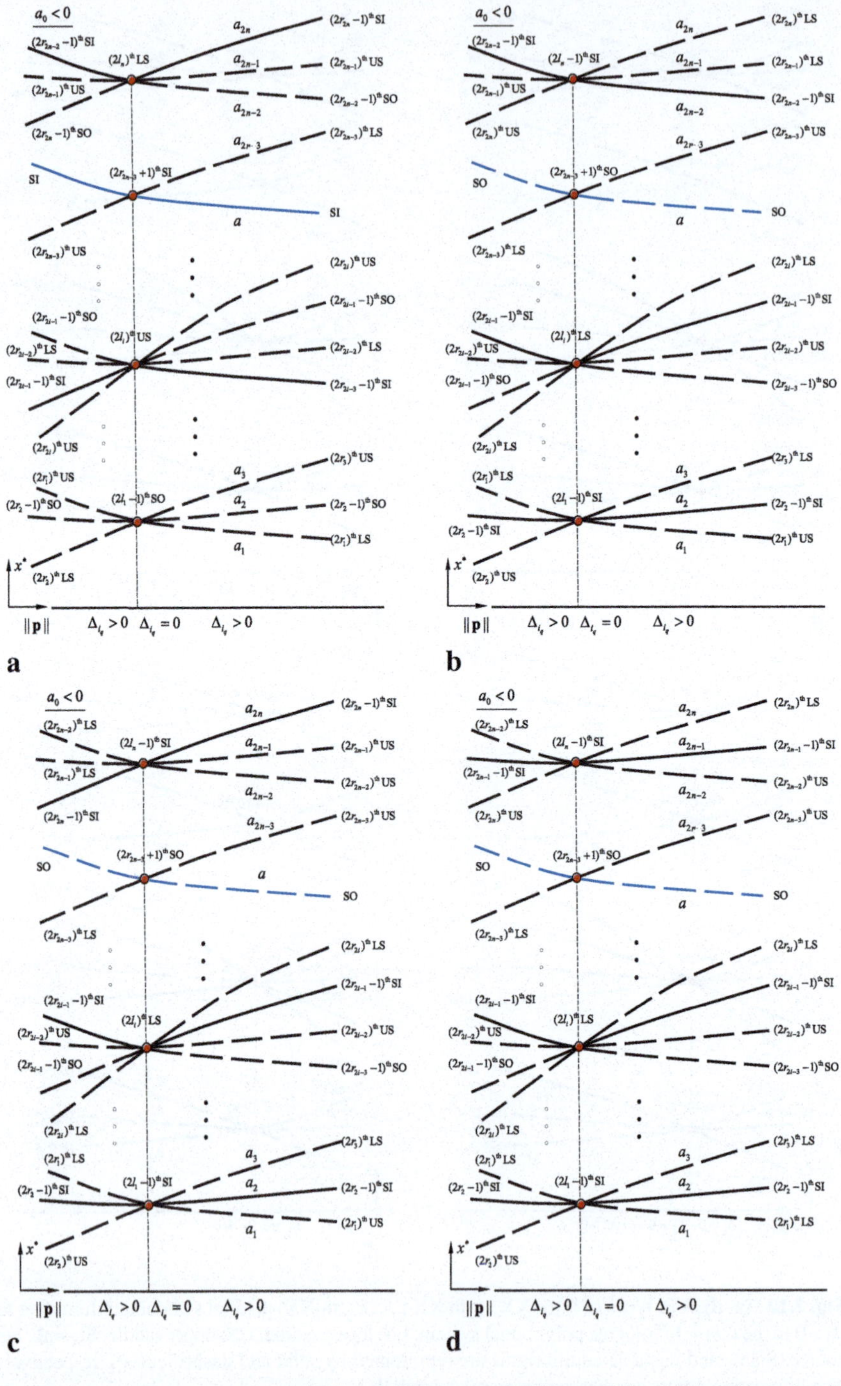

Fig. 7.13 **(a–d)** Four types of (r_1th XX : r_2th XX : . . . : r_nth XX) parallel switching bifurcation for $a_0 < 0$ in the $(2m+1)^{\text{th}}$-degree polynomial system. LS: lower saddle, US: upper saddle, SI: sink, SO: source. Stable and unstable equilibriums are represented by solid and dashed curves, respectively. The bifurcation points are marked by circular symbols

$$
\begin{aligned}
&\{a_1^-, a_2^-, \ldots, a_{2n_1+1}^-\} = \mathrm{sort}\{c_1, c_2 \ldots, c_{2n_1}, a\}, a_s^- \leq a_{s+1}^- \text{ before bifurcation} \\
&\{a_1^+, a_2^+, \ldots, a_{2n_3+1}^+\} = \mathrm{sort}\{c_1, \ldots, c_{2n_1}, a; b_1^{(1)}, b_2^{(1)}, \ldots, b_1^{(n_2)}, b_2^{(n_2)}\}, \\
&a_s^+ \leq a_{s+1}^+, n_3 = n_1 + n_2 \text{ after bifurcation}
\end{aligned}
\tag{7.235}
$$

or

$$
\begin{aligned}
&\{a_1^-, a_2^-, \ldots, a_{2n_3+1}^-\} = \mathrm{sort}\{c_1, c_2 \ldots, c_{2n_1}, a; b_1^{(1)}, b_2^{(1)}, \ldots, b_1^{(n_2)}, b_2^{(n_2)}\}, \\
&a_s^- \leq a_{s+1}^-, n_3 = n_1 + n_2 \text{ before bifurcation;} \\
&\{a_1^+, a_2^+, \ldots, a_{2n_1+1}^+\} = \mathrm{sort}\{c_1, \ldots, c_{2n_1}, a\}, a_s^+ \leq a_{s+1}^+ \text{ after bifurcation;}
\end{aligned}
\tag{7.236}
$$

and

$$
\left.
\begin{aligned}
&B_{j_1} = B_{j_2} = \ldots = B_{j_s} \,(j_{k_1} \in \{1, 2, \ldots, n\}; j_{k_1} \neq j_{k_2}) \\
&(k_1, k_2 \in \{1, 2, \ldots, s\}; k_1 \neq k_2) \\
&\Delta_j = 0 \,(j \in U \subset \{1, 2, \ldots, n_2\}) \\
&c_i \neq -\frac{1}{2} B_j (i = 1, 2, \ldots, 2n_1, j = 1, 2, \ldots, n_2)
\end{aligned}
\right\} \text{at bifurcation.}
\tag{7.237}
$$

Consider a just before bifurcation of $((\beta_1^-)^{\text{th}}\mathrm{XX}_1^- : (\beta_2^-)^{\text{th}}\mathrm{XX}_2^- : \ldots : (\beta_{s_1}^-)^{\text{th}}\mathrm{XX}_{s_1}^-)$ for higher order equilibriums. For $\beta_i^- = 2l_i^- - 1$, $\mathrm{XX}_i^- \in \{\mathrm{SO}, \mathrm{SI}\}$ and for $\alpha_i^- = 2l_i^-$, $\mathrm{XX}_i^- \in \{\mathrm{US}, \mathrm{LS}\}$ $(i = 1, 2, \ldots, s_1)$. The detailed structures are as follows:

$$
\left.
\begin{aligned}
&(r_s^{(i)-})^{\text{th}} \text{ order } \mathrm{XX}_s^{(i)-}, x^* = a_{k_i}^{(i)-}, \\
&\vdots \\
&(r_j^{(i)-})^{\text{th}} \text{ order } \mathrm{XX}_j^{(i)-}, x^* = a_j^{(i)-} \\
&\vdots \\
&(r_1^{(i)-})^{\text{th}} \text{ order } \mathrm{XX}_1^{(i)-}, x^* = a_j^{(i)-}
\end{aligned}
\right\}
\xrightarrow[\text{switching bifurcation}]{\beta_i^- = \sum_{j=1}^{k_i} r_j^{(i)-}} (\beta_i^-)^{\text{th}} \text{ order } \mathrm{XX}^{(i)-}.
\tag{7.238}
$$

The bifurcation set of $((\beta_1^-)^{\text{th}}\mathrm{XX}_1^- : (\beta_2^-)^{\text{th}}\mathrm{XX}_2^- : \ldots : (\beta_{s_1}^-)^{\text{th}}\mathrm{XX}_{s_1}^-)$ at the same parameter point is called a *left-parallel-straw-bundle* switching bifurcation.

Consider a just after bifurcation of $((\beta_1^+)^{\text{th}}\mathrm{XX}_1^+, (\beta_2^+)^{\text{th}}\mathrm{XX}_2^+ : \ldots : (\beta_{s_2}^+)^{\text{th}}\mathrm{XX}_{s_2}^+)$ for simple sources and sinks. For $\beta_i^+ = 2l_i^+ - 1$, $\mathrm{XX}_i^+ \in \{\mathrm{SO}, \mathrm{SI}\}$ and for $\beta_i^+ = 2l_i^+$, $\mathrm{XX}_i^+ \in \{\mathrm{US}, \mathrm{LS}\}$ The four detailed structures are as follows:

$$
(\beta_i^+)^{\text{th}}\,\text{order}\,\text{XX}^{(i)+} \xrightarrow[\text{switching bifurcation}]{\beta_i^-=\sum_{j=1}^{s_i} r_j^{(i)+}} \begin{cases} (r_{s_i}^{(i)+})^{\text{th}}\,\text{order}\,\text{XX}_{s_i}^{(i)+}, x^* = a_{k_i}^{(i)+}, \\ \vdots \\ (r_j^{(i)+})^{\text{th}}\,\text{order}\,\text{XX}_j^{(i)+}, x^* = a_j^{(i)+} \\ \vdots \\ (r_1^{(i)+})^{\text{th}}\,\text{order}\,\text{XX}_1^{(i)+}, x^* = a_j^{(i)+}. \end{cases} \tag{7.239}
$$

The bifurcation set of $((\beta_1^+)^{\text{th}}\text{XX}_1^+ : (\beta_2^+)^{\text{th}}\text{XX}_2^+ : \ldots : (\beta_{s_2}^+)^{\text{th}}\text{XX}_{s_2}^+)$ at the same parameter point is called a *right-parallel-straw-bundle* switching bifurcation.

(i) For the just before and after bifurcation structure, if there exists a relation of

$$
\begin{array}{l} (\beta_i^-)^{\text{th}}\text{XX}_i^- = (\beta_j^+)^{\text{th}}\text{XX}_j^+ = (\beta_j^+)^{\text{th}}\text{XX}, \text{for}\, x^* = a_i^- = a_j^+ \\ (i,j \in \{1,2,\ldots,k\}), \text{XX} \in \{\text{US}, \text{LS}, \text{SO}, \text{SI}\} \end{array} \tag{7.240}
$$

then the bifurcation is a $(\beta_j)^{\text{th}}$ *switching* bifurcation for higher order equilibriums.

(ii) Just for the just before bifurcation structure, if there exists a relation of

$$
\begin{array}{l} (2l_i^-)^{\text{th}}\text{XX}_i^- = (2l)^{\text{th}}\text{XX}, \text{for}\ x^* = a_i^- = a_i \\ (i \in \{1,2,\ldots,s_1\}, \text{XX} \in \{\text{US}, \text{LS}\}) \end{array} \tag{7.241}
$$

then the bifurcation is a $(2l)^{\text{th}}$XX *left-appearing* (or *right-vanishing*) bifurcation for higher order equilibriums.

(iii) Just for the just after bifurcation structure, if there exists a relation of

$$
\begin{array}{l} (2l_i^+)^{\text{th}}\text{XX}_i^+ = (2l)^{\text{th}}\text{XX}, \text{for}\, x^* = a_i^+ = a_i \\ (i \in \{1,2,\ldots,s_1\}), \text{XX} \in \{\text{US}, \text{LS}\} \end{array} \tag{7.242}
$$

then the bifurcation is a $(2l)^{\text{th}}$XX *right-appearing* (or *left-vanishing*) bifurcation for higher order equilibriums.

(iv) For the just before and after bifurcation structure, if there exists a relation of

$$
\begin{array}{l} (\beta_i^-)^{\text{th}}\text{XX}_i^- \neq (\beta_j^+)^{\text{th}}\text{XX}_j^+\, \text{for}\, x^* = a_i^- = a_j^+ \\ \text{XX}_i^-, \text{XX}_j^+ \in \{\text{US}, \text{LS}, \text{SO}, \text{SI}\} \\ (i \in \{1,2,\ldots,s_1\}, j \in \{1,2,\ldots,s_2\}), \end{array} \tag{7.243}
$$

then two *flower-bundle* switching bifurcations of higher order equilibriums are as follows.

(iv$_1$) For $\beta_j = \beta_i + 2l$, the bifurcation is called a $(\beta_j^+)^{\text{th}}\text{XX}_j^+$ right *flower-bundle* switching bifurcation for the $(\beta_i^-)^{\text{th}}\text{XX}_i^-$ to $(\beta_j^+)^{\text{th}}\text{XX}_j^+$ switching of *higher order* equilibriums with the appearance (or birth) of $(2l)^{\text{th}}\text{XX}$ right-appearing (or left-vanishing) bifurcation.

(iv$_2$) For $\beta_j = \beta_i - 2l$, the bifurcation is called a $(\beta_i^-)^{\text{th}}\text{XX}_i^-$ left *flower-bundle* switching bifurcation for the $(\beta_i^-)^{\text{th}}\text{XX}_i^-$ to $(\beta_j^+)^{\text{th}}\text{XX}_j^+$ switching of higher order equilibriums with the vanishing (or death) of $(2l)^{\text{th}}\text{XX}$ *left-appearing* (or right-vanishing) bifurcation.

A general parallel switching bifurcation is

$$\begin{aligned}&((\beta_1^-)^{\text{th}}\text{XX}_1^- : (\beta_2^-)^{\text{th}}\text{XX}_2^- : \ldots : (\beta_{s_1}^-)^{\text{th}}\text{XX}_{s_1}^-) \xrightarrow[\text{bifucation}]{\text{switching}}\\ &((\beta_1^+)^{\text{th}}\text{XX}_1^+ : (\beta_2^+)^{\text{th}}\text{XX}_2^+ : \ldots : (\beta_{s_2}^+)^{\text{th}}\text{XX}_{s_2}^+).\end{aligned} \tag{7.244}$$

Such a general, parallel switching bifurcation consists of the *left* and *right* parallel-bundle switching bifurcations for higher order equilibriums.

If the *left* and *right* parallel-bundle switching bifurcations are same in a parallel *flower-bundle* switching bifurcation, i.e.,

$$\begin{aligned}&(\beta_i^-)^{\text{th}}\text{XX}_i^- = (\beta_i^+)^{\text{th}}\text{XX}_i^+ = \beta^{\text{th}}\text{XX},\\ &\text{for } x^* = a_i^- = a_i^+ \; (i = 1, 2, \ldots, s)\end{aligned} \tag{7.245}$$

then the parallel *flower-bundle* switching bifurcation becomes a parallel *straw-bundle switching* bifurcation of $((\alpha_1)^{\text{th}}\text{XX} : (\beta_2)^{\text{th}}\text{XX} : \ldots : (\beta_s)^{\text{th}}\text{XX})$.

If the *left* and *right* parallel-bundle switching bifurcations are different in a parallel *flower-bundle* switching bifurcation, i.e.,

$$\begin{aligned}&(\alpha_i^-)^{\text{th}}\text{XX}_i^- = (2l_i^-)^{\text{th}}\text{XX}, (\alpha_j^+)^{\text{th}}\text{XX}_j^+ = (2l_j^+)^{\text{th}}\text{YY},\\ &\text{for} x^* = a_i^- \neq a_j^+ \; (i = 1, 2, \ldots, s_1; j = 1, 2, \ldots, s_2),\\ &\text{XX} \in \{\text{US}, \text{LS}\}, \text{YY} \in \{\text{US}, \text{LS}\}\end{aligned} \tag{7.246}$$

then the parallel *flower-bundle* switching bifurcation for higher order equilibriums becomes a combination of two independent left and right parallel appearing bifurcations:

(i) $((2l_1^-)^{\text{th}}\text{XX}_1^- : (2l_2^-)^{\text{th}}\text{XX}_2^- : \ldots : (2l_{s_1}^-)^{\text{th}}\text{XX}_{s_1}^-)$-*left parallel sprinkler-spraying-appearing (or right-vanishing)* bifurcation and

(ii) $((2l_1^+)^{\text{th}}\text{XX}_1^+ : (2l_2^+)^{\text{th}}\text{XX}_2^+ : \ldots : (2l_{s_2}^+)^{\text{th}}\text{XX}_{s_2}^+)$-*right parallel sprinkler-spraying-appearing (or left-vanishing)* bifurcation.

The parallel switching and appearing bifurcations for higher order equilibriums are presented in Fig. 7.14a–d. The *waterfall* appearing bifurcations and the *flower-bundle* switching bifurcations for higher order equilibriums are presented.

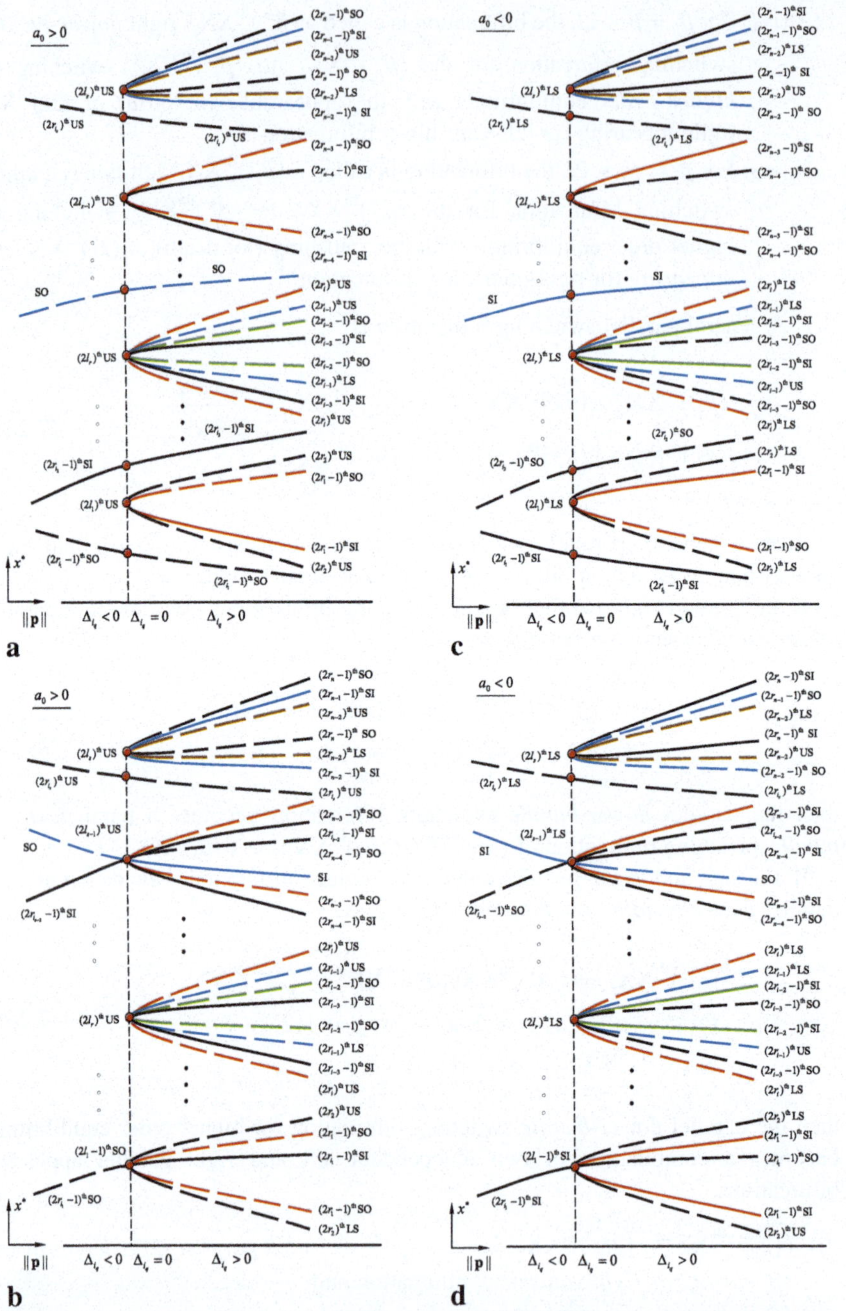

Fig. 7.14 (r_1th XX : r_2th XX : … : r_nth XX) parallel bifurcation ($a_0 > 0$): (**a**) without switching, and (**b**) with switching. The (r_1th XX : r_2th XX : … : r_nth XX) parallel bifurcation ($a_0 < 0$): (**c**) without switching, and (**d**) with switching. LS: lower saddle, US: upper saddle, SI sink, SO: source. Stable and unstable equilibriums are represented by solid and dashed curves, respectively. The bifurcation points are marked by circular symbols

Reference

Luo, A.C.J., 2019, The stability and bifurcation of the $(2m + 1)^{th}$-degree polynomial systems, *Journal of Vibration Testing and System Dynamics*, in press.

Chapter 8
Infinite-Equilibrium Systems

In this chapter, dynamical systems with infinite equilibriums are discussed through the local analysis. A method for equilibriums in nonlinear dynamical systems is developed. The generalized normal forms of nonlinear dynamical systems at equilibriums are presented for a better understanding of singularity in nonlinear dynamical systems. The dynamics of infinite-equilibrium dynamical systems is discussed for the complexity and singularity of nonlinear dynamical systems. A few examples are presented for complexity and singularity in infinite-equilibrium systems.

8.1 Equilibrium Computations

In this section, a method for equilibriums in nonlinear dynamical systems is presented, which is the extension of the Newton-Raphson method. Using the local analysis, the existence of equilibriums is discussed as in Luo (2019).

Definition 8.1 Consider an n-dimensional, dynamical system

$$\dot{\mathbf{x}} = \mathbf{f}(\mathbf{x}, \mathbf{p}). \tag{8.1}$$

For a given point $\mathbf{x}_0$ with parameter $\mathbf{p}_0$, there is a neighborhood $U(\mathbf{x}_0)$. Suppose for parameter $\mathbf{p}_0$, the dynamical system in Eq. (8.1) has an equilibrium $\mathbf{x}^*$. The vector field in $U(\mathbf{x}_0)$ is expanded as

$$\mathbf{f}(\mathbf{x}^*, \mathbf{p}_0) \approx \mathbf{f}(\mathbf{x}_0, \mathbf{p}_0) + D\mathbf{f}(\mathbf{x}_0, \mathbf{p}_0)(\mathbf{x}^* - \mathbf{x}_0) = \mathbf{0}. \tag{8.2}$$

A. C. J. Luo, *Bifurcation and Stability in Nonlinear Dynamical Systems*, Nonlinear Systems and Complexity 28, https://doi.org/10.1007/978-3-030-22910-8_8

For the Jacobian matrix of the vector field at the given point $(\mathbf{x}_0, \mathbf{p}_0)$, the corresponding distinct eigenvalues are given by

$$|D\mathbf{f}(\mathbf{x}_0, \mathbf{p}_0) - \lambda \mathbf{I}| = 0, \tag{8.3}$$

and the eigenvectors $\mathbf{v}_k$ ($k = 1, 2, \ldots, n$) for simple eigenvalues are computed by

$$(D\mathbf{f}(\mathbf{x}_0, \mathbf{p}_0) - \lambda_k \mathbf{I})\mathbf{v}_k = \mathbf{0}. \tag{8.4}$$

The covariant matrix for the Jacobian matrix $D\mathbf{f}(\mathbf{x}_0, \mathbf{p}_0)$ is

$$\mathbf{Q} = (\mathbf{v}_1, \mathbf{v}_2, \ldots, \mathbf{v}_n) = (a_{ij})_{n\times n}, \tag{8.5}$$

and the contravariant matrix of the Jacobian matrix $D\mathbf{f}(\mathbf{x}_0, \mathbf{p}_0)$ is

$$\mathbf{P} = (\mathbf{v}^1, \mathbf{v}^2, \ldots, \mathbf{v}^n) = (a^{ij})_{n\times n} = \mathbf{Q}^{-1}. \tag{8.6}$$

Suppose a new transform is

$$\mathbf{x} = c^k \mathbf{v}_k, \mathbf{x}_0 = c_0^k \mathbf{v}_k, \tag{8.7}$$

and the covariant component is

$$c^k = (\mathbf{v}^k)^{\mathrm{T}} \cdot \mathbf{x} = (\mathbf{v}^k)^{\mathrm{T}} \cdot c^j \mathbf{v}_j = c^j \delta_j^k. \tag{8.8}$$

Thus, the component of the vector field on the covariant direction of $\mathbf{v}_k$ is expanded by

$$\begin{aligned} f^k =& (\mathbf{v}^k)^{\mathrm{T}} \cdot \mathbf{f}(\mathbf{x}, \mathbf{p}) \\ =& (\mathbf{v}^k)^{\mathrm{T}} \cdot \mathbf{f}(\mathbf{x}_0, \mathbf{p}_0) + G_{k(j_1)} z^{j_1} \\ &+ \sum_{q=2}^{m_k} \frac{1}{q!} G^{(q)}_{k(j_1 j_2 \cdots j_q)} (z^{j_1} z^{j_2} \ldots z^{j_q}) \\ &+ \frac{1}{(m_k+1)!} G^{(m_k+1)}_{k(j_1 j_2 \cdots j_{m_k+1})} (\theta_1 z^{j_1})(\theta_2 z^{j_2}) \ldots (\theta_{m_k+1} z^{j_{m_k+1}}) \end{aligned} \tag{8.9}$$

where

$$
\begin{aligned}
z^j &= (\mathbf{v}^j)^{\mathrm{T}} \cdot \mathbf{y} = c^j - c_0^j \quad (j = 1, 2, \ldots, n), \\
G_{k(j_1)} &= (\mathbf{v}^k)^{\mathrm{T}} \cdot \partial_{c^{j_1}} \mathbf{f}(\mathbf{x}, \mathbf{p})|_{(\mathbf{x}_0, \mathbf{p}_0)} \\
&= (\mathbf{v}^k)^{\mathrm{T}} \cdot \partial_{\mathbf{x}} \mathbf{f}(\mathbf{x}, \mathbf{p})|_{(\mathbf{x}_0, \mathbf{p}_0)} \mathbf{v}_{j_1}, \\
G^{(r)}_{k(j_1 j_2 \cdots j_r)} &= (\mathbf{v}^k)^{\mathrm{T}} \cdot \partial^{(r)}_{c^{j_1} c^{j_2} \ldots c^{j_r}} \mathbf{f}(\mathbf{x}, \mathbf{p})|_{(\mathbf{x}_0, \mathbf{p}_0)} \\
&= (\mathbf{v}^k)^{\mathrm{T}} \cdot \partial^{(r)}_{\mathbf{x}} \mathbf{f}(\mathbf{x}, \mathbf{p})|_{(\mathbf{x}_0, \mathbf{p}_0)} \mathbf{v}_{j_1} \mathbf{v}_{j_2} \ldots \mathbf{v}_{j_r} \\
&= G^{(r)}_{k(j_1 j_2 \cdots j_r)}(\mathbf{x}_0, \mathbf{p}_0).
\end{aligned}
\tag{8.10}
$$

Definition 8.2 Consider an n-dimensional, dynamical system in Eq. (8.1). For a given point $\mathbf{x}_0$ with parameter $\mathbf{p}_0$, there is a neighborhood $U(\mathbf{x}_0)$. Suppose for parameter $\mathbf{p}_0$, the dynamical system in Eq. (8.1) has an equilibrium $\mathbf{x}^*$. The Jacobian matrix of the vector field $\mathbf{f}(\mathbf{x}, \mathbf{p})$ in $U(\mathbf{x}_0)$ has distinct eigenvalues λ_k $(k = 1, 2, \ldots, n)$ and the corresponding covariant and contravariant eigenvectors are $\mathbf{v}_k$ and $\mathbf{v}^k$ $(k = 1, 2, \ldots, n)$, respectively. The covariant component of the vector field on the covariant direction of $\mathbf{v}_k$ is

$$
f^k = (\mathbf{v}^k)^{\mathrm{T}} \cdot \mathbf{f}(\mathbf{x}, \mathbf{p}). \tag{8.11}
$$

For a specific $\mathbf{p} = \mathbf{p}_0$, if

$$
\begin{aligned}
&\lambda_k \neq 0 \ (k = 1, 2, \ldots, n) \\
&| G^{(2)}_{k(j_1 j_2)}(\mathbf{x}_0, \mathbf{p}_0) z^{j_1} z^{j_2} | << | \lambda_k z^{j_1} \delta^k_{j_1} | \\
&(j_1, j_2 = 1, 2, \ \ldots, n),
\end{aligned}
\tag{8.12}
$$

then the covariant component of the vector field is expanded as

$$
f^k \approx (\mathbf{v}^k)^{\mathrm{T}} \cdot \mathbf{f}(\mathbf{x}_0, \mathbf{p}_0) + (\mathbf{v}^k)^{\mathrm{T}} \cdot D\mathbf{f}(\mathbf{x}_0, \mathbf{p}_0) \mathbf{v}_j z^j = 0. \tag{8.13}
$$

Thus, with $D\mathbf{f}(\mathbf{x}_0, \mathbf{p}_0)\mathbf{v}_j = \lambda_j \mathbf{v}_j$, there is an approximate solution for a specific $\mathbf{p} = \mathbf{p}_0$,

$$
z^k \approx -(\lambda_k)^{-1} (\mathbf{v}^k)^{\mathrm{T}} \cdot \mathbf{f}(\mathbf{x}, \mathbf{p}_0), \text{for } k = 1, 2, \ \ldots, n, \tag{8.14}
$$

and

$$\begin{aligned}
&\mathbf{x}^{(i)} = \mathbf{x}^{(i-1)} + \mathbf{Q}^{(i-1)}\mathbf{z}^{(i-1)}, (i = 1, 2, \ldots),\\
&\text{with } \mathbf{x}^{(0)} \equiv \mathbf{x}_0, \mathbf{z}^{(0)} \equiv \mathbf{z}_0, \mathbf{Q}^{(0)} = \mathbf{Q} \text{ for } i = 1,\\
&\mathbf{Q}^{(i-1)} = (\mathbf{v}_1^{(i-1)}, \mathbf{v}_2^{(i-1)}, \ldots, \mathbf{v}_n^{(i-1)}) \text{ for}\\
&(D\mathbf{f}(\mathbf{x}^{(i-1)}, \mathbf{p}_0) - \lambda_k^{(i-1)}\mathbf{I})\mathbf{v}_k^{(i-1)} = \mathbf{0}\\
&(k = 1, 2, \ldots, n).
\end{aligned} \tag{8.15}$$

(a) If

$$\|\mathbf{f}(\mathbf{x}^{(i)}, \mathbf{p}_0)\| \leq \varepsilon, \tag{8.16}$$

then $\mathbf{x}^{(i)} \approx \mathbf{x}^*$ is called an approximate solution of equilibrium for $\mathbf{f}(\mathbf{x}^*, \mathbf{p}_0) \approx \mathbf{0}$ in the sense of ε.

(b) If

$$\|\mathbf{f}(\mathbf{x}^{(i)}, \mathbf{p}_0)\| > \varepsilon, \tag{8.17}$$

then the point $\mathbf{x}^{(i)}$ is not an approximate solution of equilibrium for $\mathbf{f}(\mathbf{x}^*, \mathbf{p}_0) \approx \mathbf{0}$ in the sense of ε. Thus, let $\mathbf{x}_0 = \mathbf{x}^{(i)}$ and repeat the same procedure until Eq. (8.16) is satisfied.

Definition 8.3 Consider an n-dimensional, dynamical system in Eq. (8.1). For a given point $\mathbf{x}_0$ with parameter $\mathbf{p}_0$, there is a neighborhood $U(\mathbf{x}_0)$. Suppose for parameter $\mathbf{p}_0$, the dynamical system in Eq. (8.1) has an equilibrium $\mathbf{x}^*$. The Jacobian matrix of the vector field $\mathbf{f}(\mathbf{x}, \mathbf{p})$ in $U(\mathbf{x}_0)$ has distinct eigenvalues λ_k $(k = 1, 2, \ldots, n)$ and the corresponding covariant and contravariant eigenvectors are $\mathbf{v}_k$ and $\mathbf{v}^k$ $(k = 1, 2, \ldots, n)$, respectively. The covariant component of the vector field on the covariant direction of $\mathbf{v}_k$ is in Eq. (8.11).

(i) For a specific $\mathbf{p} = \mathbf{p}_0$, if

$$\begin{aligned}
&\lambda_s \neq 0 \; (s = 1, 2, \ldots, n \text{ but } s \neq k)\\
&| G^{(2)}_{k(j_1 j_2)}(\mathbf{x}_0, \mathbf{p}_0) z^{j_1} z^{j_2} | << | \lambda_k z^j \delta_j^k |\\
&(j, j_1, j_2 = 1, 2, \ldots, n),
\end{aligned} \tag{8.18}$$

then the covariant component of the vector field is expanded as

$$f^s \approx (\mathbf{v}^s)^{\mathrm{T}} \cdot \mathbf{f}(\mathbf{x}_0, \mathbf{p}_0) + (\mathbf{v}^s)^{\mathrm{T}} \cdot D\mathbf{f}(\mathbf{x}_0, \mathbf{p}_0)\mathbf{v}_j z^j = 0. \tag{8.19}$$

Thus, with $D\mathbf{f}(\mathbf{x}_0, \mathbf{p}_0)\mathbf{v}_j = \lambda_j \mathbf{v}_j$, there is an approximate solution of equilibrium for a specific $\mathbf{p} = \mathbf{p}_0$,

$$z^s \approx -(\lambda_s)^{-1}(\mathbf{v}^s)^{\mathrm{T}} \cdot \mathbf{f}(\mathbf{x}, \mathbf{p}_0), \text{ for } s = 1, 2, \ldots, n \text{ but } s \neq k. \tag{8.20}$$

(ii) For a specific $\mathbf{p} = \mathbf{p}_0$, on the covariant direction of $\mathbf{v}_k$

$$\begin{aligned}
&\lambda_s \neq 0 \; (s = 1, 2, \ldots, n \text{ but } s \neq k) \\
&| G^{(s_k)}_{k(j_1 j_2 \cdots j_{s_k})}(\mathbf{x}_0, \mathbf{p}_0) z^{j_1} z^{j_2} \ldots z^{j_{s_k}} | \\
&>> | G^{(s_k+1)}_{k(j_1 j_2 \cdots j_{s_k+1})}(\mathbf{x}_0, \mathbf{p}_0) z^{j_1} z^{j_2} \ldots z^{j_{s_k+1}} | \leq \varepsilon \\
&(j_1, j_2, \ldots, j_{s_k+1} = 1, 2, \ldots, n);
\end{aligned} \tag{8.21}$$

then the covariant component of the vector field is expanded as

$$f^k \approx (\mathbf{v}^k)^{\mathrm{T}} \cdot \mathbf{f}(\mathbf{x}_0, \mathbf{p}_0) + \lambda_k z^k + \sum_{q=2}^{s_k} \frac{1}{q!} G^{(q)}_{k(j_1 j_2 \cdots j_q)} (z^{j_1} z^{j_2} \ldots z^{j_q}) = 0. \tag{8.22}$$

(ii$_1$) Equation (8.22) is equivalent to

$$\begin{aligned}
&f^k \approx A_0 (z^k)^{2m_k} + A_1 (z^k)^{2m_k - 1} + \cdots + A_{2m_k - 1} z^k + A_{2m_k} = 0 \\
&A_i = A_i(z^1, z^2, \cdots, z^{k-1}, z^{k+1}, \ldots, z^n) \; (i = 0, 1, 2, \ldots, 2m_k),
\end{aligned} \tag{8.23}$$

and the standard form for simple solution is

$$f^k \approx a_0 \prod_{i=1}^{m_k} [(z^k)^2 + B_i z^k + C_i] = 0. \tag{8.24}$$

(a) If

$$\Delta_i = B_i^2 - 4C_i < 0 \text{ for } i = 1, 2, \ldots, m_k, \tag{8.25}$$

then Eq. (8.22) does not have any solution. Thus, the equilibrium does not exist in the neighborhood $U(\mathbf{x}_0)$.

(b) If

$$\begin{aligned}
&\Delta_i = B_i^2 - 4C_i > 0 \text{ for } i \in \{i_1, \ldots, i_{r_1}\} \subset \{1, 2, \ldots, m_k\}, \\
&\Delta_j = B_j^2 - 4C_j < 0 \text{ for } j \in \{j_1, \ldots, j_{r_2}\} \subset \{1, 2, \ldots, m_k\}, \\
&r_1 + r_2 = m_k,
\end{aligned} \tag{8.26}$$

then Eq. (8.22) has r_1-pairs of solutions in the neighborhood $U(\mathbf{x}_0)$. Thus, there are $2r_1$-approximate solutions of equilibriums for a specific $\mathbf{p} = \mathbf{p}_0$,

$$\begin{aligned}&z_{1,2}^{k(i)}=-\frac{1}{2}(B_i\pm\sqrt{\Delta_i})\text{ for }i\in\{i_1,\ldots,i_{r_1}\}\subset\{1,2,\ldots,m_k\},\\&\{z_1^{(k)},z_2^{(k)},\ldots,z_{2r_1}^{(k)}\}=\text{sort}\{z_1^{k(i_1)},z_2^{k(i_1)},\ldots,z_1^{k(i_{r_1})},z_2^{k(i_{r_2})}\}\\&\text{with }z_s^{(k)}\le z_{s+1}^{(k)},\end{aligned}\tag{8.27}$$

and

$$\begin{aligned}&\mathbf{x}_j^{(i)}=\mathbf{x}_j^{(i-1)}+\mathbf{Q}_j^{(i-1)}\mathbf{z}_j^{(i-1)},(i=1,2,\ldots;j=1,2,\ldots,2r_1),\\&\text{with }\mathbf{x}_j^{(0)}\equiv\mathbf{x}_{j0},\mathbf{z}_j^{(0)}\equiv\mathbf{z}_{j0},\mathbf{Q}^{(0)}=\mathbf{Q}\text{ for }i=1,\\&\mathbf{Q}_j^{(i-1)}=(\mathbf{v}_{j1}^{(i-1)},\mathbf{v}_{j2}^{(i-1)},\ldots,\mathbf{v}_{jn}^{(i-1)})\text{ for}\\&(D\mathbf{f}(\mathbf{x}_j^{(i-1)},\mathbf{p}_0)-\lambda_{jk}^{(i-1)}\mathbf{I})\mathbf{v}_{jk}^{(i-1)}=\mathbf{0}\\&(k=1,2,\ldots,n).\end{aligned}\tag{8.28}$$

If

$$\|\mathbf{f}(\mathbf{x}_j^{(i)},\mathbf{p}_0)\|\le\varepsilon,\tag{8.29}$$

then $\mathbf{x}_j^{(i)}\approx\mathbf{x}_j^*$ ($j=1,2,\ldots,2r_1$) is called a set of approximate solutions of equilibrium for $\mathbf{f}(\mathbf{x}^*,\mathbf{p}_0)\approx\mathbf{0}$ in the sense of ε.

(c) If

$$\begin{aligned}&\Delta_i=B_i^2-4C_i=0\text{ for }i\in\{i_1,\ldots,i_{r_1}\}\subset\{1,2,\ldots,m_k\},\\&\Delta_j=B_j^2-4C_j>0\text{ for }j\in\{j_1,\ldots,j_{r_2}\}\subset\{1,2,\ldots,m_k\},\\&\Delta_l=B_l^2-4C_l<0\text{ for }l\in\{l_1,\ldots,l_{r_3}\}\subset\{1,2,\ldots,m_k\},\\&r_1+r_2+r_3=m_k,\end{aligned}\tag{8.30}$$

then Eq. (8.22) has r_1-pairs of repeated solutions and r_2-pairs of simple solutions in the neighborhood $U(\mathbf{x}_0)$. Thus, there are (r_1 + $2r_2$)-approximate solutions of equilibriums for a specific $\mathbf{p}=\mathbf{p}_0$,

$$\begin{aligned}&z_{1,2}^{k(i)}=-\frac{1}{2}B_i,\Delta_i=0\text{ for }i\in\{i_1,\ldots,i_{r_1}\}\subset\{1,2,\ldots,m_k\}\\&z_{1,2}^{k(j)}=-\frac{1}{2}(B_j\pm\sqrt{\Delta_j})\text{ for }j\in\{j_1,\ldots,j_{r_2}\}\subset\{1,2,\ldots,m_k\},\\&\{z_1^{(k)},z_2^{(k)},\ldots,z_{2(r_1+r_2)}^{(k)}\}=\text{sort}\{z_1^{k(i_1)},z_2^{k(i_1)},\ldots,z_2^{k(i_{r_1})};z_1^{k(j_1)},z_2^{k(j_1)},\ldots,z_2^{k(j_{r_2})}\}\\&\text{with }z_s^{(k)}\le z_{s+1}^{(k)},\end{aligned}\tag{8.31}$$

and

$$\begin{aligned}
&\mathbf{x}_j^{(i)} = \mathbf{x}_j^{(i-1)} + \mathbf{Q}^{(i-1)}\mathbf{z}_j^{(i-1)}, (i = 1, 2, \ldots; j = 1, 2, \ldots, r_1 + 2r_2)\\
&\text{with } \mathbf{x}_j^{(0)} \equiv \mathbf{x}_{j0}, \mathbf{z}_j^{(0)} \equiv \mathbf{z}_{j0}, \mathbf{Q}^{(0)} = \mathbf{Q} \text{ for } i = 1,\\
&\mathbf{Q}_j^{(i-1)} = (\mathbf{v}_{j1}^{(i-1)}, \mathbf{v}_{j2}^{(i-1)}, \ldots, \mathbf{v}_{jn}^{(i-1)}) \text{ for}\\
&(D\mathbf{f}(\mathbf{x}_j^{(i-1)}, \mathbf{p}_0) - \lambda_{jk}^{(i-1)}\mathbf{I})\mathbf{v}_{jk}^{(i-1)} = \mathbf{0}\\
&(k = 1, 2, \ldots, n).
\end{aligned} \tag{8.32}$$

If

$$\|\mathbf{f}(\mathbf{x}_j^{(i)}, \mathbf{p}_0)\| \leq \varepsilon, \tag{8.33}$$

then $\mathbf{x}_j^{(i)} \approx \mathbf{x}_j^*$ $(j = 1, 2, \ldots, r_1 + 2r_2)$ is called a set of approximate solutions of equilibrium for $\mathbf{f}(\mathbf{x}^*, \mathbf{p}_0) \approx \mathbf{0}$ in the sense of ε.

(ii$_2$) Equation (8.22) is equivalent to

$$\begin{aligned}
&f^k \approx A_0(z^k)^{2m_k+1} + A_1(z^k)^{2m_k} + \cdots + A_{2m_k}z^k + A_{2m_k+1} = 0\\
&A_i = A_i(z^1, z^2, \ldots, z^{k-1}, z^k, \ldots, z^n)\ (i = 0, 1, 2, \ldots, 2m_k + 1),
\end{aligned} \tag{8.34}$$

and the standard form for simple solutions is

$$f^k \approx a_0(z^k - a)\prod_{i=1}^{m_k}[(z^k)^2 + B_i z^k + C_i] = 0. \tag{8.35}$$

(a) If

$$\Delta_i = B_i^2 - 4C_i < 0 \text{ for } i = 1, 2, \ldots, m_k, \tag{8.36}$$

then Eq. (8.22) has a solution. Thus, the equilibrium has one solution of equilibrium in the neighborhood $U(\mathbf{x}_0)$.

(b) If

$$\begin{aligned}
&\Delta_i = B_i^2 - 4C_i > 0 \text{ for } i \in \{i_1, \ldots, i_{r_1}\} \subset \{1, 2, \ldots, m_k\},\\
&\Delta_j = B_j^2 - 4C_j < 0 \text{ for } j \in \{j_1, \ldots, j_{r_2}\} \subset \{1, 2, \ldots, m_k\},\\
&r_1 + r_2 = m_k,
\end{aligned} \tag{8.37}$$

then Eq. (8.22) has $(2r_1 + 1)$-simple solutions in the neighborhood $U(\mathbf{x}_0)$. Thus, there exist $(2r_1 + 1)$-approximate solutions of equilibriums for a specific $\mathbf{p} = \mathbf{p}_0$ as

$$
\begin{aligned}
&z_{1,2}^{k(i)} \approx -\frac{1}{2}(B_i \pm \sqrt{\Delta_i}) \text{ for } i \in \{i_1, \ldots, i_{r_1}\} \subset \{1,2, \ldots, m_k\}, \\
&\{z_1^{(k)}, z_2^{(k)}, \ldots, z_{2r_1+1}^{(k)}\} = \text{sort}\{a, z_1^{k(i_1)}, z_2^{k(i_1)}, \ldots, z_1^{k(i_{r_1})}, z_2^{k(i_{r_2})}\} \\
&\text{with } z_s^{(k)} \leq z_{s+1}^{(k)},
\end{aligned} \tag{8.38}
$$

and

$$
\begin{aligned}
&\mathbf{x}_j^{(i)} = \mathbf{x}_j^{(i-1)} + \mathbf{Q}^{(i-1)}\mathbf{z}_j^{(i-1)}, \\
&(i = 1, 2, \ldots; j = 1, 2, \ldots, 2r_1 + 1) \\
&\text{with } \mathbf{x}_j^{(0)} \equiv \mathbf{x}_{j0}, \mathbf{z}_j^{(0)} \equiv \mathbf{z}_{j0} \text{ for } i = 1.
\end{aligned} \tag{8.39}
$$

If

$$
\|\mathbf{f}(\mathbf{x}_j^{(i)}, \mathbf{p}_0)\| \leq \varepsilon, \tag{8.40}
$$

then $\mathbf{x}_j^{(i)} \approx \mathbf{x}_j^*$ ($j = 1, 2, \ldots, 2r_1 + 1$) is called a set of approximate solutions of equilibrium for $\mathbf{f}(\mathbf{x}^*, \mathbf{p}_0) \approx \mathbf{0}$ in the sense of ε.

(c) If

$$
\begin{aligned}
&\Delta_i = B_i^2 - 4C_i = 0 \text{ for } i \in \{i_1, \ldots, i_{r_1}\} \subset \{1,2, \ldots, m_k\}, \\
&\Delta_j = B_j^2 - 4C_j > 0 \text{ for } j \in \{j_1, \ldots, j_{r_2}\} \subset \{1,2, \ldots, m_k\}, \\
&\Delta_l = B_l^2 - 4C_l < 0 \text{ for } l \in \{j_1, \ldots, j_{r_3}\} \subset \{1,2, \ldots, m_k\}, \\
&r_1 + r_2 + r_3 = m_k,
\end{aligned} \tag{8.41}
$$

then Eq. (8.22) has r_1-pairs of repeated solutions and r_2-pairs of simple solutions in the neighborhood $U(\mathbf{x}_0)$. Thus, there are $(r_1 + 2r_2 + 1)$-approximate solutions of equilibriums for a specific $\mathbf{p} = \mathbf{p}_0$,

$$
\begin{aligned}
&z_{1,2}^{k(i)} \approx -\frac{1}{2}B_i, \Delta_i = 0 \text{ for } i \in \{i_1, \ldots, i_{r_1}\} \subset \{1,2, \ldots, m_k\} \\
&z_{1,2}^{k(j)} \approx -\frac{1}{2}(B_j \pm \sqrt{\Delta_j}) \text{ for } j \in \{j_1, \ldots, j_{r_2}\} \subset \{1,2, \ldots, m_k\}, \\
&\{z_1^{(k)}, z_2^{(k)}, \ldots, z_{2(r_1+r_2)+1}^{(k)}\} \\
&= \text{sort}\{a, z_1^{k(i_1)}, z_2^{k(i_1)}, \ldots, z_2^{k(i_{r_1})}; z_1^{k(j_1)}, z_2^{k(j_1)}, \ldots, z_2^{k(j_{r_2})}\} \\
&\text{with } z_s^{(k)} \leq z_{s+1}^{(k)}
\end{aligned} \tag{8.42}
$$

and

$$
\begin{aligned}
&\mathbf{x}_j^{(i)} = \mathbf{x}_j^{(i-1)} + Q^{(i-1)}\mathbf{z}_j^{(i-1)},\\
&(i = 1, 2, \ldots; j = 1, 2, \ldots, r_1 + 2r_2 + 1)\\
&\text{with } \mathbf{x}_j^{(0)} \equiv \mathbf{x}_{j0}, \mathbf{z}_j^{(0)} \equiv \mathbf{z}_{j0}, Q^{(0)} = Q \text{ for } i = 1\\
&Q_j^{(i-1)} = (\mathbf{v}_{j1}^{(i-1)}, \mathbf{v}_{j2}^{(i-1)}, \ldots, \mathbf{v}_{jn}^{(i-1)})\\
&\text{for } (D\mathbf{f}(\mathbf{x}_j^{(i-1)}, \mathbf{p}_0) - \lambda_{jk}^{(i-1)}\mathbf{I})\mathbf{v}_{jk}^{(i-1)} = \mathbf{0}.\\
&(k = 1, 2, \ldots, n).
\end{aligned}
\tag{8.43}
$$

If

$$
\|\mathbf{f}(\mathbf{x}_j^{(i)}, \mathbf{p}_0)\| \le \varepsilon, \tag{8.44}
$$

then $\mathbf{x}_j^{(i)} \approx \mathbf{x}_j^*$ ($j = 1, 2, \ldots, r_1 + 2r_2 + 1$) is called a set of approximate solutions of equilibrium for $\mathbf{f}(\mathbf{x}^*, \mathbf{p}_0) \approx \mathbf{0}$ in the sense of ε.

The afore-presented method is an extension of the Newton-Raphson method. For such an extension, the solutions of equilibriums of nonlinear dynamical systems are determined through the eigenvector space of the linearized systems at the initial guessed solutions rather than the state space of original nonlinear systems. In fact, such a new method is completely different from the Newton-Raphson method. The singularity of nonlinear systems in eigenvector space is employed to determine solution existence and multiplicity.

8.2 Normal Forms

In this section, a general normal form of dynamical systems is presented through the eigenvector space of equilibriums. As in Luo (2019), the eigenvector space is developed, and the corresponding normal forms are developed by the Taylor series expansion.

Nonlinear dynamical systems experiencing the simple eigenvalues are discussed first, and the eigenvector space is developed through the following definition.

Definition 8.4 Consider an n-dimensional, dynamical system

$$
\dot{\mathbf{x}} = \mathbf{f}(\mathbf{x}, \mathbf{p}). \tag{8.45}
$$

There is an equilibrium point $\mathbf{x}^*$ with parameter $\mathbf{p}_0$,

$$
\mathbf{f}(\mathbf{x}^*, \mathbf{p}_0) = \mathbf{0}. \tag{8.46}
$$

In a neighborhood $U(\mathbf{x}^*)$ at parameter $\mathbf{p}_0$, the vector field in Eq. (8.45) is linearized by

$$\mathbf{f}(\mathbf{x},\mathbf{p}_0) = D\mathbf{f}(\mathbf{x}^*,\mathbf{p}_0)(\mathbf{x}-\mathbf{x}^*). \tag{8.47}$$

For the Jacobian matrix, the corresponding eigenvalues are determined by

$$|D\mathbf{f}(\mathbf{x}^*,\mathbf{p}_0) - \lambda\mathbf{I}| = 0. \tag{8.48}$$

Thus, the foregoing equation gives

$$f(\lambda) = (-1)^n\lambda^n + a_1\lambda^{n-1} + \cdots + a_{n-1}\lambda + a_n = 0. \tag{8.49}$$

(i) For $n = 2m$, consider simple eigenvalues of Eq. (8.48) as

$$\left.\begin{array}{l} f(\lambda) = \lambda^{2m} + a_1\lambda^{2n-1} + \cdots + a_{2m-1}\lambda + a_{2m} = 0 \\ \text{equivalent to} \\ f(\lambda) = \prod_{i=1}^{m}(\lambda^2 + B_i\lambda + C_i) = 0. \end{array}\right\} \tag{8.50}$$

(i$_1$) If

$$\left.\begin{array}{l} \Delta_{i_1} = B_{i_1}^2 - 4C_{i_1} > 0 \\ (i_1 \in \{s_1^{(1)}, s_2^{(1)}, \ldots, s_{n_1}^{(1)}\} \subset \{1,2,\ldots,2m\}), \end{array}\right\} \tag{8.51}$$

then

$$\left.\begin{array}{l} \lambda_{1,2}^{(i_1)} = -\frac{1}{2}B_{i_1} \pm \frac{1}{2}\sqrt{B_{i_1}^2 - 4C_{i_1}}, \\ (D\mathbf{f}(\mathbf{x}^*,\mathbf{p}_0) - \lambda_j^{(i_1)}\mathbf{I})\mathbf{v}_j^{(i_1)} = \mathbf{0}\ (j = 1,2), \\ (i_1 \in \{s_1^{(1)}, s_2^{(1)}, \ldots, s_{n_1}^{(1)}\} \subset \{1,2,\ldots,m\}). \end{array}\right\} \tag{8.52}$$

(i$_2$) If

$$\left.\begin{array}{l} \Delta_{i_2} = B_{i_2}^2 - 4C_{i_2} < 0 \\ (i_2 \in \{s_1^{(2)}, s_2^{(2)}, \ldots, s_{n_2}^{(2)}\} \subset \{1,2,\ldots,m\}), \end{array}\right\} \tag{8.53}$$

then

$$
\begin{aligned}
&\lambda_{1,2}^{(i_2)} = -\frac{1}{2}B_{i_2} \pm \frac{1}{2}\sqrt{4C_{i_2} - B_{i_2}^2}\,\mathbf{i} \equiv \alpha^{(i_2)} \pm \beta^{(i_2)}\mathbf{i}, \\
&D\mathbf{f}(\mathbf{x}^*, \mathbf{p}_0)\mathbf{u}_1^{(i_2)} = \alpha^{(i_2)}\mathbf{u}_1^{(i_2)} + \beta^{(i_2)}\mathbf{v}_1^{(i_2)}, \\
&D\mathbf{f}(\mathbf{x}^*, \mathbf{p}_0)\mathbf{v}_1^{(i_2)} = -\beta^{(i_2)}\mathbf{u}_1^{(i_2)} + \alpha^{(i_2)}\mathbf{v}_1^{(i_2)}, \\
&\mathbf{i} = \sqrt{-1}, (i_2 \in \{s_1^{(2)}, s_2^{(2)}, \ldots, s_{n_2}^{(2)}\} \subset \{1, 2, \ldots, m\}).
\end{aligned} \tag{8.54}
$$

(i_3) If

$$
\begin{aligned}
&\Delta_{i_3} = B_{i_3}^2 - 4C_{i_3} = 0, \;\; B_{i_3} \neq B_{j_3} \\
&(i_3, j_3 \in \{s_1^{(3)}, s_2^{(3)}, \ldots, s_{n_3}^{(3)}\} \subset \{1, 2, \ldots, m\}),
\end{aligned} \tag{8.55}
$$

then

$$
\begin{aligned}
&\lambda_{1,2}^{(i_3)} = -\frac{1}{2}B_{i_3} = \lambda^{(i_3)}, \\
&D\mathbf{f}(\mathbf{x}^*, \mathbf{p}_0)\mathbf{v}_1^{(i_3)} = \lambda^{(i_3)}\mathbf{v}_1^{(i_3)} \\
&D\mathbf{f}(\mathbf{x}^*, \mathbf{p}_0)\mathbf{v}_2^{(i_3)} = \lambda^{(i_3)}\mathbf{v}_2^{(i_3)} + \delta^{(i_3)}\mathbf{v}_1^{(i_3)}, \\
&(D\mathbf{f}(\mathbf{x}^*, \mathbf{p}_0) - \lambda^{(i_3)}\mathbf{I})\mathbf{v}_j^{(i_3)} = \mathbf{0} \text{ for } j = 1, 2 \text{ and } \delta^{(i_3)} = 0; \\
&(D\mathbf{f}(\mathbf{x}^*, \mathbf{p}_0) - \lambda^{(i_3)}\mathbf{I})^2\mathbf{v}_j^{(i_3)} = \mathbf{0} \text{ for } j \in \{1, 2\} \text{ and } \delta^{(i_3)} = 1 \\
&(i_3 \in \{s_1^{(3)}, s_2^{(3)}, \ldots, s_{n_3}^{(3)}\} \subset \{1, 2, \ldots, m\}).
\end{aligned} \tag{8.56}
$$

(ii) For $n = 2m + 1$, consider simple eigenvalues of Eq. (8.48) as

$$
\begin{aligned}
&f(\lambda) = -\lambda^{2m+1} + a_1\lambda^{2m} + \cdots + a_{2m}\lambda + a_{2m+1} = 0 \\
&\text{equivalent to} \\
&f(\lambda) = -(\lambda - \lambda_1)\prod_{i=1}^{m}(\lambda^2 + B_i\lambda + C_i) = 0.
\end{aligned} \tag{8.57}
$$

(ii_1) For $\lambda = \lambda_1$

$$
(D\mathbf{f}(\mathbf{x}^*, \mathbf{p}_0) - \lambda_1\mathbf{I})\mathbf{v}_1 = \mathbf{0}. \tag{8.58}
$$

(ii_2) If

$$
\begin{aligned}
&\Delta_{i_1} = B_{i_1}^2 - 4C_{i_1} > 0 \\
&(i_1 \in \{s_1^{(1)}, s_2^{(1)}, \ldots, s_{n_1}^{(1)}\} \subset \{1, 2, \ldots, m\}),
\end{aligned} \tag{8.59}
$$

then

$$\begin{aligned}&\lambda_{1,2}^{(i_1)} = -\frac{1}{2}B_{i_1} \pm \frac{1}{2}\sqrt{B_{i_1}^2 - 4C_{i_1}},\\&(D\mathbf{f}(\mathbf{x}^*, \mathbf{p}_0) - \lambda_j^{(i_1)}\mathbf{I})\mathbf{v}_j^{(i_1)} = \mathbf{0}\ (j = 1, 2),\\&(i_1 \in \{s_1^{(1)}, s_2^{(1)}, \ldots, s_{n_1}^{(1)}\} \subset \{1, 2, \ldots, m\}).\end{aligned} \tag{8.60}$$

(ii$_3$) If

$$\begin{aligned}&\Delta_{i_2} = B_{i_2}^2 - 4C_{i_2} < 0\\&(i_2 \in \{s_1^{(2)}, s_2^{(2)}, \ldots, s_{n_2}^{(2)}\} \subset \{1, 2, \ldots, m\}),\end{aligned} \tag{8.61}$$

then

$$\begin{aligned}&\lambda_{1,2}^{(i_2)} = -\frac{1}{2}B_{i_2} \pm \frac{1}{2}\sqrt{4C_{i_2} - B_{i_2}^2}\,\mathbf{i} \equiv \alpha^{(i_2)} \pm \beta^{(i_2)}\mathbf{i},\\&D\mathbf{f}(\mathbf{x}^*, \mathbf{p}_0)\mathbf{u}_1^{(i_2)} = \alpha^{(i_2)}\mathbf{u}_1^{(i_2)} + \beta^{(i_2)}\mathbf{v}_1^{(i_2)},\\&D\mathbf{f}(\mathbf{x}^*, \mathbf{p}_0)\mathbf{v}_1^{(i_2)} = -\beta^{(i_2)}\mathbf{u}_1^{(i_2)} + \alpha^{(i_2)}\mathbf{v}_1^{(i_2)},\\&\mathbf{i} = \sqrt{-1}, (i_2 \in \{s_1^{(2)}, s_2^{(2)}, \ldots, s_{n_2}^{(2)}\} \subset \{1, 2, \ldots, 2m\}).\end{aligned} \tag{8.62}$$

(ii$_4$) If

$$\begin{aligned}&\Delta_{i_3} = B_{i_3}^2 - 4C_{i_3} = 0,\ \ B_{i_3} \neq B_{j_3}\\&(i_3, j_3 \in \{s_1^{(3)}, s_2^{(3)}, \ldots, s_{n_3}^{(3)}\} \subset \{1, 2, \ldots, m\}),\end{aligned} \tag{8.63}$$

then

$$\begin{aligned}&\lambda_{1,2}^{(i_3)} = -\frac{1}{2}B_{i_3} = \lambda^{(i_3)},\\&D\mathbf{f}(\mathbf{x}^*, \mathbf{p}_0)\mathbf{v}_1^{(i_3)} = \lambda^{(i_3)}\mathbf{v}_1^{(i_3)}\\&D\mathbf{f}(\mathbf{x}^*, \mathbf{p}_0)\mathbf{v}_2^{(i_3)} = \lambda^{(i_3)}\mathbf{v}_2^{(i_3)} + \delta^{(i_3)}\mathbf{v}_1^{(i_3)},\\&(D\mathbf{f}(\mathbf{x}^*, \mathbf{p}_0) - \lambda^{(i_3)}\mathbf{I})\mathbf{v}_j^{(i_3)} = \mathbf{0}\\&\text{for } j = 1, 2 \text{ and } \delta^{(i_3)} = 0;\\&(D\mathbf{f}(\mathbf{x}^*, \mathbf{p}_0) - \lambda^{(i_3)}\mathbf{I})^2\mathbf{v}_j^{(i_3)} = \mathbf{0} \text{ for } j \in \{1, 2\} \text{ and } \delta^{(i_3)} = 1\\&(i_3 \in \{s_1^{(3)}, s_2^{(3)}, \ldots, s_{n_3}^{(3)}\} \subset \{1, 2, \ldots, m\}).\end{aligned} \tag{8.64}$$

Nonlinear dynamical systems experiencing the repeated eigenvalues are discussed first. The eigenvector space is developed through the following definition.

Definition 8.5 Consider an n-dimensional, dynamical system of $\dot{\mathbf{x}} = \mathbf{f}(\mathbf{x}, \mathbf{p})$ in Eq. (8.45). An equilibrium point $\mathbf{x}^*$ with parameter $\mathbf{p}_0$ is given by $\mathbf{f}(\mathbf{x}^*, \mathbf{p}_0) = \mathbf{0}$ in Eq. (8.46). In a neighborhood $U(\mathbf{x}^*)$ at parameter $\mathbf{p}_0$, the vector field in Eq. (8.45) is linearized, i.e., $\mathbf{f}(\mathbf{x}, \mathbf{p}_0) = D\mathbf{f}(\mathbf{x}^*, \mathbf{p}_0)(\mathbf{x} - \mathbf{x}^*)$ in Eq. (8.47). For the Jacobian matrix of $D\mathbf{f}(\mathbf{x}^*, \mathbf{p}_0)$, the corresponding eigenvalues are determined by Eq. (8.48). Thus, the eigenvalue equation is given in Eq. (8.49). Consider a general case for simple and repeated eigenvalues of

$$\begin{aligned} f(\lambda) =&(-1)^n\lambda^n + a_1\lambda^{n-1} + \cdots + a_{n-1}\lambda + a_n = 0 \\ &\text{equivalent to} \\ f(\lambda) =&(-1)^n \prod_{i_1=1}^{l_i} (\lambda - \lambda_{i_1}^{(1)}) \prod_{i_2=1}^{l_2} (\lambda - \alpha_{i_2}^{(2)} - \beta_{i_2}^{(2)}\mathbf{i})(\lambda - \alpha_{i_2}^{(2)} + \beta_{i_2}^{(2)}\mathbf{i}) \\ &\times \prod_{i_3=1}^{l_3} (\lambda - \lambda_{i_3}^{(3)})^{q_{i_3}} \prod_{i_4=1}^{l_4} ((\lambda - \alpha_{i_4}^{(4)} - \beta_{i_4}^{(4)}\mathbf{i})(\lambda - \alpha_{i_4}^{(4)} + \beta_{i_4}^{(4)}\mathbf{i}))^{q_{i_4}} \\ =& 0. \end{aligned} \tag{8.65}$$

(i) For simple real eigenvalues, if

$$\begin{aligned} &\lambda = \lambda_{i_1}^{(1)} \\ &(i_1 \in \{s_1^{(1)}, s_2^{(1)}, \ldots, s_{l_1}^{(1)}\} \subset \{1, 2, \ldots, n\}), \end{aligned} \tag{8.66}$$

then

$$\begin{aligned} &(D\mathbf{f}(\mathbf{x}^*, \mathbf{p}_0) - \lambda_{i_1}^{(1)}\mathbf{I})\mathbf{v}^{(i_1)} = \mathbf{0} \\ &(i_1 \in \{s_1^{(1)}, s_2^{(1)}, \ldots, s_{l_1}^{(1)}\} \subset \{1, 2, \ldots, n\}). \end{aligned} \tag{8.67}$$

(ii) For complex eigenvalues with simple pairs, if

$$\begin{aligned} &\lambda_1 = \alpha_{i_2}^{(2)} + \beta_{i_2}^{(2)}\mathbf{i}, \lambda_2 = \alpha_{i_2}^{(2)} - \beta_{i_2}^{(2)}\mathbf{i}, \\ &\mathbf{i} = \sqrt{-1}, (i_2 \in \{s_1^{(2)}, s_2^{(2)}, \ldots, s_{l_2}^{(2)}\} \subset \{1, 2, \ldots, n\}), \end{aligned} \tag{8.68}$$

then

$$
\begin{aligned}
&D\mathbf{f}(\mathbf{x}^*, \mathbf{p}_0)\mathbf{v}_1^{(i_2)} = \alpha_{i_2}^{(2)}\mathbf{v}_1^{(i_2)} + \beta_{i_2}^{(2)}\mathbf{v}_2^{(i_2)},\\
&D\mathbf{f}(\mathbf{x}^*, \mathbf{p}_0)\mathbf{v}_1^{(i_2)} = -\beta_{i_2}^{(2)}\mathbf{v}_1^{(i_2)} + \alpha_{i_2}^{(2)}\mathbf{v}_2^{(i_2)},\\
&(i_2 \in \{s_1^{(2)}, s_2^{(2)}, \ldots, s_{l_2}^{(2)}\} \subset \{1, 2, \ldots, n\}).
\end{aligned} \tag{8.69}
$$

(iii) For repeated real eigenvalues, if

$$
\begin{aligned}
&\lambda = \lambda_{i_3}^{(3)} \text{ with } q_{i_3}\text{-repeated}\\
&(i_3 \in \{s_1^{(3)}, s_2^{(3)}, \ldots, s_{l_3}^{(3)}\} \subset \{1, 2, \ldots, n\}),
\end{aligned} \tag{8.70}
$$

then

$$
\begin{aligned}
&D\mathbf{f}(\mathbf{x}^*, \mathbf{p}_0)\mathbf{v}_1^{(i_3)} = \lambda_{i_3}^{(3)}\mathbf{v}_1^{(i_3)}\\
&D\mathbf{f}(\mathbf{x}^*, \mathbf{p}_0)\mathbf{v}_j^{(i_3)} = \lambda_{i_3}^{(3)}\mathbf{v}_j^{(i_3)} + \delta_j^{(i_3)}\mathbf{v}_{j-1}^{(i_3)},\\
&(j = 2, 3, \cdots, N_{i_3}), \delta_j^{(i_3)} \in \{0, 1\}\\
&(i_3 \in \{s_1^{(3)}, s_2^{(3)}, \ldots, s_{l_3}^{(3)}\} \subset \{1, 2, \ldots, n\}).
\end{aligned} \tag{8.71}
$$

(iv) For complex eigenvalues with repeated pairs, if

$$
\begin{aligned}
&\lambda_1 = \alpha_{i_4}^{(4)} + \beta_{i_4}^{(4)}\mathbf{i}, \lambda_2 = \alpha_{i_4}^{(4)} - \beta_{i_4}^{(4)}\mathbf{i}, \text{ with } q_{i_4}\text{-repeated.}\\
&\mathbf{i} = \sqrt{-1}, (i_4 \in \{s_1^{(4)}, s_2^{(4)}, \ldots, s_{l_4}^{(4)}\} \subset \{1, 2, \ldots, n\}),
\end{aligned} \tag{8.72}
$$

then

$$
\begin{aligned}
&\left.\begin{aligned}
&D\mathbf{f}(\mathbf{x}^*, \mathbf{p}_0)\mathbf{u}_1^{(i_4)} = \alpha_{i_4}^{(4)}\mathbf{u}_1^{(i_4)} + \beta_{i_4}^{(4)}\mathbf{v}_1^{(i_4)},\\
&D\mathbf{f}(\mathbf{x}^*, \mathbf{p}_0)\mathbf{v}_1^{(i_4)} = -\beta_{i_4}^{(4)}\mathbf{u}_1^{(i_4)} + \alpha_{i_4}^{(4)}\mathbf{v}_1^{(i_4)},
\end{aligned}\right\}\\
&\left.\begin{aligned}
&D\mathbf{f}(\mathbf{x}^*, \mathbf{p}_0)\mathbf{u}_j^{(i_4)} = \alpha_{i_4}^{(4)}\mathbf{u}_j^{(i_4)} + \beta_{i_4}^{(4)}\mathbf{v}_j^{(i_4)} + \delta_j^{(i_4)}\mathbf{u}_{j-1}^{(i_4)},\\
&D\mathbf{f}(\mathbf{x}^*, \mathbf{p}_0)\mathbf{v}_j^{(i_4)} = -\beta_{i_4}^{(4)}\mathbf{u}_j^{(i_4)} + \alpha_{i_4}^{(4)}\mathbf{v}_j^{(i_4)} + \delta_j^{(i_4)}\mathbf{v}_{j-1}^{(i_4)},
\end{aligned}\right\}\\
&\delta_j^{(i_3)} \in \{0, 1\}, (j = 2, 3, \ldots, q_{i_4})\\
&(i_4 \in \{s_1^{(4)}, s_2^{(4)}, \ldots, s_{l_4}^{(4)}\} \subset \{1, 2, \ldots, n\}).
\end{aligned} \tag{8.73}
$$

The covariant matrix for the Jacobian matrix $D\mathbf{f}(\mathbf{x}_0, \mathbf{p}_0)$ is

$$\begin{aligned}
&\mathbf{Q} = (\mathbf{v}_1, \mathbf{v}_2, \ldots, \mathbf{v}_n) = (a_{ij})_{n\times n},\\
&\{\mathbf{v}_1, \mathbf{v}_2, \ldots, \mathbf{v}_n\} = \text{inorder}\{\mathbf{v}_1^{(s_1^{(1)})}, \ldots, \mathbf{v}_{l_1}^{(s_{l_1}^{(1)})};\\
&\qquad \mathbf{u}_1^{(s_1^{(2)})}, \mathbf{v}_1^{(s_1^{(2)})}, \ldots, \mathbf{u}_{l_2}^{(s_{l_2}^{(2)})}, \mathbf{v}_{l_2}^{(s_{l_2}^{(2)})};\\
&\qquad \underbrace{\mathbf{v}_1^{(i_3)}, \ldots, \mathbf{v}_{q_{i_3}}^{(i_3)}; (i_3 = s_1^{(3)}, \ldots, s_{l_3}^{(3)});}_{l_3\text{-group}}\\
&\qquad \underbrace{\mathbf{u}_1^{(i_4)}, \mathbf{v}_1^{(i_4)}, \ldots, \mathbf{u}_{q_{i_4}}^{(i_4)}, \mathbf{v}_{q_{i_4}}^{(i_4)}; (i_4 = s_1^{(4)}, \ldots, s_{l_3}^{(4)})\},}_{l_4\text{-group}}
\end{aligned} \tag{8.74}$$

and the contravariant matrix of the Jacobian matrix $D\mathbf{f}(\mathbf{x}_0, \mathbf{p}_0)$ is

$$\mathbf{P} = (\mathbf{v}^1, \mathbf{v}^2, \ldots, \mathbf{v}^n) = (a^{ij})_{n\times n} = \mathbf{Q}^{-1}. \tag{8.75}$$

Thus

$$\begin{aligned}
\mathbf{P}D\mathbf{f}(\mathbf{x}^*, \mathbf{p}_0)\mathbf{Q} &= \mathbf{Q}^{-1}D\mathbf{f}(\mathbf{x}^*, \mathbf{p}_0)\mathbf{Q}\\
&= diag\ (\mathbf{A}^{(1)}, \mathbf{A}^{(2)}, \mathbf{A}^{(3)}, \mathbf{A}^{(4)}),
\end{aligned} \tag{8.76}$$

where

$$\mathbf{A}^{(1)} = diag\ (\lambda_1^{(1)}, \lambda_2^{(1)}, \ldots, \lambda_{l_1}^{(1)}); \tag{8.77}$$

$$\begin{aligned}
&\mathbf{A}^{(2)} = diag\ (\mathbf{A}_1^{(2)}, \mathbf{A}_2^{(2)}, \ldots, \mathbf{A}_{l_2}^{(2)}),\\
&\mathbf{A}_r^{(2)} = \begin{bmatrix} \alpha_{i_2}^{(2)} & \beta_{i_2}^{(2)} \\ -\beta_{i_2}^{(2)} & \alpha_{i_2}^{(2)} \end{bmatrix} (i_2 = s_r^{(2)}; r = 1, 2, \ldots, l_2);
\end{aligned} \tag{8.78}$$

$$\begin{aligned}
&\mathbf{A}^{(3)} = diag\ (\mathbf{B}_1^{(3)}, \mathbf{B}_2^{(3)}, \ldots, \mathbf{B}_{l_3}^{(3)}),\\
&\mathbf{B}_r^{(3)} = \begin{bmatrix} \lambda_{i_3}^{(3)} & \delta_1^{(i_3)} & \cdots & 0 & 0 \\ 0 & \lambda_{i_3}^{(3)} & \cdots & 0 & 0 \\ \vdots & \vdots & \ddots & \vdots & \vdots \\ 0 & 0 & \cdots & \lambda_{i_3}^{(3)} & \delta_{q_{i_3}-1}^{(i_3)} \\ 0 & 0 & \cdots & 0 & \lambda_{i_3}^{(3)} \end{bmatrix}_{q_{i_3}\times q_{i_3}}\\
&(i_3 = s_r^{(3)}; r = 1, 2, \ldots, l_3),\\
&\delta_s^{(i_3)} \in \{0, 1\}\ (s = 1, 2, \ldots, q_{i_3} - 1);
\end{aligned} \tag{8.79}$$

$$\mathbf{A}^{(4)} = diag\,(\mathbf{B}_1^{(4)}, \mathbf{B}_2^{(4)}, \ldots, \mathbf{B}_{l_4}^{(4)}),$$

$$\mathbf{B}_r^{(4)} = \begin{bmatrix} \mathbf{A}_{i_4}^{(4)} & \boldsymbol{\delta}_1^{(i_4)} & \cdots & \mathbf{0} & \mathbf{0} \\ \mathbf{0} & \mathbf{A}_{i_4}^{(4)} & \cdots & \mathbf{0} & \mathbf{0} \\ \vdots & \vdots & \ddots & \vdots & \vdots \\ \mathbf{0} & \mathbf{0} & \cdots & \mathbf{A}_{i_4}^{(4)} & \boldsymbol{\delta}_{q_{i_4}-1}^{(i_4)} \\ \mathbf{0} & \mathbf{0} & \cdots & 0 & \mathbf{A}_{i_4}^{(4)} \end{bmatrix}_{(2q_{i_4}\times 2q_{i_4})} \tag{8.80}$$

$$(i_4 = s_r^{(4)}, r = 1, 2, \ldots, l_4),$$

$$\mathbf{A}_{i_4}^{(4)} = \begin{bmatrix} \alpha_{i_4}^{(4)} & \beta_{i_4}^{(4)} \\ -\beta_{i_4}^{(4)} & \alpha_{i_4}^{(4)} \end{bmatrix}_{2\times 2}, \boldsymbol{\delta}_s^{(i_4)} = \begin{bmatrix} \delta_s^{(i_4)} & 0 \\ 0 & \delta_s^{(i_4)} \end{bmatrix}_{2\times 2} \text{ and}$$

$$\delta_s^{(i_4)} \in \{0, 1\}\ (s = 1, 2, \ldots, q_{i_4} - 1).$$

From the eigenvector space of the linearized Jacobian matrix of equilibriums, the corresponding normal forms are developed through the following definition.

Definition 8.6 Consider an n-dimensional, dynamical system of $\dot{\mathbf{x}} = \mathbf{f}(\mathbf{x}, \mathbf{p})$ in Eq. (8.45). An equilibrium point $\mathbf{x}^*$ with parameter $\mathbf{p}_0$ is given by $\mathbf{f}(\mathbf{x}^*, \mathbf{p}_0) = \mathbf{0}$ in Eq. (8.46). In a neighborhood $U(\mathbf{x}^*)$ at parameter $\mathbf{p}_0$, the vector field in Eq. (8.45) is linearized, i.e., $\mathbf{f}(\mathbf{x}, \mathbf{p}_0) = D\mathbf{f}(\mathbf{x}^*, \mathbf{p}_0)(\mathbf{x} - \mathbf{x}^*)$ in Eq. (8.47). For the Jacobian matrix of $D\mathbf{f}(\mathbf{x}^*, \mathbf{p}_0)$, the corresponding eigenvalues are determined by Eq. (8.48). Thus, the eigenvalue equation is given in Eq. (8.49). The covariant matrix for the Jacobian matrix $D\mathbf{f}(\mathbf{x}_0, \mathbf{p}_0)$ is

$$\mathbf{Q} = (\mathbf{v}_1, \mathbf{v}_2, \ldots, \mathbf{v}_n) = (a_{ij})_{n\times n}, \tag{8.81}$$

and the contravariant matrix of the Jacobian matrix $D\mathbf{f}(\mathbf{x}_0, \mathbf{p}_0)$ is

$$\mathbf{P} = (\mathbf{v}^1, \mathbf{v}^2, \ldots, \mathbf{v}^n) = (a^{ij})_{n\times n} = \mathbf{Q}^{-1}. \tag{8.82}$$

Suppose a new transform is

$$\mathbf{x} = c^k \mathbf{v}_k, \mathbf{x}^* = c^{*k} \mathbf{v}_k, \tag{8.83}$$

and the covariant component is

$$c^k = (\mathbf{v}^k)^{\mathrm{T}} \cdot \mathbf{x} = (\mathbf{v}^k)^{\mathrm{T}} \cdot c^j \mathbf{v}_j = c^j \delta_j^k. \tag{8.84}$$

Thus, the component of the vector field on the covariant direction of $\mathbf{v}_k$ is expanded by

$$\begin{aligned} f^k =&(\mathbf{v}^k)^{\mathrm{T}} \cdot \mathbf{f}(\mathbf{x},\mathbf{p}) \\ =&(\mathbf{v}^k)^{\mathrm{T}} \cdot \mathbf{f}(\mathbf{x}^*,\mathbf{p}_0) + G_{k(j_1)} z^{j_1} \\ &+\sum_{q=2}^{m_k} \frac{1}{q!} G^{(q)}_{k(j_1 j_2 \cdots j_q)} (z^{j_1} z^{j_2} \ldots z^{j_q}) \\ &+\frac{1}{(m_k+1)!} G^{(m_k+1)}_{k(j_1 j_2 \cdots j_{m_k+1})} (\theta_1 z^{j_1})(\theta_2 z^{j_2}) \ldots (\theta_{m_k+1} z^{j_{m_k+1}}), \end{aligned} \tag{8.85}$$

where

$$\begin{aligned} z^j &= (\mathbf{v}^j)^{\mathrm{T}} \cdot \mathbf{y} = c^j - c^{*j} \quad (j = 1,2,\ldots,n), \\ G_{k(j_1)} &= (\mathbf{v}^k)^{\mathrm{T}} \cdot \partial_{c^{j_1}} \mathbf{f}(\mathbf{x},\mathbf{p})|_{(\mathbf{x}^*,\mathbf{p}_0)} \\ &= (\mathbf{v}^k)^{\mathrm{T}} \cdot \partial_{\mathbf{x}} \mathbf{f}(\mathbf{x},\mathbf{p})|_{(\mathbf{x}^*,\mathbf{p}_0)} \mathbf{v}_{j_1}, \\ G^{(r)}_{k(j_1 j_2 \cdots j_r)} &= (\mathbf{v}^k)^{\mathrm{T}} \cdot \partial^{(r)}_{c^{j_1} c^{j_2} \cdots c^{j_r}} \mathbf{f}(\mathbf{x},\mathbf{p})|_{(\mathbf{x}^*,\mathbf{p}_0)} \\ &= (\mathbf{v}^k)^{\mathrm{T}} \cdot \partial^{(r)}_{\mathbf{x}} \mathbf{f}(\mathbf{x},\mathbf{p})|_{(\mathbf{x}^*,\mathbf{p}_0)} \mathbf{v}_{j_1} \mathbf{v}_{j_2} \ldots \mathbf{v}_{j_r} \\ &= G^{(r)}_{k(j_1 j_2 \cdots j_r)} (\mathbf{x}^*,\mathbf{p}_0). \end{aligned} \tag{8.86}$$

(i) For simple real eigenvalues, consider

$$\begin{aligned} &\lambda = \lambda^{(1)}_{i_1} \\ &(i_1 \in \{s^{(1)}_1, s^{(1)}_2, \ldots, s^{(1)}_{l_1}\} \subset \{1,2,\ldots,n\}) \end{aligned} \tag{8.87}$$

for a specific $\mathbf{p} = \mathbf{p}_0$ on the covariant direction of $\mathbf{v}_{i_1}$. If

$$\begin{aligned} &\lambda^{(1)}_k \neq 0 \; (k = i_1 \in \{s^{(1)}_1, s^{(1)}_2, \ldots, s^{(1)}_{l_1}\} \subset \{1,2,\ldots,n\}) \\ &|G^{(s_k)}_{k(j_1 j_2 \cdots j_{s_{i_1}})} (\mathbf{x}_0,\mathbf{p}_0) z^{j_1} z^{j_2} \ldots z^{j_{s_k}}| \\ &>>|G^{(s_k+1)}_{k(j_1 j_2 \cdots j_{s_{i_1}+1})} (\mathbf{x}_0,\mathbf{p}_0) z^{j_1} z^{j_2} \ldots z^{j_{s_k+1}}| \leq \varepsilon \\ &(j_1, j_2, \ldots j_{s_k+1} = 1,2,\ldots,n), \end{aligned} \tag{8.88}$$

then the normal form on the covariant direction of $\mathbf{v}_{i_1}$ is

$$\dot{z}^{i_1} \approx \lambda_k^{(1)} z^{i_1} + \sum_{q=2}^{s_{i_1}} \frac{1}{q!} G^{(q)}_{i_1(j_1 j_2 \cdots j_q)} (z^{j_1} z^{j_2} \ldots z^{j_q})$$
$$(i_1 \in \{s_1^{(1)}, s_2^{(1)}, \ldots, s_{l_1}^{(1)}\} \subset \{1, 2, \ldots, n\}). \tag{8.89}$$

(ii) For complex eigenvalues with simple pairs, consider

$$\lambda_1^{(i_2)} = \alpha_{i_2}^{(2)} + \beta_{i_2}^{(2)} \mathbf{i}, \lambda_2^{(i_2)} = \alpha_{i_2}^{(2)} + \beta_{i_2}^{(2)} \mathbf{i},$$
$$(i_2 \in \{s_1^{(2)}, s_2^{(2)}, \ldots, s_{l_2}^{(2)}\} \subset \{1, 2, \ldots, n\}) \tag{8.90}$$

for a specific $\mathbf{p} = \mathbf{p}_0$ on the covariant eigenvector plane of $\mathbf{u}_1^{(i_2)}$ and $\mathbf{v}_1^{(i_2)}$. If

$$\begin{gathered}
|G^{(s_{(k)_1})}_{k(j_1 j_2 \cdots j_{s_{(k)_1}})}(\mathbf{x}_0, \mathbf{p}_0) z^{j_1} z^{j_2} \ldots z^{j_{s_{(k)_1}}}| \\
>> |G^{(s_k+1)}_{k(j_1 j_2 \cdots j_{s_{(k)_1}+1})}(\mathbf{x}_0, \mathbf{p}_0) z^{j_1} z^{j_2} \ldots z^{j_{s_{(k)_1}+1}}| = \varepsilon, \\
|G^{(s_k)}_{k(j_1 j_2 \cdots j_{s_{(k)_2}})}(\mathbf{x}_0, \mathbf{p}_0) z^{j_1} z^{j_2} \ldots z^{j_{s_{(k)_2}}}| \\
>> |G^{(s_k+1)}_{k(j_1 j_2 \cdots j_{s_{(k)_2}+1})}(\mathbf{x}_0, \mathbf{p}_0) z^{j_1} z^{j_2} \ldots z^{j_{s_{(k)_2}+1}}| = \varepsilon, \\
(j_1, j_2, \ldots, j_q = 1, 2, \ldots, n), \\
(k = i_2 \in \{s_1^{(2)}, s_2^{(2)}, \ldots, s_{n_2}^{(2)}\} \subset \{1, 2, \ldots, n\}),
\end{gathered} \tag{8.91}$$

then the normal form on the covariant eigenvector plane of $\mathbf{u}_1^{(i_2)}$ and $\mathbf{v}_1^{(i_2)}$ is

$$\begin{aligned}
\dot{z}^{(i_2)_1} &\approx \alpha_{i_2}^{(2)} z^{(i_2)_1} - \beta_{i_2}^{(2)} z^{(i_2)_2} + \sum_{q=2}^{s_{(i_2)_1}} \frac{1}{q!} G^{(q)}_{(i_2)_1(j_1 j_2 \cdots j_q)} (z^{j_1} z^{j_2} \ldots z^{j_q}), \\
\dot{z}^{(i_2)_2} &\approx \beta_{i_2}^{(2)} z^{(i_2)_1} + \alpha_{i_2}^{(2)} z^{(i_2)_2} + \sum_{q=2}^{s_{(i_2)_2}} \frac{1}{q!} G^{(q)}_{(i_2)_2(j_1 j_2 \cdots j_q)} (z^{j_1} z^{j_2} \ldots z^{j_q}),
\end{aligned} \tag{8.92}$$

with $(i_2 \in \{s_1^{(2)}, s_2^{(2)}, \ldots, s_{l_2}^{(2)}\} \subset \{1, 2, \ldots, n\})$,
$(j_1, j_2, \ldots, j_q = 1, 2, \ldots, n)$.

(iii) For repeated real eigenvalues, consider

$$
\begin{aligned}
&\lambda = \lambda_{i_3}^{(3)} \text{ with } q_{i_3}\text{-repeated} \\
&(i_3 \in \{s_1^{(3)}, s_2^{(3)}, \ldots, s_{l_3}^{(3)}\} \subset \{1, 2, \ldots, n\})
\end{aligned}
\tag{8.93}
$$

for a specific $\mathbf{p} = \mathbf{p}_0$ on the covariant space of $\mathbf{v}_j^{(i_3)}$ $(j = 2, 3, \ldots, q_{i_3})$. If

$$
\begin{aligned}
&\lambda_{i_3}^{(3)} \neq 0 \; (k = i_3 \in \{s_1^{(3)}, s_2^{(3)}, \ldots, s_{l_3}^{(3)}\} \subset \{1, 2, \ldots, n\}) \\
&|G_{(k)_j (j_1 j_2 \cdots j_{s_{(k)_j}})}^{(s_{(k)_j})}(\mathbf{x}_0, \mathbf{p}_0) z^{j_1} z^{j_2} \ldots z^{j_{s_{(k)_j}}}| \\
&>> |G_{(k)_j (j_1 j_2 \cdots j_{s_{(k)_j}+1})}^{(s_{(k)_j}+1)}(\mathbf{x}_0, \mathbf{p}_0) z^{j_1} z^{j_2} \ldots z^{j_{s_{(k)_j}+1}}| \leq \varepsilon \\
&(j_1, j_2, \ldots, j_{s_{(k)_j}+1} = 1, 2, \ldots, n),
\end{aligned}
\tag{8.94}
$$

then the normal form on the covariant space of $\mathbf{v}_j^{(i_3)}$ $(j = 1, 2, 3, \ldots, q_{i_3})$ is

$$
\begin{aligned}
&\dot{z}^{(k)_1} \approx \lambda_{i_3}^{(3)} z^{(k)_1} + \sum_{q=2}^{s_{k_1}} \frac{1}{q!} G_{(k)_1 (j_1 j_2 \cdots j_q)}^{(q)} (z^{j_1} z^{j_2} \ldots z^{j_q}) \\
&\left.
\begin{aligned}
\dot{z}^{(k)_j} &\approx \delta_j^{(i_3)} z^{(k)_{j-1}} + \lambda_{i_3}^{(3)} z^{(k)_j} \\
&\quad + \sum_{q=2}^{s_{(k)_j}} \frac{1}{q!} G_{(k)_j (j_1 j_2 \cdots j_q)}^{(q)} (z^{j_1} z^{j_2} \ldots z^{j_q})
\end{aligned}
\right\} (j = 2, 3, \ldots, q_{i_3}), \\
&\delta_j^{(i_3)} \in \{0, 1\} \; (k = i_3 \in \{s_1^{(3)}, s_2^{(3)}, \ldots, s_{l_3}^{(3)}\} \subset \{1, 2, \ldots, n\}).
\end{aligned}
\tag{8.95}
$$

(iv) For complex eigenvalues with repeated pairs, consider

$$
\begin{aligned}
&\lambda_1 = \alpha_{i_4}^{(4)} + \beta_{i_4}^{(4)} \mathbf{i}, \lambda_2 = \alpha_{i_4}^{(4)} - \beta_{i_4}^{(4)} \mathbf{i}, \text{ with } q_{i_4}\text{-repeated} \\
&(i_4 \in \{s_1^{(4)}, s_2^{(4)}, \ldots, s_{l_4}^{(4)}\} \subset \{1, 2, \ldots, n\})
\end{aligned}
\tag{8.96}
$$

on the covariant eigenvector space of $\mathbf{u}_j^{(i_2)}$ and $\mathbf{v}_j^{(i_2)}$ $(j = 1, 2, 3, \ldots, q_{i_4})$. If

$$
\begin{aligned}
&|G^{(s_{(k)j1})}_{(k)_{j1}(j_1 j_2 \cdots j_{s_{(k)j1}})}(\mathbf{x}_0, \mathbf{p}_0) z^{j_1} z^{j_2} \ldots z^{j_{s_{(k)j1}}}| \\
&>>|G^{(s_{(k)j1}+1)}_{(k)_{j1}(j_1 j_2 \cdots j_{s_{(k)j1}+1})}(\mathbf{x}_0, \mathbf{p}_0) z^{j_1} z^{j_2} \ldots z^{j_{s_{(k)j1}+1}}| \leq \varepsilon; \\
&|G^{(s_{(k)j2})}_{(k)_{j2}(j_1 j_2 \cdots j_{s_{(k)j2}})}(\mathbf{x}_0, \mathbf{p}_0) z^{j_1} z^{j_2} \ldots z^{j_{s_{(k)j2}}}| \\
&>>|G^{(s_{(k)j2}+1)}_{(k)_{j2}(j_1 j_2 \cdots j_{s_{(k)j2}+1})}(\mathbf{x}_0, \mathbf{p}_0) z^{j_1} z^{j_2} \ldots z^{j_{s_{(k)j2}+1}}| \leq \varepsilon; \\
&(j_1, j_2, \cdots j_q = 1, 2, \ldots, n), \\
&(k = i_4 \in \{s_1^{(4)}, s_2^{(4)}, \ldots, s_{l_2}^{(4)}\} \subset \{1, 2, \ldots, n\}), \\
&(j = 1, 2, \ldots, q_{i_4})
\end{aligned}
\tag{8.97}
$$

then

$$
\left.
\begin{aligned}
\dot{z}^{(k)_{11}} &\approx \alpha_{i_4}^{(4)} z^{(k)_{11}} - \beta_{i_4}^{(4)} z^{(k)_{12}} \\
&\quad + \sum_{q=2}^{s_{(k)_{11}}} \frac{1}{q!} G^{(q)}_{(k)_{11}(j_1 j_2 \cdots j_q)} (z^{j_1} z^{j_2} \ldots z^{j_q}), \\
\dot{z}^{(k)_{12}} &\approx \beta_{i_4}^{(2)} z^{(k)_{11}} + \alpha_{i_4}^{(4)} z^{(k)_{12}} \\
&\quad + \sum_{q=2}^{s_{(k)_{12}}} \frac{1}{q!} G^{(q)}_{(k)_{12}(j_1 j_2 \cdots j_q)} (z^{j_1} z^{j_2} \ldots z^{j_q});
\end{aligned}
\right\}
$$

$$
\left.
\begin{aligned}
\dot{z}^{(k)_{j1}} &\approx \delta_j^{(i_4)} z^{(k)_{(j-1)1}} + \alpha_{i_4}^{(4)} z^{(k)_{j1}} - \beta_{i_4}^{(4)} z^{(k)_{j2}} \\
&\quad + \sum_{q=2}^{s_{(k)_{j1}}} \frac{1}{q!} G^{(q)}_{(k)_{j1}(j_1 j_2 \cdots j_q)} (z^{j_1} z^{j_2} \ldots z^{j_q}), \\
\dot{z}^{(k)_{12}} &\approx \delta_j^{(i_4)} z^{(k)_{(j-1)2}} + \beta_{i_4}^{(2)} z^{(k)_{j1}} + \alpha_{i_4}^{(4)} z^{(k)_{j2}} \\
&\quad + \sum_{q=2}^{s_{(k)_{j2}}} \frac{1}{q!} G^{(q)}_{(k)_{j2}(j_1 j_2 \cdots j_q)} (z^{j_1} z^{j_2} \ldots z^{j_q});
\end{aligned}
\right\}
\tag{8.98}
$$

$(j = 2, 3, \ldots, q_{i_4})$,

with $(k = i_4 \in \{s_1^{(4)}, s_2^{(4)}, \ldots, s_{l_4}^{(4)}\} \subset \{1, 2, \ldots, n\})$

$(j_1, j_2, \ldots, j_q = 1, 2, \ldots, n)$.

The normal forms of nonlinear dynamical systems presented herein are based on the eigenvector space of the corresponding linearized systems at equilibriums in general. From simple and repeated eigenvalues of the linearized systems at equilibriums, the corresponding normal forms of the original nonlinear systems are developed through the differential geometry representation. The infinite-equilibrium systems in the following section can be introduced from the normal forms of nonlinear dynamical systems at bifurcation points via the local analysis.

8.3 Infinite-Equilibrium Systems

In this section, using singularity analysis, the stability of equilibriums in the infinite-equilibrium system will be discussed as in Luo (2019).

Definition 8.7 Consider an autonomous dynamical system as

$$\begin{aligned} \dot{x}_i &= g_{i1}(\mathbf{x}, \mathbf{p}_{i1}) g_{i2}(\mathbf{x}, \mathbf{p}_{i2}) \ldots g_{ir_i}(\mathbf{x}, \mathbf{p}_{ir_i}) \\ &(i = 1, 2, \ldots, n), r_i \geq 1 \end{aligned} \tag{8.99}$$

where

$$g_i(\mathbf{x}, \mathbf{p}) \equiv g_{i1}(\mathbf{x}, \mathbf{p}_{i1}) g_{i2}(\mathbf{x}, \mathbf{p}_{i2}) \ldots g_{ir_i}(\mathbf{x}, \mathbf{p}_{ir_i}). \tag{8.100}$$

Such a system is called an infinite-equilibrium system if the following conditions exist

$$\begin{aligned} & g_i(\mathbf{x}, \mathbf{p}) = 0 (i = 1, 2, \ldots, n), \\ & g_{i_1 j_1}(\mathbf{x}, \mathbf{p}_{i_1 j_1}) \equiv g_{i_2 j_2}(\mathbf{x}, \mathbf{p}_{i_2 j_2}) = \varphi(\mathbf{x}, \mathbf{p}) = 0 \\ & \text{for } i_1, i_2 \in \{1, 2, \ldots, n\}, i_1 \neq i_2; \\ & j_1 \in \{1, 2, \ldots, r_{i_1}\}, j_2 \in \{1, 2, \ldots, r_{i_2}\}. \end{aligned} \tag{8.101}$$

The corresponding surface of $\varphi(\mathbf{x}, \mathbf{p}) = 0$ is called an infinite-equilibrium surface in such a dynamical system.

Definition 8.8 Consider an autonomous dynamical system as

$$\begin{aligned} \dot{x}_i &= \varphi(\mathbf{x}, \mathbf{p}_1) g_i(\mathbf{x}, \mathbf{p}) \\ &(i = 1, 2, \ldots, n). \end{aligned} \tag{8.102}$$

(i) An $(n-1)$-dimensional surface of $\varphi(\mathbf{x}, \mathbf{p}_1) = 0$ is called an $(n-1)$-dimensional infinite-equilibrium surface if equilibrium $\mathbf{x}^*$ in Eq. (8.102) satisfies the following condition:

$$\varphi(\mathbf{x}^*, \mathbf{p}_1) = 0. \tag{8.103}$$

(ii) The equilibrium $\mathbf{x}^*$ is called a simple equilibrium if

$$\begin{array}{l} \mathbf{x}^* \neq \mathbf{x}_\alpha^* \text{ for } \mathbf{x}^*, \mathbf{x}_\alpha^* \in E \text{ and } \alpha = 1, 2, \ldots \\ \text{with } E = \{\mathbf{x}^* | g_i(\mathbf{x}^*, \mathbf{p}) = 0, i = 1, 2, \ldots, n\}. \end{array} \tag{8.104}$$

(iii) The equilibrium $\mathbf{x}^*$ is called a r-repeated equilibrium (or r^{th}-order equilibrium) if

$$\begin{array}{l} \mathbf{x}^* = \mathbf{x}_1^* = \mathbf{x}_2^* = \cdots = \mathbf{x}_{r-1}^* \\ \text{for } \mathbf{x}^*, \mathbf{x}_\alpha^* \in E \text{ and } \alpha = 1, 2, \ldots, r-1 \\ \text{with } E = \{\mathbf{x}^* | g_i(\mathbf{x}^*, \mathbf{p}) = 0, i = 1, 2, \ldots, n\}. \end{array} \tag{8.105}$$

(iv) The equilibrium $\mathbf{x}^*$ is called an intersection equilibrium on the infinite-equilibrium surface if

$$\left.\begin{array}{l} \varphi(\mathbf{x}^*, \mathbf{p}_1) = 0, \\ g_i(\mathbf{x}^*, \mathbf{p}) = 0, i = i_1, i_2, \ldots, i_{n-1} \end{array}\right\} \\ i_\alpha \in \{1, 2, \ldots, n\} \text{ and } \alpha = 1, 2, \ldots, n-1. \tag{8.106}$$

8.3.1 One-Infinite-Equilibrium Systems

Consider a simple dynamical system with one infinite equilibrium as

$$\dot{x}_1 = x_2, \dot{x}_2 = x_1 x_2. \tag{8.107}$$

The equilibriums are

$$\begin{array}{l} x_1^* \in (-\infty, \infty), x_2^* = 0; \\ x_1^* = 0, x_2^* = 0. \end{array} \tag{8.108}$$

The first integral manifold of Eq. (8.107) is

$$x_2 = \frac{1}{2} x_1^2 + C. \tag{8.109}$$

Consider the first integral manifolds going through the following points:

$$
\begin{aligned}
x_2 &= \frac{1}{2}x_1^2 \ \text{ for } (x_1^*, x_2^*) = (0,0),\\
x_2 &= \frac{1}{2}(x_1^2 - a^2) \ \text{ for } (x_1^*, x_2^*) = (\pm a, 0),\\
x_2 &= \frac{1}{2}(x_1^2 + b^2) \ \text{ for } (x_1, x_2) = (0, \frac{1}{2}b^2)
\end{aligned}
\tag{8.110}
$$

where $a > 0$.

From Eq. (8.107), the variational equations at the equilibriums are

$$
\begin{aligned}
\Delta\dot{x}_1 &= \Delta x_2 = G^{(1)}_{1(j_1)} z^{j_1},\\
\Delta\dot{x}_2 &= x_2^*\Delta x_1 + x_1^*\Delta x_2 + \Delta x_1\Delta x_2\\
&= G^{(1)}_{2(j_1)} z^{j_1} + \frac{1}{2!}A^{(2)}_{2(j_1 j_2)} z^{j_1} z^{j_2}
\end{aligned}
\tag{8.111}
$$

where

$$
\begin{aligned}
&z^j = \Delta x_j \ (j = 1,2),\\
&G^{(1)}_{1(1)} = 0, G^{(1)}_{1(2)} = 1; G^{(1)}_{2(1)} = x_2^*, G^{(1)}_{1(2)} = x_1^*,\\
&G^{(2)}_{2(11)} = G^{(2)}_{2(22)} = 0, G^{(2)}_{2(12)} = G^{(2)}_{2(21)} = 1.
\end{aligned}
\tag{8.112}
$$

For equilibrium of $(x_1^*, x_2^*) = (0,0)$, the variational equation is

$$
\Delta\dot{x}_1 = \Delta x_2 \text{ and } \Delta\dot{x}_2 = \Delta x_1\Delta x_2. \tag{8.113}
$$

On the x_1-direction, $\Delta\dot{x}_1 = \Delta x_2 > 0$ if $\Delta x_2 > 0$. The equilibrium of $(0,0)$ is the first-order upper saddle. However, $\Delta\dot{x}_1 = \Delta x_2 < 0$ if $\Delta x_2 < 0$. So the equilibrium of $(0,0)$ is the first-order lower saddle. On the x_2-direction, for $\Delta x_2 > 0$, $\Delta\dot{x}_2 = \Delta x_1\Delta x_2 < 0$ if $\Delta x_1 < 0$, and $\Delta\dot{x}_2 = \Delta x_1\Delta x_2 < 0$ if $\Delta x_1 > 0$. Thus, the equilibrium of $(0,0)$ is the second-order upper saddle for $\Delta x_2 > 0$ on the x_2-direction. For $\Delta x_2 < 0$, $\Delta\dot{x}_2 = \Delta x_1\Delta x_2 > 0$ if $\Delta x_1 < 0$, and $\Delta\dot{x}_2 = \Delta x_1\Delta x_2 < 0$ if $\Delta x_1 > 0$. Thus, the equilibrium of $(0,0)$ is the second-order lower saddle for $\Delta x_2 > 0$ on the x_2-direction. Therefore, the equilibrium of $(0,0)$ on the first integral manifold is the second-order upper saddle.

For $(x_1^*, x_2^*) = (a,0)$, the variational equation is

$$
\Delta\dot{x}_1 = \Delta x_2 \text{ and } \Delta\dot{x}_2 = a\Delta x_2. \tag{8.114}
$$

On the x_1-direction, $\Delta\dot{x}_1 = \Delta x_2 > 0$ if $\Delta x_2 > 0$ and $\Delta\dot{x}_1 = \Delta x_2 < 0$ if $\Delta x_2 < 0$. The equilibrium of $(a,0)$ is the first-order upper saddle for $\Delta x_2 > 0$ and the first-order lower saddle for $\Delta x_2 < 0$ on the x_1-direction. On the x_2-direction, for $a > 0$, $\Delta\dot{x}_2 = a\Delta x_2 < 0$ if $\Delta x_2 < 0$ and $\Delta\dot{x}_2 = a\Delta x_2 > 0$ if $\Delta x_2 > 0$. Thus, the equilibrium of $(a,0)$ is the second-order source on the x_2-direction. Therefore, the equilibrium of $(a,0)$ is a source.

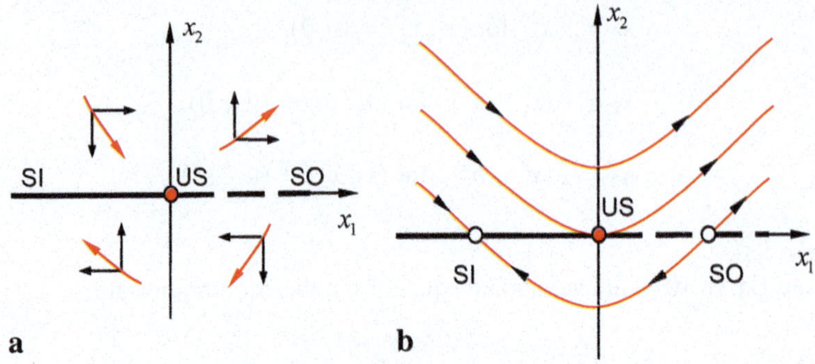

Fig. 8.1 Equilibrium stability for an infinite-equilibrium system: (**a**) local analysis for equilibriums, (**b**) the first integral manifolds. The solid and dashed thick lines are for sink and source infinite equilibriums, respectively. The parabolic curves are for the first integral manifolds. The filled circular symbol is for the upper saddle (US). The hollow circular symbols are for sink (SI) and source (SO) on the infinite-equilibrium line

For $(x_1^*, x_2^*) = (-a, 0)$, the variational equation is

$$\Delta\dot{x}_1 = \Delta x_2 \text{ and } \Delta\dot{x}_2 = -a\Delta x_2. \tag{8.115}$$

On the x_1-direction, $\Delta\dot{x}_1 = \Delta x_2 > 0$ if $\Delta x_2 > 0$ and $\Delta\dot{x}_1 = \Delta x_2 < 0$ if $\Delta x_2 < 0$. The equilibrium of $(-a, 0)$ is the first-order upper saddle for $\Delta x_2 > 0$ and the first-order lower saddle for $\Delta x_2 < 0$ on the x_1-direction. On the x_2-direction, for $a > 0$, $\Delta\dot{x}_2 = -a\Delta x_2 > 0$ if $\Delta x_2 < 0$ and $\Delta\dot{x}_2 = -a\Delta x_2 < 0$ if $\Delta x_2 > 0$. Thus, the equilibrium of $(-a, 0)$ is the first-order sink on the x_2-direction. Therefore, the equilibrium of $(-a, 0)$ is a sink.

The infinite sink of $(-a, 0)$ and the infinite source of $(a, 0)$ for $a \in (0, \infty)$ with the second-order upper-saddle switching point $(0, 0)$ form the entire infinite equilibrium. The phase portrait with the first integral manifolds is presented in Fig. 8.1. The solid and dashed thick lines are for sink and source infinite equilibriums, respectively. The parabolic curves are for the first integral manifolds. The filled circular symbol is for the upper saddle (US) on the first integral manifold. The hollow circular symbols are for sink (SI) and source (SO) on the infinite-equilibrium line of $x_2 = 0$.

8.3.2 Two-Infinite-Equilibrium Systems

Consider a dynamical system with two infinite-equilibriums

$$\begin{aligned} \dot{x}_1 &= x_2(x_1 - a_1)(x_2 - b_1), \\ \dot{x}_2 &= -x_1(x_1 - a_1)(x_2 - b_1) \end{aligned} \tag{8.116}$$

where $a_1 > 0$, $b_1 > 0$. The one simple equilibrium and two infinite equilibriums are

$$\left.\begin{array}{l}x_1^* = a_1, x_2^* \in (-\infty,\infty),\\ x_1^* \in (-\infty,\infty), x_2 = b_1;\end{array}\right\} \text{for two infinite equilbriums,}$$
$$x_1^* = a_1, x_2 = b_1;\ \text{for intersection equilbriums,} \tag{8.117}$$
$$x_1^* = 0, x_2^* = 0\ \text{for simple eqilibrium,}$$
$$\left.\begin{array}{l}x_1^* = 0, x_2^* = b_1;\\ x_1^* = a_1, x_2^* = 0\end{array}\right\} \text{for simple eqilibriums.}$$

The stability of equilibriums is determined by the local analysis. The corresponding G-functions are as follows:

$$\left.\begin{array}{l}G_{1(1)}^{(1)} = x_2^*(x_2^* - b_1),\\ G_{1(2)}^{(1)} = (x_1^* - a_1)(x_2^* - b_1) + x_2^*(x_1^* - a_1);\\ G_{2(1)}^{(1)} = -(x_1^* - a_1)(x_2^* - b_1) - x_1^*(x_2^* - b_1),\\ G_{2(2)}^{(1)} = -x_1^*(x_1^* - a_1);\end{array}\right\} \text{for the first order,}$$
$$\left.\begin{array}{l}G_{1(11)}^{(2)} = 0,\\ G_{1(12)}^{(2)} = G_{1(21)}^{(2)} = 2[(x_2^* - b_1) + x_2^*],\\ G_{1(22)}^{(2)} = 2(x_1^* - a_1),\\ G_{2(11)}^{(2)} = -2(x_2^* - b_1),\\ G_{2(12)}^{(2)} = G_{2(21)}^{(2)} = -[(x_1^* - a_1) + x_1^*],\\ G_{2(22)}^{(2)} = 0.\end{array}\right\} \text{for the second order.} \tag{8.118}$$

Consider the infinite equilibriums of $x_1^* = a_1$, $x_2^* \in (-\infty, \infty)$. The corresponding G-functions and variational equations are given by

$$\left.\begin{array}{l}G_{1(1)}^{(1)} = x_2^*(x_2^* - b_1),\\ G_{1(2)}^{(1)} = 0;\\ G_{2(1)}^{(1)} = -a_1(x_2^* - b_1),\\ G_{2(2)}^{(1)} = 0;\end{array}\right\}$$
$$\Delta\dot{x}_1 = G_{1(1)}^{(1)}\Delta x_1 = x_2^*(x_2^* - b_1)\Delta x_1, \tag{8.119}$$
$$\Delta\dot{x}_2 = G_{2(1)}^{(1)}\Delta x_1 = -a_1(x_2^* - b_1)\Delta x_1.$$

Thus, for $x_2^* \in (b_1, \infty)$, there exists

$$\left.\begin{array}{l}\Delta\dot{x}_1 > 0, \Delta\dot{x}_2 < 0 \text{ for } \Delta x_1 > 0,\\ \Delta\dot{x}_1 < 0, \Delta\dot{x}_2 > 0 \text{ for } \Delta x_1 < 0.\end{array}\right\} \tag{8.120}$$

Such an infinite-equilibrium interval of $x_2^* \in (b_1, \infty)$ is a source of the first order. For $x_2^* \in (0, b_1)$, we have

$$\left.\begin{array}{l}\Delta\dot{x}_1 < 0, \Delta\dot{x}_2 > 0 \text{ for } \Delta x_1 > 0,\\ \Delta\dot{x}_1 > 0, \Delta\dot{x}_2 < 0 \text{ for } \Delta x_1 < 0.\end{array}\right\} \tag{8.121}$$

Such an infinite-equilibrium interval of $x_2^* \in (0, b_1)$ is a sink of the first order. For $x_2^* \in (-\infty, 0)$, we have

$$\left.\begin{array}{l}\Delta\dot{x}_1 > 0, \Delta\dot{x}_2 > 0 \text{ for } \Delta x_1 > 0,\\ \Delta\dot{x}_1 < 0, \Delta\dot{x}_2 > 0 \text{ for } \Delta x_1 < 0.\end{array}\right\} \tag{8.122}$$

Such an infinite-equilibrium interval of $x_2^* \in (-\infty, 0)$ is a source of the first order.

Consider the infinite equilibriums of $x_1^* \in (-\infty, \infty)$, $x_2^* = b_1$. The corresponding G-functions and variational equations are

$$\begin{array}{l}\left.\begin{array}{l}G_{1(1)}^{(1)} = 0,\\ G_{1(2)}^{(1)} = b_1(x_1^* - a_1);\\ G_{2(1)}^{(1)} = 0,\\ G_{2(2)}^{(1)} = -x_1^*(x_1^* - a_1);\end{array}\right\}\\ \Delta\dot{x}_1 = G_{1(2)}^{(1)}\Delta x_1 = b_1(x_1^* - a_1)\Delta x_2,\\ \Delta\dot{x}_2 = G_{2(2)}^{(1)}\Delta x_1 = -x_1^*(x_1^* - a_1)\Delta x_2.\end{array} \tag{8.123}$$

Thus, for $x_1^* \in (a_1, \infty)$, we have

$$\left.\begin{array}{l}\Delta\dot{x}_1 > 0, \Delta\dot{x}_2 < 0 \text{ for } \Delta x_2 > 0,\\ \Delta\dot{x}_1 < 0, \Delta\dot{x}_2 > 0 \text{ for } \Delta x_2 < 0.\end{array}\right\} \tag{8.124}$$

Such an infinite-equilibrium interval of $x_1^* \in (a_1, \infty)$ is a sink of the first order. For $x_1^* \in (a_1, 0)$, there exists

$$\left.\begin{array}{l}\Delta\dot{x}_1 < 0, \Delta\dot{x}_2 > 0 \text{ for } \Delta x_2 > 0,\\ \Delta\dot{x}_1 > 0, \Delta\dot{x}_2 < 0 \text{ for } \Delta x_2 < 0.\end{array}\right\} \tag{8.125}$$

Such an infinite-equilibrium interval of $x_2^* \in (0, b_1)$ is a source of the first order. For $x_1^* \in (-\infty, 0)$, there exists

$$\left.\begin{array}{l}\Delta\dot{x}_1 < 0, \Delta\dot{x}_2 < 0 \text{ for } \Delta x_2 > 0,\\ \Delta\dot{x}_1 > 0, \Delta\dot{x}_2 > 0 \text{ for } \Delta x_2 < 0.\end{array}\right\} \tag{8.126}$$

Such an infinite-equilibrium interval of $x_1^* \in (-\infty, 0)$ is a sink of the first order.

For the equilibrium point of $x_1^* = 0, x_2^* = b_1$, the variational equation is

$$\begin{array}{l}\Delta\dot{x}_1 = G_{1(2)}^{(1)}\Delta x_2 = -a_1 b_1 \Delta x_2,\\ \Delta\dot{x}_2 = \frac{1}{2!}[G_{2(12)}^{(2)} + G_{2(21)}^{(2)}]\Delta x_1 \Delta x_2 = a_1 \Delta x_1 \Delta x_2;\\ \text{where}\\ G_{1(1)}^{(1)} = 0, G_{1(2)}^{(1)} = -a_1 b_1; G_{2(1)}^{(1)} = 0, G_{2(2)}^{(1)} = 0;\\ G_{2(11)}^{(1)} = 0, G_{2(12)}^{(1)} = 2a_1, G_{2(22)}^{(1)} = 0.\end{array} \tag{8.127}$$

On the x_1-direction, $\Delta\dot{x}_1 > 0$ for $\Delta x_2 < 0$ and $\Delta\dot{x}_1 < 0$ for $\Delta x_2 > 0$. Thus, there is a lower saddle of the first order for $\Delta x_2 > 0$, and there is an upper saddle of the first order for $\Delta x_2 < 0$. However, on the x_2-direction, as $\Delta x_1 > 0$, we have $\Delta\dot{x}_2 > 0$ for $\Delta x_2 < 0$ and $\Delta\dot{x}_2 < 0$ for $\Delta x_2 > 0$. The second-order source exists. As $\Delta x_1 < 0$, we have $\Delta\dot{x}_2 < 0$ for $\Delta x_2 < 0$ and $\Delta\dot{x}_2 > 0$ for $\Delta x_2 > 0$. The second-order sink exists. Thus, the second-order upper saddle is for the equilibrium point of $x_1^* = 0, x_2^* = b_1$.

For the equilibrium point of $x_1^* = a_1, x_2^* = 0$, the variational equation is

$$\begin{array}{l}\Delta\dot{x}_1 = \frac{1}{2!}[G_{1(12)}^{(1)} + G_{1(21)}^{(1)}]\Delta x_1 \Delta x_2 = -b_1 \Delta x_1 \Delta x_2,\\ \Delta\dot{x}_2 = G_{2(1)}^{(1)}\Delta x_1 = a_1 b_1 \Delta x_1;\\ \text{where}\\ G_{1(1)}^{(1)} = 0, G_{1(2)}^{(1)} = 0;\\ G_{1(11)}^{(1)} = 0, G_{1(12)}^{(1)} = G_{1(21)}^{(1)} = -b_1, G_{1(22)}^{(1)} = 0;\\ G_{2(1)}^{(1)} = a_1 b_1, G_{2(2)}^{(1)} = 0.\end{array} \tag{8.128}$$

On the x_1-direction, as $\Delta x_2 > 0$, we have $\Delta\dot{x}_1 > 0$ for $\Delta x_1 < 0$ and $\Delta\dot{x}_1 < 0$ for $\Delta x_1 > 0$. There is the second-order source. As $\Delta x_2 < 0$, we have $\Delta\dot{x}_1 < 0$ for $\Delta x_1 < 0$ and $\Delta\dot{x}_1 > 0$ for $\Delta x_1 > 0$. There is a second-order source. On the x_2-direction, $\Delta\dot{x}_2 > 0$ for $\Delta x_1 > 0$ and $\Delta\dot{x}_2 < 0$ for $\Delta x_1 < 0$. Thus, there is an upper saddle of the first order for $\Delta x_1 > 0$, and there is a lower saddle of the first order for $\Delta x_1 < 0$. Thus, the second-order lower saddle in the two directions is for the equilibrium point of $x_1^* = a_1, x_2^* = b_1$.

For the equilibrium point of $x_1^* = a_1, x_2^* = b_1$, the variational equation is

$$\begin{aligned}
&\Delta\dot{x}_1 = \frac{1}{2!}[G^{(2)}_{1(12)} + G^{(2)}_{1(21)}]\Delta x_1 \Delta x_2 = b_1 \Delta x_1 \Delta x_2,\\
&\Delta\dot{x}_2 = \frac{1}{2!}[G^{(2)}_{2(12)} + G^{(2)}_{a(21)}]\Delta x_1 \Delta x_2 = -a_1 \Delta x_1 \Delta x_2;\\
&\text{where}\\
&G^{(1)}_{1(1)} = 0, G^{(1)}_{1(2)} = 0; G^{(1)}_{2(1)} = 0, G^{(1)}_{2(2)} = 0;\\
&G^{(2)}_{1(11)} = 0, G^{(2)}_{1(12)} = G^{(2)}_{1(21)} = b_1, G^{(2)}_{1(22)} = 0;\\
&G^{(2)}_{2(11)} = 0, G^{(2)}_{2(12)} = G^{(2)}_{2(21)} = -a_1, G^{(2)}_{2(22)} = 0.
\end{aligned} \tag{8.129}$$

On the x_1-direction, as $\Delta x_2 > 0$, we have $\Delta\dot{x}_1 < 0$ for $\Delta x_1 < 0$ and $\Delta\dot{x}_1 > 0$ for $\Delta x_1 > 0$. There is the second-order source. As $\Delta x_2 < 0$, we have $\Delta\dot{x}_1 > 0$ for $\Delta x_1 < 0$ and $\Delta\dot{x}_1 < 0$ for $\Delta x_1 > 0$. There is a second-order sink. On the x_2-direction, as $\Delta x_1 > 0$, we have $\Delta\dot{x}_2 > 0$ for $\Delta x_2 < 0$ and $\Delta\dot{x}_2 < 0$ for $\Delta x_2 > 0$. The second-order source exists. As $\Delta x_1 < 0$, we have $\Delta\dot{x}_2 < 0$ for $\Delta x_2 < 0$ and $\Delta\dot{x}_2 > 0$ for $\Delta x_2 > 0$. The second-order sink exists. Thus, the second-order upper saddle for the diagonal direction is for the equilibrium point of $x_1^* = a_1, x_2^* = b_1$.

For equilibrium of $x_1^* = 0, x_2^* = 0$, we have

$$\begin{aligned}
&\Delta\dot{x}_1 = G^{(1)}_{1(2)}\Delta x_2 = a_1 b_1 \Delta x_2,\\
&\Delta\dot{x}_2 = G^{(1)}_{2(1)}\Delta x_1 = -a_1 b_1 \Delta x_1;\\
&\text{where}\\
&G^{(1)}_{1(1)} = 0, G^{(1)}_{1(2)} = a_1 b_1;\\
&G^{(1)}_{2(1)} = -a_1 b_1, G^{(1)}_{2(2)} = 0.
\end{aligned} \tag{8.130}$$

Thus, this equilibrium is a center.

From the foregoing local analysis of equilibriums, the stability of equilibriums is summarized in Table 8.1. The equilibrium of $(0,0)$ is a center, which is not included. The rich dynamics in the infinite dynamical systems can be obtained. Such equilibriums possess higher order singularity, which cannot be analyzed by the traditional eigenvalue analysis.

The first integral manifold of Eq. (8.116) is

$$x_1^2 + x_2^2 = C. \tag{8.131}$$

Consider the manifolds going through the simple and intersection equilibriums and the corresponding manifolds are

Table 8.1 Summarization of equilibrium stability for Eq. (8.116)

Infinite	$x_1^* \in (a_1, \infty), x_2^* = b_1$	$x_1^* \in (0, a_1), x_2^* = b_1$	$x_1^* \in (-\infty, 0), x_2^* = b_1$
	SI	SO	SI
Infinite	$x_1^* = a_1, x_2^* \in (b_1, \infty)$,	$x_1^* = a_1, x_2^* \in (0, b_1)$	$x_1^* = a_1, x_2^* \in (-\infty, 0)$
	SO	SI	SO
Simple	$x_1^* = a_1, x_2^* = 0$,	$x_1^* = 0, x_2^* = b_1$	$x_1^* = a_1, x_2^* = b_1$
	LS	US	US

Infinite Infinite equilibrium, Simple: Simple equilibrium, SI: sink, SO: source, US: upper saddle, LS: lower saddle

Table 8.2 Signs of $(\dot{x}_1, \dot{x}_2)$ in different domains for Eq. (8.116)

$(\dot{x}_1, \dot{x}_2)$	$x_1 \in (a_1, \infty)$	$x_1 \in (0, a_1)$	$x_1 \in (-\infty, 0)$
$x_2 \in (b_1, \infty)$	$(+, -)$	$(-, +)$	$(-, -)$
$x_2 \in (0, b_1)$	$(-, +)$	$(+, -)$	$(+, +)$
$x_2 \in (-\infty, 0)$	$(+, +)$	$(-, -)$	$(-, +)$

$$
\begin{aligned}
&x_1^2 + x_2^2 = a_1^2 \text{ for } x_1^* = a_1, x_2^* = 0;\\
&x_1^2 + x_2^2 = b_1^2 \text{ for } x_1^* = 0, x_2^* = b_1;\\
&x_1^2 + x_2^2 = a_1^2 + b_1^2 \text{ for } x_1^* = a_1, x_2^* = b;\\
&x_1^2 + x_2^2 = 0 \text{ for } x_1^* = 0, x_2^* = 0;\\
&x_1^2 + x_2^2 = C \text{ for others.}
\end{aligned}
\tag{8.132}
$$

From the differential equation in Eq. (8.116), in the different domains separated by the equilibriums, the signs of $(\dot{x}_1, \dot{x}_2)$ are presented in Table 8.2, which indicate the directions of the first-order integral manifolds. The illustrations for the signs in phase plane are presented in Fig. 8.2.

From the local analysis of equilibrium stability, the complete picture of equilibriums with the first integral manifolds is presented in Fig. 8.3. The solid and dashed thick lines are for sink and source infinite equilibriums, respectively. The circles are for the first integral manifolds. The filled circular symbol is for the upper saddle of the second order in two directions. The hollow circular symbols are for the center and two intersection points between infinite-equilibrium lines and coordinates.

8.3.3 *Higher Order Infinite-Equilibrium Systems*

Consider a dynamical system with an infinite equilibrium of the second-order singularity as

$$\dot{x}_1 = -x_1 x_2^2, \dot{x}_2 = x_1 x_2 \sin x_1 \tag{8.133}$$

where $a_1 > 0$, $b_1 > 0$.

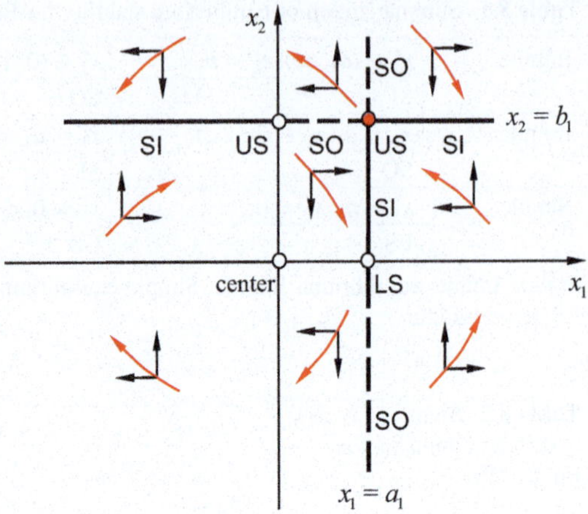

Fig. 8.2 Local analysis of equilibriums for the first integral manifolds. The solid and dashed thick lines are for sink and source infinite equilibriums, respectively. The filled circular symbol is for the intersection equilibrium. The hollow circular symbols are for the center and intersection equilibriums between infinite-equilibriums and coordinates. SI: sink, SO: source, US: upper saddle, LS: lower saddle

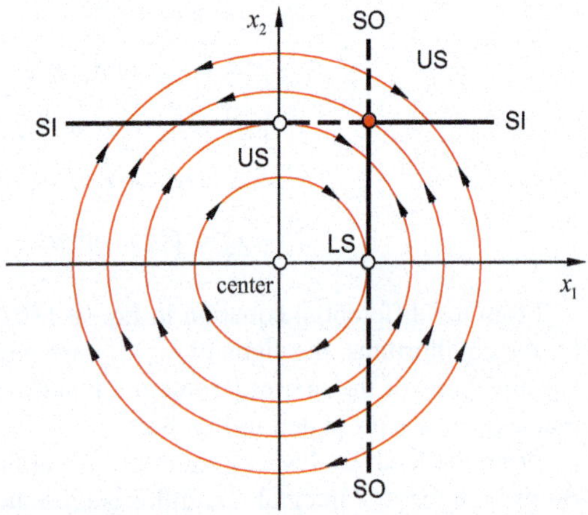

Fig. 8.3 Equilibrium stability for an infinite-equilibrium system in Eq. (8.116). The solid and dashed thick lines are for sink and source infinite equilibriums, respectively. The circles are for the first integral manifolds. The filled circular symbol is for the intersection equilibrium. The hollow circular symbols are for the center and intersection equilibriums between infinite-equilibriums and coordinates. SI: sink, SO: source, US: upper saddle, LS: lower saddle

The simple equilibriums and infinite equilibriums of the nonlinear dynamical system in Eq. (8.133) are

$$\left.\begin{array}{l} x_1^* = 0, x_2^* \in (-\infty, \infty); \\ x_1^* \in (-\infty, \infty), x_2^* = 0 \end{array}\right\} \text{for infintite-equilibriums,}$$
$$x_1^* = 0, x_2^* = 0 \;\; \text{for intersection equilbriums,}$$
$$x_1^* = \pm k\pi \; (k = 1, 2, \ldots), x_2^* = 0 \;\; \text{for simple equilbriums.} \tag{8.134}$$

For the infinite equilibriums of $x_2^* = 0, x_1^* \sin x_1^* \neq 0$, the variational equation is

$$\begin{aligned} \Delta\dot{x}_1 &= -x_1^* \Delta x_2 \Delta x_2, \\ \Delta\dot{x}_2 &= x_1^* \sin x_1^* \Delta x_2. \end{aligned} \tag{8.135}$$

On the x_1-direction, $\Delta\dot{x}_1 < 0$ for $x_1^* > 0$ which implies the second-order lower saddle, and $\Delta\dot{x}_1 > 0$ for $x_1^* < 0$ which implies the second-order upper saddle. On the x_2-direction, the equilibrium is the first-order source for $x_1^* \sin x_1^* > 0$, and sink for $x_1^* \sin x_1^* < 0$. Thus, the infinite equilibrium of $x_2^* = 0, x_1^* \sin x_1^* \neq 0$ is named a sink or source.

For the infinite equilibrium of $x_1^* = 0, x_2^* \neq 0$, the variational equation is

$$\begin{aligned} \Delta\dot{x}_1 &= -(x_2^*)^2 \Delta x_1, \\ \Delta\dot{x}_2 &= x_2^* \cos x_1^* \Delta x_1 \Delta x_1 = x_2^* \Delta x_1 \Delta x_1. \end{aligned} \tag{8.136}$$

On the x_1-direction, the equilibrium is the first-order sink. On the x_2-direction, $\Delta\dot{x}_2 > 0$ for $x_2^* > 0$ which implies the second-order upper-saddle, and $\Delta\dot{x}_2 < 0$ for $x_2^* < 0$ which implies the second-order lower saddle. Thus, the infinite equilibrium of $x_1^* = 0, x_2^* \neq 0$ is named the sink with different type on the x_2-direction.

For the simple equilibriums of $x_2^* = 0, x_1^* = \pm m\pi$ $(m = 1, 2, \ldots)$, the variational equation is

$$\begin{aligned} \Delta\dot{x}_1 &= -x_1^* \Delta x_2 \Delta x_2, \\ \Delta\dot{x}_2 &= x_1^* \cos x_1^* \Delta x_1 \Delta x_2. \end{aligned} \tag{8.137}$$

On the x_1-direction, $\Delta\dot{x}_1 < 0$ for $x_1^* > 0$ which implies the second-order lower saddle, and $\Delta\dot{x}_1 > 0$ for $x_1^* < 0$ which implies the second-order upper saddle. On the x_2-direction,

(i) $\Delta\dot{x}_2 > 0$ for $x_1^* \cos x_1^* > 0$ with $\Delta x_1 > 0$, which implies the second-order source.
(ii) $\Delta\dot{x}_2 < 0$ for $x_1^* \cos x_1^* > 0$ with $\Delta x_1 < 0$, which implies the second-order sink.
(iii) $\Delta\dot{x}_2 < 0$ for $x_1^* \cos x_1^* < 0$ with $\Delta x_1 > 0$, which implies the second-order sink.
(iv) $\Delta\dot{x}_2 > 0$ for $x_1^* \cos x_1^* < 0$ with $\Delta x_1 < 0$, which implies the second-order source.

Thus, the equilibrium based on the x_1-direction is called the second-order lower saddle or upper saddle.

For the equilibrium of $x_1^* = 0, x_2^* = 0$, the variational equation is

$$\begin{aligned} \Delta\dot{x}_1 &= -\Delta x_1 \Delta x_2 \Delta x_2, \\ \Delta\dot{x}_2 &= \Delta x_1 \Delta x_1 \Delta x_2. \end{aligned} \tag{8.138}$$

Table 8.3 Summarization of equilibrium stability for Eq. (8.133)

$x_2^*=0$	$x_1^*\in(4\pi,3\pi)$	$x_1^*\in(3\pi,2\pi)$	$x_1^*\in(2\pi,\pi)$	$x_1^*\in(\pi,0)$
	SI	SO	SI	SO
$x_2^*=0$	$x_1^*\in(-\pi,0)$	$x_1^*\in(-2\pi,-\pi)$	$x_1^*\in(-3\pi,-2\pi)$	$x_1^*\in(-4\pi,-3\pi)$
	SO	SI	SO	SI
$x_1^*=0$	$x_2^*\in(0,\infty)$		$x_2^*\in(-\infty,0)$	
	SI		SI	
$x_2^*=0$	$x_1^*=2m\pi$	$x_1^*=(2m-1)\pi$	$x_1^*=-2m\pi$	$x_1^*=-(2m-1)\pi$
	Second-order LS	Second-order LS	Second-order US	Second-order US
$x_2^*=0$	$x_1^*=0$			
	Center with third-order SI and SO			

SI: sink, SO: source, US: upper saddle, LS: lower saddle

On the x_1-direction, the equilibrium is the third-order sink. On the x_2-direction, the equilibrium is the third-order source. The equilibrium of $(x_1^*,x_2^*)=(0,0)$ is named the third-order center herein.

From the foregoing local analysis, the stability of equilibriums is tabulated in Table 8.3.

The first integral manifold of Eq. (8.133) is

$$\frac{1}{2}x_2^2-\cos x_1=C. \tag{8.139}$$

If the manifolds are going through the intersection equilibriums, the corresponding manifolds are given by

$$\begin{aligned}\frac{1}{2}x_2^2-\cos x_1&=1\text{ for }x_1^*=\pm(2m-1)\pi,x_2^*=0;\\ \frac{1}{2}x_2^2-\cos x_1&=-1\text{ for }x_1^*=(2m)\pi,x_2^*=0;\\ \frac{1}{2}x_2^2-\cos x_1&=C\text{ for others.}\end{aligned} \tag{8.140}$$

From the local analysis of equilibrium stability, equilibriums with the first integral manifolds are presented in Fig. 8.4. The solid and dashed thick lines are for sink and source infinite-equilibriums, respectively. The curves are for the first integral manifolds. The hollow circular symbols are for switching saddles. The filled circular symbol is for the center with third-order sink and source.

8.4 Network-Infinite-Equilibrium Systems

In this section, nonlinear dynamical systems with a network of infinite equilibriums are discussed as in Luo (2019).

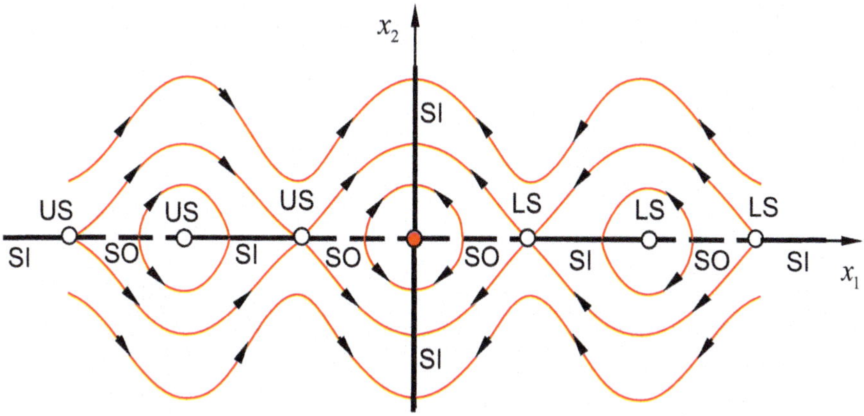

Fig. 8.4 Equilibrium stability for an infinite-equilibrium system. The solid and dashed thick lines are for sink and source infinite-equilibriums, respectively. The filled circular symbol is for the center with third-order sink and source. The hollow circular symbols are for upper-saddle (US) and lower-saddle (LS) on the infinite-equilibrium line

Definition 8.9 Consider an autonomous dynamical system as

$$\begin{array}{l} \dot{x}_i = \varphi_1(\mathbf{x}, \mathbf{p}_1)\varphi_2(\mathbf{x}, \mathbf{p}_2)\ldots\varphi_l(\mathbf{x}, \mathbf{p}_l)g_i(\mathbf{x}, \mathbf{p}) \\ (i = 1, 2, \ldots, n), r_i \geq 1. \end{array} \tag{8.141}$$

(i) An $(n-1)$-dimensional surface network of $\varphi_j(\mathbf{x}, \mathbf{p}_j) = 0$ $(j = 1, 2, \ldots, l)$ in phase space is called an infinite-equilibrium network of dynamical system in Eq. (8.141).

(ii) An $(n-1)$-dimensional surface of $\varphi_j(\mathbf{x}, \mathbf{p}_j) = 0$ $(j \in \{1, 2, \ldots, l\})$ is called an $(n-1)$-dimensional infinite-equilibrium surface if the equilibrium $\mathbf{x}^*$ of Eq. (8.141) satisfies the following condition, i.e.,

$$\varphi_j(\mathbf{x}^*, \mathbf{p}_j) = 0 \text{ for } j \in \{1, 2, \ldots, l\}. \tag{8.142}$$

(iii) An $(n-2)$-dimensional intersection edge of $\varphi_{j_1}(\mathbf{x}, \mathbf{p}_{j_1}) = 0$ and $\varphi_{j_2}(\mathbf{x}, \mathbf{p}_{j_2}) = 0$ is called an $(n-2)$-dimensional infinite-equilibrium edge if the equilibrium $\mathbf{x}^*$ of Eq. (8.141) satisfies the following conditions, i.e.,

$$\begin{array}{l} \varphi_{j_1}(\mathbf{x}^*, \mathbf{p}_{j_1}) = 0, \varphi_{j_2}(\mathbf{x}^*, \mathbf{p}_{j_2}) = 0 \\ \varphi_{j_1}(\mathbf{x}, \mathbf{p}_{j_1}) \neq \varphi_{j_2}(\mathbf{x}, \mathbf{p}_{j_2}) \text{ and } D\varphi_{j_1}(\mathbf{x}, \mathbf{p}_{j_1}) \neq D\varphi_{j_2}(\mathbf{x}, \mathbf{p}_{j_2}) \\ \text{for } j_1, j_2 \in \{1, 2, \ldots, l\}. \end{array} \tag{8.143}$$

(iv) An $(n-r)$-dimensional intersection edge of $\varphi_{j_\alpha}(\mathbf{x}, \mathbf{p}_{j_\alpha}) = 0$ $(\alpha = 1, 2, \ldots, r)$ is called an $(n-r)$-dimensional infinite-equilibrium edge if the equilibrium $\mathbf{x}^*$ of Eq. (8.141) satisfies the following condition, i.e.,

$$\begin{aligned} &\varphi_{j_1}(\mathbf{x}^*, \mathbf{p}_{j_1}) = 0, \varphi_{j_2}(\mathbf{x}^*, \mathbf{p}_{j_2}) = 0, \ldots, \varphi_{j_r}(\mathbf{x}^*, \mathbf{p}_{j_r}) = 0, \\ &\varphi_{j_\alpha}(\mathbf{x}, \mathbf{p}_{j_\alpha}) \neq \varphi_{j_\beta}(\mathbf{x}, \mathbf{p}_{j_\beta}) \text{ and } D\varphi_{j_\alpha}(\mathbf{x}, \mathbf{p}_{j_\alpha}) \neq D\varphi_{j_\beta}(\mathbf{x}, \mathbf{p}_{j_\beta}) \\ &\text{for } j_1, j_2, \ldots, j_r \in \{1, 2, \ldots, l\}; \alpha, \beta \in \{1, 2, \ldots, r\}, \alpha \neq \beta. \end{aligned} \tag{8.144}$$

(v) An intersection vertex of $\varphi_{j_\alpha}(\mathbf{x}, \mathbf{p}_{j_\alpha}) = 0$ $(\alpha = 1, 2, \ldots, n)$ is called an infinite-equilibrium vertex if the equilibrium $\mathbf{x}^*$ of Eq. (8.141) satisfies the following condition, i.e.,

$$\begin{aligned} &\varphi_{j_1}(\mathbf{x}^*, \mathbf{p}_{j_1}) = 0, \varphi_{j_2}(\mathbf{x}^*, \mathbf{p}_{j_2}) = 0, \ldots, \varphi_{j_n}(\mathbf{x}^*, \mathbf{p}_{j_n}) = 0, \\ &\varphi_{j_\alpha}(\mathbf{x}, \mathbf{p}_{j_\alpha}) \neq \varphi_{j_\beta}(\mathbf{x}, \mathbf{p}_{j_\beta}) \text{ and } D\varphi_{j_\alpha}(\mathbf{x}, \mathbf{p}_{j_\alpha}) \neq D\varphi_{j_\beta}(\mathbf{x}, \mathbf{p}_{j_\beta}) \\ &\text{for } j_1, j_2, \ldots, j_n \in \{1, 2, \ldots, l\}, \alpha, \beta \in \{1, 2, \ldots, n\}, \alpha \neq \beta. \end{aligned} \tag{8.145}$$

(vi) An $(n-1)$-dimensional surface of $\varphi_j(\mathbf{x}, \mathbf{p}_j) = 0$ $(j \in \{1, 2, \ldots, l\})$ in phase space is called a rth-order, $(n-1)$-dimensional, infinite equilibrium if

$$\begin{aligned} &\varphi_j(\mathbf{x}, \mathbf{p}_j) \equiv \varphi_{j_1}(\mathbf{x}, \mathbf{p}_{j_1}) = \cdots = \varphi_{j_r}(\mathbf{x}, \mathbf{p}_{j_r}) \\ &\text{for } j_1, j_2, \ldots, j_r \in \{1, 2, \ldots, l\}. \end{aligned} \tag{8.146}$$

8.4.1 A Network-Infinite-Equilibrium System

To demonstrate the infinite-equilibrium network systems, consider a dynamical system of four infinite equilibriums, i.e.,

$$\begin{aligned} \dot{x}_1 &= x_2(x_1 - a_1)(x_2 - b_1)(x_1 + a_1)(x_2 + b_1), \\ \dot{x}_2 &= -x_1(x_1 - a_1)(x_1 - b_1)(x_1 + a_1)(x_2 + b_1) \end{aligned} \tag{8.147}$$

where $a_1 > 0$, $b_1 > 0$. Such four infinite equilibriums form a network of the infinite equilibrium in phase space. The dynamical behaviors of the dynamical systems are separated in different subdomains.

The simple equilibriums and infinite equilibriums are

Table 8.4 Summarization of equilibrium stability for Eq. (8.147)

$x_2^* = b_1$	$x_1^* \in (a_1, \infty)$,	$x_1^* \in (0, a_1)$	$x_1^* \in (-a_1, 0)$,	$x_1^* \in (-\infty, -a_1)$
	SI	SO	SI	SO
$x_2^* = -b_1$	$x_1^* \in (a_1, \infty)$,	$x_1^* \in (0, a_1)$	$x_1^* \in (-a_1, 0)$,	$x_1^* \in (-\infty, -a_1)$
	SO	SI	SO	SI
$x_1^* = a_1$	$x_2^* \in (b_1, \infty)$,	$x_2^* \in (0, b_1)$	$x_2^* \in (-b_1, 0)$	$x_2^* \in (-\infty, -b_1)$
	SO	SI	SO	SI
$x_1^* = -a_1$	$x_2^* \in (b_1, \infty)$,	$x_2^* \in (0, b_1)$	$x_2^* \in (-b_1, 0)$	$x_2^* \in (-\infty, -b_1)$
	SI	SO	SI	SO
$x_1^* = a_1$	$x_2^* = b_1$,	$x_2^* = 0$	$x_2^* = -b_1$	
	US	LS	US	
$x_1^* = 0$	$x_2^* = b_1$,	$x_2^* = 0$	$x_2^* = -b_1$	
	US	Center	LS	
$x_1^* = -a_1$	$x_2^* = b_1$,	$x_2^* = 0$	$x_2^* = -b_1$	
	LS	US	LS	

SI: sink, SO: source, US: upper saddle, LS: lower saddle

$$\left.\begin{array}{l}\left.\begin{array}{l}x_1^* = a_1, x_2^* \in (-\infty, \infty);\\ x_1^* = -a_1, x_2^* \in (-\infty, \infty);\\ x_1^* \in (-\infty, \infty), x_2^* = b_1;\\ x_1^* \in (-\infty, \infty), x_2^* = -b_1\end{array}\right\}\text{for infintite equilibriums,}\\ \left.\begin{array}{l}x_1^* = a_1, x_2^* = b_1; x_1^* = -a_1, x_2^* = b_1;\\ x_1^* = a_1, x_2^* = -b_1; x_1^* = -a_1, x_2^* = -b_1\end{array}\right\}\text{for intersection equilbriums,}\\ x_1^* = 0, x_2^* = 0, \text{for a simple equilbrium,}\\ \left.\begin{array}{l}x_1^* = 0, x_2^* = b_1; x_1^* = 0, x_2^* = -b_1;\\ x_1^* = a_1, x_2^* = 0; x_1^* = -a_1, x_2^* = 0\end{array}\right\}\text{for simple equilbriums.}\end{array}\right. \tag{8.148}$$

The similar local analysis of equilibriums can be completed, and the stability of equilibriums is summarized in Table 8.4. The first integral manifold of Eq. (8.147) is

$$x_1^2 + x_2^2 = C. \tag{8.149}$$

If the first integral manifolds go through the simple and intersection equilibrium, the corresponding first integral manifolds are given by

$$\begin{array}{l}x_1^2 + x_2^2 = a_1^2 \text{ for } x_1^* = \pm a_1, x_2^* = 0;\\ x_1^2 + x_2^2 = b_1^2 \text{ for } x_1^* = 0, x_2^* = \pm b_1;\\ x_1^2 + x_2^2 = a_1^2 + b_1^2 \text{ for } x_1^* = \pm a_1, x_2^* = \pm b;\\ x_1^2 + x_2^2 = 0 \text{ for } x_1^* = 0, x_2^* = 0;\\ x_1^2 + x_2^2 = C \text{ for others.}\end{array} \tag{8.150}$$

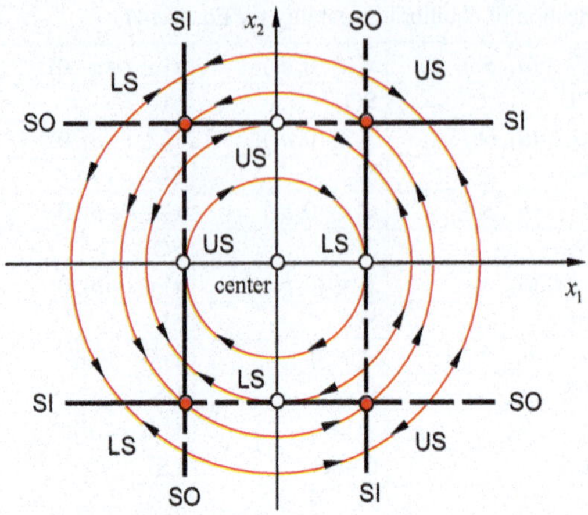

Fig. 8.5 Equilibrium stability for an infinite-equilibrium system in Eq. (8.147). The solid and dashed thick lines are for sink and source infinite equilibriums, respectively. The circles are for the first integral manifolds. The filled circular symbol is for the intersection equilibrium between two infinite equilibrium lines. The hollow circular symbols are for a simple equilibrium and the intersection points between infinite-equilibrium lines and coordinates. SI: Sink, SO: Source, US: upper saddle, LS: lower saddle

From the local analysis of equilibrium stability, equilibriums with the first integral manifolds are presented in Fig. 8.5. The solid and dashed thick lines are for sink and source infinite equilibriums, respectively. The circles are for the first integral manifolds. The filled circular symbol is for the upper saddle of the second order in two directions. The hollow circular symbols are for lower-saddles and upper-saddles in one direction and the center. The network of four infinite equilibriums has nine (9) subdomains separated by the four infinite equilibriums.

8.4.2 *Circular Infinite-Equilibrium Systems*

In this section, a few dynamical systems of circular infinite-equilibrium systems will be presented.

A. *Harmonic motions*: Consider a dynamical system of a circular infinite equilibrium with a harmonic motion, i.e.,

$$\begin{aligned}\dot{x}_1 &= (x_1^2 + x_2^2 - R^2)^m x_2,\\ \dot{x}_2 &= -(x_1^2 + x_2^2 - R^2)^m x_1\end{aligned} \tag{8.151}$$

where $R > 0$. The simple equilibriums and infinite equilibriums are

$$
\begin{aligned}
& x_1^{*2} + x_2^{*2} = R^2 \text{ for the infintite equilibrium;} \\
& \left.\begin{aligned} & x_1^* = \pm R, x_2^* = 0; \\ & x_1^* = 0, x_2^* = \pm R \end{aligned}\right\} \text{ for intersection equilibrium;} \\
& x_1^* = 0, x_2^* = 0 \text{ for simple equilbrium.}
\end{aligned} \tag{8.152}
$$

Consider

$$
\begin{aligned}
& x_1 = \rho \cos\theta, x_2 = \rho \sin\theta; \\
& \dot{x}_1 = \dot{\rho} \cos\theta - \rho\dot{\theta} \sin\theta, \\
& \dot{x}_2 = \dot{\rho} \sin\theta + \rho\dot{\theta} \cos\theta.
\end{aligned} \tag{8.153}
$$

Thus, Eq. (8.151) becomes

$$
\begin{aligned}
& \dot{\rho} = 0 \Rightarrow \rho = C, \\
& \dot{\theta} = -(\rho - R)^m (\rho + R)^m.
\end{aligned} \tag{8.154}
$$

The infinite equilibrium is at $\rho^* = R$. From Eq. (8.154), we have

$$
\begin{aligned}
& \dot{\theta} = 0 \text{ for } \rho = R, \\
& \left.\begin{aligned} & \dot{\theta} = -(\rho - R)^m (\rho + R)^m > 0 \text{ for } \rho < R, \\ & \dot{\theta} = -(\rho - R)^m (\rho + R)^m < 0 \text{ for } \rho > R \end{aligned}\right\} \\
& \text{for } m = 2l - 1 \text{ with } l = 1, 2, \ldots; \\
& \left.\begin{aligned} & \dot{\theta} = -(\rho - R)^m (\rho + R)^m < 0 \text{ for } \rho < R, \\ & \dot{\theta} = -(\rho - R)^m (\rho + R)^m < 0 \text{ for } \rho > R \end{aligned}\right\} \\
& \text{for } m = 2l \text{ with } l = 1, 2, \ldots.
\end{aligned} \tag{8.155}
$$

The motions of $\rho > R$ and $\rho < R$ for $m = 2l - 1$ in phase space are clockwise and counterclockwise, respectively. With $\rho = R$, the system is static, which is an infinite equilibrium. With decreasing ρ, the rotation is from negative to positive direction. Thus, the infinite equilibrium of $\rho = R$ is the $(2l - 1)^{\text{th}}$-order sink (SI).

The motions of $\rho > R$ and $\rho < R$ for $m = 2l$ in phase space are clockwise and counterclockwise, respectively. With $\rho = R$, the system is static, which is an infinite equilibrium. With decreasing ρ, the rotation is in the negative direction only. Thus, the infinite equilibrium of $\rho = R$ is the $(2l)^{\text{th}}$-order lower saddle (LS).

From Eq. (8.155), we have

$$\begin{aligned}\theta &= C_1 \ \text{for}\ \rho = R,\\ \theta &= -(\rho^2 - R^2)^m t + \theta_0 \ \text{for}\ \rho < R,\\ \theta &= -(\rho^2 - R^2)^m t + \theta_0 \ \text{for}\ \rho > R.\end{aligned} \tag{8.156}$$

The first integral manifold of Eq. (8.151) is

$$x_1^2 + x_2^2 = C. \tag{8.157}$$

If the first integral manifolds go through the simple and intersection equilibrium, the corresponding first integral manifolds are given by

$$\begin{aligned}&x_1^{*2} + x_2^{*2} = R^2 \ \text{for the infinite-equilibrium};\\ &x_1^2 + x_2^2 = C_1^2 \ \text{for}\ C_1 < R \ \text{with counterclockwise rotation};\\ &x_1^2 + x_2^2 = C_2^2 \ \text{for}\ C_2 > R \ \text{with clockwise rotation}.\end{aligned} \tag{8.158}$$

From the local analysis of equilibrium stability, equilibriums with the first integral manifolds are presented in Fig. 8.6. The thick circle is for the infinite-equilibrium. The thin circles are for the first integral manifolds. The hollow circular symbols are for the center and intersection points between the infinite-equilibrium circle and coordinates.

B. *Linear motions*: Consider a dynamical system of a circular infinite equilibrium with linear motions, i.e.,

$$\begin{aligned}\dot{x}_1 &= (x_1^2 + x_2^2 - R^2)^m x_1,\\ \dot{x}_2 &= (x_1^2 + x_2^2 - R^2)^m x_2\end{aligned} \tag{8.159}$$

where $R > 0$. The simple equilibriums and infinite equilibriums are

$$\begin{aligned}&x_1^{*2} + x_2^{*2} = R^2 \ \text{for the infintite-equilibrium};\\ &\left.\begin{aligned}&x_1^* = \pm R, x_2^* = 0;\\ &x_1^* = 0, x_2^* = \pm R\end{aligned}\right\} \text{for intersection equilibrium};\\ &x_1^* = 0, x_2^* = 0 \ \text{for simple equilbrium}.\end{aligned} \tag{8.160}$$

Consider

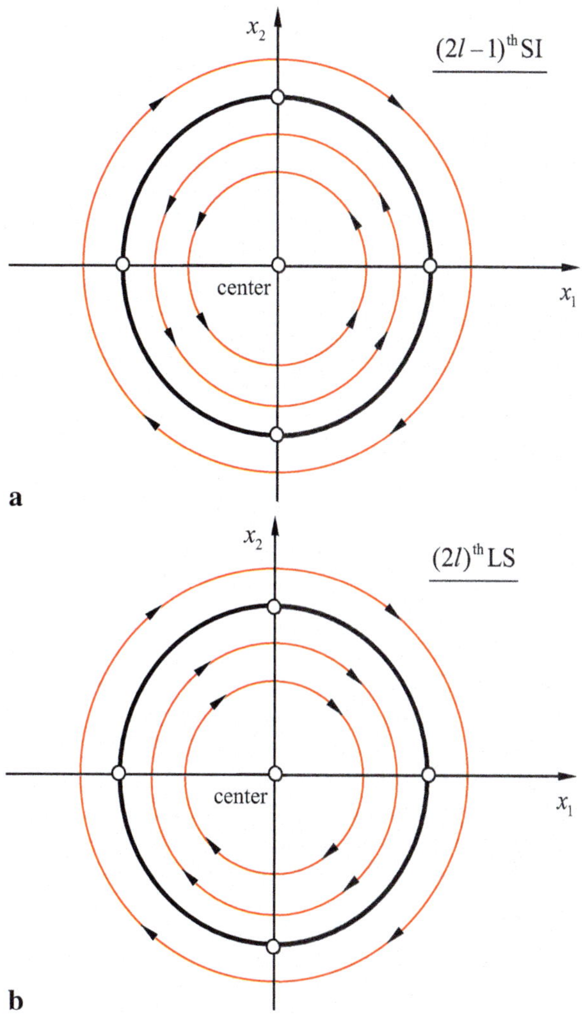

Fig. 8.6 Equilibrium stability for an infinite-equilibrium system in Eq. (8.151). (**a**) The opposite directions of motion rotation and (**b**) the same directions of motion rotation on both sides of the infinite equilibrium. The solid thick circle is for lower-saddle infinite equilibriums. The thin circles are for the first integral manifolds. The hollow circular symbols are for the center and the intersection points between infinite-equilibrium circle and coordinates. LS: lower saddle, SI: sink

$$\begin{aligned} x_1 &= \rho\cos\theta, x_2 = \rho\sin\theta; \\ \dot{x}_1 &= \dot{\rho}\cos\theta - \rho\dot{\theta}\sin\theta, \\ \dot{x}_2 &= \dot{\rho}\sin\theta + \rho\dot{\theta}\cos\theta. \end{aligned} \tag{8.161}$$

Thus, Eq. (8.159) becomes

$$\begin{aligned} \dot{\rho} &= \rho(\rho^2 - R^2)^m = \rho(\rho - R)^m(\rho + R)^m, \\ \dot{\theta} &= 0 \Rightarrow \theta = C. \end{aligned} \tag{8.162}$$

The equilibriums are $\rho^* = 0$ and $\rho^* = R$ for a specific $\theta = C \in [0, 2\pi)$. We have

$$\begin{aligned}
&\dot{\rho} = 0 \text{ for } \rho = R,\\
&\left.\begin{aligned}
&\dot{\rho} = \rho(\rho - R)^m(\rho + R)^m < 0 \text{ for } \rho < R,\\
&\dot{\rho} = \rho(\rho - R)^m(\rho + R)^m > 0 \text{ for } \rho > R
\end{aligned}\right\}\\
&\text{for } m = 2l - 1 \text{ with } l = 1, 2, \ldots;\\
&\left.\begin{aligned}
&\dot{\rho} = \rho(\rho - R)^m(\rho + R)^m > 0 \text{ for } \rho < R,\\
&\dot{\rho} = \rho(\rho - R)^m(\rho + R)^m > 0 \text{ for } \rho > R
\end{aligned}\right\}\\
&\text{for } m = 2l \text{ with } l = 1, 2, \ldots;
\end{aligned} \tag{8.163}$$

$$\begin{aligned}
&\dot{\rho} = 0 \text{ for } \rho = 0,\\
&\Delta\dot{\rho} = (-R^2)^m \Delta\rho < 0 \text{ for } \Delta\rho > 0,\\
&\text{for } m = 2l - 1 \text{ with } l = 1, 2, \ldots;\\
&\Delta\dot{\rho} = (-R^2)^m \Delta\rho > 0 \text{ for } \Delta\rho > 0,\\
&\text{for } m = 2l \text{ with } l = 1, 2, \ldots.
\end{aligned} \tag{8.164}$$

The motions of $\rho > R$ and $\rho < R$ for $m = 2l - 1$ in the radial direction are outwards and inwards the center, respectively. With $\rho = R$, the system is static, which is an infinite-equilibrium. Thus, the infinite-equilibrium of $\rho* = R$ is the $(2l)^{\text{th}}$-order source (SO) in the radial direction. The equilibrium of $(0, 0)$ is a sink.

The motions of $\rho > R$ and $\rho < R$ for $m = 2l$ in the radial direction are outwards the center. With $\rho = R$, the system is static, which is an infinite equilibrium. Thus, the infinite equilibrium of $\rho^* = R$ is the $(2l)^{\text{th}}$-order upper saddle (US) in the radial direction. The equilibrium of $(0, 0)$ is a source.

The corresponding equilibriums are

$$\begin{aligned}
&\rho^* = 0 \text{ for a simple equilibrium},\\
&\rho^* = R \text{ for infinite-equilbrium},\\
&\theta^* = 0, \frac{1}{2}\pi, \pi, \frac{3}{2}\pi; \rho^* = R \text{ for intersection points.}
\end{aligned} \tag{8.165}$$

The first integral manifold of Eq. (8.152) is

$$x_2 = Cx_1. \tag{8.166}$$

If the first integral manifolds go through the infinite equilibrium, the corresponding first integral manifolds are given by

$$\begin{aligned}
&x_2 = Cx_1 \text{ with } C = x_2^*/x_1^* = \tan\theta^*,\\
&x_1^{*2} + x_2^{*2} = R^2 \text{ for the infinite-equilibrium.}
\end{aligned} \tag{8.167}$$

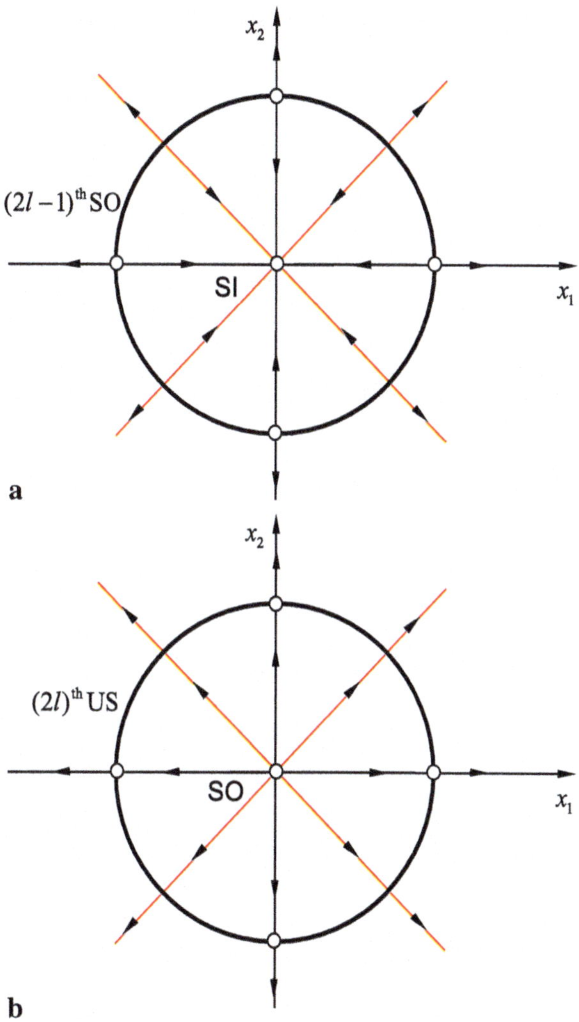

Fig. 8.7 Equilibrium stability for an infinite-equilibrium system in Eq. (8.159). (**a**) The sink of $(0, 0)$ and the $(2l-1)^{\text{th}}$-order source of infinite equilibrium, (**b**) the source of $(0, 0)$ and the $(2l)^{\text{th}}$-order upper saddle of infinite equilibrium. The solid circular curve is for infinite equilibriums. The straight lines are for the first integral manifolds. The hollow circular symbols are for a simple equilibrium and intersection points between the infinite-equilibrium circle and coordinates. SI: sink, SO: source, US: upper saddle

From the local analysis of equilibrium stability, equilibriums with the first integral manifolds are presented in Fig. 8.7. The thick circle is for the infinite-equilibrium circle. The thin lines are for the first integral manifolds. The hollow circular symbols are for the simple equilibrium and the intersection points between the infinite-equilibrium circle and coordinates.

C. *Hyperbolic motion*: Consider a dynamical system of a circular infinite equilibrium with hyperbolic motions, i.e.,

$$\begin{aligned} \dot{x}_1 &= (x_1^2 + x_2^2 - R^2)^m x_2, \\ \dot{x}_2 &= (x_1^2 + x_2^2 - R^2)^m x_1 \end{aligned} \tag{8.168}$$

where $R > 0$.The simple equilibriums and infinite equilibriums are

$$\begin{aligned}
&x_1^{*2} + x_2^{*2} = R^2 \text{ for the infintite-equilibrium;}\\
&\left.\begin{aligned} &x_1^* = \pm R, x_2^* = 0;\\ &x_1^* = 0, x_2^* = \pm R \end{aligned}\right\} \text{ for intersection equilibrium;}\\
&x_1^* = 0, x_2^* = 0 \text{ for simple equilbrium.}
\end{aligned} \tag{8.169}$$

Consider

$$\begin{aligned}
x_1 &= \rho\cos\theta, x_2 = \rho\sin\theta;\\
\dot{x}_1 &= \dot{\rho}\cos\theta - \rho\dot{\theta}\sin\theta,\\
\dot{x}_2 &= \dot{\rho}\sin\theta + \rho\dot{\theta}\cos\theta.
\end{aligned} \tag{8.170}$$

Thus, Eq. (8.168) becomes

$$\begin{aligned}
\dot{\rho} &= \rho(\rho - R)^m(\rho + R)^m \sin 2\theta,\\
\dot{\theta} &= (\rho - R)^m(\rho + R)^m \cos 2\theta.
\end{aligned} \tag{8.171}$$

The corresponding equilibriums are

$$\begin{aligned}
&\rho^* = 0, \text{for simple equilibrium,}\\
&\rho^* = R, \text{for infinite-equilbrium,}\\
&\left.\begin{aligned} &\theta^* = 0, \frac{1}{2}\pi, \pi, \frac{3}{2}\pi; \rho^* = R,\\ &\theta^* = \frac{1}{4}\pi, \frac{3}{4}\pi, \frac{3}{4}\pi, \frac{5}{4}\pi; \rho^* = R \end{aligned}\right\} \text{for intersection points.}
\end{aligned} \tag{8.172}$$

For the simple equilibrium of $(0, 0)$, the linearized equation is

$$\begin{aligned}
\Delta\dot{x}_1 &= (-R^2)^m \Delta x_2,\\
\Delta\dot{x}_2 &= (-R^2)^m \Delta x_1
\end{aligned} \tag{8.173}$$

with the corresponding eigenvalues are

$$\lambda_1 = -(-R^2)^m, \lambda_2 = (-R^2)^m. \tag{8.174}$$

For $m = 2l - 1$, we have

$$\lambda_1 = R^{2(2l-1)} > 0, \lambda_2 = -R^{2(2l-1)} < 0. \tag{8.175}$$

For $m = 2l$, we have

Table 8.5 Summarization of equilibrium stability for Eq. (8.168)

$\rho^* = R$	$\theta^* \in (0, \frac{1}{2}\pi)$,	$\theta^* \in (\frac{1}{2}\pi, \pi)$	$\theta^* \in (\pi, \frac{3}{2}\pi)$	$\theta^* \in (\frac{3}{2}\pi, 2\pi)$
$m = 2l - 1$	$(2l-1)^{\text{th}}$SO	$(2l-1)^{\text{th}}$SI	$(2l-1)^{\text{th}}$SO	$(2l-1)^{\text{th}}$SI
$m = 2l$	$(2l)^{\text{th}}$US	$(2l)^{\text{th}}$LS	$(2l)^{\text{th}}$US	$(2l)^{\text{th}}$LS
$\rho^* = R$	$\theta^* = 0$	$\theta^* = \frac{1}{2}\pi$	$\theta^* = \pi$	$\theta^* = \frac{3}{2}\pi$
$m = 2l - 1$	$(2l)^{\text{th}}$US	$(2l)^{\text{th}}$US	$(2l)^{\text{th}}$LS	$(2l)^{\text{th}}$US
$m = 2l$	$(2l+1)^{\text{th}}$US	$(2l+1)^{\text{th}}$US	$(2l+1)^{\text{th}}$LS	$(2l+1)^{\text{th}}$LS
$\rho^* = R$	$\theta^* = \frac{1}{4}\pi$	$\theta^* = \frac{3}{4}\pi$	$\theta^* = \frac{5}{4}\pi$	$\theta^* = \frac{7}{4}\pi$
$m = 2l - 1$	$(2l-1)^{\text{th}}$SO	$(2l-1)^{\text{th}}$SI	$(2l-1)^{\text{th}}$SO	$(2l-1)^{\text{th}}$SI
$m = 2l$	$(2l)^{\text{th}}$US	$(2l)^{\text{th}}$LS	$(2l)^{\text{th}}$US	$(2l)^{\text{th}}$LS

SI: sink, SO: source, US: upper saddle, LS: lower saddle

$$\lambda_1 = -R^{2(2l)} < 0, \lambda_2 = R^{2(2l)} > 0. \tag{8.176}$$

Therefore, the equilibrium of $(0, 0)$ is a saddle. From the local analysis, the behaviors of equilibriums are summarized in Table 8.5.
The first integral manifold of Eq. (8.168) is

$$\begin{aligned} x_2^2 - x_1^2 &= C_1, \\ x_1^2 - x_2^2 &= C_2. \end{aligned} \tag{8.177}$$

If the first integral manifolds go through the simple and intersection equilibrium, and the infinite equilibrium, the corresponding first integral manifolds are given by

$$\left.\begin{aligned} &x_2^2 - x_1^2 = R^2, \\ &x_1^2 - x_2^2 = R^2 \end{aligned}\right\} \text{for the intersection equilibriums,}$$
$$\left.\begin{aligned} &x_2^2 - x_1^2 = C_1 \equiv x_2^{*2} - x_1^{*2}, \\ &x_1^2 - x_2^2 = C_2 \equiv x_1^{*2} - x_2^{*2}, \\ &x_1^{*2} + x_2^{*2} = R^2 \end{aligned}\right\} \text{for the infinite equilibrium,} \tag{8.178}$$
$$\left.\begin{aligned} &x_2 = x_1, \\ &x_2 = -x_1 \end{aligned}\right\} \text{for the simple equilibrium of } (0, 0).$$

From the local analysis of equilibrium stability, equilibriums with the first integral manifolds are presented in Fig. 8.8. The solid and dashed thick lines are for sink and source infinite equilibriums, respectively. The circles are for the first integral manifolds. The hollow circular symbols are for lower-saddles and upper-saddles in one direction, related to the infinte-equilibrium circle and coordinates. The equilibrium at the center point is a saddle-node point.

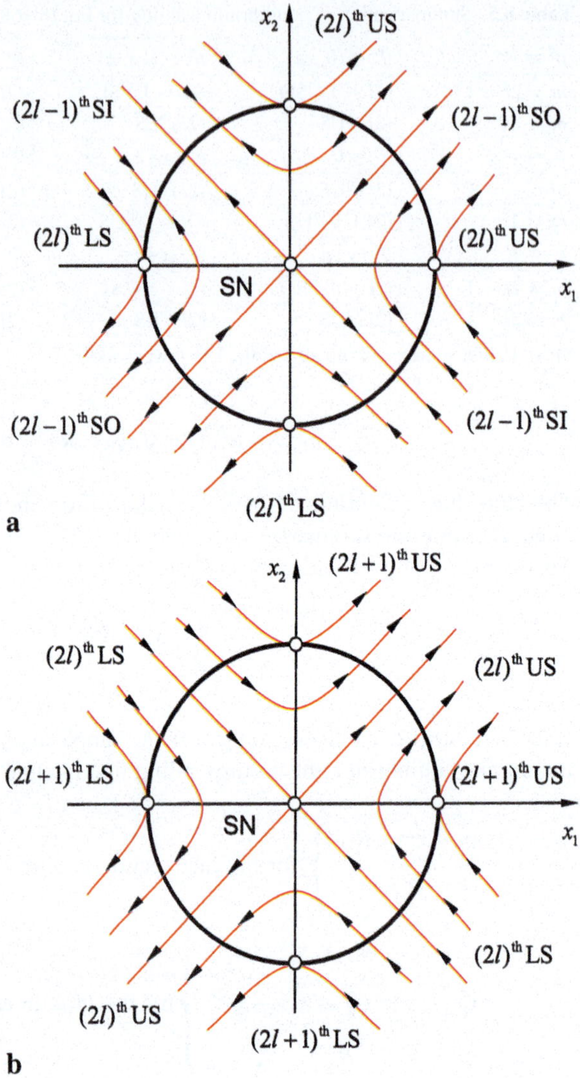

Fig. 8.8 Equilibrium stability for an infinite-equilibrium system in Eq. (8.159). (**a**) The saddle node of (0, 0) and $(2l-1)^{th}$-order sources and sinks, and (**b**) the saddle node of (0, 0) and $(2l)^{th}$-order upper saddles and lower saddles. The thick circle is for the infinite equilibrium. The hollow circular symbols are for simple equilibrium and intersection equilibriums between the infinite-equilibrium circle and coordinates. SI: sink, SO: source, US: upper saddle, LS: lower saddle, SN: saddle node

Reference

Luo, A.C.J., 2019, On dynamics of infinite-equilibrium systems, *International Journal of Dynamics and Control,* in press.

Index

A. C. J. Luo, *Bifurcation and Stability in Nonlinear Dynamical Systems*, Nonlinear Systems and Complexity 28, https://doi.org/10.1007/978-3-030-22910-8

The manufacturer's authorised representative in the EU is Springer Nature Customer Service Centre GmbH, Europaplatz 3, 69115 Heidelberg, Germany. If you have any concerns regarding our products, please contact ProductSafety@springernature.com

Printed and bound by CPI Group (UK) Ltd, Croydon, CR0 4YY
28/09/2026
02224077-0012